What special qualities of mind set the great apes apart from other
from ourselves? In this book, field and laboratory researchers show t
abilities in both social and ecological domains, including tool use, pr
consolation, teaching and culture itself. Great apes are also shown to
levels, traditionally considered to be uniquely human. Here, the mechanisms involved in building these
abilities – especially the lengthy development and 'enculturation' processes – are emphasized, showing
how new discoveries are changing views on how primate and human intelligence evolved.

This book is written for researchers interested in current research and theoretical views of great ape
cognition.

Reaching into thought

Reaching into thought
The minds of the great apes

Edited by

ANNE E. RUSSON
Associate Professor of Psychology, Glendon College of York University, Toronto

KIM A. BARD
Research Scientist, Division of Reproductive Biology, Yerkes Regional Primate Center, Emory University

SUE TAYLOR PARKER
Professor of Anthropology, Sonoma State University, Rohnert Park, California

PUBLISHED BY THE PRESS SYNDICATE OF THE UNIVERSITY OF CAMBRIDGE
The Pitt Building, Trumpington Street, Cambridge CB2 1RP, United Kingdom

CAMBRIDGE UNIVERSITY PRESS
The Edinburgh Building, Cambridge CB2 2RU, UK http://www.cup.cam.ac.uk
40 West 20th Street, New York, NY 10011-4211, USA http://www.cup.org
10 Stamford Road, Oakleigh, Melbourne 3166, Australia

© Cambridge University Press 1996

This book is in copyright. Subject to statutory exception
and to the provisions of relevant collective licensing agreements,
no reproduction of any part may take place without
the written permission of Cambridge University Press

First published 1996
First paperback edition 1998

Printed in the United Kingdom at the University Press, Cambridge

Typeset in Ehrhardt 9.5/12 pt [VN]

A catalogue record for this book is available from the British Library

Library of Congress cataloguing in publication data
Reaching into thought : the minds of the great apes / edited by Anne E. Russon,
  Kim A. Bard, Sue Taylor Parker.
    p. cm.
  Includes index.
  ISBN 0 521 47168 (hc)
  1. Apes – Psychology. 2. Animal intelligence. I. Russon, Anne E. II. Bard,
    Kim A. III. Parker, Sue Taylor.
QL737.P96R424 1996
599.88'4'0451 – dc20   95-20350 CIP

ISBN 0 521 47168 0 hardback
ISBN 0 521 64496 8 paperback

We dedicate this book to

all those great apes who have tolerated and suffered human presence, as we struggle to overcome our preconceptions to see them more clearly as they are.

Contents

List of contributors

1. Exploring the minds of the great apes: Issues and controversies 1
   ANNE E. RUSSON AND KIM A. BARD

PART ONE   The scope of great ape intelligence

2. Chimpanzees and capuchin monkeys: Comparative cognition 23
   JAMES R. ANDERSON
3. Acting and understanding: Tool use revisited through the minds of capuchin monkeys 57
   ELISABETTA VISALBERGHI AND LUCA LIMONGELLI
4. Consolation, reconciliation, and a possible cognitive difference between macaques and chimpanzees 80
   FRANS B. M. DE WAAL AND FILIPPO AURELI
5. The misunderstood ape: Cognitive skills of the gorilla 111
   RICHARD W. BYRNE
6. Ostensive behavior in great apes: The role of eye contact 131
   JUAN CARLOS GÓMEZ
7. Imitation in everyday use: Matching and rehearsal in the spontaneous imitation of rehabilitant orangutans (*Pongo pygmaeus*) 152
   ANNE E. RUSSON
8. "More is less": The elicitation of rule-governed resource distribution in chimpanzees 177
   SARAH T. BOYSEN
9. Tool-using behavior in wild *Pan paniscus*: Social and ecological considerations 190
   ELLEN J. INGMANSON

| 10 | Comparison of chimpanzee material culture between Bossou and Nimba, West Africa | 211 |

TETSURO MATSUZAWA AND GEN YAMAKOSHI

PART TWO Organization of great ape intelligence: Development, culture, and evolution

| 11 | Influences on development in infant chimpanzees: Enculturation, temperament, and cognition | 235 |

KIM A. BARD AND KATHRYN H. GARDNER

| 12 | Heterochrony and the evolution of primate cognitive development | 257 |

JONAS LANGER

| 13 | Simon says: The development of imitation in an enculturated orangutan | 278 |

H. LYN MILES, ROBERT W. MITCHELL AND STEPHEN E. HARPER

| 14 | Imitation, pretense, and mindreading: Secondary representation in comparative primatology and developmental psychology? | 300 |

ANDREW WHITEN

| 15 | Self-awareness and self-knowledge in humans, apes, and monkeys | 325 |

DANIEL HART AND MARY PAT KARMEL

| 16 | Apprenticeship in tool-mediated extractive foraging: The origins of imitation, teaching, and self-awareness in great apes | 348 |

SUE TAYLOR PARKER

| 17 | The effect of humans on the cognitive development of apes | 371 |

JOSEP CALL AND MICHAEL TOMASELLO

| 18 | Three approaches for assessing chimpanzee culture | 404 |

CHRISTOPHE BOESCH

| 19 | On the wild side of culture and cognition in the great apes | 430 |

SUE TAYLOR PARKER AND ANNE E. RUSSON

Index   451

Contributors

JAMES R. ANDERSON
Department of Psychology, University of Stirling,
Stirling FK9 4LA, Scotland

FILIPPO AURELI
Yerkes Regional Primate Center, Emory University,
Atlanta, GA 30322, USA

KIM A. BARD
Yerkes Regional Primate Research Center, Emory
University, Atlanta, GA 30322, USA

CHRISOTOPHE BOESCH
Zoologisches Institut der Universität Basel, CH-4051,
Basel, Rheinsprung 9, Switzerland

SARAH T. BOYSEN
Comparative Cognition Project, Rm. 48 Townsend
Hall, 1885 Neil Avenue, The Ohio State University,
Colombus, OH 432210-1222, USA

RICHARD W. BYRNE
Scottish Primate Research Group, Psychology
Department, University of St Andrews, St Andrews,
Fife KY16 9JU, Scotland

JOSEP CALL
Department of Psychology and Yerkes Regional
Primate Research Center, Emory University, Atlanta,
GA 30322, USA

FRANS B. M. DE WAAL
Yerkes Regional Primate Center, Emory University,
Atlanta, GA 30322, USA

KATHRYN H. GARDNER
Yerkes Regional Primate Research Center, Emory
University, Atlanta, GA 30322, USA

JUAN CARLOS GÓMEZ
Psychology Department, University of St. Andrews,
St. Andrews, Fife KY16 9JU, Scotland

STEPHEN E. HARPER
Department of Anthropology, Emory University,
Atlanta, GA 30322, USA

DANIEL HART
Department of Psychology, Rutgers University,
Camden, NJ 08102, USA

ELLEN J. INGMANSON
Department of Anthropology, Dickinson College,
Carlisle, PA 17013, USA

MARY PAT KARMEL
Department of Psychology, The Catholic University
of America, Washington, DC 20064, USA

JONAS LANGER
Department of Psychology, University of California
at Berkeley, Berkeley, CA 94928, USA

LUCA LIMONGELLI
Reparto Psicologia comparato, Instituto di Psicologia,
CNR, Via U. Aldrovandi 16b, 00197 Roma, Italy

TETSURO MATSUZAWA
Section of Language and Intelligence, Department of
Behavioral and Brain Sciences, Primate Research
Institute, Kyoto University, Inuyama City, Aichi,
484 Japan

H. LYN MILES
Department of Sociology and Anthropology,
University of Tennessee, 615 McCallie Avenue,
Chattanooga, TN 37403, USA

ROBERT W. MITCHELL
Department of Psychology, Eastern Kentucky
University, Richmond, KY 40475, USA

SUE TAYLOR PARKER
Department of Anthropology, Sonoma State
University, Rohnert Park, CA 94928, USA

ANNE E. RUSSON
Department of Psychology, Glendon College,
2275 Bayview Avenue, Toronto, Ontario,
M4N 3M6, Canada

MICHAEL TOMASELLO
Department of Psychology and Yerkes Regional
Primate Research Center, Emory University, Atlanta,
GA 30322, USA

ELISABETTA VISALBERGHI
Reparto Psicologia comparato, Instituo de Psicologia,
CNR, Via U. Aldrovandi 16b, 00197 Roma, Italy

ANDREW WHITEN
Scottish Primate Research Group, Psychology
Department, University of St Andrews, St Andrews,
Fife KY16 9JU, Scotland

GEN YAMAKOSHI
Section of Ecology, Department of Ecology and
Social Behaviour, Primate Research Institute, Kyoto
University, Inuyama City, Aichi, 484 Japan

Acknowledgements

Special thanks to two fine students who helped enormously in completing this volume: Paul Vasey, who dedicated hours of his time participating in our editing, and Tom Mitchell, who assisted greatly in organizing and preparing the manuscript.

1

Exploring the minds of the great apes: Issues and controversies

ANNE E. RUSSON AND KIM A. BARD

INTRODUCTION

Interest in intelligence has always been intense: it verges on an obsession within Western cultural and scientific traditions, where intelligence reigns as a central defining characteristic of the human species. Interest has always been strongly anthropocentric, but almost as alluring has been contemplating the possibility that there exist, in other worlds or perhaps in other worldly species, other intelligences like our own. Both interests spotlight the nonhuman species most like us, the great apes. Because they are the living species most closely related to humans, they offer important living models of the minds from which modern human intelligence evolved. Scholars ranging from philosophers to linguists, neuropsychologists, and anthropologists have invoked their capacities in developing and justifying their views of the human mind (e.g. Corballis, 1991; Donald, 1991; Gibson & Ingold, 1993; Karmiloff-Smith, 1992).

Our aim in this volume is to present current material that contributes to understanding the minds of the great apes, as they are distinguished from the minds of humans and the minds of other nonhuman primates. Although by now several decades have been devoted to careful research on great apes, laboratory and field studies are still regularly generating rich new material on the range, flexibility, and complexity of great apes' behavior and the depth of their understanding. We have invited contributions from researchers working primarily with great apes, those working with several primate species, developmental psychologists working with humans, and primatologists working primarily on monkey species, in order to highlight contrasts and similarities between great ape, human, and other nonhuman primate intelligence. We expressly invited studies of great apes across a wide range of living conditions, from wild to colony or zoo to laboratory, and the corresponding range of research methods, from observational field studies to controlled laboratory experimentation. Our contributors offer new evidence: from the field, on social cognition in bonobos, complex tool use in chimpanzees, and regional variations in the behavior of chimpanzees with characteristics reminiscent of culture; and from the laboratory, on conflict resolution processes, imitation, ostension, and numerical abilities. Theoretical issues that our contributors address include: the most complex levels of ability attained by great apes; the extent to which there are interconnections between great apes' intellectual abilities; the extent to which great apes' social traditions reflect cultural mechanisms; the extent of similarities and differences in intelligence among great ape species; the extent of the intellectual gap between great apes and humans; the extent of the gap between great apes and monkeys; and the extent to which evolutionary, socio-cultural, and developmental perspectives can contribute to understanding great apes' intellect.

Our discussions arise from the view offered by behavioral expressions of intelligence and leave two critical factors in intelligence, the brain and language, as beyond its scope. An assessment of great ape intelligence in terms of the brain would be of great value, but it would require a book in and of itself. We consider great apes' language capacities equally vital to understanding their intelligence, and leave them aside because they have been discussed recently and extensively by others (e.g. Gibson & Ingold, 1993; Parker & Gibson, 1990; Savage-Rumbaugh *et al.*, 1993).

In this opening chapter, we offer a brief historical overview of explorations into the minds of the great apes and an introduction to the two central issues we address in this volume: (a) the *scope* of great ape intellectual abilities, in their breadth and their complexity; (b) the *organization* of these abilities, for instance interdependencies among various abilities and domains, and the developmental, socio-cultural, and evolutionary mechanisms from which it derives. Current themes in exploring *scope* concern great apes' capacities for acquiring a range of specific abilities (e.g. object manipulation, tool use, classification, imitation, pretense, language, deception, mindreading) and the highest levels of complexity they can achieve in each. The *organization* of great ape intelligence is a relatively new issue emerging from findings on the broad scope of great ape intelligence. Themes receiving current attention include the relations and interdependencies among individual intellectual abilities and even domains, and the developmental and socio-cultural processes that influence how these abilities are constructed. Undergoing reevaluation are evolutionary reconstructions, especially concerning the evolutionary forces that could have formed the intelligences of great apes and humans and correspondingly, the models of intelligence that are compatible with comparative evolutionary perspectives.

HISTORICAL AND MODERN VIEWS OF GREAT APE INTELLIGENCE

In the established view, great apes are classified into three genera and four species – orangutan (*Pongo pygmaeus*), gorilla (*Gorilla gorilla*), chimpanzee (*Pan troglodytes*), and bonobo or pygmy chimpanzee (*Pan paniscus*)[1] (e.g. Tuttle, 1986).

The great apes were not clearly distinguished from other nonhuman primates by Western thinkers until the 1600s, but they, along with the other nonhuman primates, had been noticed since antiquity for their unsettling similarity to humans (whence "simians"), particularly in their intellect (French, 1994; Salisbury, 1994). The characteristics of great apes inject ambiguity into neat distinctions between "human" and "animal" and the resulting tension is reflected in varying stances about their nature – some periods and some communities have emphasized the differences, others the similarities. For instance, Aristotle knew of nonhuman primates and saw human and animal intelligence as similar except that humans had reasoning abilities beyond the sensorimotor abilities of animals (French, 1994; Gibson, 1993). The early Christians, apparently as a way of distinguishing themselves from the pagan ancients, pronounced unbreachable boundaries. Medieval thinkers returned to a sense of ambiguities from as early as the twelfth century (Janson, 1952; Salisbury, 1994). Medieval scholars likened nonhuman primates to humans because of their imitative and problem-solving abilities, but considered them as imperfect and foolish copies; they did, however, place nonhuman primates in a special category, bridging the divide between humans and animals (Salisbury, 1994). In so doing, they initiated the shift away from views of unbreachable separation between animals and humans and toward views of continuity.

The medieval sense of continuity is considered to have founded the present scientific view of humans as part of the animal kingdom and particularly closely related to great apes (Salisbury, 1994); it now has evolutionary justification. Fossil and molecular evidence show that humans and great apes are related through recent common ancestors, the Hominidae (Begun, 1992; King & Wilson, 1975). Chimpanzees and humans are each others' closest living genealogical relatives; they share a common ancestor unique to themselves more recently than either shares a common ancestor with any other living species. Zoologically, evolutionary relations as close as those between modern humans and chimpanzees would normally result in classifying them together as a single genus, separated only at the species level. Molecular methods further indicate that the relationships are so close as to say humans *are* great apes (e.g. Begun, 1992; for a recent discussion, see Byrne, 1995).

Despite great apes' importance, scientific research into great ape intelligence was sparse until the 1960s. Few volumes dedicated to great ape intelligence have appeared since the pioneering work from the beginning of the twentieth century. The twentieth century history of the social sciences offers important insights into these curious lacunae; because it also set the course of today's questions, we sketch it briefly.

Darwin's (1871) publication of *The Descent of Man and Selection in Relation to Sex* sparked scientific interest in comparative and evolutionary studies of mentality, in the spirit of studying the human mind within Darwin's evolutionary framework. The new disciplines founding much of this work were anthropology and comparative psychology, the former focusing primarily on evolutionary origins of human mentality and the latter fixing on comparative studies of nonhuman species (for discussions, see Antinucci, 1989; Parker, 1990; Tuttle, 1986). This impetus produced significant advances in understanding great ape intelligence, through the work of pioneers such as Yerkes in America, Köhler in Europe, and Kohts in Russia (e.g. Köhler, 1925; Kohts, 1921, 1928; Yerkes, 1916, 1925, 1943; Yerkes & Yerkes, 1929). For instance, Köhler and Yerkes found: great apes could solve problems by "insight"; they could construct and use complex tools; they could abstract, generalize, and infer from their experiences at rudimentary levels; and chimpanzees exhibited the makings of culture.

This pre-1930 pioneering work was almost the last to concentrate on great ape intelligence until the 1960s, despite its provocative findings. Perhaps through discomfort with the ramifications of evolutionary theory, that humans are not special creatures separated from and above animality, the twentieth century saw a resurgence of discontinuity views attesting to firm boundaries between animal and human abilities. This was manifest in a turn towards logical–positive epistemologies and specialization of scholarship within disciplinary boundaries (Gibson, 1993; Parker, 1990). Both anthropology and comparative psychology adopted these newer views, raising them to mainstream status over most of the middle part of this century (Parker, 1990). In consequence, work on great apes was severely muted. American anthropologists made a place for the study of great apes and other nonhuman primates by the 1930s but within physical rather than cultural anthropology and so out of the mainstream. Comparative psychologists continued some work with great apes but from the perspective of behaviorism. Beyond presuming a human–animal distinction, behaviorists were uninterested in the mind or species differences ("Pigeon, rat, monkey, which is which? It doesn't matter." – Skinner, 1956, p. 230) and many of their practices (reductionist approaches, concern for universal processes, tightly controlled experimentation and rearing conditions, and preference for convenient species) militated against encountering great apes' more complex capacities.

More revealing studies then tended to be maverick ones. The Hayes and the Kelloggs raised chimpanzees in their homes as members of their families to study language and motor development, coincidentally producing findings on the development of discriminatory abilities, imitation, problem-solving, and the effects of "cross-fostering" on apes' intellectual development (Hayes & Hayes, 1951, 1952, 1954; Hayes & Nissen, 1971; Kellogg & Kellogg, 1933). Carpenter (1934, 1938) and Nissen (1931) pioneered field studies of free-living nonhuman primates. This work, peripheral at the time, paved the way for future work on great ape intelligence. Overall, up to the late 1960s, evidence was insufficient to characterize great ape intelligence comparatively (Tuttle, 1986); to some, it appeared that primate learning did not differ from other vertebrate learning, except perhaps in speed (e.g. Warren, 1965). Others, near the end of this era, had a glimmering that there were special qualities to nonhuman primate intelligence: "In both variety and complexity, the problems that fall within the repertory of the primate contrast sharply with those testable on rodent or carnivore" (Harlow, 1951, p. 234).

The 1960s brought revolutionary changes. The cognitive revolution brought back the mind, legitimizing studies of cognition and language as well as reintroducing cognition to comparative psychology under the rubric of animal cognition. Increased international exchange after World War II brought significant innovations. For instance, Japanese scholars, who had developed their own primatological research tradition relatively independently, began to interact more extensively with Western counterparts. This introduced their research traditions, characterized by interests in social structures, the roles of both the individual and the group, and historical views, as well as by long-term field methods and reliance on provisioning (Asquith, 1991, 1994). European ethology emphasized evolutionary perspectives, especially the importance of considering a species' intelligence as an adaptive capacity with qualities evolved in conjunction with species-specific contexts and functions. Comparative psychology established methods or measuring complex learning at levels reflecting emergent symbolic abilities (e.g. Rumbaugh, 1969). In physical anthropology, Sherwood Washburn and Louis Leakey helped bring primatology to greater prominence by articulating the significance of the living primates to the evolutionary study of human behavior (Parker, 1990). Developmental psychology, through Jean Piaget, brought views of human intelligence as epigenetic, that is that higher level cognitive abilities are constructed sequentially on the basis of lower level ones, in accordance with physical world and social experiences.

The innovations of the 1960s had significant repercussions. The application of Piaget's developmental approach to great ape and monkey intelligence revealed striking similarities in the patterning of their cognitive development to that of humans, although the focus of study remained at low levels of sensory and motor abilities (e.g. initially through Jolly, 1972; then Mathieu et al., 1976; Chevalier-Skolnikoff, 1977, 1983, 1989 and discussions; Parker, 1977; Redshaw, 1978; Parker & Gibson, 1979; Mathieu & Bergeron, 1981; Chevalier-Skolnikoff et al., 1982; Antinucci, 1989; Russon, 1990; Bard, 1993). Challenges to traditional landmarks of human mentality began as a result of the long-term field studies of each of the three great ape genera initiated during this period (e.g. chimpanzees use and manufacture of tools – Goodall, 1963), studies of language in relation to intelligence (e.g. Miles, 1983; Premack, 1971, 1984), and experimental research suggesting great apes' potential for rudimentary symbolic level processing (e.g. Menzel, 1973, 1978; Premack & Woodruff, 1978; Rumbaugh, 1969). Japanese studies of social traditions in monkeys, which they likened to culture (e.g. Kawai, 1965), contributed to interest in the social dimensions of primate intelligence.

Views and theoretical perspectives changed significantly in light of this work. The argument appeared that primate intelligence is centrally adapted to primates' complex sociality (e.g. Jolly, 1966; Humphrey, 1976) and it subsequently took on major significance (e.g. Byrne & Whiten, 1988). Developmentally oriented findings underlined the importance of ontogeny to primate intelligence. The understanding that has accumulated on great ape cognitive development has come to attract human developmental psychologists, who see the potential of this work to illuminate human cognitive development. As landmarks of human intelligence were found in great apes, interest extended beyond the evolutionary divergence of the hominids to that of the great apes themselves, and led to hypotheses about the evolution of great ape intelligence – for instance, that intelligent tool-assisted extractive foraging on embedded foods constituted their key intellectual adaptation (Parker & Gibson, 1977, 1979). This also increased calls for continuity models of intelligence, as evidence accumulated that human intelligence probably evolved by expanding capacities already present in recent ancestors (e.g. Gibson, 1993). Such changes stimulated further field studies specifically aimed at understanding great ape intelligence, provided impetus to the study of great apes other than chimpanzees, expanded the expressions of intelligence considered researchable, altered the models, the methods and even the definitions of fields concerned with animal intelligence, brought the disciplines of physical anthropology and comparative psychology closer together, and began a reintegration in studies concerned with Tinbergen's causes of behavior – proximate mechanisms, ontogeny, phylogeny, and evolutionary function (Parker, 1990; Tinbergen, 1963).

The scope of great ape intelligence

The post-1960s research increased appreciation for great apes' capacities, but the full scope of their intelligence, both in breadth (the range of problem areas in which their intelligence is expressed) and complexity (the levels of sophistication they attain), has not yet been established.

At least three research perspectives guide explorations concerning scope. A first focuses on the constellation of intellectual abilities that may characterize the great ape–human divide, such as tool use and tool manufacture, imitation, sense of self, pedagogy, and culture. Although rudiments of virtually all the abilities traditionally considered hallmarks of human intelligence are now accepted as within great apes' scope (Gibson, 1993), scholars continue to probe these abilities in great apes for qualitative distinctions and for suggestions about the evolutionary antecedents to the human forms. A second perspective derives from researchers who study primate groups and social dynamics. It focuses attention on abilities for handling social problems, such as conflict resolution (de Waal, 1989; Kummer, 1967), theory of mind (e.g. Premack & Woodruff, 1978), imitation (Kawai, 1965; Mitchell, 1987; Whiten & Ham, 1992), social communication, deception (Byrne & Whiten, 1988; Mitchell & Thompson, 1986), and teaching (Boesch, 1991; Caro & Hauser, 1992). The third perspective is that of Piagetian developmental psychology, which considers yet another array of intellectual abilities (e.g. object concept, causal reasoning, logicomathematical reasoning). Research undertaken from this perspective explores the constructive processes and changes involved in building great ape intellectual abilities. An additional advantage of this perspective is that it conceptualizes intellectual abilities as multileveled and ordered in terms of complexity, an important view in considering our second issue, the organization of great ape intelligence.

The turn towards continuity and away from discontinuity views has had its own impact on this issue. First, great ape–human intellectual differences are being reformulated in terms of degree rather than of kind, accepting that the great apes and humans share critical abilities at rudimentary levels (Gibson, 1993), and attention is now focused on the highest levels of an ability attained by great apes or by other nonhuman primates. Secondly, human rather than animal models are increasingly considered the more appropriate as the basis for comparative assessments of great ape abilities. The ultimate benchmarks for comparison are human abilities because questions about intelligence tend to be anthropocentric; in addition, the great apes are more closely related to humans than they are to any other animal species; and only human models address great apes' higher level abilities. A rehabilitated anthropomorphism is then one of the many products of this shift to continuity views. Among its more lighthearted manifestations is an insistence for anthropomorphic descriptions of great ape behavior, as, for instance, deception (Mitchell & Thompson, 1986), "Machiavellian intelligence" (Byrne & Whiten, 1988) and "chimpanzee politics" (de Waal, 1982).

The significance of developmental approaches to understanding the scope of great apes' intellectual abilities is increasingly apparent. Clearly, for instance, it is inappropriate to compare the performance of the most frequently used great ape subjects, individuals 3–5 years old, with adult humans: the former are infants and young juveniles with immature abilities, whereas the latter possess mature intellectual abilities and skills. We also now recognize that great apes do not fully master some of their intellectually based skills until near adolescence, from 7–10 years of age in the wild (e.g. Boesch & Boesch, 1993; Matsuzawa, 1994). Further, researchers find considerable variability in great apes' performances, especially in tasks demanding high-level cognitive abilities. This variability tends to correlate with rearing conditions, prior experience, and age; indeed, some consider that great apes may not attain certain abilities, like language and symbolic level imitation, except under particular rearing conditions (e.g. Boesch, 1991; Tomasello *et al.*, 1993b; Whiten & Ham, 1992). For these reasons, developmental considerations are currently prominent in investigations into the scope of great ape intelligence (e.g. see Chapters 10, 11, 12, 13, 14, 15, 16, and 17).

In this context, we consider briefly current issues concerning the scope (breadth and complexity) of great ape intelligence.

Understanding of the *breadth* of great apes' intellectual span continues to expand, even now, with each new ability considered opening new avenues for understanding great ape intelligence. Most of our contributors delve into the intricacies of established abilities (e.g. causal reasoning as seen through tool use, object manipulation, and food processing; logicomathematical reasoning such as classification and quantification; social abilities such as imitation, pretense, deception, mindreading, mirror-self-recognition, self-awareness, teaching, and cooperation). Others discuss newer indices of sophisticated intelligence, particularly in the domain of social intelligence where the phenomena that index complex cognition are not yet clearly identified (e.g. reconciliation and consolation (Chapter 4); ostension (Chapter 6); self-related phenomena (Chapter 15); secondary representation (Chapter 14); use of tools to solve social problems (Chapter 9)).

Regarding *complexity*, various abilities initially treated as representing a single achievement (self-recognition, theory of mind, true imitation) are being reconstrued as multileveled ones. To illustrate, primates' intelligent tool use may involve several achievements reflecting multiple levels of cognitive sophistication. On the basis of the monkey species that have been tested, primarily cebus and macaques, monkeys may not understand the relationships involved in tool-mediated problem-solving whereas the great apes and humans do (Visalberghi & Trinca, 1989); great apes sometimes use sets of tools rather than single tools (see McGrew, 1992) whereas monkeys very rarely do (there exists only one report of tool set use in cebus monkeys – Westergaard & Suomi, 1993); and only great apes have been found to use a tool to modify or to make a second tool (Matsuzawa, 1991; Toth *et al.*, 1993; Wright, 1972). Each of these great ape variations in tool use is considered to reflect higher level cognitive abilities. Likewise, other shared abilities (e.g. tactical deception, imitative learning, self-awareness, mindreading) are being reconstrued as multileveled and devolving from a multiplicity of mechanisms varying in cognitive complexity (e.g. Byrne, 1995; Mitchell, 1987; Parker *et al.*, 1994;

Whiten & Byrne, 1988). Several of our contributors discuss the levels at which great apes express a given ability or identify variation in the levels at which a given ability can be expressed – causality (Chapters 3 and 12), self-concept (Chapter 15), logicomathematical abilities (Chapters 8 and 12), conflict resolution (Chapter 4), imitation-related phenomena (Chapters 3, 5, 7, 13, 14 and 16), teaching (Chapter 16), and metarepresentation (Chapter 14). Of central interest are the highest levels of an ability that great apes or other nonhuman primates attain.

Although great apes' abilities fall well short of those achieved by humans, evidence that all the great apes can handle such tasks as rudimentary language, insightful or tool-assisted problem-solving, and abstract learning is seriously challenging traditional views that their reach is bounded by symbolic level processing (e.g. Gardner *et al.*, 1989; Lethmate, 1982; McGrew, 1992; Miles, 1983; Meador *et al.*, 1987; Mignault, 1985; Parker *et al.*, 1994; Patterson & Cohn, 1990; Premack & Woodruff, 1978; Savage-Rumbaugh *et al.*, 1993). Since the mid-1980s, research has increasingly targeted great apes' abilities beyond sensorimotor and into symbolic levels. Several researchers now argue that great apes have demonstrated the capacity for rudimentary symbolic level processing in several abilities, such as imitation (Custance & Bard, 1994; Russon & Galdikas, 1993; Tomasello *et al.*, 1993b; Whiten & Custance, 1994), deception (Byrne & Whiten, 1988), tool use and object manipulation (e.g. Matsuzawa, 1994; McGrew, 1992; Russon & Galdikas, 1992), and classification (Langer, 1993). Similarly, researchers argue that the great apes achieve abilities that hinge on symbolic processes, such as self-awareness, pretense, analogical reasoning, quantification, demonstration teaching, and mindreading (e.g. Boesch, 1991; Boysen & Berntson, 1990; Gardner *et al.*, 1989; Langer, 1993; Meador *et al.*, 1987; Parker *et al.*, 1994; Povinelli *et al.*, 1992; Premack, 1984; Whiten, 1991). Among the most striking of the realizations emerging from this work is that *all* the great apes share these intellectual capacities; what differs is their tendency to express them.

Many of our contributors present further evidence and arguments concerning great ape capacities at symbolic levels – Chapters 10, 16, 18, and 19 (proto-cultural processes), Chapter 15 (self-related concepts), Chapter 8 (logicomathematical reasoning about number and quantity), Chapter 12 (logicomathematical and causal reasoning), Chapter 13 (imitation, deception, and pretense), Chapter 16 (imitation, self-awareness, and teaching), Chapters 5 and 7 (imitation), and Chapter 4 (consolation). Significantly, other nonhuman primates have not yet shown abilities beyond sensorimotor levels (Antinucci, 1989; Visalberghi & Trinca, 1989; Chapters 2, 3, and 4).

The implication is that the significant move beyond sensorimotor to symbolic thought evolved with the ancestors of the great apes, not the ancestors of humans. In Aristotle's terms, this would qualify great apes as rational and creative thinkers, like humans. This also supports the possibility that these hallmark abilities distinguish great ape from other nonhuman primate intelligence. Contention around this view is high. Our contributors do not themselves agree on the intellectual scope of great apes (compare Chapters 7, 13, 14, and 16 with Chapter 17). Some argue that great apes do not attain the highest level abilities except when raised with intensive human tutoring (e.g. Tomasello *et al.*, 1993b; Chapter 17); others, that their most complex abilities are attained in the wild (e.g. Chapter 18). Neither is the distinction between great ape and monkey intelligence well established (compare Chapter 2 with Chapter 3).

The organization of great ape intelligence

Increasing appreciation for the scope of great ape intelligence has raised the question of how to understand their particular constellation of abilities. Do the intellectual abilities that have been demonstrated by great apes represent a simple collection or are they interrelated; if they are interrelated, in what way and at what levels? Empirical evidence remains relatively limited because research has tended to consider individual abilities in isolation rather than in relation to one another.

The possibilities concerning interrelationships range between two poles: great ape intelligence comprises independent, problem-specific, module-like abilities or it constitutes some generalized capacity. Both positions have been advocated as models for human intelligence (e.g. Fodor, 1983; Gibson, 1993; Sternberg, 1990). Intermediate possibilities include interconnections among discrete abilities at some intermediate level, such as large problem domains (e.g. social, ecological). Modularity or problem-specificity models are the most strongly advocated for learning in nonhuman species (e.g. Davey, 1989; Hirschfield & Gelman, 1994a), including monkeys, based on evidence that their abilities for handling social and ecological problems are domain-specific and mutually inaccessible (Cheney & Seyfarth, 1990; Langer, 1989). Generality–modularity issues, however, are not yet resolved for either humans or nonhumans (e.g. Greenfield, 1991; Hirschfield & Gelman, 1994b; Karmiloff-Smith, 1992); and although earlier studies generalized across nonhuman primates, more recently investigators have focused on differences between monkeys and great apes. Evidence on the organization of great ape abilities could play a key role in these debates.

Research on great apes has focused on two possibilities, interconnections among abilities (e.g. between individual abilities, within domains, or between domains – see Gibson, 1993; Langer, 1989), or some overarching, generalized capacity such as hierarchical mental construction that generates individual abilities centrally (e.g. Byrne, 1994a; Galdikas & Vasey, 1992; Greenfield, 1991; Russon & Galdikas, 1994).

Those exploring interconnections assess interplay within particular subsets of abilities, such as co-occurrences, interdependencies, or developmental parallels. Evidence exists for links among some individual abilities (e.g. self-awareness and imitation co-occur in great apes (see Parker et al., 1994); language training may enhance chimpanzees' performance on analogy problems (see Premack, 1984)). Particular attention has been devoted to the relations between the large domains of social and ecological intelligence. The focus of attention shifted from ecological to social intelligence in the late 1980s. Early discussions aimed at establishing the relative primacy of social versus ecological intelligence in primate evolution (e.g. Byrne & Whiten, 1988; Dunbar, 1992). Within the primates, however, the great apes require an added consideration: great apes' feeding niche (e.g. embedded foods) is considered to have further shaped their intelligence (e.g. one capable of handling extractive foraging – see Byrne & Byrne, 1991; Parker & Gibson, 1977). While researchers continue to recognize the distinctive qualities of the intellectual abilities within each domain, they are now, more intensively, examining interconnections between the two domains (e.g. how great apes' imitative abilities contribute to their technological ones). Several of our contributors offer further demonstrations or arguments for interplay among individual abilities (e.g. see Chapters 5, 11, 13, 15, and 16), within domains (e.g. Chapters 13 and 14), and between domains (Chapters 9, 16, 18, and 19).

The ways in which abilities are constructed play a central determining role in the organizational patterns that appear. Developmental models therefore offer important perspectives on how intellectual abilities may interrelate. These models are particularly important to primates, who, with prolonged immaturity, rely heavily on joint influences of environment, genetic programming, and the socio-cultural milieu for constructing many behavioral achievements (e.g. Scarr & McCartney, 1983). Of equal importance is the possibility that something akin to culture is a contributor, a possibility that is hotly debated. Evolutionary perspectives suggest another view on how great ape intelligence is organized.

Developmental considerations

Developmental perspectives offer a variety of insights into the organization of great apes' intellectual abilities. For instance, the sequential emergence, developmentally, of a subset of abilities suggests structural interdependencies among them, in that later emerging abilities depend on the earlier ones; developmental parallels, particularly synchronized progress across abilities, suggest those abilities are coordinated or integrated, whereas asynchronous progress suggests more isolated, problem-specific organization. A number of researchers argue for particular sequentially-based interdependencies (e.g. imitation, pretense, and teaching (see Chapters 13, 14, 16, and 19); or imitation and self-awareness (see Parker *et al.*, 1994)). Synchronous development is considered to characterize the early development of human cognitive abilities. In contrast, asynchronous development has been reported for monkeys (Antinucci, 1989; Cheney & Seyfarth, 1990; Langer, 1993; Parker, 1977) and some degree of synchrony, for great apes (e.g. Mathieu & Bergeron, 1981; Russon, 1990; Langer, 1993). Langer (Chapter 12) compares this developmental synchrony across great apes, humans, and some monkeys, and argues for its significance in the evolution of complex cognitive capacities. Developmental parallels could also support centralized models, if they reflect similar construction processes for discrete abilities. The patterning underlying the parallels may determine which model is favored (e.g. in Chapter 12, Langer favors interconnectivity interpretations for his findings). More conclusive assessments await more intense study of the co-occurrences and developmental synchonies among great apes' abilities across problem areas and even across domains.

Developmental models that portray cognitive abilities as multileveled are useful for assessing great ape intelligence in comparative perspective (e.g. Antinucci, 1989; Mitchell, 1987; Parker & Gibson, 1979, 1990; Whiten, 1991). Especially useful are models portraying cognitive development as a sequential epigenetic process because they facilitate scaling intellectual levels in a fashion amenable to evolutionary analyses.

Among the significant empirical findings emerging from work based on such epigenetic models is that the development of higher level abilities is delayed in great apes in comparison with humans; for instance, great apes appear to achieve symbolic level capacities only as juveniles, whereas humans do so late in infancy. In consequence, attention is shifting from infancy to juvenile periods in considering great apes' high level abilities (e.g. see Chapters 10, 18, and 19). In this context, Piaget's logicomathematical and physical models have been favored (Piaget, 1952, 1954, 1962; and see Antinucci, 1989; Dore & Dumas, 1987; Parker & Gibson, 1990), along with Case's model of causal reasoning (Case 1985; and see Russon & Galdikas, 1992), Fischer's skill-focused model (Fischer, 1980), Langer's logicomathematical model (Langer, 1980, 1981; and see Langer, 1989, 1993), and Vygotsky's socio-cultural model (e.g. Vygotsky, 1962; and see Tomasello et al., 1993a). Developmental models are important to several of our contributors (see Chapters 8, 10, 11, 12, 13, 14, 16, 17, and 19).

Socio-cultural considerations
Determining the degree to which great apes transmit representations of knowledge socially, as humans do in their cultures, is one of the challenges emerging from research into the links between social and ecological intelligence in great apes (e.g. Kummer, 1971; Nishida, 1987). This is clearly important in establishing the links between human and great ape intelligence. Further, great apes' capacity to sustain culture-like representations is important because such representations shape the fabric of individually constructed intelligence. Because culture has been considered the exclusive province of humanity, debates are intense (e.g. Galef, 1990, 1992; Heyes, 1993; Quiatt, 1993; Tomasello et al., 1993a). Answers depend to a considerable degree on the definition of culture that is adopted. Most of our contributors touch on these issues either directly or indirectly. There is increasing empirical force to the view that great apes' proto-cultural abilities influence the ecologically related cognitive abilities of individual group members (e.g. Wrangham, et al., 1994; Chapters 18 and 19). Present debates concern what definitions of culture are suitable for comparative purposes, whether great ape social traditions merit the name of culture, what form their evolutionary precursors could have taken, and what impact these and human "cultures" have on the quality of great apes' intelligence.

Correspondingly, some are searching for theories that model great ape abilities as a function of the particular socio-cultural context in which they develop (e.g. Tomasello et al., 1993a); a number of our contributors are involved (see Chapters 10, 11, 13, 16, 17, and 19). Some are exploring mechanisms considered critical to culture, such as imitation, teaching, and other forms of social transmission (see Chapters 5, 7, 10, 13, and 16). Others identify behavioral phenomena that may represent social traditions (see Chapters 9, 10, and 18). Yet others explore potential differences in the quality of great ape cognition and performance as a function of the social context in which it develops – especially the human rearing conditions that have been called "enculturation" (see Chapters 11, 13, and 17).

Evolutionary considerations
The cognitive abilities achieved by great apes, the degree of integration-independence of their intelligence across domains, and the relative strengths of social and ecological processing remain critical to scenarios of the nature and evolution of human intelligence (Antinucci, 1989; Cheney & Seyfarth, 1990; Gibson & Ingold, 1993; Langer, 1989; Parker & Gibson, 1990).

Findings that great apes have the capacity for abilities traditionally considered as adaptations to the hominid niche are forcing reworkings of evolutionary reconstructions of higher primate intelligence. Current views lean toward the conclusion that the move from sensorimotor to rudimentary symbolic or rational thought came with the evolutionary divergence of the common great ape–human ancestor, and that divergence of the human lineage involved additional changes that created yet more advanced levels of symbolic thought (e.g. Gibson, 1993). The implications are that the evolutionary forces that shaped these landmark abilities occurred at the divergence of the common great ape–human ancestor, not the divergence of the human lineage; whatever evolutionary forces shaped these landmark abilites, they are unlikely to be the forces that shaped whatever new capacities did emerge with the hominids. Scholars are now reworking evolutionary reconstructions around both points of divergence, to reconsider the ecological pressures that could have shaped these traditional landmark abilities and what abilities emerged uniquely with human ancestors along with what selection pressures could have shaped them (e.g. Byrne, 1994b; Dunbar, 1992; Parker, 1992; Povinelli & Cant, 1992; Russon & Galdikas, 1993; Whiten, 1991).

The organization of great ape intelligence has important evolutionary implications. If intellectual abilities were shaped independently, each reflects its own particular selective pressures and timetables; if they evolved as part of a complex adaptive whole, a more unified evolutionary package and timetable are indicated. Two sets of recent findings (that great ape intelligence resembles human intelligence in showing interconnections among its discrete abilities (e.g. Gibson, 1993) and that monkey intelligence may be domain-specific (e.g. Cheney & Seyfarth, 1990)) suggest that some form of generality in intelligence came with the hominoids. The evolutionary advantages of generality, the mechanisms that effected it, and the reasons for the evolution of humans' yet greater capacity for interconnections to even higher levels, are all issues now in need of greater consideration (Diamond, 1990; Gibson, 1990, 1993; Greenfield, 1991). Some of the evolutionary reconstructions proposed in this volume concern the specification of which individual abilities may have been become interconnected and what mechanisms may have been involved (e.g. Chapters 12, 14, and 16).

Developmental models of intelligence, particularly epigenetic ones, offer special advantages to the enterprise of evolutionary reconstructions because they offer explicit and plausible models of cognitive abilities in terms of levels and they lay out sequences of steps linking higher and lower level cognitive abilities, suggesting plausible paths in evolution (e.g. Antinucci, 1989; Cheney & Seyfarth, 1990; Greenfield, 1991; Parker & Gibson, 1990; and see Chapters 12 and 16). The value of developmental models applied comparatively is further illustrated by the concept of heterochrony. Heterochrony, adjustments in ontogenetic timetables, may be one of the major mechanisms through which evolutionary change is achieved (Gould, 1977; McKinney & McNamara, 1991). Changes to the timing of the development of various abilities may then have been critical to the evolution of great ape and human intelligence (see Parker, 1995; and see Chapter 12).

A point of contention concerns the significance of individual species of great apes in understanding the evolution of higher primate intelligence. Orangutans play a pivotal role in modeling the common ancestral intelligence because their lineage was the earliest to diverge from the common great ape–human ancestor, and fossil evidence shows them to be "living relics" that have changed little in the last $12\frac{1}{2}$ million years (e.g. Schwartz, 1987). They are also important as the closest outgroup for the African hominoids. Gorillas are essential pivots in modeling the ancestor common to the African great apes and humans and the closest outgroup for humans and chimpanzees. They resemble both *Pan* species very closely anatomically, suggesting their common African ancestor as a species with capacities similar to those shared by modern gorillas and chimpanzees (Byrne, 1994b; Chapter 5). Chimpanzees are distinguished as human's closest living relatives and some researchers consider they offer the best living model of the common ancestor of the great apes (e.g. Andrews & Cronin, 1982; Begun, 1992). Which species represents the best model of the common ancestral hominoid is currently unresolved, in this volume as elsewhere. Those working with orangutans (see Chapters 7 and 13) offer the orangutan as important, as the earliest of the great apes to diverge and the one least changed since its divergence; others consider chimpanzees as a more appropriate model, based on recent evidence that the chimpanzee may most closely resemble the common ancestor (see Chapter 16).

ORGANIZATION OF THIS VOLUME

We have used these issues of scope and organization in arranging the chapters in this volume, although no simple scheme captures the complexities in the issues addressed. Many of our contributors do not treat a single issue or theme in isolation; rather, their work cuts across several themes in exploration of the multiple implications of empirical findings (e.g. socio-cultural influences on the development of a particular ability; how the developmental trajectories of discrete abilities contribute to evolutionary reconstructions). We have therefore arranged chapters according to what we consider to be their primary focus.

CONCLUSION

A balanced and thorough understanding of the minds of the great apes could never stem from the work of a single person. With three genera incorporating at least four species with life spans close to our own and four cornerstone approaches to the study of behavior (proximate mechanisms, ontogeny, phylogeny, and adaptive function; Tinbergen, 1963), it inevitably takes many researchers, many years, varied approaches, and diverse interests to capture the many facets of these complex minds. This volume represents our collective effort to draw on the understanding of those currently involved in studying these minds much like our own, to offer what we hope is a rich view of great ape thought. Research over the last 30 years has contributed to the current wave of emphasis on similarities. Researchers are regularly finding heretofore unexpected realms and degrees of similarity, which are particularly useful for evolutionary reconstructions. When we reconsider differences, these similarities are forcing changes in how the boundaries between great apes and humans are placed and articulated. Despite the intensity of research, it is only now becoming possible to address some questions. Although we now realize the importance of understanding the differences between great ape and other nonhuman primate intelligence, the potential for culture in great apes, and the highest levels of understanding they can attain, a fuller understanding awaits future research.

ACKNOWLEDGEMENTS

The authors thank Sue Taylor Parker for her helpful comments and suggestions on earlier versions of this manuscript. K.A. Bard was supported in part by NIH grants RR-00165, RR-06158, and RR-03591. A.E. Russon was supported in part by NSERC funds and a Glendon College Research Fellowship.

NOTES
1 Three of the great ape species are subdivided into several subspecies: orangutans (Bornean and Sumatran); gorillas (Western lowland, Eastern lowland, and mountain); and chimpanzees (one Eastern and two Western subspecies). Recent molecular evidence suggests that subspecies of both chimpanzees and gorillas may merit classification as separate species (Morin *et al.*, 1994; Ruvolo, 1994).

REFERENCES

Andrews, P. & Cronin, J. (1982). The relationship of *Sivapithecus* and *Ramapithecus* and the evolution of the orangutan. *Nature*, 297, 541.

Antinucci, F. (ed.) (1989). *Cognitive Structure and Development in Nonhuman Primates*. Hillsdale, NJ: Erlbaum.

Asquith, P.J. (1991). Primate research groups in Japan: Orientations and east-west differences. In *The Monkeys of Arashiyama. Thirty-five Years of Research in Japan and the West*, ed. L.M. Fedigan & P.J. Asquith, pp. 81–98. New York: State University of New York Press.

Asquith, P.J. (1994). The intellectual history of field studies in primatology: East and west. In *Strength in Diversity: A Reader in Physical Anthropology*, ed. A. Herring & L. Chan, pp. 49–75. Toronto: Canadian Scholars' Press.

Bard, K.A. (1993). Cognitive competence underlying tool use in free-ranging orang-utans. In *The Use of Tools by Human and Non-Human Primates*, ed. A. Berthelet & J. Chavaillon, pp. 103–13. Oxford: Clarendon Press.

Begun, D.R. (1992). Miocene fossil hominids and the chimp-human clade. *Science*, 257, 1929–33.

Boesch, C. (1991). Teaching in wild chimpanzees. *Animal Behaviour*, 41, 530–2.

Boesch, C. & Boesch, H. (1993). Transmission aspects of tool use in wild chimpanzees. In *Tools, Language and Cognition in Human Evolution*, ed. K.R. Gibson & T. Ingold, pp. 171–84. Cambridge: Cambridge University Press.

Boysen, S. & Berntson, G.G. (1990). The emergence of numerical competence in the chimpanzee (*Pan troglodytes*). In *"Language" and Intelligence in Monkeys and Apes: Comparative Developmental Perspectives*, ed. S.T. Parker & K.R. Gibson, pp. 435–50. New York: Cambridge University Press.

Byrne, R.W. (1994a). The problem of hierarchical structure in serial order: What description and explanations are appropriate for complex patterns of gorilla manual actions? Paper presented at the XVth Congress of the International Primatological Society, Kuta-Bali, Indonesia, 3–8 August.

Byrne, R.W. (1994b). The evolution of intelligence. In *Behaviour and Evolution*, eds. P.J.B. Slater & T.R. Halliday, pp. 223–65. Cambridge: Cambridge University Press.

Byrne, R.W. (1995). *The Thinking Ape: Evolutionary Origins of Intelligence*. Oxford: Oxford University Press.

Byrne, R.W. & Byrne, J.M.E. (1991). Hand preferences in the skilled gathering tasks of mountain gorillas (*Gorilla gorilla beringei*). *Cortex*, 27, 521–46.

Byrne, R.W. & Whiten, A. (eds.) (1988). *Machiavellian Intelligence: Social Expertise and the Evolution of Intellect in Monkeys, Apes and Humans*. Oxford: Clarendon Press.

Caro, T.M. & Hauser, M. (1992). Is there teaching in nonhuman animals? *Quarterly Review of Biology*, 67, 151–74.

Carpenter, C.R. (1934). A field study of the behavior and social relations of howling monkeys. *Comparative Psychology Monographs*, 10, 1–168.

Carpenter, C.R. (1938). A survey of wild life conditions in Atjeh, North Sumatra, with special reference to the orang-utan. *Netherlands Committee for International Nature Protection*, Amsterdam, Communications no. 12, pp. 1–34.

Case, R. (1985). *Intellectual Development: Birth to Adulthood*. New York: Academic Press.

Cheney, D.L. & Seyfarth, R.M. (1990). *How Monkeys See the World: Inside the Mind of Another Species*. Chicago: University of Chicago Press.

Chevalier-Skolnikoff, S. (1977). A Piagetian model for describing and comparing socialization in monkey, ape, and human infants. In *Primate Biosocial Development: Biological, Social and Ecological Perspectives*, ed. S. Chevalier-Skolnikoff & F. Poirier, pp. 159–87. New York: Garland Publishing, Inc.

Chevalier-Skolnikoff, S. (1983). Sensorimotor development in orangutan and other primates. *Journal of Human Evolution*, 12, 545–561.

Chevalier-Skolnikoff, S. (1989). Spontaneous tool use and sensorimotor intelligence in *Cebus* compared with other monkeys and apes. *Behavioral and Brain Sciences*, 12, 561–627.

Chevalier-Skolnikoff, S., Galdikas, B.M.F. & Skolnikoff, A.Z. (1982). The adaptive significance of higher intelligence in wild orangutans: A preliminary report. *Journal of Human Evolution*, 11, 639–52.

Corballis, M. (1991). *The Lopsided Ape: Evolution of the Generative Mind*. Chicago: University of Chicago Press.

Custance, D. & Bard, K.A. (1994). The comparative and developmental study of self-recognition and imitation: The importance of social factors. In *Self-Awareness in Animals and Humans: Developmental Perspectives*, ed. S.T. Parker, R.W. Mitchell & M.L. Boccia, pp. 207–26. New York: Cambridge University Press.

Darwin, C. (1871). *The Descent of Man and Selection in Relation to Sex*. London: John Murray.

Davey, G. (1989). *Ecological Learning Theory*. London: Routledge.

de Waal, F. (1982). *Chimpanzee Politics: Power and Sex Among Apes*. New York: Harper & Row, Publishers.

de Waal, F. (1989). *Peacemaking among Primates*. Cambridge, MA: Harvard University Press.

Diamond, A. (1990). The development of and neural bases of high cognitive functions. *Annals of the New York Academy of Sciences*, 608, 637–76.

Donald, M. (1991). *Origins of the Modern Mind*. Cambridge, MA: Harvard University Press.

Dore, F.Y. & Dumas, C. (1987). Psychology of animal cognition: Piagetian studies. *Psychological Bulletin*, **102**, 219–33.

Dunbar, R.I. (1992). Neocortex size as a constraint on group size in primates. *Journal of Human Evolution*, **20**, 469–93.

Fischer, K. (1980). A theory of cognitive development: The control and construction of hierarchies of skills. *Psychological Review*, **87**, 477–531.

Fodor, J. (1983). *Modularity of Mind*. Cambridge, MA: MIT Press.

French, R. (1994). *Ancient Natural History*. London: Routledge.

Galdikas, B.M.F. & Vasey, P. (1992). Why are orangutans so smart? In *Social Processes and Mental Abilities in Non-Human Primates*, ed. F. Burton, pp. 183–224. Queenston, Ont.: The Edward Mellon Press.

Galef, B.G. Jr (1990). Tradition in animals: Field observations and laboratory analysis. In *Interpretation and Explanation in the Study of Animal Behavior*, ed. M. Beckoff & D. Jamieson, pp. 74–95. Boulder, CO: Westview Press.

Galef, B.G. Jr (1992). The notion of culture. *Human Nature*, **3**, 157–78.

Gardner, R.A., Gardner, B.T. & Van Cantfort, T.E. (1989). *Teaching Sign Language to Chimpanzees*. Albany, NY: State University of New York Press.

Gibson, K.R. (1990). New perspectives on instincts and intelligence: Brain size and the emergence of hierarchical mental construction skills. In *"Language" and Intelligence in Monkeys and Apes: Comparative Developmental Perspectives*, ed. S.T. Parker & K.R. Gibson, pp. 97–128. New York: Cambridge University Press.

Gibson, K.R. (1993). General introduction: Animal minds, human minds. In *Tools, Language, and Cognition in Human Evolution*, ed. K.R. Gibson & T. Ingold, pp. 3–19. Cambridge: Cambridge University Press.

Gibson, K.R. & Ingold, T. (eds.) (1993). *Tools, Language, and Cognition in Human Evolution*. Cambridge: Cambridge University Press.

Goodall, J. (1963). Feeding behaviour of wild chimpanzees: A preliminary report. *Symposia of the Zoological Society of London*, **10**, 39–48.

Gould, S.J. (1977). *Ontogeny and Phylogeny*. Cambridge, MA: The Belknap Press.

Greenfield, P. (1991). Language, tools and the brain: The ontogeny and phylogeny of hierarchically organized sequential behavior. *Behavioral and Brain Sciences*, **14**, 531–95.

Hayes, K.J. & Hayes, C. (1951). The intellectual development of a home raised chimpanzee. *Proceedings of the American Philosophical Society*, **95**, 105–9.

Hayes, K.J. & Hayes, C. (1952). Imitation in a home-raised chimpanzee. *Journal of Comparative and Physiological Psychology*, **45**, 470–4.

Hayes, K.J. & Hayes, C. (1954). The cultural capacity of chimpanzee. *Human Biology*, **26**, 288–303.

Hayes, K.J. & Nissen, C. (1971). Higher mental functions in a home-reared chimpanzee. In *Behavior of Non-human Primates*, ed. H.S. Schrier & F. Stollnitz, vol. 4, pp. 59–115. New York: Academic Press.

Heyes, C. (1993). Imitation, culture and cognition. *Animal Behaviour*, **46**, 999–1010.

Hirschfeld, L.A. & Gelman, S.A. (1994a). Toward a topography of mind: An introduction to domain specificity. In *Mapping the Mind: Domain Specificity in Cognition and Culture*, ed. L.A. Hirschfeld & S.A. Gelman, pp. 3–35. Cambridge: Cambridge University Press.

Hirschfeld, L.A. & Gelman, S.A. (eds.). (1994b). *Mapping the Mind: Domain Specificity in Cognition and Culture*. Cambridge: Cambridge University Press.

Humphrey, N.K. (1976). The social function of intellect. In *Growing Points in Ethology*, ed. P.P.G. Bateson & R.A. Hinde, pp. 303–17. Cambridge: Cambridge University Press.

Janson, H.W. (1952). *Apes and Ape Lore in the Middle Ages and the Renaissance*. Studies of the Warburg Institute, vol. 20. London: Warburg Institute.
Jolly, A. (1966). Lemur social behavior and primate intelligence. *Science*, **153**, 501–6.
Jolly, A. (1972). *The Evolution of Primate Behavior*. New York: Macmillan.
Karmiloff-Smith, A. (1992). *Beyond Modularity: A Developmental Perspective on Cognitive Science*. Cambridge, MA: Bradford–MIT Press.
Kawai, M. (1965). Newly acquired pre-cultural behavior of the natural troop of Japanese monkeys on Koshima Islet. *Primates*, **6**, 1–30.
Kellogg, W. & Kellogg, L. (1933). *The Ape and the Child: A Study of Environmental Influence upon Early Behavior*. New York: McGraw-Hill.
King, M.C. & Wilson, A.C. (1975). Evolution at two levels in humans and chimpanzees. *Science*, **188**, 107–16.
Köhler, W. (1925). *The Mentality of Apes*. London: Kegan Paul, Trench & Co. Ltd.
Kohts, N.N. (1921). [*Report of the Zoopsychology Laboratory of the Darwinian Museum Moscow*]. (In Russian.)
Kohts, N. (1928). Recherches sur l'intelligence du chimpanze par la méthode de "choix d'après modèle". *Journale de la Psychologie Normale et Pathologique*, **28**, 255–75.
Kummer, H. (1967). Tripartite relations in hamadryas baboons. In *Social Communication among Primates*, ed. S.A. Altmann, pp. 63–72. Chicago: University of Chicago Press.
Kummer, H. (1971). *Primate Societies*. Chicago: Aldine.
Langer, J. (1980). *The Origins of Logic: Six to Twelve Months*. New York: Academic Press.
Langer, J. (1981). Logic in infancy. *Cognition*, **10**, 181–6.
Langer, J. (1989). Comparisons with the human child. In *Cognitive Structure and Development of Nonhuman Primates*, ed. F. Antinucci, pp. 229–42. Hillsdale, NJ: Lawrence Erlbaum.
Langer, J. (1993). Comparative cognitive development. In *Tools, Language and Cognition in Evolution*, ed. K.R. Gibson & T. Ingold, pp. 230–50. Cambridge: Cambridge University Press.
Lethmate, J. (1982). Tool-using skills of orang-utans. *Journal of Human Evolution*, **11**, 49–64.
Mathieu, M. & Bergeron, G. (1981). Piagetian assessment on cognitive development in chimpanzee (*Pan troglodytes*). In *Primate Behavior and Sociobiology*, ed. A.B. Chiarelli & R.S. Corruccini, pp. 142–7. Berlin: Springer-Verlag.
Mathieu, M., Daudelin, N., Dagenais, Y. & Decarie, T.G. (1980). Piagetian causality in two house-reared chimpanzees (*Pan troglodytes*). *Canadian Journal of Psychology*, **34**, 179–86.
Matsuzawa, T. (1991). Nesting cups and metatools in chimpanzees. *Behavioral and Brain Sciences*, **14**, 570–1.
Matsuzawa, T. (1994). Field experiments on use of stone tools in the wild. In *Chimpanzee Cultures*, ed. R.W. Wrangham, W.C. McGrew, F.B.M. de Waal & P.G. Heltne, pp. 351–70. Cambridge, MA: Harvard University Press.
McGrew, W.C. (1992). *Chimpanzee Material Culture: Implications for Human Evolution*. Cambridge: Cambridge University Press.
McKinney, M.L. & McNamara, K.J. (1991). *Heterochrony: The Evolution of Ontogeny*. New York: Plenum Press.
Meador, D., Rumbaugh, D.M., Pate, J.L. & Bard, K.A. (1987). Learning, problem solving, cognition, and intelligence. In *Comparative Primate Biology*: vol. 2B: *Behavior, Cognition, and Motivation*, ed. G. Mitchell & J. Erwin, pp. 17–83. New York: Alan R. Liss.
Menzel, E.W. (1973). Chimpanzee spatial memory organization. *Science*, **182**, 943–5.
Menzel, E.W. (1978). Cognitive mapping in chimpanzees. In *Cognitive Processes in Animal Behavior*, ed. S.H. Hulse, H. Fowler & W.K. Honig, pp. 375–422. Hillsdale, NJ: Lawrence Erlbaum.

Mignault, C. (1985). Transition between sensorimotor and symbolic activities in nursery-reared chimpanzees (*Pan troglodytes*). *Journal of Human Evolution*, 14, 747–58.

Miles, L. (1983). Apes and language: The search for communicative competence. In *Language in Apes*, ed. J. De Luce & H.T. Wilder, pp. 43–61. New York: Springer-Verlag.

Mitchell, R.W. (1987). A comparative-developmental approach to understanding imitation. In *Perspectives in Ethology*, eds. P.P.G. Bateson & P.H. Klopfer, vol. 7, pp. 183–215. New York: Plenum Press.

Mitchell, R.W. & Thompson, N.S. (eds.). (1986). *Deception: Perspectives on Human and Nonhuman Deceit*. Albany, NY: State University of New York Press.

Morin, P.A., Moore, J.J., Chakraborty, F., Jin, L., Goodall, J. & Woodruff, D.S. (1994). Kin selection, social structure, gene flow, and the evolution of chimpanzees. *Science*, 265, 1193–201.

Nishida, T. (1987). Local traditions and cultural tradition. In *Primate Societies*, ed. B.B. Smuts, D.L. Cheney, R.M. Seyfarth, R.W. Wrangham & T.T. Struhsaker, pp. 462–74. Chicago: University of Chicago Press.

Nissen, H.W. (1931). A field study of the chimpanzee. *Comparative Psychology Monographs*, 8, 1–122.

Parker, S.T. (1977). Piaget's sensorimotor series in an infant macaque: A model for comparing unstereotyped behavior and intelligence in human and nonhuman primates. In *Primate Biosocial Development: Biological, Social and Ecological Determinants*, ed. S. Chevalier-Skolnikoff & F.E. Poirier, pp. 43–113. New York: Garland Publishing, Inc.

Parker, S.T. (1990). Origins of comparative developmental evolutionary studies of primate mental abilities. In *"Language" and Intelligence in Monkeys and Apes: Comparative Developmental Perspectives*, ed. S.T. Parker & K.R. Gibson, pp. 3–64. New York: Cambridge University Press.

Parker, S.T. (1992). Imitation as an adaptation for apprenticeship in foraging and feeding. Paper presented at the XIVth Congress of the International Primatological Society, Strasbourg, France, 16–21 August.

Parker, S.T. (1995). Using cladistic analysis of comparative data to reconstruct the evolution of cognitive development in hominids. In *Phylogenies and the Comparative Method*, ed. E. Martins. New York: Oxford University Press (in press).

Parker, S.T. & Gibson, K.R. (1977). Object manipulation, tool use, and sensorimotor intelligence as feeding adaptations in cebus monkeys and great apes. *Journal of Human Evolution*, 6, 623–41.

Parker, S.T. & Gibson, K.R. (1979). A developmental model for the evolution of language and intelligence in early hominids. *Behavioral and Brain Sciences*, 2, 367–408.

Parker, S.T. & Gibson, K.R. (eds.) (1990). *"Language" and Intelligence in Monkeys and Apes: Comparative Developmental Perspectives*. New York: Cambridge University Press.

Parker, S.T., Mitchell, R.W. & Boccia, M.L. (eds.) (1994). *Self-Awareness in Animals and Humans: Developmental Perspectives*. New York: Cambridge University Press.

Patterson, F.G. & Cohn, R.H. (1990). Language acquisition in a lowland gorilla: Koko's first ten years of vocabulary development. *Word*, 41, 97–143.

Piaget, J. (1952). *The Origins of Intelligence in Childhood*. New York: International Universities Press.

Piaget, J. (1954). *The Construction of Reality in the Child*. New York: Basic Books.

Piaget, J. (1962). *Play, Dreams and Imitation in Childhood*. New York: Norton.

Povinelli, D.J. & Cant, J.G.H. (1992). Orangutan clambering and the evolutionary origins of self-conception. Paper presented at the XIVth Congress of the International Primatological Society, Strasbourg, France, 16–21 August.

Povinelli, D.J., Nelson, K.E. & Boysen, S.T. (1992). Comprehension of role reversal in chimpanzees: evidence of empathy? *Animal Behaviour*, 43, 633–40.

Premack, D. (1971). Language in a chimpanzee. *Science*, 172, 808–22.
Premack, D. (1984). Possible general effects of language training on the chimpanzee. *Human Development*, 27, 268–81.
Premack, D. & Woodruff, G. (1978). Does the chimpanzee have a theory of mind? *Behavioral and Brain Sciences*, 1, 515–26.
Quiatt, D. (1993). *Primate Behaviour: Information, Social Knowledge, and the Evolution of Culture.* Cambridge: Cambridge University Press.
Redshaw, M. (1978). Cognitive development in human and gorilla infants. *Journal of Human Evolution*, 7, 133–41.
Rumbaugh, D.M. (1969). The transfer index: An alternate measure of learning set. *Proceedings of the 2nd International Congress of Primatology, 1968, Atlanta, GA*, vol. 1, pp. 267–72. Basel: Karger.
Russon, A.E. (1990). The development of peer social interaction in infant chimpanzees: Comparative social, Piagetian, and brain perspectives. In *"Language" and Intelligence in Monkeys and Apes: Comparative Developmental Perspectives*, eds. S.T. Parker & K.R. Gibson, pp. 379–419. New York: Cambridge University Press.
Russon, A.E. & Galdikas, B.M.F. (1992). Cognitive complexity in the object manipulations of free-ranging rehabilitant orangutans. Paper presented at the XIVth Congress of the International Primatological Society, Strasbourg, France, 16–21 August.
Russon, A.E. & Galdikas, B.M.F. (1993). Imitation in free-ranging rehabilitant orangutans (*Pongo pygmaeus*). *Journal of Comparative Psychology*, 107, 147–61.
Russon, A.E. & Galdikas, B.M.F. (1994). The hierarchical organization of complex orangutan object manipulation and tool using routines. Paper presented at XVth Congress of the International Primatological Society, Kuta-Bali, Indonesia, 3–8 August.
Russon, A.E. & Galdikas, B.M.F. (1995). Constraints on great ape imitation: Model and action selectivity in rehabilitant orangutan (*Pongo pygmaeus*) imitation. *Journal of Comparative Psychology* 109, 5–17.
Ruvolo, M. (1994). Molecular evolutionary processes and conflicting gene trees: The Hominoid case. *American Journal of Physical Anthropology*, 94, 89–113.
Salisbury, J.E. (1994). *The Beast Within: Animals in the Middle Ages*. New York: Routledge.
Savage-Rumbaugh, E.S., Murphy, J., Sevcik, R., Brakke, K., Williams, S. & Rumbaugh, D.M. (1993). Language comprehension in ape and child. *Monographs of the Society for Research in Child Development*, 58, 1–122.
Scarr, S. & McCartney, K. (1983). How people make their own environments: A theory of genotype-environment effects. *Child Development*, 54, 424–35.
Schwartz, J.H. (1987). *The Red Ape: Orang-utans and Human Origins*. Boston, MA: Houghton-Mifflin.
Skinner, B.F. (1956). A case history in the scientific method. *Scientific American*, 11, 221–33.
Sternberg, R.J. (1990). *Metaphors of Mind: Conceptions of the Nature of Intelligence*. Cambridge: Cambridge University Press.
Tinbergen, N. (1963). On aims and methods of ethology. *Zeitschrift für Tierpsychologie*, 20, 410–429.
Tomasello, M., Kruger, A.C. & Ratner, H.H. (1993a). Cultural learning. *Behavioral and Brain Sciences*, 16, 495–552.
Tomasello, M., Savage-Rumbaugh, E.S. & Kruger, A.C. (1993b). Imitative learning of actions on objects by children, chimpanzees, and enculturated chimpanzees. *Child Development*, 64, 1688–705.

Toth, N., Schick, K.D., Savage-Rumbaugh, E.S., Sevcik, R. & Rumbaugh, D.M. (1993). *Pan* the tool maker: Investigations into the stone tool-making and tool-using capabilities of a bonobo (*Pan paniscus*). *Journal of Archaeological Science*, 20, 81–91.

Tuttle, R.H. (1986). *Apes of the World: Their Social Behavior, Communication, Mentality and Ecology.* Park Ridge, NJ: Noyes Publications.

Visalberghi, E. & Trinca, L. (1989). Tool use in capuchin monkeys: Distinguishing between performing and understanding. *Primates*, 30, 511–21.

Vygotsky, L. (1962). *Thought and Language*. Cambridge, MA: MIT Press.

Westergaard, G.C. & Suomi, S.J. (1993). Use of a tool-set by capuchin monkeys (*Cebus apella*). *Primates*, 34, 459–62.

Whiten, A. (ed.) (1991). *Natural Theories of Mind: Evolution, Development and Simulation of Everyday Mindreading.* Oxford: Basil Blackwell Ltd.

Whiten, A. & Byrne, R.W. (1988). Tactical deception in primates. *Behavioral and Brain Sciences*, 11, 233–73.

Whiten, A. & Custance, D. (1994). Functions and mechanisms of imitation: Studies of monkeys, apes and human children. Paper presented at the Social Learning and Tradition Workshop, Madingley, Cambridgeshire, UK, 21–26 August.

Whiten, A. & Ham, R. (1992). On the nature and evolution of imitation in the animal kingdom: Reappraisal of a century of research. In *Advances in the Study of Behavior*, eds. P.J.B. Slater, J.S. Rosenblatt, C. Beer & M. Milinski, vol. 21, pp. 239–83. New York: Academic Press.

Wrangham, R.W., McGrew, W.C., de Waal, F.B.M. & Heltne, P.G. (eds.) (1994). *Chimpanzee Cultures*. Cambridge, MA: Harvard University Press.

Wright, R.V.S. (1972). Imitative learning of a flaked-tool technology: The case of an orangutan. *Mankind*, 8, 296–306.

Yerkes, R.M. (1916). The mental life of monkeys and apes: a study of ideational behavior. *Behavior Monographs*, 3, 1–145.

Yerkes, R.M. (1925). *Almost Human*. New York: The Century Co.

Yerkes, R.M. (1943). *Chimpanzees: A Laboratory Colony*. New Haven, CT: Yale University Press.

Yerkes, R.M. & Yerkes, A.W. (1929). *The Great Apes*. New Haven, CT: Yale University Press.

PART ONE

The scope of great ape intelligence

2

Chimpanzees and capuchin monkeys: Comparative cognition

JAMES R. ANDERSON

INTRODUCTION: CHIMPANZEES AND COGNITION

The combined efforts of comparative psychologists, anthropologists, biologists, ecologists, and others have done much to elucidate the problem-solving abilities and thought processes of our nearest living relatives, chimpanzees (*Pan troglodytes*) and, to a lesser extent, pygmy chimpanzees or bonobos (*Pan paniscus*). We also have available an impressive body of knowledge about problem-solving and mental activities in capuchin monkeys (*Cebus*), the other primate genus put under the spotlight here. The aim of this chapter is to compare current knowledge regarding cognitive abilities in these two primate genera, one closely related to humans, one much more distantly related.

The two species of chimpanzees, along with gorillas (*Gorilla gorilla*) and orangutans (*Pongo pygmaeus*), belong to the great ape family, which shared a common ancestor with humans somewhere between 6 and 15 million years ago. It is generally accepted that the great apes are the most cognitively advanced nonhuman primates. The four species of *Cebus* belong to the family Cebidae, one of the two (or three) New World primate families, believed to have separated from the ancestral Old World primates about 30 million years ago. As will become clearer, capuchins merit special attention due to their many similarities with humans and great apes in morphology, behavior, and life history patterns (Fragaszy *et al.*, 1990). It has been suggested that similar object manipulatory and sensorimotor abilities arose in cebus monkeys and chimpanzees through convergent evolution, as an adaptation for the extraction of embedded food sources (Parker & Gibson, 1977).

New studies are continually refining our understanding of thought processes in these and other species, but as a useful starting point for the present comparative discussion, the following facts may be borne in mind regarding chimpanzees:

1. They progress through all six stages of sensorimotor intelligence, as well as certain preoperational capacities when tested on Piagetian tasks (Dore & Dumas, 1987).
2. They can learn a "vocabulary" of arbitrary symbols and use them to communicate intelligently with other individuals (Gardner *et al.*, 1989; Parker & Gibson, 1990).

3. They show sometimes insightful solutions to cognitive problems, including problems requiring the use of tools (Beck, 1977; Köhler, 1925).
4. They show a wide range of different forms of tool use, manufacture tools and use tool-kits. Further, when they use tools they clearly understand the nature of the task and the cause–effect relationship involved in the act (Boesch & Boesch, 1990, 1993; Goodall, 1986; McGrew, 1992; Chapter 3).
5. They show self-awareness as expressed through self-recognition in mirrors (Gallup, 1970, 1994).
6. They can imitate the actions of others (Custance & Bard, 1994; Hayes & Hayes, 1952; see also Tomasello *et al.*, 1993).
7. They engage in *at least* second-order intentionality, meaning that they can represent another individual's mental state, and use this representation to control their own behavior (Byrne & Whiten, 1988, 1992; Premack & Woodruff, 1978).

The above list is impressive, and naturally raises the question as to what extent such abilities are shared with other species. The present comparison is largely restricted to capuchin monkeys, but it should be remembered that chimpanzees may be equalled or surpassed in certain cognitive domains, including tool use and linguistic performance, by the other great apes (see Lethmate, 1982; McGrew, 1992; Miles, 1986; Parker & Gibson, 1990; Patterson, 1980; Chapters 5 and 7). The case can also be made that some nonprimates, even some nonmammals, engage in behaviors as cognitively complex as those shown by chimpanzees (Beck, 1982, 1986; Herman & Morrel-Samuels, 1990; Pepperberg, 1983; Schusterman & Gisiner, 1988). In order to facilitate the present discussion, chimpanzees are considered as being the "cognitive champions" among nonhumans, but "great apes" could probably replace "chimpanzees" in much of what follows. It is also noteworthy, however, that while gorillas may have the greatest absolute neocortical volume among nonhuman primates (see Tuttle, 1986; Chapter 5), chimpanzees score higher than the other great apes and all monkeys on encephalization and neocortical progression indices (Gibson, 1986; Parker, 1990; Stephan, 1972).

To sum up: based on the available evidence, chimpanzees are considered to be the most cognitively advanced nonhuman primates in existence, although much if not all of their cognitive prowess may be shared with the other great apes. Some of this evidence may be seen to be opening up a "cognitive gap" within the order Primates, between, on the one hand, chimpanzees along with the other great apes and humans, and, on the other hand, all other species: prosimians, monkeys, and lesser apes. It is interesting to note that such a rift had already been proposed many years ago by Yerkes & Yerkes (1929) with regard to tool use. But is the cognitive division justified? What follows is a critical, point-by-point evaluation of some of the evidence relating to the cognitive abilities of chimpanzees compared to those of capuchin monkeys.

CAPUCHINS AS COGNITIVE "CURIOSITIES"

"What is so special about capuchin monkeys?" it might be asked. While it is true that in many ways these monkeys are no more interesting or valuable than other species, in the context of a debate on cognition, they are indeed worthy of special attention. Experimental data on capuchin problem-solving and tool use were already being collected over 60 years ago (see Klüver, 1937). Since Klüver's analysis of the behavior of a single young adult female, observations in the wild and especially in captivity have accumulated to such an extent that these monkeys are now widely acknowledged as being the most prolific tool users among nonhuman primates other than great apes (Beck, 1980; Visalberghi, 1990).

The impressive capacity for tool use in itself arouses interest in the cognitive abilities of capuchins, but we are interested in cognition in a more general sense. On some of the more "traditional" tests of learning ability, capuchin monkeys do not excel. For example, in their review of the performance of different species on object discrimination learning sets (DLS), Fobes & King (1982) show capuchin monkeys as scoring higher than squirrel monkeys (over 70% correct choices on trial 2 after 500 problems, versus 800 problems), but lower than spider monkeys (over 80% correct after 500 problems) and rhesus monkeys (over 80% after 300 problems); chimpanzees outperform all monkeys (80% correct after 150 problems). However, there are wide variations in the methods used in the studies on DLS, in terms of stimulus presentation procedures and performance criteria. In another study, capuchin monkeys made fewer errors than did owl monkeys, squirrel monkeys, and gibbons on successive spatial discrimination reversals (Gossette, 1970). There were no important differences between capuchin, rhesus monkeys and chimpanzees in cross-modal (touch-to-vision) recognition using a method based on discrimination of edible and inedible food objects (Elliott, 1977). Nevertheless, capuchins solved fewer detour (bent wire) problems than macaques, resembling squirrel and woolly monkeys on these tests (Davis & Leary, 1968). Capuchin monkeys ranked below squirrel monkeys, rhesus monkeys, and chimpanzees in solving patterned strings problems, but one of the two subjects tested by Harlow & Settlage (1934) was only 1 year old, and species differences in visual acuity may also be a confounding factor (Harris & Meyer, 1971). In the domain of auditory short-term memory, Colombo & D'Amato (1986) found maximum retention intervals in capuchin monkeys that largely surpassed those of Japanese macaques, but again the studies used different methods.

Only one published study has assessed capuchins on the Transfer Index (TI), a measure based on multiple two-choice discrimination problems with reversals that yield evidence of a qualitative shift in learning among primates, with "mediational" processing taking precedence over associative learning as we cross the primate order from prosimians to great apes (chimpanzees; Rumbaugh & Pate, 1984). De Lillo & Visalberghi (1994) obtained overall TI scores for *Cebus apella* that approached those of the great apes. However, results using another paradigm designed to differentiate between associative and mediational learning (ML) modes led the same authors to conclude that "... in ML testing, the capuchin mode of learning reveals features typical of monkeys and not of apes" (De Lillo & Visalberghi, 1994, p. 285). It is difficult to draw firm conclusions from this study, however, due to the significant individual differences in patterns of performance among the four subjects. Furthermore, De Lillo & Visalberghi (1994) acknowledged that other paradigms, i.e. conditional matching to sample, and serial learning, do provide evidence of mediational processing in capuchins (D'Amato & Colombo, 1988; D'Amato & Salmon, 1984; D'Amato et al., 1985).

In contrast to the somewhat ambiguous picture that emerges from "traditional" experimental psychological methods of assessing learning and cognition, the data coming from studies focusing on sensorimotor intelligence is less ambiguous: capuchins (like great apes) surpass other monkeys. Schino et al. (1990) compared capuchin monkeys and macaques on tasks derived from the Piagetian object concept series. Adults but not juveniles responded nonrandomly on trials involving invisible displacement, but only the capuchin monkey appeared to solve the task by means of mental representation, i.e. stage 6 representational skills. Mathieu et al. (1976) obtained object permanence data indicative of stage 6 capacities in an adult *C. capucinus* and a chimpanzee, but not in a woolly monkey (*Lagothrica flavicauda*). Natale (1989) applied a Piagetian analysis to the use of a stick to rake in objects and concluded that their cebus and gorilla subjects, but not macaques, reached "a fully spatialized conception of causality" (Natale, 1989, p. 131). Chevalier-Skolnikoff (1989) suggest that object manipulation and tool use by capuchin monkeys revealed a superior comprehension of object–object relations than that found in other monkeys; this issue will be taken up later. Finally, capuchin monkeys score higher than almost all other monkeys on indices of brain development related to cognitive potential (Fragaszy, 1990; Gibson, 1986; Stephan, 1972).

To sum up: from the point of view of cognition, capuchin monkeys often appear to resemble great apes more than they resemble other monkeys, to whom they are obviously phylogenetically more closely related. The origin of this resemblance may lie in adaptation for extractive foraging (Parker & Gibson, 1977). In the remainder of this chapter, some of the parallels and contrasts between chimpanzees' and capuchin monkeys' cognitive achievements are evaluated. The overall aims are to provide an up-to-date overview of recent research, to evaluate the evidence pertaining to cognitive processes in the two species, and to point out some areas for further research.

COMPARATIVE EARLY NEURO-BEHAVIORAL DEVELOPMENT

An ontogenetic generality is that mammalian species showing advanced cognitive complexity as adults go through a relatively long period of immaturity; the prolonged periods as infants and juveniles are prime periods for learning and cognitive development (Schultz, 1969). In this context, neuro-behavioral competence at birth and shortly thereafter may serve as a useful indicator of ultimate cognitive potential. Capuchin monkeys and chimpanzees both have a long life span compared to other primate species. In captivity, chimpanzees may live to over 50 years of age, with capuchins coming not too far behind (see Bartus et al., 1982). In comparison, the maximal life span of macaques is around 35 years (Cutler, 1984). Capuchins and chimpanzees are also both behaviorally less competent at birth than many other species of nonhuman primates. Direct comparisons have been made between neonatal capuchin monkeys and another member of the family Cebidae, namely squirrel monkeys, concerning brain and body growth, perceptual, motor, cognitive, and social development. Elias (1977) reported that the weight of a squirrel monkey's brain at birth was over 60% of the adult value, whereas in neonatal capuchin monkeys it was less than 50%. Neonates were removed from their mothers at birth, nursery-reared, and their perceptual and motor development assessed weekly. Capuchin monkeys lagged 3–6 weeks behind squirrel monkeys on a number of measures including locomotor development and stage 4 object permanence. In a comparison of behavioral development in mother and group-reared infants, Fragaszy *et al.* (1991) found that capuchins lagged behind squirrel monkeys on several early behavioral criteria, often by several weeks. Excursions from the mother, social play, and object handling, among other measures, all appeared later in the capuchin monkeys. Fragaszy (1990) gave an overview of the relatively slow development of postural control, independent locomotion, and voluntary grasping of objects in capuchin monkeys compared to squirrel monkeys, macaques and baboons, and related this developmental tardiness to capuchins' ultimate advanced level of cognitive functioning.

Although neuro-behavioral development in capuchin monkeys is slow compared to that in other species of monkeys, it is fast indeed compared to that of the great apes, including chimpanzees. For example, in contrast to neonatal capuchins, chimpanzee newborns are relatively poor at clinging, and need to be supported and carried most of the time during locomotion by the mother (Plooij, 1987). Steinbacher (1941, cited in van de Rijt-Plooij & Plooij, 1987) observed one chimpanzee baby who could not maintain the ventro-ventral position on the mother by himself until nearly 3 months of age (but Nicolson, 1977, observed unaided clinging during travel in one 2 week old baby). During the neonatal period (up to 30 days), attentional state and motor performance of chimpanzees strikingly resemble those observed in humans, as measured by the Brazleton Neonatal Behavioral Assessment Scale (Bard, 1990). Throughout the first year, sensorimotor development of chimpanzees parallels that of human infants, according to assessment based on the Užgiris–Hunt scales (Hallock & Worobey, 1984), while Piagetian techniques show faster but more limited sensorimotor development in Old World monkeys than in chimpanzees and other great apes (Chevalier-Skolnikoff, 1983). A 7 month old female chimpanzee showed tertiary circular reactions typical of a stage 5 human infant (10–11 to 18 months), but her repertoire was relatively limited (Mathieu & Bergeron, 1981). According to Chevalier-Skolnikoff (1983), infant chimpanzees and orangutans complete sensorimotor stages 2 through 4 faster than humans, but require much longer to fully attain stage 6 (up to 8 years, compared to 2 years; see also Spinozzi & Poti, 1993).

There are interesting species contrasts in the development of independent locomotion, which is related to the development of early prehension and manipulation schemes, and thereby to cognitive development (Antinucci, 1989). Squirrel monkey infants may be out of contact with the mother or other individuals for at least 50% of the time at the age of 15 weeks, whereas this degree of independence is not reached by capuchin infants until 23 weeks of age (Fragaszy *et al.*, 1991). Chimpanzee infants are much slower still to start making excursions from the mother. In the wild, infants were never out of contact with the mother to a comparable extent even at 9 months of age (van de Rijt-Plooij &

Plooij, 1987). In a zoo setting, infants over 2 years of age spent two thirds of the time out of reach of the mother, rising to 80% of the time at 4 years of age (Yoshida *et al.*, 1991). Some other differences in behavioral developmental speed between mother-reared rhesus monkeys and chimpanzees were described by Dienske (1984), and Russon (1990) described the later onset of different patterns of social play in chimpanzees compared to rhesus monkeys. Poti & Spinozzi (1994) summarized the relation between early locomotion, prehension schemes, and cognitive development in several species of hand- or nursery-reared primates. Infant macaques start to locomote at stage 2 and show high levels of buccal prehension (directly taking objects in the mouth). Capuchins also start to locomote at stage 2 but show much more, albeit simple, manual exploration of objects. Chimpanzees start to locomote at stage 3, i.e. later than other primate species, but they are already relatively skilled object manipulators by this time.

OBJECT MANIPULATION AND COMBINATION

The emergence of object manipulation and object–object relations is intimately related to the development of cognitive abilities (Vauclair, 1984). Such object operations depend on functional manipulative abilities, which are themselves subject to development. Capuchin monkeys show frequent and varied precision grips (Costello & Fragaszy, 1988), starting at 13 weeks of age, which is several weeks later than in macaques and baboons but several months earlier than in chimpanzees (Fragaszy, 1990). Combinatorial or "generative" object related acts held to be precursors of tool use – appear in the second half of the first year in capuchin monkeys (Fragaszy & Adams-Curtis, 1991; Natale, 1989), and have recently been described in tool-using infant baboons (Westergaard, 1993). Use of a stick to dip for syrup first appeared towards the end of the first year of life in the capuchins studied by Westergaard & Fragaszy (1987). In a longitudinal study of the use of a stick to rake in food, Parker & Poti (1990) recorded some instances of effective tool use by a female *C. apella* after the age of 9 months, but consistent success, indicating comprehension of the stick as a detached tool, was not observed until the age of 15 months. The earliest successful tool-assisted nut-cracking in the group studied by Anderson (1990) was by a 2 year old female.

Gibson (1990a) noted that object manipulation developed slowly in chimpanzees. Combinatorial acts with objects were virtually absent during the first year in a chimpanzee and a bonobo studied by Vauclair & Bard (1983), and very rare in the baby chimpanzees studied by Poti & Spinozzi (1994). Object–object relations increase in the chimpanzee's second year, but they do not attain the levels seen in human infants (Langer, 1994; Mignault, 1985). Efficient tool use also develops slowly in chimpanzees. The stick problem is solved at about 3–4 years (Schiller, 1952), while adult-like proficiency in tool use to capture termites and driver ants is not achieved before the age of 3 years and 5 years, respectively (McGrew, 1974, 1977, 1992). Tool-assisted nut-cracking may not be perfected until about 10 years of age (Boesch & Boesch-Ackermann, 1991).

Several studies have compared object manipulation in different species of primates. In their paper on reactions to novel objects in over 200 zoo-housed mammals and reptiles, Glickman & Sroges (1966) noted that capuchin monkeys and chimpanzees showed a wider variety of manipulatory responses than most other species studied. In a comparison limited to primates, Parker (1974) found a greater diversity of actions performed on an object (a knotted piece of rope protruding into the subject's cage) in great apes than in lemurs, monkeys, and gibbons. Surprisingly, capuchin monkeys generally obtained low scores, and ranked alongside or below lemurs and langurs. However, the impoverished object-directed repertoire of the capuchin monkeys studied by Parker appears to be anomalous. Torigoe (1985, 1986) studied object manipulation in over 70 species of primates and found that capuchins were best classified alongside the apes in terms of the variety of actions observed and body parts used. In particular, capuchins frequently brought objects into contact with one another. Such combinatorial object manipulation appears to be particularly important for the emergence of tool use skills. Greenfield (1991) described different complexities of combinatorial object manipulation and provided evidence that chimpanzees show some of the same object manipulation schemes as human infants. Westergaard & Suomi (1994) showed that some capuchin monkeys also engaged in the most complex object combinatorial pattern ("subassembly") when tested with the same materials (nesting cups) as humans and chimpanzees were, but their performances were dominated by the simplest pattern ("pairing").

SKILLFUL TOOL USE

Although examples of tool use in species that are phylogenetically distant from primates deserve more analysis in terms of underlying cognitive complexity than they are usually accorded (Beck, 1986; Chapter 3), it is efficient tool use by skilled primates that arouses the greatest interest from anthropologists and psychologists. The most rigorous definition of tool use is given by Beck (1980, p. 10): "the external employment of an unattached environmental object to alter more efficiently the form, position, or condition of another object, another organism, or the user itself when the user holds or carries the tool during or just prior to use and is responsible for the proper and effective orientation of the tool." Among the nonhuman primates, chimpanzees show the broadest repertoire of tool-using acts, at least in the wild (Beck, 1980; Boesch & Boesch, 1990; Goodall, 1986; McGrew, 1992). Tool use plays no important role in any wild population of gorillas or orangutans studied so far, but in captivity these apes can match chimpanzees in terms of solving tasks requiring tool use, showing that they possess the same tool-use-relevant cognitive potential as that shown by chimpanzees.

Köhler (1925) sought to clarify the thought processes involved in tool use in captive chimpanzees, and proposed insight to characterize the sudden comprehension of a solution by internally restructuring the elements of the problem in relation to each other. The best-known example concerns the chimpanzee Sultan, who joined two sticks together to form a tool long enough to reach food lying outside the cage. Insight is considered to be an important cognitive process because it may lead to the more efficient discovery of a solution – in terms of time and energy expended – than overt (or indeed covert) trial-and-error learning. However, it is difficult to identify an insightful tool use act with any certainty (Anderson, 1989; Bernstein, 1989). A major problem is that individuals with experience of objects and their usefulness in a given situation may continue to learn more about object–object relations by engaging in sometimes apparently aimless manipulations (Baldwin, 1989), so that what appears to be an insightful solution to a task might simply involve the transfer of an already established manipulatory pattern to a new context; this raises the possibility of explanation based on transfer of learning, rather than mental recombination (see Windholz & Lamal, 1985). Schiller (1952) reported that, even in the absence of any food-related problem, 31 out of 48 chimpanzees spontaneously joined two short sticks that could be fitted together. He emphasized the interaction of unlearned motor patterns and experience in the emergence of so-called insightful solutions (for a discussion of interpretations of the problem-solving performances by Köhler's chimpanzees, see King & Forbes, 1982).

Insight is not easy to define operationally. Existing attempts, such as unsuccessful responding on the task, followed by a period of nongoal oriented activity and then the sudden appearance of the solution (e.g. Beck, 1967), appear overly restrictive. Despite these problems, a number of examples of intuitively insightful tool use by chimpanzees may be found in the literature (see, e.g. Goodall, 1986; Toth et al., 1993).

In contrast to tool use by chimpanzees, accounts of tool use in monkeys rarely give cause to consider the possibility that insight is at work. Indeed, studies of the acquisition of tool use by immature and adult monkeys tend to emphasize the role of overt trial-and-error and successive approximations to the final act, rather than cognitive restructuring. This is true for capuchin monkeys as well as other species (Anderson, 1985, 1990; Beck, 1980; Candland et al., 1978; Visalberghi, 1987). However, one instance of a tool use solution with at least some characteristics of insight (as operationalized by Beck (1967)) was observed in an adult male *C. apella*. Skilled at inserting sticks into holes in a box in order to extract honey, the capuchin was given a stick which was too thick to fit into the holes, while appropriately thin sticks were lying just out of direct reach outside the cage. On previous occasions, in the absence of thin sticks, the male obtained appropriate tools for dipping by gnawing and splintering the thick stick. On this occasion, however, after a few cursory attempts to insert the thick stick, instead of splintering it he stopped briefly, then quickly carried it across to the front of the cage near to where the thin sticks were lying and used it to rake one in; he then used the latter stick to dip for the honey. The male repeated this act upon subsequent presentations of the problem (Anderson & Henneman, 1994, see Figure 2.1), and never behaved inappropriately by using one thin stick to rake in another. He had been caught in the wild as a juvenile, and had lived with a female and their offspring on an island in a zoo for many years. Although we do not have all the details of his past experience, to our knowledge he had never before been exposed to such a problem.

The behavior of the adult male capuchin described above not only suggests insight, it also meets the definition of the use of a *tool-set*, namely two or more types of tools used sequentially to achieve a goal (Brewer & McGrew, 1990); this has been observed in free-ranging chimpanzees, who used different sticks to break open an arboreal bees' nest and then extract the honey (Brewer & McGrew, 1990), or who used a stone and then a stick to crack open nuts and extract the meat (Boesch & Boesch, 1990). Matsuzawa (1991) described an adult female chimpanzee using one stone to keep another stone (the anvil) steady while a third stone was used as a hammer. Captive capuchin monkeys have recently been shown to use a tool-set: a stone to pound open a walnut and then a stick to extract the inner meat (Westergaard & Suomi, 1993). The use of tool-sets, therefore, is clearly not beyond the ability of capuchin monkeys. One category of tool use that remains to be demonstrated in capuchin monkeys is *secondary* tool use, i.e. the use of one object as a tool for making a tool out of another object. This has been shown in captive chimpanzees (Kitahara-Frisch *et al.*, 1987) and in a pygmy chimpanzee (Toth *et al.*, 1993). However, given existing observations on use of tools and tool-sets by capuchins, it would be surprising if secondary tool use could not be demonstrated.

Tool use and mental representation

Tool use may reflect advanced cognitive abilities, or it may be based on relatively simple learning (or even innate) mechanisms. In other words, a distinction can be made between "intelligent" and "unintelligent" tool use (Parker & Gibson, 1977). The diversity of contexts in which chimpanzees use tools, and the flexibility of their behavior in selecting appropriate materials and modifying tools provide convincing evidence that these apes understand what they are doing when they use tools (Boesch & Boesch, 1993; McGrew, 1974). In other words, chimpanzees mentally represent the elements of the problem, and understand the causal relationship between the function of the tool, the necessary action to be performed, and the desired outcome (for other evidence of such comprehension in chimpanzees, see Premack and Woodruff, 1978).

Do capuchin monkeys show evidence of mental representation during tool use? Visalberghi & Trinca (1989) found that capuchin monkeys skilled at pushing food out of a transparent horizontal tube with sticks made persistent errors. For example, they used inappropriate tools and modified tools excessively, leading to the conclusion that "in tool use the cognitive capacity of capuchins does not reach the sixth Piagetian stage of representational means" (Visalberghi & Trinca, 1989, p. 520). Further evidence that capuchin monkeys have a limited understanding of cause–effect relationships during tool use comes from a report by Visalberghi & Limongelli (1994). In this study, the monkeys again had to use a stick to push food out of a transparent horizontal tube, but the tube contained a trap, such that pushing from the wrong end led to the loss of the bait. Only one out of four tool users systematically avoided the trap, but subsequent experimentation revealed that this monkey used sensory cues (visible distance) rather than mental representation of the problem when deciding how to respond. The authors concluded that capuchin monkeys do not understand the cause–effect relationships of tool use (see Chapter 3).

In contrast to the view that tool use in capuchin monkeys is not based on mental representation is that view that object manipulation and tool use by these monkeys reflects adanced levels of sensorimotor intelligence, *sensu* Piaget. This argument was made most strongly by Chevalier-Skolnikoff (1989), who described behaviors suggestive

(a)

(b)

Figure 2.1. An adult male *Cebus apella* uses one tool (a), (b) and (c) to rake in another tool, suitable for dipping (d).

(c)

(d)

of insight and representation, including tool use and possible instances of deferred imitation. However, the evidence put forward by Chevalier-Skolnikoff was weak (Anderson, 1989; Bard & Vauclair, 1989; Bernstein, 1989; Fragaszy, 1989; Visalberghi, 1989), and inadequate for concluding that "insight", "mental combination" or "representation" occurred (note also that these terms do not mean the same thing; Greenfield, 1989). In this context, spontaneous use of tool-sets by capuchin monkeys is relevant (Anderson & Henneman, 1994; Westergaard & Suomi, 1993), as they suggest that these primates may indeed mentally represent the problem and its solution. Finally, on the basis of responses to variations of the classic stick and food problem, Natale (1989) concluded that macaques never arrived at an understanding of the conditions leading to efficient use of the stick to rake in food, whereas a gorilla and a cebus monkey did. Natale also pointed out the value of analyzing how the principles and variables regulating the use of a tool are understood by the user, rather than emphasizing simple success or failure. Further data are clearly necessary for clarifying to what extent capuchins have an understanding of their tool use. Direct comparisons of the behavior of capuchin monkeys and chimpanzees at various stages of development on variations of similar tool use tasks would be valuable. Chapters 3 and 12 in this volume are of particular interest in this regard.

The question of mental representation is not limited to tool use, however, and from this point of view other experiments on capuchin monkeys are informative. For example, D'Amato (1991) reviewed a series of studies showing that cebus monkeys can extract positional information from serially ordered events and that, unlike that of pigeons, their performance appears to be based on an internal representation of the stimulus series, rather than simpler discriminative processes. In a preliminary study of puzzle-box-solving, Simons & Holtkotter (1986) found that a cebus monkey used visual control, rather than mechanical "fiddling" to undo various catches to open a box that contained food. A definitive synthesis of the relationships between the cognitive processes involved in tool use and those operating in other problem-solving situations is awaited, but Langer's work (Chapter 12 this volume) goes in this direction.

SELF-RECOGNITION

The ability to recognize one's own reflection in a mirror, or some other visual representation of self such as a photograph or televised image, is considered significant by researchers interested in cognitive development: developmental, comparative and clinical psychologists, anthropologists, and psychiatrists. Self-recognition implies that an individual is capable of becoming the object of his or her own attention (Gallup, 1982), that the individual has a representation of how he or she should appear from a different visual perspective (Anderson, 1984a; Mitchell, 1993a). It is an important phenomenon in the wider context of self-concept or self-consciousness (Parker et al., 1994).

Studies spanning over two decades have yielded the following general picture regarding the ability of nonhuman primates to self-recognize when confronted with their reflection in a mirror: Great apes show evidence of self-recognition, whereas monkeys do not (Parker *et al.*, 1994). Several criteria may be used to evaluate whether a nonverbal primate recognizes that the reflection is a representation of its own body: experimenting with unusual postures or facial expressions while self-monitoring in the mirror, spontaneously using the reflection to explore or groom normally unseen parts of the body, using the mirror to inspect visually and tactually a mark on a normally unseen body part (the mark test introduced by Gallup, 1970) (Anderson, 1993).

Like tool use, self-recognition has been studied much more extensively in chimpanzees than in the other great apes. Several confirmatory studies have appeared, including small-sample studies of orangutans and gorillas (Calhoun & Thompson, 1988; Lethmate & Dücker, 1973; Miles, 1990; Parker, 1991; Parker *et al.*, 1994; Patterson, 1986; Patterson & Cohn, 1994; Suarez & Gallup, 1981; Swartz & Evans, 1994). The early development of self-recognition in chimpanzees has similarities with that in humans, but passing the mark test appears later (Lin *et al.*, 1992; Povinelli *et al.*, 1993). An interesting recent development is the finding that some apparently normal chimpanzees may fail to show one, two, or even all three of the above-mentioned criteria of self-recognition (Povinelli *et al.*, 1993; Swartz & Evans, 1991). The implications of such variability in the extent to which self-recognition is shown remain to be fully worked out, but its discovery may explain why earlier attempts to demonstrate self-recognition in gorillas, often using small numbers of subjects in less than favorable settings, yielded largely negative results (Ledbetter & Basen, 1982; Lethmate, 1974).

As far as monkeys are concerned, however, the widespread failure to find self-recognition does not appear to be attributable to some form of dissociation among criteria or to individual differences in ability, because all three criteria are *consistently* absent in monkeys. Most studies have been done on macaques, and have revealed, instead of the emergence of behaviors indicative of self-recognition, a pattern consisting of social responding followed by habituation and (sometimes) avoidance, with a retained potential to respond socially (Anderson, 1984b, 1994; Gallup, 1975, 1987). It has been suggested that monkeys may lack a sufficiently well-integrated self-awareness to be able to self-recognize (Gallup, 1982, 1987).

The possibility that competent mirror use and tool use are related cognitively has been raised by a number of researchers; both involve the use of an intermediate object to solve a problem (Anderson, 1992; McGrew, 1992; Riviello *et al.*, 1993; Vauclair & Anderson, 1995). Given their proclivity for tool use, how capuchin monkeys respond to mirrors is of particular interest. This question has been studied recently (for a review, see Anderson & Marchal, 1994).

The results of these studies, mostly on *C. apella*, show that capuchin monkeys treat their image in a mirror much as do other monkeys, with social responses initially predominating then diminishing over time, but never being replaced by behaviors indicative of self-recognition (Anderson, 1994; Anderson & Roeder, 1989; Collinge, 1989; Riviello *et al.*, 1993). The failure of capuchin monkeys to self-recognize occurs whether the mirror is large and stationary or small and portable (Figure 2.2(*a*) and (*b*)). One of the conditions used by Anderson & Roeder (1989) deserves particular comment. In order to reduce eye-to-eye contact between the viewing subject and its reflection in the usual face-on mirror arrangement, angled mirrors were used, positioned so that eye-to-eye contact was impossible. This condition dramatically reduced social responses directed towards the mirror, but did not lead to any signs of self-recognition. Thus, it cannot be argued that the emergence of self-recognition was impeded by the subjects' being locked into an aggressive "stare-out" with the mirror image.

The most recent study of capuchin monkeys and mirror images (Marchal & Anderson, 1993), on *C. capucinus*, confirms that these primates fail to show signs of self-recognition, and that like macaques (Anderson, 1986; Itakura, 1987) they can make use of the reflection of their own bodies and the environment to find objects that are not directly visually accessible. In this experiment, subjects were given the opportunity to locate and pick off raisins stuck onto a target surface facing away from the home cage and invisible except in a mirror which was sometimes placed directly opposite, reflecting back the target area and the subjects (Figure 2.2(c)). Two out of three subjects showed clearly better performance, in terms of the number of reaches required to find the raisin, when the mirror was present compared to when a nonreflecting surface was present.

In chimpanzees (see Menzel *et al.*, 1985), mirror-guided reaching appears to be an extension of the capacity to recognize visual representations of parts of one's own body. At the present time, the mental operations underlying mirror-mediated finding of hidden objects by monkeys have not been satisfactorily determined. Might successful performance indicate some degree of *partial* self-recognition (of hand and arm), or is the

Figure 2.2. (*a*) Social reaction towards a large mirror in *Cebus apella*. (*b*) Manipulation of a small, portable mirror. (*c*) Use of the reflection to find a hidden raisin by an adult male *Cebus capucinus*.

Cognition in chimpanzees and capuchin monkeys

(b)

(c)

location of the reflected limb in relation to the bait simply used as an environmental cue for where to start searching manually? The latter hypothesis invokes a simple associative learning process rather than self-recognition, and it must be remembered that the mirror-using capuchin monkeys showed no spontaneous signs of self-recognition. Further studies are indicated to shed light on the processes controlling the mirror-mediated location of hidden objects by apparently nonself-recognizing monkeys. One interesting methodological extension of the basic paradigm with monkeys would be to replace the mirror with televised images, to allow greater stimulus control.

Some authors have drawn attention to the fact that mirror-self-recognition in chimpanzees taps the ability of these apes to respond to the physical or somatic aspect of self, but that mirror-induced self-awareness in humans goes beyond simply responding to one's body as an object (Mitchell, 1993a; Morin & DeBlois, 1989; Parker *et al.*, 1994). Indeed, social psychologists have found that the presence of a mirror may induce a wide range of sometimes subtle behavioral and cognitive effects in humans, explainable in terms of self-concept (Buss, 1980; Duval & Wicklund, 1972). Similar mirror-induced self-awareness effects remain to be demonstrated in nonhuman primates (Mitchell, 1993a); no one has yet *tried* to test self-recognizing apes for any of the objective self-awareness effects observed in humans (Anderson, 1993; see also Chapter 15). Nevertheless, following Gallup (1982) and Povinelli (1993), there is a growing consensus that the ability to recognize oneself in a mirror is intimately bound up with self-awareness in a more general sense – an ability to become aware of one's own physical *and mental state* and an accompanying ability to hypothesize about the mental states of *other individuals*, i.e. theory of mind, which takes us into the domain of social cognition.

SOCIAL COGNITION

Increasing importance is being attributed to social relationships as being at the source of the highly evolved cognitive capacities in many species of primates (Byrne & Whiten, 1988; Cheney *et al.*, 1986; Humphrey, 1976; Jolly, 1966). For individuals living in complex groups, it is advantageous to be able to assess rapidly and accurately the signals given by others, and to recognize not only social relationships, but *relationships between relationships* (Seyfarth & Cheney, 1988). Laboratory and field studies both contribute to revealing the psychological skills individuals may call upon in dealing with others. One example of observational and experimental approaches converging to provide evidence for a particular social skill concerns tactical deception. Byrne & Whiten (1992, p. 612) defined tactical deception as "acts from the normal repertoire of the agent, deployed such that another individual is likely to misinterpret what the acts signify, to the advantage of the agent." The basic phenomenon of deception is widespread among animals, but it can be categorized according to the complexity of the mental processes involved, with higher levels being produced with greater cognitive control than lower levels (Mitchell, 1986, 1993b). The most interesting levels of deception, from the point of view of the thought processes involved, are those involving at least "second-order intentionality" or "mindreading" (Byrne & Whiten, 1988, 1992; Dennett, 1988), in which the actor attributes goals or intentions to others.

There exist many descriptions of tactical deception in chimpanzees, ranging in complexity from simple concealment or witholding of information, to the use of a social tool in which both the tool and the target are deceived, to the understanding of being deceived and the taking of counterdeceptive measures (Byrne & Whiten, 1992; de Waal, 1986, 1992; Menzel, 1974; Nishida, 1990). Such accounts have been reported by careful and experienced observers, sometimes with control over the context in which the acts occur (e.g. Menzel, 1974), but confirmation from experimental situations apt to increase the production of deceptive acts is desirable (Bernstein, 1988; de Waal, 1992). The best experimental data come from Woodruff & Premack (1979). Individually tested chimpanzees could obtain food by correctly signalling its location to a cooperative trainer who gave the food, or by misinforming an uncooperative trainer who kept any food found on the basis of the ape's signals. "Honest" signalling to the cooperative trainer, usually by pointing an outstretched arm or leg towards the baited object, appeared quickly in the four chimpanzees tested. Deception by witholding information from the uncooperative trainer appeared more gradually. Eventually, two of the chimpanzees systematically sent the uncooperative trainer to the wrong location, and were thus allowed to obtain the food. This study does not appear to have been replicated.

Although several studies have documented the basics about group life in capuchin monkeys (the importance of kin relations, dominance, etc.), there exist almost no reports bearing directly on the socio-cognitive abilities of capuchins. Cebidae are underrepresented in records of deception (Byrne & Whiten, 1992). Further, the four records of deception retained by Byrne & Whiten were classified at the lowest levels; in fact, there were no records of "higher order" (second order and above) tactical deception, i.e. involving mental or visual perspective-taking, by any New World primates. In contrast, great apes, particularly chimpanzees, are well represented in this category. Gibson (1990b) gave a few examples of low-level deceptive acts in a pet capuchin monkey, such as giving false alarm calls to attract attention, or looking away from a desired object.

Are capuchin monkeys capable of tactical deception in an experimental situation? Mitchell & Anderson (1997) tested two adult male *C. apella* in a protocol similar to that of Woodruff & Premack (1979), but over a much shorter time span and with more trials each day. The capuchins received over 300 trials in each of the two conditions: "cooperative" trainer and "uncooperative" trainer. Figure 2.3 illustrates a trial in the uncooperative trainer condition, and Table 2.1 shows the percentage of the last 50 trials in which the trainer chose the baited or the unbaited object on the basis of the subjects' signals (or lack thereof). It can be seen that both subjects indicated the baited object to the cooperative trainer on most trials. One subject behaved such that the uncooperative trainer chose the baited object slightly less frequently than chance. More interestingly, however, the second subject *systematically sent the uncooperative trainer to the wrong object*, and thereby obtained the food, because the uncooperative trainer left empty handed and the experimenter who baited the objects returned, praised the subject, and gave him the food.

Table 2.1. *Choice of the baited and unbaited object by cooperative and uncooperative trainers, based on information given by two individually tested* Cebus apella

	Cooperative trainer		Uncooperative trainer	
Subject	Baited object	Unbaited object	Baited object	Unbaited object
Churchill	42	8	21	29
Coluche	49	1	8	42

Data represent the last 50 trials. In cooperative trainer trials, if the trainer chooses the baited object the subject gets the food; in uncooperative trainer trials, the subject gets the food only if the trainer chooses the unbaited object.

(a)

Figure 2.3. A trial in the uncooperative trainer condition. (*a*) and (*b*) The uncooperative trainer pays no attention while another trainer shows the subject which one of two objects is baited. (*c*) and (*d*) The uncooperative trainer arrives and chooses the object he thinks is baited, based on the subject's signals. (*e*) The uncooperative trainer has chosen the correct object so eats the food. If the uncooperative trainer chooses the unbaited object, he leaves empty handed and the first trainer comes back and gives the subject the food (not shown).

(d)

(e)

It would be premature to conclude from the above experiment that tactical deception based on higher order socio-cognitive or intentional processes had been demonstrated in a capuchin monkey. A simpler explanation would be that the "deceiving" individual had simply mastered a conditional discrimination: if A (cooperative trainer approaches) do X (indicate "correct" object), if B (uncooperative trainer approaches) do Y (indicate "incorrect" object). Finer-grained analyses of the monkey's behavior in this context and in variants of the protocol will be necessary for teasing out the processes underlying the performance. (Note that the same cautionary remark holds for the single experimental study of deception in chimpanzees by Woodruff & Premack, 1979.)

As mentioned above, there is a dearth of records of tactical deception involving perspective-taking by capuchin monkeys. The potential of capuchin monkeys to make use of another's visual perspective was recently addressed in a study by Anderson et al. (1995). Three adult C. apella were given series of object-choice tasks under two conditions: (1) the experimenter provided no overt cues as to which was the correct (i.e., baited) object, (2) the experimenter indicated the correct object by staring at it, pointing to it, or both. Figure 2.4 illustrates the different conditions. None of the subjects used the experimenter's gaze as a cue to increase their frequency of finding the food, but all three fared significantly better when the experimenter "pointed," i.e. laid a hand with extended index finger near to the correct object. This finding, which we have recently replicated in rhesus monkeys, supports other experimentally obtained data suggesting that monkeys may be limited in their ability to take the visual (and/or mental) perspective of others; in this respect they appear less capable than chimpanzees and other great apes (Byrne & Whiten, 1992; Cheney & Seyfarth, 1990; Gómez, 1991; Povinelli et al., 1990, 1991). Of course, such data do not show that monkeys *cannot* engage in perspective-taking, only that the ability may not be easily demonstrated experimentally.

Imitation
One behavioral phenomenon that appears intimately related to self-recognition and visual perspective-taking, and thereby to self-awareness and social cognition, is imitation (Hart & Fegley, 1994; Meltzoff & Moore, 1992; Mitchell, 1993a; Parker, 1991). The ability to imitate is clearly advantageous; like insight, it allows acquisition of a novel behavior more efficiently than trial-and-error learning; in fact it enables "no-trial learning" (Meltzoff, 1988). Imitation is also important in discussions of self-awareness and social cognition because, when shown by very young human infants, it precedes other important cognitive phenomena such as self-recognition and person permanence (but for contradictory evidence in the chimpanzee, see Custance & Bard, 1994). Meltzoff & Moore (1992) suggested that very early imitation serves as the base upon which social cognition develops.

There is good evidence that chimpanzees are able to imitate motor acts. For example, Hayes & Hayes (1952) reported spontaneous and induced imitation of gestures and object-related acts in a home-raised chimpanzee. Sign-language trained chimpanzees imitate each other and humans in the contexts of signing and using implements (Gardner et al., 1989). Recently, Custance & Bard (1994) and Custance et al. (1994)

Figure 2.4. Conditions used in the study of use of experimenter-given cues by *Cebus apella*.
(*a*) Baseline condition – the experimenter remains immobile and stares at a neutral, midline point.
(*b*) Experimenter's head and gaze are oriented towards the baited object. (*c*) Experimenter stares at

(c)

(d)

and points to the baited object. (d) Experimenter points to the baited object, while staring at neutral, midline point. All three subjects scored significantly better than chance in conditions (c) and (d) only.

showed that young chimpanzees can learn to imitate both familiar and novel gestures shown by a human. The ability of chimpanzees to imitate novel use of a tool has been called into question by Tomasello et al. (1987) and Nagell et al. (1993), who observed that laboratory-reared chimpanzees did not increase efficiency by reproducing the precise tool use technique of a skilled model (another chimpanzee or a human). "Enculturated" chimpanzees, i.e. those reared by humans in relatively enriched and responsive environments, appear to show greater imitative propensities than their nonenculturated counterparts (Tomasello et al., 1993). In general, however, the particular conditions of the laboratory setting should be borne in mind when assessing both successes and failures in such experiments (McGrew, 1992); great apes in more enriched, free-ranging settings appear to imitate object-related acts more readily than those in captivity (Russon & Galdikas, 1993).

Evaluations of capuchin monkeys' ability to imitate have been largely limited to observations of the role of the social context in the dissemination of novel behaviors, such as food-washing (Visalberghi & Fragaszy, 1990a), and tool use (Fragaszy & Visalberghi, 1990) in groups containing skilled and naive individuals. These studies demonstrate the importance of interindividual proximity, and local and stimulus enhancement in the spread of a novel behavior within a group, but there is no evidence for imitation (Visalberghi & Fragasy, 1990b; see also Fragaszy et al., 1994). Stimulus enhancement could also account for the "imitation" reported in an early study by Warden et al. (1940), in which the monkeys were presented with a demonstrator monkey solving a puzzle device in a neighboring cage before being given access to an identical puzzle. Parker & Poti (1990) noted the lack of imitation in the development of the use of a stick as a rake in *C. apella*, but they suggested that the monkeys might be able to learn about object–object cause–effect relationships through observation. One adult male in the group studied by Anderson (1990) used a log to break open a hazelnut for the first time immediately after seeing another individual doing so, but the male was already skilled at opening nuts using hammer stones. Chevalier-Skolnikoff (1989) observed capuchin monkeys reproduce object-directed acts observed in another individual several months earlier, and interpreted these observations as evidence of deferred imitation, which is accepted as being cognitively more complex than simultaneous imitation (Meltzoff, 1988), but Chevalier-Skolnikoff's interpretations have been heavily criticized (see Open Peer Commentary in Chevalier-Skolnikoff, 1989). Finally, Gibson (1990b) noted stimulus enhancement and social facilitation effects, but no imitation of motor acts in a tame capuchin monkey. There appear to have been no attempts to train capuchin monkeys to imitate gestures using methods similar to those employed by Custance & Bard (1994). One attempt to train a long-tailed macaque to imitate the act of scratching failed when the subject had to transfer scratching to novel body parts (Mitchell & Anderson, 1993), but it might be worthwhile to attempt a similar procedure with capuchin monkeys.

CONCLUDING REMARKS

The main conclusion to be drawn from this review is that we lack sufficient comparable data to be sure whether or not there are important qualitative differences in the respective cognitive abilities of chimpanzees and capuchin monkeys. The strongest evidence to date for such a difference concerns self-recognition; in contrast to chimpanzees, no capuchin monkey has shown any sign that it recognizes its own reflection. Certain social correlates of self-recognition (or markers of mind, in Gallup's (1982) terms) have been documented in chimpanzees, including the ability to imitate and to intentionally deceive; specific investigations of these skills in capuchin monkeys are needed. Another example of a potentially interesting research question concerns "linguistic" ability. The extent to which capuchins could master elements of an artificial system of communication is unknown, but D'Amato (1991) has drawn attention to the similarities between knowledge of ordinal position from sequential events, as shown in *Cebus*, and comprehension of language by apes (e.g. Savage-Rumbaugh, 1988).

The question of mental representation or comprehension of cause–effect relations in the contexts of object manipulation and tool use remains open. Some experiments indicate that capuchin monkeys are limited in this area, whereas other observations suggest that at least some individuals may progress from a simple "instrumental" understanding, based on successes and failures of their actions, to a "representational" understanding, based on a conception of how things work (see Langer, 1994). Further, some paradigms give evidence of mediational thought processes in capuchin monkeys, whereas others suggest only simpler associative processes. This overall fragmentary picture recalls the hypothesis of "domain-specific" intelligence, or "modularity of mind" (Cheney & Seyfarth, 1990). Langer (1989) suggested that the cognitive structures of macaques and cebus monkeys are disorganized and asynchronic, in contrast to the situation in humans. It may be that not only the development of, but also the degree of organization among, various cognitive structures contribute to some of the observable chimpanzee–capuchin differences in behavior.

ACKNOWLEDGEMENTS

My own research described in this chapter was conducted at the Centre de Primatologie of the Université Louis Pasteur. I thank the Director, Nicolas Herrenschmidt, and the personnel at the "Fort" for encouragement and support. Thanks also to Hélène Neveu (who appears in Figure 2.3), Fabienne Aujard, and Elisabeth Ludes for help on the "experimental deception" study.

REFERENCES

Anderson, J.R. (1984a). The development of self-recognition: A review. *Developmental Psychobiology*, 17, 35–49.

Anderson, J.R. (1984b). Monkeys with mirrors: Some questions for primate psychology. *International Journal of Primatology*, 5, 81–98.

Anderson, J.R. (1985). Development of tool-use to obtain food in a captive group of *Macaca tonkeana*. *Journal of Human Evolution*, **14**, 637–45.
Anderson, J.R. (1986). Mirror-mediated finding of hidden food by monkeys (*Macaca tonkeana* and *M. fascicularis*). *Journal of Comparative Psychology*, **100**, 237–42.
Anderson, J.R. (1989). On the contents of capuchins' cognitive tool-kit. *Behavioral and Brain Sciences*, **12**, 588–9.
Anderson, J.R. (1990). Use of objects as hammers to open nuts by capuchin monkeys (*Cebus apella*). *Folia Primatologica*, **54**, 138–45.
Anderson, J.R. (1992). L'outil et le miroir: leur rôle dans l'étude des processus cognitifs chez les primates non humains. *Psychologie Française*, **37**, 81–90.
Anderson, J.R. (1993). To see ourselves as others see us: A response to Mitchell. *New Ideas in Psychology*, **11**, 339–46.
Anderson, J.R. (1994). The monkey in the mirror: A strange conspecific. In *Self-awareness in Animals and Humans: Developmental Perspectives*, ed. S.T. Parker, R.W. Mitchell & M.L. Boccia pp. 315–329. New York: Cambridge University Press.
Anderson, J.R. & Henneman, M.-C. (1994). Solutions to a tool-use problem in a pair of *Cebus apella*. *Mammalia* **58**, 351–61.
Anderson, J.R. & Marchal, P. (1994). Capuchin monkeys and confrontations with mirrors. In *Current Primatology*, vol. II: *Social Development, Learning and Behaviour*, ed. J.-J. Roeder, B. Thierry, J.R. Anderson & N. Herrenschmidt, pp. 371–380. Strasbourg: Université Louis Pasteur.
Anderson, J.R. & Roeder, J.-J. (1989). Responses of capuchin monkeys (*Cebus apella*) to different conditions of mirror-image stimulation. *Primates*, **30**, 581–7.
Anderson, J.R., Salleberry, P. & Barbier, H. (1995). Use of experimenter-given cues during object-choice tasks by capuchin monkeys. *Animal Behaviour*, **49**, 201–208.
Antinucci, F. (ed.) (1989). *Cognitive Structure and Development in Nonhuman Primates*. Hillsdale, NJ: Lawrence Erlbaum.
Baldwin, J.D. (1989). Does "spontaneous" behavior require "cognitive special creation"? *Behavioral and Brain Sciences*, **12**, 589–90.
Bard, K.A. (1990). A brief report on neurobehavioral integrity in chimpanzee and human neonates. *Primate Foundation of Arizona Newsletter*, **2**, 3–5
Bard, K.A. & Vauclair, J. (1989). What's the tool and where's the goal? *Behavioral and Brain Sciences*, **12**, 590–1.
Bartus, R.T., Dean, R.L. III, Beer, B. & Lippa, A.S. (1982). The cholinergic hypothesis of geriatric memory dysfunction. *Science*, **217**, 408–17.
Beck, B.B. (1967). A study of problem solving by gibbons. *Behaviour*, **28**, 95–109.
Beck, B.B. (1977). Köhler's chimpanzees: How did they really perform? *Zoologischer Garten N.F. Jena*, **47**, 352–60.
Beck, B.B. (1980). *Animal Tool Behavior: The Use and Manufacture of Tools by Animals*. New York: Garland STPM Press.
Beck, B.B. (1982). Chimpocentrism: Bias in cognitive ethology. *Journal of Human Evolution*, **11**, 3–17.
Beck, B.B. (1986). Tools and intelligence. In *Animal Intelligence: Insights into the Animal Mind*, ed. R.J. Hoage & L. Goldman, pp. 135–47. Washington, DC: Smithsonian Institution Press.
Bernstein, I.S. (1988). Metaphor, cognitive belief, and science. *Behavioral and Brain Sciences*, **11**, 247–8.
Bernstein, I.S. (1989). Cognitive explanations: Plausibility is not enough. *Behavioral and Brain Sciences*, **12**, 593–4.

Boesch, C. & Boesch, H. (1990). Tool use and tool making in wild chimpanzees. *Folia Primatologica*, **54**, 86–99.

Boesch, C. & Boesch, H. (1993). Diversity of tool use and tool-making in wild chimpanzees. In *The Use of Tools by Human and Non-human Primates*, ed. A. Berthelet & J. Chavaillon, pp. 158–68. Oxford: Clarendon Press.

Boesch, C. & Boesch-Ackermann, H. (1991). Les chimpanzés et l'outil. *La Recherche*, **22**, 724–31.

Brewer, S.M. & McGrew, W.C. (1990). Chimpanzee use of a tool-set to get honey. *Folia Primatologica*, **54**, 100–4.

Buss, A.H. (1980). *Self-consciousness and Social Anxiety*. San Francisco: Freeman.

Byrne, R.W. & Whiten, A. (eds.) (1988). *Machiavellian Intelligence: Social Expertise and the Evolution of Intellect in Monkeys, Apes, and Humans*. Oxford: Clarendon Press.

Byrne, R.W. & Whiten, A. (1992). Cognitive evolution in primates: Evidence from tactical deception. *Man* (N.S.), **27**, 609–27.

Calhoun, S. & Thompson, R.L. (1988). Long-term retention of self-recognition by chimpanzees. *American Journal of Primatology*, **15**, 361–5.

Candland, D.K., French, J.A. & Johnson, C.N. (1978). Object-play: Test of a categorized model by the genesis of object-play in *Macaca fuscata*. In *Social Play in Primates*, ed. E.O. Smith, pp. 259–96. New York: Academic Press.

Cheney, D.L. & Seyfarth, R.M. (1990). *How Monkeys See the World*. Chicago: University of Chicago Press.

Cheney, D.L., Seyfarth, R.M. & Smuts, B.B. (1986). Social relationships and social cognition in nonhuman primates. *Science*, **234**, 1361–6.

Chevalier-Skolnikoff, S. (1983). Sensorimotor development in orang-utans and other primates. *Journal of Human Evolution*, **12**, 545–61.

Chevalier-Skolnikoff, S. (1989). Spontaneous tool use and sensorimotor intelligence in *Cebus* compared with other monkeys and apes. *Behavioral and Brain Sciences*, **12**, 561–627.

Collinge, N.E. (1989). Mirror reactions in a zoo colony of *Cebus* monkeys. *Zoo Biology*, **8**, 89–98.

Colombo, M. & D'Amato, M.R. (1986). A comparison of visual and auditory short-term memory in monkeys (*Cebus apella*). *Quarterly Journal of Experimental Psychology*, **38B**, 425–48.

Costello, M.B. & Fragaszy, D.M. (1988). Prehension in *Cebus* and *Saimiri*: I. Grip type and hand preference. *American Journal of Primatology*, **15**, 235–45.

Custance, D. & Bard, K.M. (1994). The comparative and developmental study of self-recognition and imitation: The importance of social factors. In *Self-awareness in Animals and Humans: Developmental Perspectives*, ed. S.T. Parker, R.W. Mitchell & M.L. Boccia, pp. 207–26. New York: Cambridge University Press.

Custance, D.M., Whiten, A. & Bard, K.A. (1994). The development of gestural imitation and self-recognition in chimpanzees (*Pan troglodytes*) and children. In *Current Primatology*, vol. 2: *Social Development, Learning and Behaviour*, ed. J.-J. Roeder, B. Thierry, J.R. Anderson & N. Herrenschmidt, pp. 381–7. Strasbourg: Université Louis Pasteur.

Cutler, R.G. (1984). Antioxidants, aging, and longevity. In *Free Radicals in Biology*, vol. VI, ed. W.A. Pryor, pp. 371–428. New York: Academic Press.

D'Amato, M.R. (1991). Comparative cognition: Processing of serial order and serial pattern. In *Current Topics in Animal Learning: Brain, Emotion, and Cognition*, ed. L. Dachowski & C.F. Flaherty, pp. 165–85. Hillsdale, NJ: Lawrence Erlbaum.

D'Amato, M.R. & Colombo, M. (1988) Representation of a serial order in monkeys (*Cebus apella*). *Journal of Experimental Psychology, Animal Behavior Processes*, **14**, 131–9.

D'Amato, M.R. & Salmon, D.P. (1984). Cognitive processes in cebus monkeys. In *Animal Cognition*, ed. H. L. Roitblat, T. G. Bever & H. S. Terrace, pp. 149–64. Hillsdale, NJ: Lawrence Erlbaum.

D'Amato, M.R., Salmon, D., Loukas E. & Tomie, A. (1985). Symmetry and transitivity of conditional relations in monkeys (*Cebus apella*) and pigeons (*Columbia livia*). *Journal of the Experimental Analysis of Behavior*, **44**, 35–47.

Davis, R.T. & Leary, R.W. (1968). Learning of detour problems by lemurs and seven species of monkeys. *Perceptual and Motor Skills*, **27**, 1031–4.

De Lillo, C. & Visalberghi, E. (1994). Transfer Index and mediational learning in tufted capuchins (*Cebus apella*). *International Journal of Primatology*, **15**, 275–87.

Dennett, C. (1988). The intentional stance in theory and practice. In *Machiavellian Intelligence: Social Expertise and the Evolution of Intellect in Monkeys, Apes, and Humans*, ed. R.W. Byrne & A. Whiten, pp. 180–202. Oxford: Clarendon Press.

de Waal, F.B.M. (1986). Deception in the natural communication of chimpanzees. In *Deception: Perspectives on Human and Nonhuman Deceit*, ed. R.W. Mitchell & N.S. Thompson, pp. 221–44. Albany, NY: State University of New York Press.

de Waal, F.B.M. (1992). Intentional deception in primates. *Evolutionary Anthropology*, **1**, 86–92.

Dienske, H. (1984). Early development of motor abilities, daytime sleep and social interactions in the rhesus monkey, chimpanzee and man. *Clinics in Developmental Medicine*, **94**, 126–43.

Dore, F.Y. & Dumas, C. (1987). Psychology of animal cognition: Piagetian studies. *Psychological Bulletin*, **102**, 219–33.

Duval, S. & Wicklund, R.A. (1972). *A Theory of Objective Self-awareness*. New York: Academic Press.

Elias, M.F. (1977). Relative maturity of cebus and squirrel monkeys at birth and during infancy. *Developmental Psychobiology*, **10**, 519–28.

Elliott, R.C. (1977). Cross-modal recognition in three primates. *Neuropsychologia*, **15**, 183–6.

Fobes, J.L. & King, J.E. (1982). Measuring primate learning abilities. In *Primate Behavior*, ed. J.L. Fobes & J.E. King, pp. 289–326. New York: Academic Press.

Fragaszy, D.M. (1989). Tool use, imitation, and insight: Apples, oranges, and conceptual pea soup. *Behavioral and Brain Sciences*, **12**, 596–8.

Fragaszy, D.M. (1990). Early behavioral development in capuchins (*Cebus*). *Folia Primatologica*, **54**, 119–28.

Fragaszy, D.M. & Adams-Curtis, L.E. (1991). Generative aspects of manipulation in tufted capuchin monkeys (*Cebus apella*). *Journal of Comparative Psychology*, **105**, 387–97.

Fragaszy, D.M., Baer, J. & Adams-Curtis, L. (1991). Behavioral development and maternal care in tufted capuchins (*Cebus apella*) and squirrel monkeys (*Saimiri sciureus*) from birth through seven months. *Developmental Psychobiology*, **24**, 375–93.

Fragaszy, D.M. & Visalberghi, E. (1990). Social processes affecting the appearance of innovative behaviors in capuchin monkeys. *Folia Primatologica*, **54**, 155–65.

Fragaszy, D.M., Visalberghi, E. & Robinson, J.G. (1990). Variability and adaptability in the genus *Cebus*. *Folia Primatologica*, **54**, 114–8.

Fragaszy, D.M., Vitale, A.F. & Ritchie, B. (1994). Variation among juvenile capuchins in social influences on exploration. *American Journal of Primatology*, **32**, 249–60.

Gallup, G.G., Jr (1970). Chimpanzees: Self-recognition. *Science*, **167**, 86–7.

Gallup, G.G., Jr (1975). Towards an operational definition of self-awareness. In *Socioecology and Psychology of Primates*, ed. R.H. Tuttle, pp. 309–41. The Hague: Mouton.

Gallup, G.G., Jr (1982). Self-awareness and the emergence of mind in primates. *American Journal of Primatology*, **2**, 237–48.

Gallup, G.G., Jr (1987). Self-awareness. In *Comparative Primate Biology*, vol. 2B: *Behavior, Cognition, and Motivation*, ed. G. Mitchell & J. Erwin, pp. 3–16. New York: Alan R. Liss.

Gallup, G.G., Jr (1994). Self-recognition: Research strategies and experimental design. In *Self-awareness in Animals and Humans: Developmental Perspectives*, ed. S.T. Parker, R.W. Mitchell & M.L. Boccia, pp. 35–50. New York: Cambridge University Press.

Gardner, R.A., Gardner, B.T. & Van Cantfort, T.E. (eds.) (1989). *Teaching Sign Language to Chimpanzees*. Albany, NY: State University of New York Press.

Gibson, K.R. (1986). Cognition, brain size and the extraction of embedded food resources. In *Primate Ontogeny, Cognition and Social Behaviour*, ed. J.G. Else & P.C. Lee, pp. 93–103. Cambridge: Cambridge University Press.

Gibson, K.R. (1990a). New perspectives on instincts and intelligence: Brain size and the emergence of hierarchical mental constructional skills. In *"Language" and Intelligence in Monkeys and Apes: Comparative Developmental Perspectives*, ed. S.T. Parker & K.R. Gibson, pp. 97–128. New York: Cambridge University Press.

Gibson, K.R. (1990b). Tool use, imitation, and deception in a captive cebus monkey. In *"Language" and Intelligence in Monkeys and Apes: Comparative Developmental Perspectives*, ed. S.T. Parker & K.R. Gibson, pp. 205–18. New York: Cambridge University Press.

Glickman, S.E. & Sroges, R.W. (1966). Curiosity in zoo animals. *Behaviour*, 26, 151–88.

Gómez, J.C. (1991). Visual behavior as a window for reading the mind of others in primates. In *Natural Theories of Mind: Evolution, Development and Simulation of Everyday Mindreading*, ed. A. Whiten, pp. 195–207. Oxford: Basil Blackwell Ltd.

Goodall, J. (1986). *The Chimpanzees of Gombe: Patterns of Behavior*. Cambridge, MA: The Belknap Press.

Gossette, R.L. (1970). Comparisons of SDR performance of gibbons and three species of New World monkeys on a spatial task. *Psychonomic Science*, 19, 301–3.

Greenfield, P.M. (1989). *Cebus* uses tools, but what about representation? Comparative evidence for generalized cognitive structures. *Behavioral and Brain Sciences*, 12, 559–60.

Greenfield, P.M. (1991). Language, tools, and the brain: The ontogeny and phylogeny of hierarchically organized sequential behavior. *Behavioral and Brain Sciences*, 14, 531–95.

Hallock, M.B. & Worobey, J. (1984). Cognitive development in chimpanzee infants (*Pan troglodytes*). *Journal of Human Evolution*, 13, 441–7.

Harlow, H.F. & Settlage, P.H. (1934). Comparative behavior of primates. VII. Capacity of monkeys to solve patterned strings tests. *Journal of Comparative Psychology*, 18, 423–35.

Harris, D.G. & Meyer, M.E. (1971). The relationship between visual acuity and performance on patterned string problems by infrahuman primates. *Psychonomic Science*, 22, 60.

Hart, D. & Fegley, S. (1994). Social imitation and the emergence of a mental model of self. In *Self-awareness in Animals and Humans: Developmental Perspectives*, ed. S.T. Parker, R.W. Mitchell & M.L. Boccia, pp. 149–65. New York: Cambridge University Press.

Hayes, K.J. & Hayes, C. (1952). Imitation in a home-raised chimpanzee. *Journal of Comparative and Physiological Psychology*, 45, 450–9.

Herman, L.M. & Morrel-Samuels, P. (1990). Knowledge acquisition and asymmetry between language comprehension and production: Dolphins and apes as general models for animals. In *Interpretation and Explanation in the Study of Animal Behavior*, vol. 1: *Interpretation, Intentionality, and Communication*, ed. M. Bekoff & D. Jamieson, pp. 283–312. Boulder, CO: Westview Press.

Humphrey, N.K. (1976). The social funtion of intellect. In *Growing Points in Ethology*, ed. P.P.G. Bateson & R.A. Hinde, pp. 303–17. Cambridge: Cambridge University Press.

Itakura, S. (1987). Use of a mirror to guide their responses in Japanese monkeys (*Macaca fuscata fuscata*). *Primates*, **28**, 343–52.
Jolly, A. (1966). Lemur social behavior and primate intelligence. *Science*, **153**, 501–6.
King, J.E. & Fobes, J.L. (1982). Complex learning by primates. In *Primate Behavior*, ed. J.L. Fobes & J.E. King, pp. 327–60. New York: Academic Press.
Kitahara-Frisch, J., Norikoshi, K. & Hara, K. (1987). Use of a bone fragment as a step toward secondary tool use in captive chimpanzees. *Primate Report*, **18**, 33–7.
Klüver, H. (1937). Re-examination of implement-using behavior in a cebus monkey after an interval of three years. *Acta Psychologia*, **2**, 347–97.
Köhler, W. (1925). *The Mentality of Apes*. New York: Harcourt, Brace & Co. Ltd.
Langer, J. (1989). Comparison with the human child. In *Cognitive Structure and Development in Nonhuman Primates*, ed. F. Antinnuci, pp. 229–42. Hillsdale, NJ: Lawrence Erlbaum.
Langer, J. (1994). From acting to understanding: The comparative development of meaning. In *The Nature and Ontogenesis of Meaning*, ed. W.F. Overton & D.S. Palermo, pp. 191–213. Hillsdale, NJ: Lawrence Erlbaum.
Ledbetter, D.H. & Basen, J.A. (1982). Failure to demonstrate self-recognition in gorillas. *American Journal of Primatology*, **2**, 307–10.
Lethmate, J. (1974). Selbst-kenntnis bei Menschenaffen. *Umschau*, **74**, 486–7.
Lethmate, J. (1982). Tool-using skills of orang-utans. *Journal of Human Evolution*, **11**, 49–64.
Lethmate, J. & Dücker, G. (1973). Untersuchungen zum Selbsterkennen im Spiegel bei Orang-Utans und einigen anderen Affenarten. *Zeitschrift für Tierpsychologie*, **33**, 248–69.
Lin, A.C., Bard, K.A. & Anderson, J.R. (1992). Development of self-recognition in chimpanzees (*Pan troglodytes*). *Journal of Comparative Psychology*, **106**, 120–7.
Marchal, P. & Anderson, J.R. (1993). Mirror-image responses in capuchin monkeys (*Cebus capucinus*): Social responses and use of reflected environmental information. *Folia Primatologica*, **61**, 165–73.
Mathieu, M. & Bergeron, G. (1981). Piagetian assessment of cognitive development in chimpanzees (*Pan troglodytes*). In *Primate Behavior and Sociobiology*, ed. A. B. Chiarelli & R. S. Corruccini, pp. 142–7. Berlin: Springer-Verlag.
Mathieu, M., Bouchard, M.A., Granger, L. & Herscovitch, J. (1976). Piagetian object-permanence in *Cebus capucinus*, *Lagothrica flavicauda* and *Pan troglodytes*. *Animal Behaviour*, **24**, 585–8.
Matsuzwa, T. (1991). Nesting cups and metatools in chimpanzees. *Behavioral and Brain Sciences*, **14**, 570–1.
McGrew, W.C. (1974). Tool use by wild chimpanzees in feeding upon driver ants. *Journal of Human Evolution*, **3**, 501–8.
McGrew, W.C. (1977). Socialization and object manipulation of wild chimpanzees. In *Primate Biosocial Development*, ed. S. Chevalier-Skolnikoff & F. E. Poirier, pp. 261–88. New York: Garland STPM Press.
McGrew, W.C. (1992). *Chimpanzee Material Culture: Implications for Human Evolution*, Cambridge: Cambridge University Press.
Meltzoff, A.N. (1988). Imitation, objects, tools, and the rudiments of language in human ontogeny. *Human Evolution*, **3**, 45–64.
Meltzoff, A.N. & Moore, M.K. (1992). Early imitation within a functional framework: The importance of person identity, movement, and development. *Infant Behavior and Development*, **15**, 479–505.
Menzel, E.W. (1974). A group of young chimpanzees in a one-acre field. In *Behavior of Nonhuman Primates*, vol. 5, ed. A. M. Schrier & F. Stollnitz, pp. 83–153. New York: Academic Press.

Menzel, E.W., Savage-Rumbaugh, E.S. & Lawon, J. (1985). Chimpanzee (*Pan troglodytes*) spatial problem solving with the use of mirrors and televised equivalents of mirrors. *Journal of Comparative Psychology*, 99, 211–7.

Mignault, C. (1985). Transition between sensorimotor and symbolic activities in nursery-reared chimpanzees (*Pan troglodytes*). *Journal of Human Evolution*, 14, 747–58.

Miles, H.L.W. (1986). How can I tell a lie? Apes, language, and the problem of deception. In *Deception: Perspectives on Human and Nonhuman Deceit*, ed. R.W. Mitchell & N.S. Thompson, pp. 245–66. Albany: State University of New York Press.

Miles, H.L.W. (1990). The cognitive foundations for reference in a signing orangutan. In *"Language" and Intelligence in Monkeys and Apes: Comparative Developmental Perspectives*, ed. S.T. Parker & K.R. Gibson, pp. 511–39. New York: Cambridge University Press.

Mitchell, R.W. (1986). A framework for discussing deception. In *Deception: Perspectives on Human and Nonhuman Deceit*, ed. R.W. Mitchell & N.S. Thompson, pp. 3–40. Albany, NY: State University of New York Press.

Mitchell, R.W. (1993a). Mental models of self-recognition: Two theories. *New Ideas in Psychology*, 11, 295–325.

Mitchell, R.W. (1993b). Animals as liars: The human face of nonhuman duplicity. In *Lying and Deception in Everyday Life*, ed. M. Lewis & C. Saarni, pp. 59–89. New York: Guilford Press.

Mitchell, R.W. & Anderson, J.R. (1993). Discrimination learning of scratching, but failure to obtain imitation and self-recognition in a long-tailed macaque. *Primates*, 34, 301–9.

Mitchell, R.W. & Anderson, J.R. (1997). Pointing, witholding information, and deception in capuchin monkeys (*Cebus apella*). *Journal of Comparative Psychology*, 111, 351–61.

Morin, A. & DeBlois, S. (1989). Gallup's mirrors: More than an operationalization of self-awareness in primates? *Psychological Reports*, 65, 287–91.

Nagell, K., Olguin, R.S. & Tomasello, M. (1993). Processes of social learning in the tool use of chimpanzees (*Pan troglodytes*) and human children (*Homo sapiens*). *Journal of Comparative Psychology*, 107, 174–86.

Natale, F. (1989). Causality II: The stick problem. In *Cognitive Structure and Development in Nonhuman Primates*, ed. F. Antinucci, pp. 121–33. Hillsdale, NJ: Lawrence Erlbaum.

Nicolson, N.A. (1977). A comparison of early behavioral development in wild and captive chimpanzees. In *Primate Biosocial Development*, ed. S. Chevalier-Skolnikoff & F.E. Poirier, pp. 529–50. New York: Garland STPM Press.

Nishida, T. (1990). Deceptive behavior in young chimpanzees: An essay. In *The Chimpanzees of the Mahale Mountains: Sexual and Life History Strategies*, ed. T. Nishida, pp. 285–89. Tokyo: University of Tokyo Press.

Parker, C.E. (1974). Behavioral diversity in ten species of non-human primates. *Journal of Comparative and Physiological Psychology*, 87, 930–7.

Parker, S.T. (1990). Why big brains are so rare: Energy costs of intelligence and brain size in anthropoid primates. In *"Language" and Intelligence in Monkeys and Apes: Comparative Developmental Perspectives*, ed. S.T. Parker & K.R. Gibson, pp. 129–54. New York: Cambridge University Press.

Parker, S.T. (1991). A developmental approach to the origins of self-recognition in great apes. *Human Evolution*, 6, 435–49.

Parker, S.T. & Gibson, K.R. (1977). Object manipulation, tool use and sensorimotor intelligence as feeding adaptations in cebus monkeys and great apes. *Journal of Human Evolution*, 6, 623–41.

Parker, S.T. & Gibson, K.R. (eds.) (1990). *"Language" and Intelligence in Monkeys and Apes: Comparative Developmental Perspectives*. New York: Cambridge University Press.

Parker, S.T., Mitchell, R.W. & Boccia, M.L. (eds.) (1994). *Self-awareness in Animals and Humans: Developmental Perspectives.* New York: Cambridge University Press.

Parker, S.T. & Poti, P. (1990). The role of innate motor patterns in ontogenetic and experiential development of intelligence use of sticks in cebus monkeys. In *"Language" and Intelligence in Monkeys and Apes: Comparative Developmental Perspectives,* ed. S. T. Parker & K. R. Gibson, pp. 219–43. New York: Cambridge University Press.

Patterson, F.G.P. (1980). Innovative uses of language by a gorilla: A case study. In *Children's Language,* vol. 2, ed. K. Nelson, pp. 497–561. New York: Gardner.

Patterson, F.G.P. (1986). The mind of the gorilla: Conversation and conservation. In *Primates: The Road to Self-sustaining Populations,* ed. K. Benirschke, pp. 933–47. New York: Springer-Verlag.

Patterson, F.G.P. & Cohn, R.H. (1994). Self-recognition and self-awareness in lowland gorillas. In *Self-awareness in Animals and Humans: Developmental Perspectives,* ed. S.T. Parker, R.W. Mitchell & M.L. Boccia, pp. 273–90. New York: Cambridge University Press.

Pepperberg, I.M. (1983). Cognition in the African grey parrot: Preliminary evidence for auditory†vocal comprehension of the class concept. *Animal Learning and Behavior,* 11, 179–85.

Plooij, F. (1987). Infant-ape behavioral development, the control of perception, types of learning and symbolism. In *Symbolism and Knowledge,* ed. J. Montangero, A. Tryphon & S. Dionnet, pp. 35–64. Geneva: Fondation Archives Jean Piaget.

Poti, P. & Spinozzi, G. (1994). Early sensorimotor development in chimpanzees (*Pan troglodytes*). *Journal of Comparative Psychology,* 108, 93–103.

Povinelli, D.J. (1993). Reconstructing the evolution of mind. *American Psychologist,* 48, 493–509.

Povinelli, D.J., Nelson, K.E. & Boysen, S.T. (1990). Inferences about guessing and knowing by chimpanzees (*Pan troglodytes*). *Journal of Comparative Psychology,* 104, 203–10.

Povinelli, D.J., Parks, K.A. & Novak, M.A. (1991). Do rhesus monkeys (*Macaca mulatta*) attribute knowledge and ignorance to others? *Journal of Comparative Psychology,* 105, 318–25.

Povinelli, D.J., Rulf, A.B., Landau, K.R. & Bierschwale, D.T. (1993). Self-recognition in chimpanzees (*Pan troglodytes*): Distribution, ontogeny, and patterns of emergence. *Journal of Comparative Psychology,* 107, 347–72.

Premack, D. & Woodruff, G. (1978). Does the chimpanzee have a theory of mind? *Behavioral and Brain Sciences,* 1, 515–26.

Riviello, M.C., Visalberghi, E. & Blasetti, A. (1993). Individual differences in responses toward a mirror by captive tufted capuchin monkeys (*Cebus apella*). *Hystrix,* 4, 35–44.

Rumbaugh, D.M. & Pate, J.L. (1984). Primates' learning by levels. In *Behavioral Evolution and Integrative Levels,* ed. G. Greenberg & E. Tobach, pp. 221–40. Hillsdale, NJ: Lawrence Erlbaum.

Russon, A.E. (1990). The development of peer social interaction in infant chimpanzees: Comparative social, Piagetian, and brain perspectives. In *"Language" and Intelligence in Monkeys and Apes: Comparative Developmental Perspectives,* ed. S.T. Parker & K.R. Gibson, pp. 379–419. New York: Cambridge University Press.

Russon A.E. & Galdikas, B.M.F. (1993). Imitation in free-ranging rehabilitant orangutans (*Pongo pygmaeus*). *Journal of Comparative Psychology,* 107, 147–61.

Savage-Rumbaugh, S. (1988). A new look at ape language: Comprehension of vocal speech and syntax. In *Nebraska Symposium on Motivation,* vol. 35, ed. D.W. Leger, pp. 201–55. Lincoln: University of Nebraska Press.

Schiller, P.H. (1952). Innate constituents of complex responses in primates. *Psychological Review,* 59, 177–91.

Schino, G., Spinozzi, G. & Berlinguer, L. (1990). Object concept and mental representation in *Cebus*

apella and *Macaca fascicularis*. *Primates*, **31**, 537–44.
Schultz, A.H. (1969). *The Life of Primates*. London: Weidenfeld & Nicolson.
Schusterman, R.J. & Gisiner, R. (1988). Artificial language comprehension in dolphins and sea lions: The essential cognitive skills. *Psychological Record*, **38**, 311–48.
Seyfarth, R.M. & Cheney, D.L. (1988). Do monkeys understand their relations? In *Machiavellian Intelligence: Social Expertise and the Evolution of Intellect in Monkeys, Apes, and Humans*, ed. R.W. Byrne & A. Whiten, pp. 69–84. Oxford: Clarendon Press.
Simons, D. & Holtkotter, M. (1986). Cognitive processes in cebus monkeys (*Cebus apella*) when solving problem-box tasks. *Folia Primatologica*, **46**, 149–63.
Spinozzi, G. & Poti, P. (1993). Piagetian stage 5 in two infant chimpanzees (*Pan troglodytes*): The development of permanence of objects and the spatialization of causality. *International Journal of Primatology*, **14**, 905–17.
Stephan, H. (1972). Evolution of primate brains: A comparative anatomical investigation. In *The Functional and Evolutionary Biology of Primates*, ed. R. Tuttle, pp. 155–74. Chicago: Aldine.
Suarez, S.D. & Gallup, G.G., Jr (1981). Self-recognition in chimpanzees and orangutans, but not gorillas. *Journal of Human Evolution*, **10**, 175–88.
Swartz, K.B. & Evans, S. (1991). Not all chimpanzees (*Pan troglodytes*) show self-recognition. *Primates*, **32**, 483–96.
Swartz, K.B. & Evans, S. (1994). Social and cognitive factors in chimpanzee and gorilla mirror behavior and self-recognition. In *Self-awareness in Animals and Humans: Developmental Perspectives*, ed. S.T. Parker, R.W. Mitchell & M.L. Boccia, pp. 187–206. New York: Cambridge University Press.
Tomasello, M., Davis-Dasilva, M., Camak, L. & Bard, K.A. (1987). Observational learning of tool-use by young chimpanzees. *Human Evolution*, **2**, 175–83.
Tomasello, M., Savage-Rumbaugh, S. & Kruger, A.C. (1993). Imitative learning of actions on objects by children, chimpanzees, and enculturated chimpanzees. *Child Development*, **64**, 1688–705.
Torigoe, T. (1985). Comparison of object-manipulation among 74 species of non-human primates. *Primates*, **26**, 182–94.
Torigoe, T. (1986). Object manipulation in primates: A comparative psychological approach to human behavior. *Hiroshima Forum for Psychology*, **11**, 89–99.
Toth, N., Schick, K.D., Savage-Rumbaugh, E.S., Sevcik, R.A. & Rumbaugh, D.M. (1993). Pan the tool maker: Investigations into the stone tool-making and tool-using capabilities of a bonobo (*Pan paniscus*). *Journal of Archaeological Science*, **20**, 81–91.
Tuttle, R.H. (1986). *Apes of the World*. Park Ridge, NJ: Noyes Publications.
Van de Rijt-Plooij, H.H.C. & Plooij, F.X. (1987). Growing independence, conflict and learning in mother–infant relations in free-ranging chimpanzees. *Behaviour*, **101**, 1–86.
Vauclair, J. (1984). Phylogenetic approach to object manipulation in human and ape infants. *Human Development*, **27**, 321–8.
Vauclair, J. & Anderson, J.R. (1995). Object manipulation, tool use and the social context in human and nonhuman primates. *Techniques et Culture* (in press).
Vauclair, J. & Bard, K.A. (1983). Development of manipulations with objects in ape and human infants. *Journal of Human Evolution*, **12**, 631–45.
Visalberghi, E. (1987). Acquisition of nut-cracking behaviour by 2 groups of capuchin monkeys (*Cebus apella*). *Folia Primatologica*, **49**, 168–81.
Visalberghi, E. (1989). Primate tool use: Parsimonious explanations make better science. *Behavioral and Brain Sciences*, **12**, 608–9.

Visalberghi, E. (1990). Tool use in *Cebus*. *Folia Primatologica*, **54**, 146–54.
Visalberghi, E. & Fragaszy, D.M. (1990a). Food-washing behavior in tufted capuchin monkeys, *Cebus apella*, and crabeating macaques, *Macaca fascicularis*. *Animal Behaviour*, **40**, 829–36.
Visalberghi E. & Fragaszy, D. M. (1990b). Do monkeys ape? In *"Language" and Intelligence in Monkeys and Apes: Comparative Developmental Perspectives*, ed. S.T. Parker & K.R. Gibson, pp. 247–73. New York: Cambridge University Press.
Visalberghi E. & Limongelli, L. (1994). Lack of comprehension of cause–effect relations in tool-using capuchin monkeys (*Cebus apella*). *Journal of Comparative Psychology*, **108**, 15–22.
Visalberghi, E. & Trinca, L. (1989). Tool use in capuchin monkeys: Distinguishing between performing and understanding. *Primates*, **30**, 511–21.
Warden, C.J., Fjeld, H.A. & Koch, A.M. (1940). Imitative behavior in cebus and rhesus monkeys. *Journal of Genetic Psychology*, **56**, 311–22.
Westergaard, G.C. (1993). Development of combinatorial manipulation in infant baboons (*Papio cynocephalus anubis*). *Journal of Comparative Psychology*, **107**, 34–8.
Westergaard, G.C. & Fragaszy, D.M. (1987). The manufacture and use of tools by capuchin monkeys (*Cebus apella*). *Journal of Comparative Psychology*, **101**, 159–68.
Westergaard, G.C. & Suomi, S.J. (1993). Use of a tool-set by capuchin monkeys (*Cebus apella*). *Primates*, **34**, 459–62.
Westergaard, G.C. & Suomi, S.J. (1994). Hierarchical complexity of combinatorial manipulation in capuchin monkeys (*Cebus apella*). *American Journal of Primatology*, **32**, 171–6.
Windholz, G. & Lamal, P.A. (1985). Köhler's insight revisited. *Teaching of Psychology*, **12**, 165–7.
Woodruff, G. & Premack, D. (1979). Intentional communication in the chimpanzee: The development of deception. *Cognition*, **7**, 333–62.
Yerkes, R.M. & Yerkes, A.W. (1929). *The Great Apes*. New Haven, CT: Yale University Press.
Yoshida, H., Norikoshi, K., Kitahara, T. & Yoshihara, K. (1991). A study of the mother–infant relationships of chimpanzees in Tokyo Tama Zoological Park. In *Primatology Today*, ed. A. Ehara, T. Kimura, O. Takenaka & M. Iwamoto, pp. 243–6. Amsterdam: Elsevier.

3

Acting and understanding: Tool use revisited through the minds of capuchin monkeys

ELISABETTA VISALBERGHI AND
LUCA LIMONGELLI

A tool is not an object but a mental program which can interrelate that object with others to implement anticipated external effects. Tools, in other words, are as much mental as material, and their description is not a photograph of the material object itself but an empirically verifiable characterization of the mental knowledge and behavioral programs which allow the object to be produced and used.

(Reynolds, 1982, p. 377).

Do nonhuman primates understand what makes one object effective as a tool, and another object ineffective? Do nonhuman primates take into account the outcome of their action when they are using a tool? By exploring these aspects of tool use in primates we hope to capture the features that make the use of tools in our species, and to a lesser extent in other species, so powerful, adaptable, and pervasive.

The capacity for using objects as tools provides humans with a master key for solving the many kinds of problem they might encounter. By using tools, humans can overcome physical and morphological limitations and expand the range of their achievements. Moreover, humans also use tools when they are not strictly needed and use them in complex ways that are characterized by the monitoring and comprehension of the consequences of their action with the tool. Their extremely flexible tool use is characterized by an appreciation of invariant relations between the tools' properties and the transformations they produce in other objects (Caron *et al.*, 1988).

The acquisition of means–ends relations and tool-using behaviors is an important achievement in infant and child development (Piaget, 1952, 1954) and is one of the few measures of cognitive development that correlate with early language acquisition (Bates *et al.*, 1979; Gopnik & Meltzoff, 1994). In children, cognitive development brings radical changes in what they are able to do with objects, including using them as tools. They gradually acquire the ability to analyze the problem they face, to represent the outcome(s) of their action(s) mentally and to plan their behavior accordingly (Piaget, 1952, 1954; Connolly & Dalgleish, 1989, 1993).

Because the capacity to use tools is considered to be one of the major achievements of our species both at the ontogenetic and phylogenetic levels, the study of tool-using behaviors in apes and monkeys has always been compelling. In general, research on the use of tools in nonhuman primates has given a privileged status to the description of performance over analysis of the underlying cognitive processes. Notable exceptions were the pivotal experimental studies by Kluver (1933, 1937), Köhler (1976), and Yerkes (1927a, b, 1943) with captive apes and monkeys. In those studies performance per se was not equated to comprehension, and careful attempts were made to assess the aspects of the task the subjects could handle cognitively. Such an approach is to be found also in the most recent detailed analysis of nut-cracking behavior in wild chimpanzees (Boesch & Boesch, 1983, 1990, 1993a; Sakura & Matsuzawa, 1991). The aim of this chapter is to go beyond documenting the similarities in tool use performance in primates noted by various authors (e.g. Beck, 1980), and to discuss the extent to which a primate that uses a tool understands its functioning along the lines suggested by Reynolds' (1982) quotation at the beginning of this chapter.

Tool use has been defined and investigated in many different ways (Beck, 1980). For example, Goodall (1986) described tools and tool use functionally, by stressing what tool use behavior allowed an actor to accomplish. This straightforward definition has led to listing the different types of tool-using behaviors performed by different species. According to Goodall, "to be classified as a tool, an object must be held in the hand (or foot, or mouth) and used in such a way as to enable the operator to attain an immediate goal" (Goodall, 1986, p. 536). On the basis of this descriptive (operational) definition, animals as different as mammals, fish, birds, and some species of invertebrates may be labeled as tool users (for an extensive review, see Beck, 1980). Furthermore, species that show very different levels of cognitive complexity and which are phylogenetically very distantly related show similar performances with tools: for example, the use of stones by Egyptian vultures (*Neophron percnopterus*) to crack open ostrich eggs resembles the use of stones to crack open nuts observed in capuchin monkeys and chimpanzees. However, in most of the species in which individuals perform tool use, this behavior does not work as a master key to open many sorts of lock, but as a key that fits a single lock. It is when tool use represents a flexible and intelligent behavioral response to various kinds of problem that it should be analyzed from a cognitive perspective.

A more productive approach to the study of tool use that took into account its cognitive aspects was proposed by Parker & Gibson (1977). In their pivotal work, they argued that inquiry into tool use behavior suffered from the lack of a good definition and an appropriate theoretical framework and proposed a typology for classifying tool use that was inspired by the work of Piaget. Parker & Gibson distinguished between four different categories of tool use (context-specific proto tool use, context-specific tool use, intelligent proto tool use, and intelligent tool use). We report their definition of intelligent tool use because it concerns the kind of tool use behaviors we are going to discuss.

Intelligent tool use involves trial and error application of several complex object manipulation schemata such as aimed throwing, using a lever, banging with a tool, raking in with a stick, probing with a stick, in different contexts, such as opening objects, raking in out-of-reach objects, extracting objects from a container without opening it, using a variety of objects such as sticks, rocks, leaves. Intelligent tool use involves accommodation to the specific situation and exploring and manipulating physical causality in a generalized manner.
(Parker & Gibson, 1977, p. 628; see also Parker & Poti, 1990).

Parker & Gibson (1977) also argued that intelligent tool use emerges with tertiary sensorimotor intelligence (diagnostic of the fifth and sixth Piagetian stages) during which the subject discovers, through purposeful trial-and-error variation, the special properties of the objects and explores and manipulates causality through accommodation to these properties. It is by performing these activities that new means to solve a task are discovered. Capuchin monkeys and chimpanzees are considered by Parker & Gibson (1977) as having both intelligent tool use and tertiary sensorimotor intelligence (see also a more recent article by Chevalier-Skolnikoff (1989) suggesting a similar view).

Finally, we want to mention Reynolds' (1982) approach to goal-directed behaviors. According to his view, a goal is the intended effect of an action. Goal-oriented actions presuppose: "(a) the ability to mentally represent the consequences of action prior to the execution of the action itself; (b) the ability to compare the actual consequences to the anticipated consequences; (c) the ability to alter subsequent action on the basis of experience with prior executions" (Reynolds, 1982, p. 343). These abilities characterize tool use as it occurs in our species: adult humans respond proficiently to the feedback deriving from their own actions and plan subsequent actions accordingly. Because these abilities capture the dynamic essence of tool use, their assessment in nonhuman primates is fundamental.

We will analyze performance in tool-using tasks in nonhuman primates by assessing the extent to which the actor plans its behavior by taking into account the features of the task and the outcome of its action. Reynold's approach to tool use has been our heuristic tool and will be the Leitmotiv throughout this chapter.

TOOL USE REVISITED THROUGH THE MINDS OF CAPUCHIN MONKEYS

During our 17 years of research experience with cebus monkeys . . ., we have alternated between marvelling at their cognitive accomplishments and being plunged to the depths of despair over their inability or extraordinary reluctance to learn a variety of apparently simple tasks.
(D'Amato & Salmon, 1984, p. 164)

In recent years, tool use behavior in capuchin monkeys has been studied in several laboratories (Anderson, 1990; Antinucci & Visalberghi, 1986; Fragaszy & Visalberghi, 1989; Natale 1989; Parker & Poti, 1990; Visalberghi, 1993a; Visalberghi & Limongelli, 1994; Westergaard, 1994; Westergaard & Fragaszy, 1987; Visalberghi & Trinca, 1989;

for a review see Visalberghi, 1990). All these studies have shown that this South American monkey genus has amazing propensities for using a variety of tools in many different tasks. For example, capuchins spontaneously use stones to crack open nuts; sticks to dip, rake, push, and extract; paper towels to soak up liquids. In terms of the variety of tools and the variety of contexts in which tools are used, capuchins are similar to chimpanzees and other ape species. However, as we will argue, this general level of description of tool use behavior is almost meaningless when comparing the species' intelligence.

The urge for a deeper appreciation of how capuchin monkeys use tools emerged when we were confronted (as were D'Amato and Salmon) with puzzling observations in which successful use of tools was interspersed with awkward and "nonsense" behaviors. How could the following episodes be explained? An adult capuchin tool-user, Hansje, had successfully learned to use straws to dip for syrup through holes in the sides of an opaque box. More than 5 years later, this monkey was presented with a new dipping task (a transparent box with holes in the top). In this new situation, she repeatedly tried to dip into the sides of the new box (E. Visalberghi & D. Fragaszy, personal observation), although it was obvious that this new box had no opening on the sides! On another occasion when a group of capuchin monkeys was presented with a nut-board and pounding tools, the adult female Erna adopted a behavior that had previously led her to success in a completely different task: Erna repeatedly tried to use straws to dip the nuts! (Fragaszy & Visalberghi, 1989; see Figure 3.1). In both these cases, the capuchins seemed to generalize (with no success, whatsoever) from previous experiences rather than to analyze the characteristics of the new task they were facing.

The above descriptions are not isolated cases. Other similar examples are provided by Klüver (1933) and by a more recent experiment by Westergaard & Suomi (1993) in which a few capuchins cracked open walnuts with stones and probed with sticks to extract the nut meat. These monkeys used these tools in both appropriate and inappropriate contexts. In fact, in 43–49% of the bouts the capuchins used sticks on intact nuts (!) and in 23–42% of the bouts they used stones on already opened nuts (the percentages correspond to the highest and lowest values scored).

Some of the capuchins' "errors" were so surprising that we decided to study them systematically. This type of awkward and inappropriate behavior could result from a lack of attention to the features of the task and/or from a lack of understanding of the requirements of the task. Similar instances of "errors" have been described in chimpanzees and gorillas using tools (e.g. Köhler, 1976; Yerkes, 1927a, b) and cannot, therefore, be considered unique to tool use in capuchins. Humans may also make "stupid errors" that resemble those we have just described. However, they possess (or develop) the capacity to figure out what is wrong and try to remove the cause of failure. Is this also the case for capuchin monkeys or are the capuchins' awkward behaviors the result of a different way of understanding what caused failure?

A fair comparison across species requires a common scale for testing different species. Because existing literature on tool use in human and nonhuman primates lacks comparable experiments, we designed and carried out a series of experiments using the

Figure 3.1. Nut-cracking board. Instead of using one of the provided pounding tools (a pulley, a wooden block), Erna uses a straw as a tool for the nuts.

same experimental paradigms across species.

As indicated in Table 3.1, a total of 12 capuchin monkeys (*Cebus apella*), 16 chimpanzees (*Pan troglodytes*), 4 bonobos (*Pan paniscus*), 1 orangutan (*Pongo pygmaeus*), and 88 children (*Homo sapiens*) were tested in one or more of the three problems described below (for more information about these experiments see references in Table 3.1). In this chapter, our main focus is the comparison between capuchins and chimpanzees. Data on children will be reported mainly to provide an indication of the levels of difficulty of the tasks presented.

The three tool problems required the subjects to push a reward out of a transparent tube by means of a stick. Problem 1 (tube task and imitation), was aimed at investigating the extent to which naive subjects took advantage of models. Problem 2 (tube task and complex conditions), was aimed at investigating the ways in which subjects dealt with misshapen tools that required modification. Problem 3 (trap-tube task), was aimed at investigating the ability to comprehend the effect caused by the use of the tool. The description of the tool task itself and the design of each related problem is reported below. Our discussion of each problem is accompanied by a puzzling episode of a tool-using capuchin monkey performing some awkward behavior. It was these episodes that prompted us to undertake the systematic investigation of capuchin tool use that we report below.

Table 3.1. *Details of tube task studies*

Studies in which tube task and imitation (Problem 1), tube task and complex conditions (Problem 2), trap-tube task (Problem 3) (see Figure 3.2) were used to test tufted capuchin monkeys (*Cebus apella*), chimpanzees (*Pan troglodytes*), bonobos (*Pan paniscus*), an orangutan (*Pongo pygmaeus*), and children (*Homo sapiens*). The number of subjects is indicated for each species. References are indicated for each study.

Study	Problems	Species	Age	References
1	Problem 1	*Cebus apella* (n = 6)	2–9 years	Visalberghi & Trinca (1989)
	Problem 2	4 of the above subjects	3–9 years	Visalberghi (1993a)
				Visalberghi (1993b)
2	Problem 1	*Pan troglodytes* (n = 6)	2–4 years	Bard *et al.* (1995)
	Problem 2	same as above		
3	Problem 2	*Cebus apella* (n = 6)	2.5–17 years	Visalberghi *et al.* (1995)
		Pan troglodytes (n = 5)		
		Pan paniscus (n = 4)		
		Pongo pygmaeus (n = 1)		
4	Problem 2 (longitudinal)	Children (*Homo sapiens*) (n = 8)	13–23 months	A. Troise & E. Visalberghi, unpublished results
5	Problem 1	Children (*Homo sapiens*) (n = 57)	12–24 months	E. Visalberghi, unpublished results
6	Problem 3	*Cebus apella* (n = 4)	3–8 years	Visalberghi & Limongelli (1994)
7	Problem 3	*Pan troglodytes* (n = 5)	5–34 years	Limongelli *et al.* (1995)
8	Problem 3	Children (*Homo sapiens*) (n = 23)	27–66 months	L. Limongelli & E. Visalberghi, unpublished results

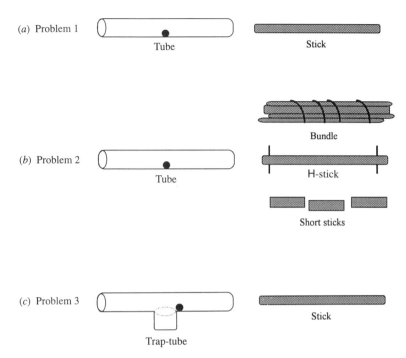

Figure 3.2. Experimental apparatuses presented as tasks: (a) Problem 1, tube task; (b) Problem 2, complex conditions; (c) Problem 3, trap-tube task.

THE TUBE TASK

The three above-mentioned problems involved the use of a piece of apparatus, hereafter called the tube task. This experimental apparatus is a transparent horizontal tube baited in the center with a food treat (see Figure 3.2(a)). The subject is provided with a stick that can be used to push the treat out of the tube. The use of such a transparent tube has a significant advantage in that it provides both the experimenter and the subject with continuous monitoring: it allows the subject to observe the position of the reward in the tube and to monitor the effects of its own behavior when using the stick in the tube.

This possibility of continuous monitoring of the effects produced by the tool is often not possible in other tasks presented in captivity (e.g. the dipping task presented to capuchin monkeys by Westergaard & Fragaszy, 1987) or in naturally occurring tasks faced by chimpanzees. For example, the depth at which termites are located in a termite mound cannot be assessed by human observers by merely watching what the chimpanzees are doing (Goodall, 1986; McGrew, 1992). However, it is worth mentioning that there are some naturally occurring tool tasks in which monitoring is possible. This is the case, for example, of the ant-dipping observed at Gombe by McGrew (1974, 1992), in which the ants are visible throughout the performance and efficient performance depends on the visual monitoring of the progress of the ants up the wand.

The tube task was solved by individuals of all the species we tested. In chimpanzees (study 2) and in children (study 4) age influenced the way in which the solution was first achieved. When first presented with the task, young children, young chimpanzees, and most capuchins engaged in a variety of attempts (including positioning the stick transversally to the opening and holding it parallel to the tube). Eventually, by trial-and-error the stick was inserted into the tube and the reward was obtained. In our studies, children over 2 years of age and chimpanzees over 5 tried to insert the stick into the opening of the tube from the very beginning of the first presentation. In contrast, older capuchins were not necessarily more proficient than younger ones in solving the tube task (studies 1 and 2). The youngest capuchin to spontaneously solve the task was 24 months old. The youngest child to spontaneously solve the task was 13 months old (Benjamin) and the youngest chimpanzee (Mercury) 42 months old (study 3).

The results of a longitudinal study (study 4) in which eight children were tested approximately every 3 months showed that children achieve a solution sometime during the second year. In this study, the youngest subject to achieve solution was 13 months old and the oldest 23 months. Another study (study 8) found that children over 2 years of age solved the tube task on their first presentation ($n = 23$, age $= 27$–66 months).

Problem 1: Tube task and imitation

Brahms, a 5 year old female capuchin, was sitting on the tube watching as Carlotta slowly pushed a reward out of the opposite end of the tube with a stick. Carlotta walked around the tube, gently inserted her fingers into the tube and pulled out her reward. Brahms continued to watch, then got off the tube and retrieved the stick that Carlotta had left inside the tube.

(F. Visalberghi, personal observation)

Later, when tested by herself, Brahms did not attempt to use the stick as a tool, not even after having witnessed a total of 57 instances of solution performed by her cagemates (see Figure 3.3). Is Brahms a "dumb" monkey? Not at all. Her lack of learning a complex behavior such as tool use through the process of imitation is indeed the rule, not the exception (Visalberghi 1993; Visalberghi & Fragaszy, 1990).

In the very first experiments (study 1) which began in 1987, three (Brahms, Roberta, and Pepe) of the six capuchin monkeys tested did not solve the tube task. Therefore, we developed an imitation task in which nonsolvers watched repeated solutions of the task performed by their successful companions (models). These so-called "lessons" were aimed at facilitating the achievement of solution by nonsolvers. Brahms received 57 lessons, Pepe and Roberta 75. The results showed that, although the observers had ample opportunity to watch the model(s) and learn from them, none of the observers acquired tool use by imitation.

These findings initiated comparable studies on chimpanzees (study 2) and children (study 5). In both these studies, subjects were assigned *a priori* to the experimental

Figure 3.3. The tube task. A capuchin monkey attentively observes a cagemate solving the task by using a stick.

group or the control group. Only the subjects of the experimental group watched the experimenter (model) solve the task.

Our study of human children (study 5, still in progress) is aimed at assessing the age at which children's acquisition of tool use is facilitated by the actions of a model. To date, a total of 57 children have been tested at different ages (12, 15, 18, 21, and 24 months) in this cross-sectional study. Children are first presented with one 3-min trial of the tube task. If the child is successful, testing ends and his or her performance serves as a data point for assessment of the percentage of children who spontaneously solve the task at any given age. Children who are not successful are randomly assigned to the control group or the model group and are presented with three more trials of the tube task. The results show that modeling does not improve the performance of 12 month old children, whereas it does improve that of 15 and 18 month old children. For example, at 15 months of age, four out of eight children exposed to the model solved the task, whereas none of the eight children of the control group solved the tube task. In older children, experience with the task, that is, the presentation of three more trials, improved performance almost as effectively as modeling (see Figure 3.4).

The noneffectiveness of modeling in this tool task contrasts with Meltzoff's (1985, 1988) findings that 12 month old children are able to imitate. However, the actions to be imitated in Meltzoff's tasks were simple and not goal-directed, i.e. they did not require problem-solving skills. More recently, Nagell *et al.* (1993) reported that 18 month old children lacked the ability to imitate the action required for solving a tool problem.

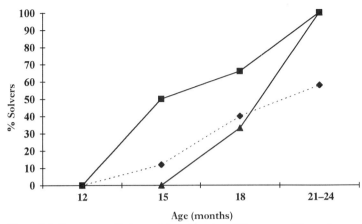

Figure 3.4. Percentages of solvers of the tube task at different ages. Diamonds correspond to the percentage of children who spontaneously solved the task after one presentation. Squares correspond to the percentage of children who solved the task in one to three trials with a Model. Triangles correspond to the percentage of children who solved the task in one to three trials with no model.

Our study of the effect of modeling in young chimpanzees (study 2) confirms the importance of age for subjects' receptivity to modeling. Despite the limited number of chimpanzees tested ($n = 6$), the results suggest that the presence of a model is not effective for 2 year old chimpanzees but may lead to quicker acquisition of tool-using behavior in 3 and 4 year old chimpanzees, as in humans. Limongelli et al. (1995, and unpublished results) reported that a 5 year old chimpanzee that did not use a stick to push a reward out of a tube during two trials, immediately did so after only one instance of modeling by the experimenter.

In short, whereas adult and juvenile capuchin monkeys did not learn how to solve the tube task by watching a model, chimpanzees and children (over a certain age) did. The failure to learn by witnessing solutions performed by models can be interpreted as a lack of understanding of what the model is doing. The observer can see but does not understand that pushing the stick inside the tube causes the displacement of the reward so as to make it available.

Elsewhere we have argued that the ability to imitate a tool-using skill relates to the ability to understand the cause–effect relation between action and outcome (Visalberghi & Fragaszy, 1990, 1995; see also Köhler, 1976). Brownell & Carriger (1990) argued that an understanding of the task is also necessary when children are required to cooperate for its solution and have found that this capacity develops at the end of the second year. R. Chalmeau, A. Gallo & E. Visalberghi (unpublished results) recently tested capuchin monkeys in a task requiring cooperation; results show that, although success often emerges from co-occurrent activities, cooperation *sensu strictu* is not present. A similar experiment using chimpanzees (R. Chalmeau & A. Gallo, unpublished results; see also Crawford, 1937) resulted in a better performance, which to some extent paralleled that of children at an age in which cooperation is emerging (approximately 18 months of age;

Acting and understanding: Tool use revisited

Figure 3.5. The tube task, complex conditions. Carlotta has received a bundle of reeds kept together by masking tape. The picture shows an example of error: the capuchin inserts the tape instead of a reed into the tube.

Brownell & Carriger, 1990). Therefore, our results on imitation of tool-using skills in capuchins, chimpanzees, and children are consistent with those on cooperation.

Problem 2: Tube task and complex conditions

Carlotta, a 3 year old female capuchin, was given a bundle of reeds held together with masking tape. (She had already solved the tube task in several trials). She bit the tape and freed the reeds, then discarded the reeds and very carefully (*sic*) inserted a piece of tape inside the tube.
(Visalberghi 1993a, and personal observation; see Figure 3.5).

Why did she do so? Why discard suitable tools (the reeds) and try an unsuitable one (the tape)? Why try the impossible?

In this problem the subject is presented with the transparent tube and tools that are not effective unless modified (See Figure 3.2(b)). There are three different tool conditions. The Bundle condition consists of several sticks firmly held together. The intact bundle is too large to fit into the tube. The H-stick condition consists of a dowel with two smaller sticks placed transversely near the end; the transverse sticks block the insertion of the dowel into the tube. The Short-Sticks condition consists of three short sticks at least two of which must be inserted into the tube one behind the other to displace the reward.

This problem was aimed at investigating what a tool user "thinks" an effective tool should look like. Repeated presentation of these conditions allowed us to assess whether the subject modified the tool before attempting a solution and whether its performance improved across trials. We presented these complex tool tasks to capuchin monkeys, chimpanzees, and children who had previously solved the tube task (studies 1–4). Data on the chimpanzees of study 3 refer only to the Bundle and H-stick conditions. Nonhuman primates received 10 trials in each complex condition and their performance was assessed by comparing the first block of 5 trials with the second block of 5 trials. Children were tested longitudinally every 3 months. On average, children underwent 10 trials in the Bundle condition, 25 trials in the Short-Sticks condition, and 14 trials in the H-stick condition. For each condition, the children's performance was evaluated by dividing into two the number of trials of the experimental session in which they reliably solved the problem and comparing the first block of trials with the last block.

We argued that if the subject understood the properties of an object that enabled it to be effective as a tool and, vice versa, the properties that made it ineffective, then the subject would modify the object before trying to use it. The subject may try to insert into the tube unmodified unsuitable tools, modified unsuitable tools, modified suitable tools or inappropriate objects. The extent to which a subject modified an unsuitable tool into a suitable one before using it provided an index of the subject's knowledge of the requirements of the task. A subject could try to use the whole Bundle or the H-stick as tools, or it could use one of the transverse sticks of the H-stick (much too short), or it could insert one short stick in one side of the tube and the other in the opposite side, etc. All these behaviors (hereafter referred to as errors) did not prevent solution, but indicated a lack of appreciation of the properties required for the tool to be effective. The subject could (a) correctly modify the complex tool before even attempting to use it, exhibiting comprehension; (b) perform errors mainly in the first trials and improve its performance in later trials; (c) perform errors at any time. These performances indicate decreasing degrees of comprehension of the problem.

When presented with the complex conditions, all the capuchins and chimpanzees over 3–4 years of age quickly solved them on the first presentation. In contrast, most of the children began to solve the complex conditions in later sessions. Whereas the performance of capuchins (as indicated by the number of errors in the first and second block of trials) did not significantly improve across trials, those of chimpanzees and children did. Overall, chimpanzees judged width or length reasonably accurately and modified the tool in an appropriate way when they were presented with tools that were too thick (Bundle) or misshapen (H). They did not often insert sticks that were visibly too short (such as the transverse sticks), or objects that were inappropriate for the task (the whole bundle or peanut shells). In capuchin monkeys these kinds of error were more frequent and the use of inappropriate objects persisted even after many vain attempts to solve the task with them.

Problem 3: Trap-tube task

Cammello, an adult male capuchin, introduced a peanut shell into the left opening of the tube (there were 10 cm between this unlikely tool and the reward!). He then went to the right opening to check whether the reward had become available.
(E. Visalberghi, personal observation)

Why does Cammello expect the reward to have moved? How can one object displace another without even touching it?

In Problem 3 (trap-tube or T-tube task) the subject was presented with a tube that had a trap-hole in the middle: the reward was placed on one of the two sides of the hole. In order to solve the problem, the subject had to insert the stick into the side of the tube from which it could push the reward out of the tube and not into the trap (see Figure 3.2(c)). Depending on the side in which the subject inserted the tool, it could either push the reward into the trap or push the reward out of the tube and obtain it. This task allowed the investigator to assess whether the subject had the ability to foresee the outcome of its action (i.e. the effect of the stick on the displacement of the reward).

Whereas the complex condition task explored the extent to which a subject took into account the requirements necessary for a tool to be effective, the T-tube problem explored the extent to which the subject took into account the relation between the action performed with the tool and the outcome resulting from this action. It should be noted that whereas in the complex conditions task error delays solution, in the T-tube task, the subject is penalized by loss of the reward (see Figure 3.6).

Problem 3 consisted of 14 ten-trial blocks. In the trap-tube task (T-tube) a subject can be 50% successful in every trial simply by systematically inserting the stick into the same side of the tube or by inserting it into one of the two sides by chance. In contrast, rates of success higher than chance can be obtained by avoiding the trap.

To avoid the trap, the subject can (a) insert the stick into the correct side of the tube straight away or (b) perform multiple attempts/insertions. In the latter case, the subject inserts the tool in one side of the tube without pushing it through to the end, withdraws it, and inserts it again. Multiple insertions are therefore a way of correcting previous wrong insertions; moreover, it can be argued that the more multiple insertions performed the less the subject represents the outcome of its action in advance and the less it is able to plan it correctly. Single insertion trials would imply an *a priori* representation of the action that will fulfill the requirements of the task (representational strategy), whereas multiple insertion trials would imply an appreciation of the outcome of the ongoing action (anticipatory strategy). (Note that representational elements also play a role in the anticipatory strategy. Therefore the above characterization aims mainly at distinguishing performances in which the entire correct strategy is planned in advance from performances in which it is not.)

Furthermore, in the T-tube task it is possible to investigate the process(es) that lead the subject to an above chance level solution, i.e. whether the subject develops a rule of action (associative process) or understands that some key features of the task, such as the trap, affect the outcome of the pushing action (comprehension of causality process).

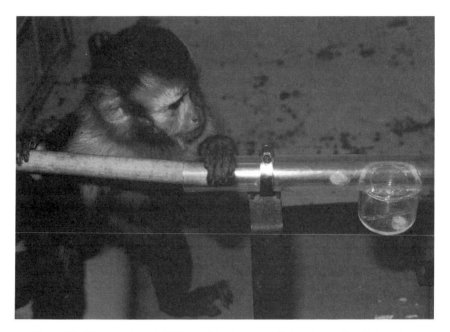

Figure 3.6. The trap-tube task. While watching the reward, Roberta inserts the stick into the wrong side of the tube. A reward lost in a previous trial is already inside the trap.

The results of our study on capuchins (study 6) showed that the rate of success of three out of four capuchins was at chance level, whereas that of the fourth subject, Roberta (3 years old), was significantly higher than chance. In particular, Roberta became 85.7% successful in the second half of the experiment (blocks 8–14). Careful analysis of her behavior and a series of control tests revealed that she was choosing the side of insertion on the basis of the distance of the reward from the openings of the tube. By looking inside the tube through its openings, Roberta evaluated the spatial configuration among stick, tube, and reward, and adopted the rule of inserting the stick in the opening farthest from the reward. The successful performance of Roberta was anticipatory (see definition above) and was based on an associative process. On the grounds of the performance of this successful subject and of those of the other three unsuccessful ones, it was argued that our capuchins did not understand the causal consequences of their actions and that they did not learn to do so after repeated experiences.

Five common chimpanzees (*Pan troglodytes*) were also tested in the T-tube task (study 7). Throughout the experiment, the success rate of three of the five subjects tested, Sarah, Kermit, and Bobby (34, 12, and 5 years old, respectively) did not increase above chance level. The remaining two subjects, Sheba and Darrell (11 and 13 years old, respectively) were successful above chance level. In the second half of the experiment, the rates of success of the latter two subjects were 98.6% and 90%, respectively. A final study (study 8), which is still in progress, has presented the same task to children between 27 and 66 months of age ($n = 23$). The preliminary results show that children under 3 years of age did not figure out a successful strategy; in contrast, children over 3 years of age fully succeeded in the task.

The behavior of the successful children and chimpanzees was qualitatively very different from that of the capuchin monkey Roberta. Typically, children and chimpanzees looked at the set-up (and not into the opening of the tube, as Roberta did) and evaluated the situation. Then they inserted the stick into the correct side of the tube and pushed it through to the end. In short, they were able to mentally take into account whether their action was going to push the reward into the trap or not. Moreover, children verbally explained the reason for their choice and what would have happened if they had chosen the other side of the tube for insertion.

In order to analyze the performance of the experimental subjects in terms of representational and anticipatory strategy, we considered the mean number of insertions per successful trial performed during the second half of the experiment, i.e. when the subjects (both the two chimpanzees and the capuchin) were successful above chance level. These values were 2.1 per trial for Roberta, the capuchin, and 1.1 for Sheba and 1.2 for Darrell, the chimpanzees. Only one child (out of ten successful ones over 3 years of age) performed two insertions in one trial. In all the other trials children performed only one insertion. This analysis shows that whereas successful children and chimpanzees essentially selected the correct side for insertion right away, the successful capuchin did not. Typically, Roberta, the one successful capuchin, reached the solution by performing multiple insertions.

In short, understanding of the causal relations involved in the T-tube is within the cognitive capabilities of chimpanzees and children, but not within those of capuchins. It should be noted that in terms of rate of success the successful capuchin monkey performed in a manner similar to that of the successful chimpanzees and children. The importance of distinguishing between understanding and performing is the topic of the following section.

TOOL USE AND THE UNDERSTANDING OF CAUSALITY

> Capuchins seem to live in a world where almost every possible attempt that can be tried to solve a task may be successful. Men, and apes, live in a world where some possibilities are discarded beforehand, because they are incompatible with their mental representation of possible solutions.
> *(Visalberghi, 1990, p. 152)*

Early in life humans have a "perceptual mechanism operating automatically and incorrigibly upon the spatio-temporal properties of events yet producing abstract descriptions of their causal structure" (Leslie, 1988, p. 187).

A tool task is a constrained experimental situation in which the goal is set *a priori* by the experimenter and the subject performs actions in order to obtain the goal. The subject's strategy to reach the goal and to deal with possible variations of the task is related to its understanding of the task and of the effects of its own behavior. By analyzing the subject's strategy of solution, the experimenter can infer how the subject "sees" the problem it is facing.

Tool use may derive from the understanding of cause–effect relations or from the formation of associations between success and actions with objects. It should be noted that comprehension of causality is a multifaceted ability that a species may possess to different extents in relation to different physical principles, or in different domains (see Piaget, 1952, 1954; and Chapter 12). The comprehension of cause–effect relations allows an individual to connect events and to produce further hypotheses of how things might be affected by the same causes. An analysis in terms of cause–effect relations that incorporates the available information and suggests a temporal and causal link between elements makes it possible to formulate possible explanations when new conditions are presented. In contrast, knowledge deriving from specific associations is often not sufficient for behaving adequately when new conditions are presented. In the T-tube experiment, causal comprehension was considered to be the subject's knowledge of the causal relation between its pushing action with the stick and the consequent movement of the reward into the trap or out of the tube.

Typically, adult humans assess the requirements of the problem, the tools available (or potentially available) for its solution, and then try to solve the task. If the same task is presented again, humans take their previous experience into account and reevaluate the possible strategies for solution. Throughout this process, causal relations are taken into account and used to choose the best strategy. But success in using a tool to solve a problem may also be based on trial and error strategies and practice (Parker & Poti, 1990). An individual can accidentally solve the task and, gradually through repeated attempts, learn the sequence of behaviors necessary for solution.

There is a major difference between these two ways of solving a tool task. In the first, the individual performs actions and evaluates why one action leads to success and another leads to failure. In the second, the individual associates a given behavior with an (unsuccessful or successful) outcome, without understanding why these outcomes are produced. In the latter case, variations of the task and/or variations of the objects available as tools are likely to worsen performance until the individual has once again formed the necessary associations.

The comparative data we have presented concern several aspects of tool use behavior. In all the species tested, tool use emerges spontaneously through adaptive and flexible interactions with objects and the frequent production of combinatorial actions. Intrinsic behavioral characteristics of chimpanzees, capuchins, and children are sufficient to produce initial discoveries of the forms of tool use that involve insertion and/or relation of an object to a substrate (for a discussion of the emergence of tool use in capuchins, see Fragaszy & Adams-Curtis, 1991; Visalberghi, 1987; in chimpanzees, see Schiller, 1952). Yet, while the capuchin monkeys, chimpanzees, and young children in our experiments were all able to use sticks to push a reward out of a tube, their comprehension of the problems involved was different. Differences were present at three levels of analysis.

First, the results of Problem 1 (tube task and imitation) show that when unable to solve a task, the three species differ in their ability to profit from observing a model. Chimpanzees over 3–4 years of age and children from about 15–18 months were able to learn to solve the task by observing a model performing the solution. In contrast capuchins, regardless of age, did not learn a tool task by watching a model. Capuchins' lack of imitation has also been reported for other problem-solving skills (Visalberghi & Fragaszy, 1990). In contrast, the results of study 2 and those of Nagell et al. (1993) show that chimpanzees exposed to a model solving a tool task improved their performance. According to Köhler (1976) and Piaget (1954) imitation in a tool-using task is related to an understanding of the causal relations between the actions performed by the model and the outcome. We believe that this is exactly what capuchins lack. On the other hand, our data show that at a certain age, chimpanzees and children develop the ability to "read" the behavior of the model, i.e. they show the ability "to understand and intelligently grasp what the action of the other means" (Köhler, 1976, p. 221).

Second, the results of the complex conditions task show that subjects of all species tested may try to use inappropriately shaped tools. However, while chimpanzees (over 5 years of age), and children (over 2 years of age) rather quickly learned what the tool should be like to be effective, capuchins did not. Chimpanzees and children apparently came to understand that a certain shape of a tool made it unsuitable for insertion into the tube (because it was too big), or for the displacement of the reward (because it was too short to come into contact with the reward). Our findings suggest that capuchins know that a tool is needed, but that they do not know what features allow one object to fit into the tube and displace the reward and what causes another to be ineffective.

Third, the results of the T-tube task suggest that, in contrast to children (over 3 years of age) and some chimpanzees, capuchins do not know how and why their actions cause certain effects. It seems that successful chimpanzees and children possess the ability to represent mentally the consequences of their action with the tool. Nevertheless the only capuchin to solve this task used a nonanticipatory strategy, which consisted of a distance-based rule of action. It is evident that capuchins were not able to acquire information about the cause–effect relation between their action with the tool and the action's outcome.

An understanding of causal relations is necessary for success in each of the above three problems (imitation, complex conditions, and T-tube). The nature, or the extent, of causal understanding required for success apparently varied across the three problems. Children and chimpanzees solved Problem 1 at a younger age than they did Problem 2, and Problem 2 at a younger age (or better) than Problem 3. Our data suggest that children and chimpanzees first learn to appreciate that a stick can be used as a tool, then they understand what makes an object effective as a tool, and only later do they understand the outcome of their action with the tool. Thus, causal understanding is not a yes/no phenomenon in humans or in chimpanzees. Rather, it is a capacity that gradually increases in power and generality throughout individual development.

Our results parallel some of the new experimental findings in the domain of social cognition that show that, unlike chimpanzees, monkeys lack other behaviors indicative of complex social intelligence, such as attribution skills (e.g. Cheney & Seyfarth, 1990a;

Povinelli et al., 1991), intentional deception (Mitchell, 1986; Byrne & Whiten, 1988), self-recognition (Parker et al., 1994; Chapter 16), teaching (Caro & Hauser, 1992; Visalberghi & Fragaszy, 1995) and cooperation (Chalmeau, 1994).

Recent experiments have demonstrated that chimpanzees are able to attribute knowledge about the location of a food item to an individual who had witnessed it being hidden and lack of knowledge to another individual that had not witnessed it (Povinelli et al., 1990). Therefore, chimpanzees seem to understand that in this circumstance access to visual information is a prerequisite for knowledge (see also Heyes, 1993). Visual access as source of knowledge is appreciated by 4 year old children (Wimmer et al., 1988).

A similar experiment carried out with macaques (*Macaca mulatta*) demonstrated that monkeys fail to attribute knowledge to others (Povinelli et al., 1991; see also Cheney & Seyfarth, 1990a). These findings refer to macaques and should not be directly generalized to other monkey species, such as capuchins. Nevertheless, since capuchins lack the ability for self-recognition (Collinge, 1989; Riviello et al., 1993; see also, Anderson, 1994), and no cases of intentional deception or teaching have been reported, it is reasonable to also expect a lack of attributional abilities in this monkey species.

As we have seen, attribution may also depend upon an appreciation of some kinds of cause–effect relations. Similarly, it can be argued that intentional deception and teaching require an understanding of what causes knowledge in another individual, so that an individual can be provided with false or appropriate information. Teaching in a nonhuman primate species is clearly a rare event, even if recent observations testify that the phenomenon does occur. Nut-cracking by chimpanzees has been observed closely for many years in the Taï forest in Côte d'Ivoire by Boesch & Boesch (1983, 1990, 1993b). The Boeschs have observed a few cases in which the mother chimpanzee appears to have actively intervened while her offspring was attempting (ineffectively) to crack nuts (Boesch, 1991; see Chapter 18). In order to teach how to crack open nuts, the teacher must understand the task and why (and when) the pupil has performed a wrong behavior so that the teaching effort can be directed to the specific segment of the task where failure occurred.

Chalmeau (1994) tested chimpanzees and capuchin monkeys in a "cooperative task" in which two handles had to be pulled simultaneously for a reward to be obtained; the handles were too far away for a single individual to pull them both. Despite the fact that in both species there were pairs that solved the task, in contrast to chimpanzees, capuchins did not understand the role of the partner in the achievement of success.

Finally, mirror-self-recognition can also be related to an understanding of the causal relation between behavior in front of the mirror and its effects in the mirror. Lewis & Brooks-Gunn (1979) reported that, before children recognize themselves in a mirror, they often go through a systematic exploration of the effects caused by their behavior in the mirror. The finding that monkeys do not achieve self-recognition (Eglash & Snowdon, 1983; Riviello et al., 1993), even when they perform apparently similar explorations, whereas apes do, is in accordance with the possibility that monkeys lack the ability to relate events in terms of causal relations.

CONCLUSIONS

The aim of our research was to compare the tool use behavior of capuchin monkeys, chimpanzees, and young children from a cognitive perspective. Our approach to the study of tool use was aimed at distinguishing performance from comprehension by teasing apart the different ways in which the solution to a task can be achieved (Visalberghi & Mason, 1983). In this effort we were helped immensely by the possibility of studying several species. Comparison prompted experiments and analyses which would never have come to mind otherwise. In particular, we wanted to investigate whether similar tool-using behaviors emerge from the same levels of comprehension of causal relations. In order to do this, we designed and presented to subjects of these primate species a set of simple experimental tasks meant to probe into the subjects' understanding of causality.

Results showed that capuchins, apes, and children were all capable of using tools despite major differences in their understanding of means–end relations, i.e. of the links between their action and its outcome. In short, unlike children and at least some of the chimpanzees, capuchins do not "know" why one tool is effective and another is not. Regardless of age, capuchins are unable to foresee the consequences of their action and to analyze a model's action with a tool in a way that might facilitate their acquisition of the same tool action. In contrast, during the course of their development apes and children acquire such abilities. Furthermore, only the latter species acquire the ability to plan correctly the sequence of actions leading to success.

We argue that these achievements of chimpanzees and children are made possible by their development of an ability to relate events causally, i.e. to extract knowledge from experience and to make hypotheses about the outcome of future actions. It is the comprehension of causality that transforms tool use – an adaptive behavior present in many different species – into the master key that we humans use to change the world.

Finally, causal understanding seems to be involved in some of the tasks presented in the experiments on social cognition of monkeys and apes (see Cheney & Seyfarth, 1990b; Povinelli *et al.*, 1990). Therefore, our hypothesis of a hiatus between monkeys and apes in the comprehension of causal relations involved in tool use is confirmed by convergent evidence from studies in the domain of social intelligence.

ACKNOWLEDGEMENTS

We are extremely grateful to K. Bard, S. Boysen, D. Fragaszy and S. Savage-Rumbaugh for their invaluable collaboration in the studies with apes. We are also grateful to the Italian students who helped collect the data on children. Finally, we want to thank our capuchin monkeys; without their successful use and limited understanding of tools, the experiments discussed here would never even have been conceived, let alone carried out.

REFERENCES

Anderson, J.R. (1990). Use of objects as hammers to open nuts by capuchin monkeys (*Cebus apella*). *Folia primatologica*, 54, 138–45.

Anderson, J.R. (1994). The monkey in the mirror: A strange conspecific. In *Self-awareness in Animals and Humans: Developmental Perspectives*, ed. S.T. Parker, R.W. Mitchell & M.L. Boccia, pp. 315–29. New York: Cambridge University Press.

Antinucci, F. & Visalberghi, E. (1986). Tool use in *Cebus apella*. A case study. *International Journal of Primatology*, 7, 351–63.

Bard, K.A., Fragaszy, D. & Visalberghi, E. (1995). Acquisition and comprehension of tool-using behavior by young chimpanzees (*Pan troglodytes*): Effects of age and modeling. *International Journal of Comparative Psychology*, 8, 1–22.

Bates, E., Benigni, L., Bretherton, I., Camaioni, L. & Volterra, V. (1979). *The Emergence of Symbols: Cognition and Communication in Infancy.* New York: Academic Press.

Beck, B.B., (1980). *Animal Tool Behavior: The Use and Manufacture of Tools by Animals.* New York: Garland STPM Press.

Boesch, C. (1991). Teaching among wild chimpanzees. *Animal Behaviour*, 41, 530–2.

Boesch, C. & Boesch, H. (1983). Optimization of nut-cracking with natural hammers by wild chimpanzees. *Behaviour*, 83, 265–85.

Boesch, C. & Boesch, H. (1990). Tool use and tool making in wild chimpanzees. *Folia Primatologica*, 54, 86–99.

Boesch, C. & Boesch, H. (1993a). Different hand postures for pounding nuts with natural hammers by wild chimpanzees. In *Hands of Primates*, ed. H. Preuschoft & D. Chivers, pp. 31–43. New York, Vienna: Springer-Verlag.

Boesch, C. & Boesch, H. (1993b). Transmission of tool use in wild chimpanzees. In *Tools, Language and Cognition in Human Evolution*, ed. K.R. Gibson & T. Ingold, pp. 171–83. Cambridge: Cambridge University Press.

Brownell, C.A. & Carriger, M.S. (1990). Changes in cooperation and self-other differentiation during the second year. *Child Development*, 61, 1164–74.

Caro, T. & Hauser, M.D. (1992). Teaching in nonhuman animals. *Quarterly Review of Biology*, 67, 151–74.

Caron, A.J., Caron, R.F. & Antell, S.E. (1988). Infant understanding of containment: an affordance perceived or a relationship conceived?. *Developmental Psychology*, 24, 620–7.

Chalmeau, R. (1994). Apprentissage en situation sociale: La cooperation chez les Primates. Ph.D. thesis. Université Paul Sabatier de Toulouse.

Cheney, D. & Seyfarth, R. (1990a). Attending to behavior versus attending to knowledge: Examining monkeys' attribution of mental states. *Animal Behaviour*, 40, 742–53.

Cheney, D. & Seyfarth, R. (1990b). *How Monkeys See the World.* Chicago: University of Chicago Press.

Chevalier-Skolnikoff, S. (1989). Spontaneous tool use and sensorimotor intelligence in *Cebus* compared with other monkeys and apes. *Behavioral and Brain Sciences*, 12, 561–627.

Collinge, N.E. (1989). Mirror reactions in a zoo colony of cebus monkeys. *Zoo Biology*, 8, 89–98.

Connolly, K. & Dalgleish, M. (1989). The emergence of a tool-using skill in infancy. *Developmental Psychology*, 25, 894–912.

Connolly, K. & Dalgleish, M. (1993). Individual patterns of tool use by infants. In *Motor Development in Early and Later Childhood: Longitudinal Approaches*, ed. A. Kalverboer, B. Hopkins & R. Geuze, pp.174–204. Cambridge: Cambridge University Press.

Crawford, M.P. (1937). The cooperative solving of problems by young chimpanzees. *Comparative Psychological Monographs*, **14**, 1–88.

D'Amato, M.R. & Salmon, D.P. (1984). Cognitive processes in cebus monkeys. In *Animal Cognition*, ed. H.L. Roitblat, T.G. Bever & H.S. Terrace, pp. 149–68. Hillsdale, NJ: Lawrence Erlbaum.

Eglash, A.R. & Snowdon, C.T. (1983). Mirror-image responses in pygmy marmosets (*Cebuella pygmaea*). *American Journal of Primatology*, **5**, 211–9.

Fragaszy, D.M. & Adams-Curtis, L. (1991). Generativity aspects of manipulation in tufted capuchin monkeys. *Journal of Comparative Psychology*, **105**, 387–97.

Fragaszy, D.M. & Visalberghi, E. (1989). Social influences on the acquisition of tool-using behaviors in tufted capuchin monkeys (*Cebus apella*). *Journal of Comparative Psychology*, **103**, 159–70.

Goodall, J. (1986). *The Chimpanzees of Gombe: Patterns of Behaviour*. Cambridge, MA: The Belknap Press.

Gopnik, A. & Meltzoff, A.N. (1994). Minds, bodies and persons: Young children's understanding of the self and others as reflected in imitation and theory of mind research. In *Self-awareness in Animals and Humans: Developmental Perspectives*, ed. S.T. Parker, R.W. Mitchell & M.L. Boccia, pp. 166–86. New York: Cambridge University Press.

Heyes, C.M. (1993). Anecdotes, training, trapping and triangulating: Do animals attribute mental states? *Animal Behaviour*, **46**, 177–88.

Klüver, H. (1933). *Behavior Mechanisms in Monkeys*. Chicago: University of Chicago Press.

Klüver, H. (1937). Re-examination of implement using behaviour in a cebus monkey after an interval of three years. *Acta Psychologica*, **2**, 347–97.

Köhler, W. (1976). *The Mentality of Apes*. 2nd edn. New York: Liveright. (First published in 1925).

Leslie, A.M. (1988). The necessity of illusion: Perception and thought in infancy. In *Thought without Language*, ed. L. Weiskrantz, pp. 185–210. Oxford: Oxford University Press.

Lewis, M. & Brooks-Gunn, J. (1979). *Social Cognition and the Acquisition of Self*. New York: Plenum Press.

Limongelli, L., Visalberghi, E. & Boysen, S.T. (1995). The comprehension of cause–effect relations in a tool-using task by chimpanzees (*Pan troglodytes*). *Journal of Comparative Psychology*. **109**, 18–26.

McGrew, W.C. (1974). Tool-use by wild chimpanzees in feeding upon driver ants. *Journal of Human Evolution*, **3**, 501–8.

McGrew, W.C. (1992). *Chimpanzee Material Culture: Implications for Human Evolution*. Cambridge: Cambridge University Press.

Meltzoff, A.N. (1985). Immediate and deferred imitation in fourteen- and twenty-four-month-old infants. *Child Development*, **56**, 62–72.

Meltzoff, A.N. (1988). Infant imitation and memory: Nine-month-olds in immediate and deferred tests. *Child Development*, **59**, 217–25.

Mitchell, R.W. (1986). A framework for discussing deception. In *Deception: Perspectives on Human and Nonhuman Deceit*, ed. R.W. Mitchell & N.S. Thompson, pp. 3–40. Albany, NY: State University of New York Press.

Nagell, K., Olguin, R.S. & Tomasello, M. (1993). Processes of social learning in the tool use of chimpanzees (*Pan troglodytes*) and human children (*Homo sapiens*). *Journal of Comparative Psychology*, **107**, 174–86.

Natale, F. (1989). Causality II: The stick problem. In *Cognitive Structure and Development in Nonhuman Primates*, ed. F. Antinucci, pp. 121–33. Hillsdale, NJ: Lawrence Erlbaum.

Parker, S.T. & Gibson, K.R. (1977). Object manipulation, tool use, and sensorimotor intelligence as feeding adaptations in cebus monkeys and great apes. *Journal of Human Evolution*, **6**, 623–41.

Parker, S. T. & Poti, P. (1990). The role of innate motor patterns in ontogenetic and experiential development of intelligent use of sticks in cebus monkeys. In *"Language" and Intelligence in Monkeys and Apes: Comparative Developmental Perspectives*, ed. S.T. Parker & K.R. Gibson, pp. 219–43. New York: Cambridge University Press.

Parker, S.T., Mitchell, R.W. & Boccia, M.L. (eds.) (1994). *Self-awareness in Animals and Humans: Developmental Perspectives*. New York: Cambridge University Press.

Piaget, J. (1952). *The Origins of Intelligence in Childhood*. New York: International Universities Press.

Piaget, J. (1954). *The Construction of Reality in the Child*. New York: Basic Books.

Povinelli, D.J., Nelson, K.E. & Boysen, S.T. (1990). Inferences about guessing and knowing by chimpanzees (*Pan troglodytes*). *Journal of Comparative Psychology*, 104, 203–10.

Povinelli, D.J., Parks, K.A. & Novak, M.A. (1991). Do rhesus monkeys (*Macaca mulatta*) attribute knowledge and ignorance to others? *Journal of Comparative Psychology*, 105, 318–25.

Reynolds, P.C. (1982). The primate constructional system: The theory and description of instrumental tool use in humans and chimpanzees. In *The Analysis of Action*, ed. M. Van Cranach & R. Harré, pp. 243–385. Cambridge: Cambridge University Press.

Riviello, M.C., Visalberghi, E. & Blasetti, A. (1993). Individual differences in responses toward a mirror by captive tufted capuchin monkeys (*Cebus apella*). *Hystrix*, 4, 35–44.

Sakura, O. & Matsusawa, T. (1991). Flexibility of wild chimpanzees nut-cracking behavior using stone hammers and anvils: An experimental analysis. *Ethology*, 87, 237–48.

Schiller, P.H. (1952). Innate constituents of complex responses in primates. *Psychological Review*, 59, 177–91.

Visalberghi, E. (1987). Acquisition of nut-cracking behavior by 2 groups of capuchin monkeys (*Cebus apella*). *Folia Primatologica*, 49, 168–81.

Visalberghi, E. (1990). Tool use in *Cebus*. *Folia Primatologica*, 54, 154–64.

Visalberghi, E. (1993a). Tool use in a South American monkey species. An overview of characteristics and limits of tool use in *Cebus apella*. In *The Use of Tools by Human and Nonhuman Primates*, ed. A. Berthelet & J. Chavaillon, pp. 118–31. Oxford: Clarendon Press.

Visalberghi, E. (1993b). Capuchin monkeys. A window into tool use activities by apes and humans. In *Tools, Language and Cognition in Human Evolution*, ed. K. Gibson & T. Ingold, pp.138–50. Cambridge: Cambridge University Press

Visalberghi, E. & Fragaszy, D. (1990). Do monkeys ape?. In *"Language" and Intelligence in Monkeys and Apes: Comparative Developmental Perspectives*, ed. S.T. Parker & K.R. Gibson, pp. 247–73. New York: Cambridge University Press.

Visalberghi, E. & Fragaszy, D. (1995). Pedagogy and imitation in monkeys: Yes, no, or maybe?. In *Handbook of Psychology in Education. New Models of Learning, Teaching, and Schooling*, ed. D. Olson. Cambridge, MA: Blackwell. (in press).

Visalberghi, E. & Limongelli, L. (1994). Lack of comprehension of cause-effect relations in tool-using capuchin monkeys (*Cebus apella*). *Journal of Comparative Psychology*, 108, 15–22.

Visalberghi, E. & Mason, W.A. (1983). Determinants of problem-solving success in *Saimiri* and *Callicebus*. *Primates*, 24, 385–96.

Visalberghi, E. & Trinca, L. (1989). Tool use in capuchin monkeys: Distinguishing between performing and understanding. *Primates*, 30, 511–21.

Visalberghi, E., Fragaszy, D.M. & Savage-Rumbaugh, S. (1995). Performance in a tool-using task by common chimpanzees (*Pan troglodytes*), bonobos (*Pan paniscus*), an orangutan (*Pongo pygmaeus*), and capuchin monkeys (*Cebus apella*). *Journal of Comparative Psychology*. 109, 52–60.

Westergaard, G.C. (1994). The subsistence technology of capuchins. *International Journal of Primatology*. 15, 521.

Westergaard, G.C. & Fragaszy, D.M. (1987). The manufacture and use of tools by capuchin monkeys (*Cebus apella*). *Journal of Comparative Psychology*, **2**, 159–68.

Westergaard, G.C. & Suomi, S.J. (1993). Use of a tool-set by capuchin monkeys (*Cebus apella*). *Primates*, **34**, 459–62.

Whiten, A. & Byrne, R.W. (1988). Tactical deception in primates. *Behavioral and Brain Sciences*, **11**, 233–73.

Wimmer, H., Hogrefe, J. & Perner, J. (1988). Children's understanding of informational access as a source of knowledge. *Child Development*, **59**, 386–96.

Yerkes, R.M. (1927a). The mind of a gorilla. Part I. *Genetic Psychology Monographs*, **2**, 1–193.

Yerkes, R.M. (1927b). The mind of a gorilla. Part II. *Genetic Psychology Monographs*, **2**, 375–551.

Yerkes, R.M. (1943). *Chimpanzees*. New Haven, CT: Yale University Press.

4

Consolation, reconciliation, and a possible cognitive difference between macaques and chimpanzees

FRANS B. M. DE WAAL AND FILIPPO AURELI

INTRODUCTION

In an attempt to understand how animal societies maintain their integrity, primatologists and ethologists have begun to pay attention to events following aggressive competition. Competition is inevitable, yet a purely competitive society would never hold together, so the reasoning goes. Hence social animals must have ways to overcome conflict, or at least to reduce its disturbing and stressful consequences.

Postconflict behavior seems to serve intragroup harmony, and to allow each member of the group to "undo" those negative consequences of aggression that immediately affect him or her. The most prominent postconflict behavior is *reconciliation*, defined by de Waal & van Roosmalen (1979) as a friendly reunion between former opponents not long after their confrontation. Many of these contacts strike the human observer as high in emotional content. In chimpanzees (*Pan troglodytes*), for example, an individual may approach his or her opponent with an outstretched hand, to which the other may respond with a kiss and embrace.

Here, we will use postconflict behavior as a window on primate cognition by analyzing variation in this behavioral category between chimpanzees and macaques. This is no simple task, because apart from reflecting social cognition, this variation may also reflect interspecific differences in temperament, emotionality, or social organization. These various sources of behavioral variation are probably interdependent, making it hard – if at all possible – to isolate the contribution of cognitive processes. Nevertheless, attempts in this direction need to be undertaken in order to develop a functional perspective on social cognition, i.e. a perspective that positions cognitive abilities in the environment to which they are adapted.

In this chapter, we introduce two main hypotheses to account for variation in postconflict behavior (see next section), and discuss the social background against which this variation needs to be evaluated (third section). This is followed by a review of available data on postconflict behavior itself (fourth section), and the pros and cons of each hypothesis (fifth and sixth sections).

TWO HYPOTHESES

A discussion of the dependency of reconciliation on social cognition first took place when Gallup (1982) proposed that reconciliation represents a "marker of mind" that requires self-awareness. This led him to predict that this process should be restricted to great apes and humans as these are the only species to recognize themselves in a mirror.

Apart from the fact that this prediction rests on a rather narrow definition of self-awareness (i.e. recognition of contingencies between one's own body and its reflection), de Waal & Yoshihara (1983) documented reconciliation in monkeys, thus contradicting Gallup's prediction. Since then, reconciliation has been investigated in many different primate species, both in captivity and in the field, without any indications that monkeys and apes differ fundamentally in this regard (for reviews, see de Waal, 1989c, 1993; Kappeler & van Schaik, 1992).

A second category of postconflict behavior exists, however, that does seem to reveal important differences between apes and monkeys: *consolation*. Consolation differs from reconciliation in that the postconflict contact is not between the opponents but between a participant in a conflict and an uninvolved third party (de Waal & van Roosmalen, 1979). An illustration is provided by the following account of how two adult female chimpanzees in the Arnhem Zoo colony used to console each other after fights (see Figure 4.1):

> Not only do they often act together against attackers, they also seek comfort and reassurance from each other. When one of them has been involved in a painful conflict, she goes to the other to be embraced. They then literally scream in each other's arms *(de Waal, 1982, p. 67)*.

In terms of function and motivation, and perhaps also cognition, a few important similarities and distinctions between reconciliation and consolation should be noted. There exist indications that reconciliation reduces the anxiety in recipients of aggression (Aureli & van Schaik, 1991b; Aureli *et al.*, 1989), and the most likely function of consolation is to calm recipients of aggression. Thus, reconciliation and consolation probably share a distress-alleviating function. Yet, whereas both parties may experience similar emotions during reconciliation due to their direct involvement in the preceding conflict and their shared interest in restoration of the relationship, parties to consolation find themselves in quite different situations. One of the two just experienced a stressful event, whereas the other merely looked on from the sidelines. This individual probably could walk away from the scene without negative repercussions.

If consolation serves to alleviate another's stress, it essentially provides a service, and may thus be categorized as altruistic behavior, or more specifically as "succorant behavior," i.e. helping and caregiving directed to distressed individuals (Scott, 1972). In

Figure 4.1. Two adult female chimpanzees, Mama and Gorilla, at Arnhem Zoo scream in each others' arms after an aggressive encounter involving one of them. Mama (left) embraces Gorilla, while Gorilla screams in a close face-to-face to Mama's infant. Photograph by Frans de Waal.

its simplest form, succorant behavior is an automatic response to the distress signals of kin and close associates. Most likely, its evolutionary origin is nurturance of the young, because succorant behavior and other signs of affection between adults are virtually absent in animals without parental care (Eibl-Eibesfeldt, 1970).

One possible mechanism underlying succorance may be that the perception of distress in another individual vicariously distresses the observer, resulting in similar internal states in the observer and the originally distressed individual. If so, the initial dissimilarity in arousal state has been replaced by a similarity. Identification with, and sensitivity to, the emotions of others emerges in humans well before caring responses are developed, such as when the rest of the infants in a nursery room burst out crying in response to the cries of one amongst them. This process, known as "empathic distress" or "emotional contagion" (Hatfield *et al.*, 1993; Hoffman, 1977, 1987), provides the ontogenetic basis for a cognitively more advanced form of succorance in which the actor understands the other's situation, distinguishes the other's distress from his or her own feelings, and acts out of genuine concern about the other's well-being. Hence, dependent on the precise mechanism involved, *empathy*, or the capacity to be emotionally affected by someone else's feelings, can be cognitively simple or complex (Hoffman, 1981, 1987; Mercer, 1972).

What we measure in our study is not so much empathy, however, but the active response to assist another, known as *sympathy*, which most likely rests on empathy (Wispé, 1991). Because our data suggest interspecific variation in the occurrence of consolation (pp. 93-9) we discuss the possibility that this reflects emotional and/or cognitive capacities present in one taxonomic group but not the other. In order to do so we compare the postconflict behavior of chimpanzees and several members of the genus *Macaca*. It is unclear at this point how our findings might generalize to other pongids, such as gorillas and orangutans, and other monkey species, including neotropical primates. Ours is a very limited comparison, but one that hopefully will inspire further research.

Macaques and chimpanzees are among the best studied nonhuman primates, both characterized by a complex yet quite different social organization. Differences in social organization are important to our analysis as they provide the wider context within which to understand postconflict behavior. Furthermore, the social organization of a species may be an important correlate of its cognitive capacity. An increasing number of studies confirm that many primates live in societies that require complex information processing for successful functioning (see e.g. Cheney & Seyfarth, 1990b; de Waal, 1982). This is partly due to the high degree of intragroup cooperation, in the form of coalitions and alliances, characteristic of the primate order (Harcourt, 1992; Harcourt & de Waal, 1992b). Several authors have tied the evolution of primate intelligence to this social complexity (e.g. Humphrey, 1976; Jolly, 1966; Kummer, 1967).

One of the greatest challenges for students of cognition in the social domain is that the species-typical context in which cognition is expressed (i.e. group life under natural or naturalistic conditions) does not lend itself easily for controlled experimentation. Conclusive demonstrations often require controls and manipulations possible only in the laboratory; interesting attempts of this kind include experiments on deception (Woodruff & Premack, 1979), attribution (Povinelli et al., 1990), self-recognition (Gallup, 1991), and imitation (Nagell et al., 1993). Laboratory studies, however, also have their limitations: they rarely shed light on the social or biological function of a capacity. What do animals use it for? What advantage might individuals that possess the capacity have over individuals that do not? Even if comparative in nature, laboratory research traditionally has failed to adopt a truly evolutionary perspective, i.e. it has paid scant attention to the adaptive significance of cognitive capacities.

Study of the natural behavior of monkeys and apes remains necessary both as an external validation of experimental findings and in order to illuminate the evolutionary reasons for particular abilities (de Waal, 1991a). To analyze cognition in the context of the social organization of a species may be called the *in vivo* approach as distinguished from a laboratory approach, which isolates cognitive capacities from the species-typical social environment. The *in vivo* approach is fraught with problems and complexities, yet essential for a functional understanding.

Laboratory research suggests a dichotomy in cognitive capacities between on the one hand, pongids and humans, and on the other hand, the rest of the primate order (e.g. Gallup, 1982; Povinelli, 1993; Rumbaugh & Pate, 1984; Visalberghi & Fragaszy, 1990; Whiten & Ham, 1992). According to our studies, these differences may be paralleled in certain aspects of the spontaneous social behavior of these species (pp. 89–99). Even if it is tempting to attribute these behavioral differences to the assumed cognitive gap we cannot exclude the alternative that they relate to differences in social organization. In other words, we will be comparing two hypotheses to explain the presence of particular postconflict behavior in chimpanzees but not macaques.

Social Constraints Hypothesis: Certain postconflict behavior is more advantageous or less risky in chimpanzee society than in macaque society.

Reconciliation and consolation are expected only in species with high levels of cooperation and strong mutual attachments. Both macaques and chimpanzees are such species. These primates maintain long-term relationships between mother and offspring, and establish strong bonds with a subset of group members, including nonrelatives. It is likely, therefore, that social mechanisms to resolve conflicts of interest, to restore relationships after aggressive encounters, and to alleviate stress in others do not occur randomly but vary with the nature and quality of the social network in which individuals operate. Chimpanzee society differs from that of macaques in having (a) a less strictly hierarchical organization, and rather high levels of social tolerance, and (b) considerable influence (through alliances) of lower echelons of the hierarchy on higher ranking individuals. Interspecific differences in reconciliation and consolation may be related to these differences in the plasticity of social organization.

Social Cognition Hypothesis: Differences in postconflict behavior reflect the higher cognitive capacities of great apes as also indicated by studies of tool use, symbol learning, mirror-self-recognition, and so on.

Human empathy may be mediated by various psychological mechanisms, including motor mimicry (emotional contagion), classical conditioning, memory-based nonverbal processes, and verbal processes. Very young children pay attention to and are emotionally disturbed by the distress of others, and try to make contact with them (Zahn-Waxler et al., 1992; Zahn-Waxler & Smith, 1992). Yet, these children may lack the distinction between self and other required for a full appreciation of the other's situation and needs. It is generally assumed that in the course of development, empathic distress becomes associated with the emergence of a cognitive sense of self and others (Hoffman, 1981, 1987). Cognitively mediated empathic responses emerge at approximately the same time during development as mirror-self-recognition (Bischof-Köhler, 1988; Johnson, 1982).

This developmental perspective is relevant in connection to a comparative one because of apparent interspecific variation in self-knowledge. The work of Gallup (1982), Povinelli (1987), and others on self-recognition in mirrors suggest that humans and great apes differ in this regard from all monkey species. Only in the former species has mirror-self-recognition been demonstrated. Thus far, all members of the genus *Macaca* have failed the critical test (Anderson, 1983; Gallup, 1970; Straumann & Anderson, 1991; for reviews and discussion, see Mitchell, 1993; Parker et al., 1994). Yet, even if we assume that mirror-tests demonstrate self-awareness, this does not mean that failing the test proves its absence. Mirror-tests provide a rather limited measure of self-knowledge.

At the most basic level the self represents the interface between the animal and the environment upon which it acts (Gibson, 1979; Gopnik & Meltzoff, 1994). Understanding one's surroundings equals understanding oneself. This means that no animal is devoid of self-knowledge, and that differences between species are probably gradual rather than absolute. The more complex their interaction with the environment, the more complete their self-knowledge and differentiation between self and other. The social environment may be particularly important in this regard: knowledge of the self in the social field is subsumed under some of the more basic forms of self-knowledge distinguished by contemporary cognitive psychologists (e.g. Neisser, 1988).

There are many indications that nonhuman primates possess an exquisite knowledge of their position relative to others. To our knowledge, Kummer (1971) was the first to speak of the *social field*, and of the need of nonhuman primates to process large amounts of information for successful social action:

> In turning from one to another context at a rapid rate, the individual primate constantly adapts to the equally versatile activities of the group members around him. Such a society requires two qualities in its members: a highly developed capacity for releasing or suppressing their own motivations according to what the situation permits and forbids; and an ability to evaluate complex social situations, that is, to respond not to specific social stimuli but to a social field.
> *(Kummer, 1971, p. 36).*

To place self-knowledge in this broader perspective is not to deny a possible connection with mirror-self-recognition, yet it does provide an argument for gradual rather than qualitative interspecific differences. Differences in postconflict behavior discussed in this chapter may represent one more indication of a greater emotional sensitivity to others in apes as compared to monkeys, hence of a greater awareness in apes of their position in the social field.

SOCIAL CONTEXT

To put variation in postconflict behavior in perspective, we start with an overview of the most important similarities and differences in the social organization of macaques and chimpanzees.

Macaque matrilines

Macaques live in social groups that usually consist of more adult females than adult males, and that move as an integral unit. Females remain in their natal group throughout their lives, whereas males generally transfer to neighboring groups at around the time of sexual maturity. The females in a macaque group are therefore genealogically related to one another and form the core of the social system. Closely related females – mothers, daughters, sisters, grandmothers, grand-daughters, aunts, and nieces – tend to associate with one another, forming kin-based clusters of individuals, named matrilines. This pattern of association with kin also holds for immature males before they leave the group. Members of the same matriline frequently exchange friendly behavior, such as grooming, and support one another in aggressive encounters, particularly against other matrilines. Support among kin clearly affects the pattern of aggressive interactions and dominance relationships. Mothers and adult daughters typically have similar ranks, so that entire matrilines can be ordered with reference to one another. Female dominance relationships are clear-cut, linear, and relatively stable over the years. By contrast, adult male dominance relationships cannot be predicted from maternal rank, but are determined by age, size, tenure in the group, and fighting ability, but rarely by alliances. Therefore, dominance relationships among adult males are less stable than those among adult females. Adult males are usually dominant over most adult females, but females can form successful coalitions against them. For more details on macaque social organization, see Eaton (1976), Lindburg (1991) and Melnick & Pearl (1987).

Chimpanzee fission and fusion

Chimpanzees live in communities that contain all age–sex classes and number anywhere from 20 to over 100 individuals. Adult females generally outnumber adult males. Unlike macaques, chimpanzees do not travel in permanent groups but associate in temporary "parties" lasting a few minutes to several days. Party membership changes constantly in a pattern of fission and fusion of parties, with the exception of the tie between mother and dependent offspring. Most parties are small, consisting of less than six individuals. Adult males are more often found in parties than adult females, the latter spending longer periods alone. Males associate more frequently with one another than with females, except when one of the females is in estrus: then they follow the female. In contrast to macaques, it is the female who migrates between communities, whereas males stay in their natal group. As a result males are genetically related, which may explain the evolution of cooperation among them (Morin et al., 1994). Given the inherent uncertainty of paternity, however, kinship among male chimpanzees is not easily recognizable to the individuals themselves (except for maternal brothers). For this reason, kinship does not affect relationships among male chimpanzees with anything near the predictability with which matrilineal relatedness determines relationships among female macaques.

Male chimpanzees, unlike male or female macaques, "patrol" the boundaries of a community range. Interactions between males of neighboring communities are typically hostile: intercommunity encounters may have lethal consequences. It is of paramount importance, therefore, that male chimpanzees maintain unity among themselves: male–male relationships within a community are characterized by a high degree of tolerance and cooperation. At the same time, male chimpanzees are particularly known for a strong motivation to defend or improve dominance positions. Early in life, males may be supported by their mothers in aggressive interactions, but such support seems unimportant in interactions among adult males. Coalitions among males are often critical in determining the outcome of power struggles.

The hierarchy among adult females is rather vague compared to that among males, but older females are generally dominant over younger ones and newly immigrated females occupy the lowest ranks. Adult males are dominant over all females. Coalitions among females are not very common in the wild but are a typical pattern of defense against male aggression in captivity. Details on chimpanzee social organization are given by de Waal (1982), Goodall (1971, 1986), Nishida (1979), Nishida & Hiraiwa-Hasegawa (1987), and Wrangham (1986).

Social relationships as investments

To understand why nonhuman primates engage in postconflict behavior, it is crucial to place this behavior in the context of social relationships, and of social organization in general. If social organization is the pattern of relationships emerging from repeated interaction among group members who individually recognize one another (cf. Hinde, 1976), interactions do not only have immediate consequences but also serve long-term functions: interactions *shape* social relationships. Naturally, we expect each individual to shape relationships to its own advantage, and to interact with others in order to gain benefits from subsequent interaction.

From an evolutionary perspective, benefits to interactants are most evident in relationships with a reproductive function, such as male–female and mother–offspring relationships. Relationships may also produce non-reproductive benefits, however, such as when two individuals protect one another against attack, tolerate one another around resources, provide vigilance against predators, or cooperate during intra- or intergroup competition. With regards to the evolution of such cooperation it is, of course, assumed here that it pays off in the long run in terms of survival and reproduction.

Kummer (1978) considered the benefits that individual A provides to B as A's *value* to B. Any individual will try to improve this value: B will select the best available A, predict A's behavior, and modify A's behavior to its own advantage. In other words, B will *invest* in the relationship with A. Whereas most of B's investments may not lead to quick profits, such as immediately useful actions by A, they may help to cultivate a relationship that is beneficial to both A and B in the long term. A good example of such investment is social grooming. One primate may groom another for over an hour without any immediate reciprocation or return-favor; after the session the two may simply separate, each going their own way. There are indications that grooming is altruistic in that it entails costs for the groomer such as reduced energy and time available for other activities (Kurland, 1977; but see Dunbar, 1988) and decreased vigilance (Maestripieri, 1993), whereas it provides benefits to the groomee in terms of hygiene (Hutchins & Barash, 1976) and tension reduction (Schino *et al.*, 1988), probably involving a pleasant sensation (Keverne *et al.*, 1989). Why would individuals provide such services to others – grooming is one of the most common social activities in primate groups – if not to foster future beneficial exchanges?

Inspired by Trivers' (1971) model of stable long-term relationships within which altruistic acts are reciprocally exchanged, Seyfarth (1977) predicted that grooming produces delayed benefits, often of a different "currency." A currency he specifically referred to is coalition formation; that is, one individual intervening in a confrontation between two others on behalf of one and against the other party (for a review, see Harcourt & de Waal, 1992a). There is indeed evidence that grooming partners support one another in fights (de Waal & Luttrell, 1988; Hemelrijk & Ek, 1991; Seyfarth, 1980), and share food with one another (de Waal, 1989b). This suggests that grooming serves as an investment in reciprocal exchange relationships.

The principle of reciprocal exchange may hold for both macaques and chimpanzees, yet major differences exist with regard to partner selection. In macaques, the matrilineal system predicts to a large extent which individuals will establish social ties, and which will not. The best general predictors of preferential association and grooming in cercopithecine monkeys (to which the macaques belong) are, in this order, matrilineal kinship and similarity in dominance rank, variables that are or tend to be, respectively, stable for life, at least among the females (de Waal, 1991b; Seyfarth, 1977, 1980). These rules are less applicable or relevant to male macaques because in the wild, the majority of males leaves their natal group.

In the chimpanzee, it is the females who migrate between communities, and hence establish ties with unrelated individuals and these ties are rather loose compared to those of female macaques. In the Arnhem Zoo colony, relationships between unrelated females are durable, and their affiliative preferences strongly overlap with coalitions during aggressive encounters. This female pattern contrasts with the remarkable opportunism observed among unrelated males in the same colony (de Waal, 1984) as well as between males in the wild (Nishida, 1983). Although male chimpanzees are fairly consistent in their support preferences from one incident to the next, they sometimes change allies slowly but radically over longer periods of time. These changes seem to be related to which partner offers the best prospects for a successful alliance during dominance struggles among the males; changes are particularly marked during periods of hierarchical instability. During such periods, male coalitions become dissociated from affiliative preferences (de Waal, 1984; Hemelrijk & Ek, 1991), and males who used to support one another may begin to support one another's opponents, i.e. turn from allies into major rivals (de Waal, 1978).

So, adult male chimpanzees rely on one another during intergroup competition. They are also prone to shift allegiance, often ignoring existing affiliative ties in favor of new, more profitable partnerships when striving for high status. This is a more complex and unpredictable pattern than that of macaques, the more so because at the same time male chimpanzees may increase the exchange of grooming with rivals, probably as a mechanism of tension reduction (de Waal, 1986b). A system in which the value attached to relationships and, consequently, the investment in them vary across time and context may require more fine-tuned processes of postconflict behavior than a relatively stable, kin-based system.

POSTCONFLICT BEHAVIOR

Reconciliation

Contrary to the traditional view that aggression causes dispersal, de Waal & van Roosmalen (1979) demonstrated that in the chimpanzee colony of Arnhem Zoo aggression led to an increase in contact. Opponents were more often within 2 m of one another and engaged in affiliative contact immediately after a conflict compared to the period before a conflict. Moreover chimpanzees preferred to contact the opponent rather than uninvolved individuals (interopponent contacts were found to constitute 30% of all postconflict contacts compared with chance expectations of 5.6%). These results demonstrated a pronounced conciliatory tendency in the chimpanzee, later confirmed by reports in the wild (Goodall, 1986).

Whereas the occurrence of peacemaking strategies in the chimpanzee comes perhaps as no surprise, few people would have predicted that a rather intolerant primate, such as the rhesus macaque (*Macaca mulatta*), would follow similar strategies. Introducing a new, more controlled methodology that compared contact rates following aggression with those during control periods, de Waal & Yoshihara's (1983) study of captive rhesus macaques produced evidence for reconciliation among monkeys. Similar results have been reported for other groups of the same species (Balcomb et al., 1993; Demaria & Thierry, 1992) as well as for other captive macaques (*M. fascicularis* (Aureli et al., 1989; Cords, 1988) *M. arctoides* (de Waal & Ren, 1988) *M. nemestrina* (Judge, 1991) *M. tonkeana* (Demaria & Thierry, 1992) *M. fuscata* (Aureli et al., 1993) *M. sylvanus* (Aureli et al., 1994) *M. nigra* (Petit & Thierry, 1995)). A field study confirmed the occurrence of reconciliation for long-tailed macaques (*M. fascicularis*) in a Sumatran forest (Aureli, 1992). In macaques, as in chimpanzees, affiliative contact between former opponents is more likely during the first few minutes after a conflict (Kappeler & van Schaik, 1992) and postconflict attractiveness is highly selective, i.e. the contact increase does not occur indiscriminately with all possible partners but specifically concerns the former opponents (e.g. de Waal & Yoshihara, 1983).

Even though both chimpanzees and macaques reconcile, they show some differences in the form of postconflict reunions. Chimpanzees are more likely to exchange behavior patterns rarely used in other contexts. According to de Waal & van Roosmalen's (1979) and de Waal's (1989b) data on captive chimpanzees, and Goodall's (1986) descriptions of chimpanzees in the wild, reconciliation in this species involves distinct behavior patterns, such as embracing, gentle touching, and mouth-to-mouth kissing (Figure 4.2). In fact, chimpanzee opponents kiss one another 10 times more often during the first postconflict contact than during subsequent contacts (de Waal & van Roosmalen, 1979). In contrast, most macaque species studied thus far do not show any behavioral distinction between postconflict and control contacts. This absence of specific conciliatory behavior is not general in the macaque genus, however, as stump-tailed macaques (*M. arctoides*) use a conspicous behavior pattern rarely observed outside the reconciliation context: the "hold-bottom ritual," in which one individual, usually the recipient of aggression, presents the hindquarters, and the other clasps the presenter's haunches. This ritual occurs in more than 30% of first postconflict contacts, a 20-fold increase compared to control contacts (de Waal & Ren, 1988). Another possible exception occurs in the Tonkean macaque (*M. tonkeana*), which shows special clasping gestures during reconciliation (Thierry, 1984).

Stump-tailed macaques are among the most conciliatory macaques: after correction for both the number of aggressive incidents and the species' normal contact tendencies, these monkeys reconcile four times more frequently than rhesus macaques (for new correction procedures and a reanalysis of existing data see Veenema *et al.*, 1994). The frequency of reconciliation among captive chimpanzees is probably closer to that of stump-tailed macaques than that of rhesus macaques. Because of this similarity between stump-tailed macaques and chimpanzees, and the shared behavioral distinctness of postconflict reunions, we believe that the greater behavioral specificity of chimpanzee reconciliations compared to those in most macaque species is not so much a function of taxonomic position, i.e. pongids versus Old World monkeys, but rather of species-typical conciliatory tendencies. This is confirmed by the fact that Tonkean macaques, which show special gestures during reconciliation, are also characterized by high levels of reconciliation (Demaria & Thierry, 1992). Rapid transitions from aggressive to affiliative motivation during reconciliation are probably facilitated by the exchange of ritualized signals that both test the attitude of the other and communicate the actor's intentions. The kiss of chimpanzees, the hold-bottom ritual of stump-tailed macaques, and the clasping gestures of Tonkean macaques all require a high level of intimacy and coordination. They assist the reconciliation process by making both the context and meaning of the contact more *explicit*. This is in contrast to the *implicit* reconciliations of rhesus macaques, which often reestablish relationships through a brief, inconspicuous brushing contact that is quite meaningful given the risks involved in mere proximity in this short-tempered species (de Waal, 1989c).

Both macaques and chimpanzees follow what seems a general rule among primates;

Figure 4.2. The most typical conciliatory gesture of the chimpanzee is the mouth-to-mouth kiss, here shown by a young adult female to the dominant male of the colony at the Yerkes Field Station. Photograph by Frans de Waal.

that is, reconciliation aims to restore valuable relationships (for a review, see Kappeler & van Schaik, 1992). In macaques, in which matrilineal kin relationships are particularly valuable, related individuals reconcile conflicts more often than unrelated ones, particularly, of course, in species with a pronounced kin-bias (Aureli et al., 1993a; Demaria & Thierry, 1992; cf. Thierry, 1990). In chimpanzees, in which males form stronger intrasexual bonds than females, conflicts among males are more often reconciled than those among females in both captivity and the wild (de Waal, 1986b; Goodall, 1986).

In order to experimentally investigate the role of relationship value, Cords & Thurnheer (1993) trained pairs of long-tailed macaques to cooperate during feeding, thus enhancing dependency between partners. They found that the probability of reconciliation following conflict increased dramatically once the partner had become a social tool to obtain food, thus confirming the hypothesis that reconciliation occurs in proportion to the value of the relationship.

Whereas the above data indicate that chimpanzees and macaques do not differ substantially in the occurrence and behavioral specificity of reconciliation, and its role in the maintenance of valuable relationships, they still may differ in other areas, such as the variability or complexity of the process. For example, there are anecdotal indications for deceptive reconciliation in chimpanzees. On six occasions, a dominant female who had been unable to catch a fleeing opponent was observed to approach this individual later on with a friendly appearance, holding her hand out in a gesture of invitation, only to suddenly change her behavior into outright attack when the other came within reach (de Waal, 1986a). Similar cases have not been reported for macaques.

Another aspect of the peacemaking process that may distinguish these taxonomic groups is *mediated reconciliation*, in which a bystander takes the initiative in bringing two opponents together. About 20 such mediations were observed during 2 months when there were frequent confrontations between male chimpanzees at Arnhem Zoo. An adult female would approach one of two males after a confrontation between them, groom this male, then walk slowly towards the second, immediately followed by the first male. She might pull this male's arm if he failed to follow. She would then groom the second male for a brief while after which she would walk off and leave the two males grooming each other (de Waal & van Roosmalen, 1979).

No such cases of mediation have ever been described for macaques. However, this does not mean that macaques entirely lack the required skills. Once, after an adult rhesus macaque male had chased a juvenile male, the juvenile's mother went to groom the adult male. After a while the juvenile went to sit less than a meter behind the two. As soon as the mother noticed him, she stepped aside to let her son take her place against the adult male's back (de Waal, 1989c, p. 238). Although she did not actually lead the second protagonist to the first, the mother seemed to at least make room for contact. This suggests that the mediation skills of chimpanzees may not have been totally without antecedent in the common ancestor with Old World monkeys.

In captive pig-tailed macaques (*M. nemestrina*), Judge (1991) found that relatives of the recipient of aggression increased contact with the aggressor following the incident. Thus, the recipient's kin seemed aware of the previous conflict and reconciled "for" their relative possibly so as to divert the aggressor's attention, or to reduce further aggression (for other data on "triadic reconciliations," see Cheney & Seyfarth, 1989). These observations indicate that both chimpanzees and macaques act as if they understand relationships in which they are not directly involved themselves, and follow both conflict and reconciliation among others with close attention. Understanding of the process of reconciliation is probably greater in chimpanzees, in which others may gather around two reconciling opponents or participate directly in the event through mediation, yet the differences appear to be mostly of degree.

This finding contradicts Gallup's (1982) prediction that chimpanzees, unlike macaques, would be capable of reconciliation because this interaction pattern is part of a set of cognitive capacities that includes mirror-self-recognition. It is, of course, still possible that reconciliation in macaques is what Gallup (1982) calls a "hard-wired analog" rather than a "marker of mind," yet reconciliation in macaques seems a rather flexible behavior. The experiment on the value of the relationship reported above (Cords & Thurnheer, 1993) is a clear example of strategic changes in the frequency of reconciliation in response to changed circumstances. In addition, de Waal & Johanowicz (1993) recently showed that juvenile rhesus macaques can learn from juvenile stump-tailed macaques, through social experience, to reconcile at a higher frequency. Such modification of reconciliation behavior strongly argues against a "hard-wired" nature.

Consolation

Typically, aggressive encounters among chimpanzees are accompanied by affiliative interactions between participants and third parties. Some of these contacts serve to recruit allies: for example, a losing individual will approach a potential supporter and stretch out an open hand in begging fashion, inviting the other to join the confrontation. Other contacts serve to prevent harmful interventions by individuals who tend to support the opponent, and to seek reassurance or protection from harm. De Waal & van Hooff (1981), who analyzed this behavioral category in detail, speak of *side-directed behavior*, meaning that the main exchange of behavior during aggressive encounters is between the combatants themselves, whereas appeals to and contacts with bystanders are directed "to the side" of the scene of conflict. Towards the end of the aggressive episode, side-directed behavior grades into consolation. The conflict being over, affiliative contact with bystanders is not made for recruitment or protection, but probably for the alleviation of distress. This functional distinction is also clear from the following description:

> An adult male challenged by another male often runs, screaming, to a third and establishes contact with him. Often both will then scream, embrace, mount, or groom each other while looking toward the original aggressor. This (. . .) is how a victim tries to enlist the help of an ally. There are occasions, however, when it seems that the primary goal is to establish reassurance contact – as when fourteen-year-old Figan, after being attacked by a rival, went to hold hands with his mother.
> *(Goodall, 1986, p. 361)*

Consolation is defined as a contact between a recipient of aggression and an uninvolved bystander not long after the former was involved in a conflict. In relation to the issue of empathy, it is important to specify whether such contact is initiated by the bystander (offering consolation) or the recipient of aggression (seeking consolation). Both kinds of contact may alleviate the distress of a victim of aggression, yet only the first kind requires empathy and sympathy, i.e. an active response to a distressed individual. Consolation was first recognized by de Waal & van Roosmalen (1979) based on observations of the Arnhem Zoo chimpanzee colony. These contacts occurred more often in the very first minute following a conflict than in subsequent minutes. Moreover, first-minute contacts included three times more hugging and gentle touching than contacts during subsequent minutes. The high rate of embraces makes consolation behaviorally distinct from reconciliation, which is typified by kissing (Table 4.1; Figure 4.3).

Because of the brevity of the postconflict observation window (i.e. 5 minutes), and the lack of control data, the above study needed replication. For this purpose, we analyzed data collected on a colony of 17 chimpanzees in a large outdoor compound at the Field Station of the Yerkes Regional Primate Research Center. The data stem from records of all social interactions during 230 lots of 90 minute observations (i.e. 345 hours) collected from a tower with an excellent overview of the enclosure. During

Table 4.1. *Frequencies of kissing and embracing among the Arnhem Zoo chimpanzees in two different postconflict contexts*

Context	Kiss	Embrace
Reconciliation	23	8
Consolation	19	57

Data from: de Waal & van Roosmalen, 1979.

these sessions we observed 418 agonistic incidents. In the *post-hoc* analysis, each such incident was used as the marker for the beginning of 30 minute interval. The 418 intervals, some of which overlapped, were analyzed together with 903 such intervals following semi-aggressive incidents. Following van Hooff's (1974) and de Waal and van Hooff's (1981) criteria, strictly aggressive interactions include particular vocalizations and facial expressions, brusque chases, biting, etc. and thus differ from incidents of milder intensity that lack these behaviors. The latter so-called semi-aggressive incidents are chases, threatening gestures, and pushing and shoving unaccompanied by agonistic vocalizations or facial expressions.

Control data concerned 30 minute intervals starting arbitrarily in the middle (i.e. at 45 minutes) of each 90 minute observation session. The advantage is that these intervals are representative for the general activity level of the group, which might not have been the case if aggression-free intervals had been selected. The hypothesis that postconflict affiliative contact with bystanders serves to alleviate stress or distress gives rise to the following testable predictions:

1. More affiliative contacts occur within a few minutes of the conflict than after longer time intervals or during control periods.
2. Because the distress of the recipient of aggression is supposedly proportional to aggression intensity, there will be more contacts following strictly aggressive rather than semi-aggressive incidents.
3. Because recipients of aggression are probably more distressed than the aggressors themselves, there will be more contacts with recipients than with aggressors.

Here we limit the analysis to affiliative contacts (e.g. kissing, embracing, grooming, gentle touching, mounts) *received by* conflict participants from bystanders. This means that we ignore most instances of side-directed behavior (which are performed by participants in the conflict towards bystanders).

As illustrated by Figure 4.4, significant differences in the rates of affiliative contact were limited to the first 2 minutes following an aggressive incident; after this point, levels returned to baseline. During the first 2 minutes, the rate of contact received by recipients of strictly aggressive actions was six times higher than baseline (comparing

Figure 4.3. In a consolation at Arnhem Zoo, a juvenile male embraces a screaming adult male, Yoroen, who has just lost a confrontation with his chief rival. Note the difference in facial expression between the two parties. From de Waal (1982).

rates for each individual, Wilcoxon matched-pairs test: $n = 17$, $t = 23$, $p < 0.01$, one-tailed). Thus, prediction 1 was supported. Rates of contact following strictly aggressive actions were also significantly above the rates following semi-aggressive actions ($n = 16$, $t = 32$, $p < 0.05$, one-tailed). Thus, prediction 2 was also supported. Aggressors, too, received more contact from bystanders during the first 2 minutes, yet

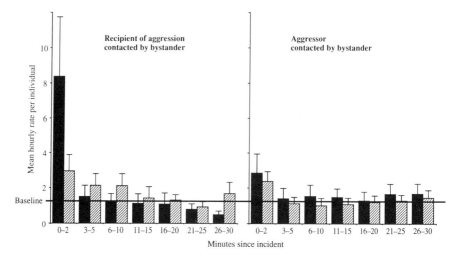

Figure 4.4. Data concerning 1321 spontaneous aggressive incidents among chimpanzees at the Yerkes Field Station. The graph shows the mean (+ SEM) hourly rate of affiliative contacts/individual received as conflict participant from a nonopponent. The 30 minute window following each incident has been divided into blocks of 2, 3, and 5 minutes. Data are presented separately: black bars for incidents involving vocalizations and/or physical contact (strictly aggressive incidents) and dashed bars for mere silent threats and lunges (semi-aggressive incidents). The mean-baseline contact rate/individual is based on arbitrarily selected control intervals in the same data set (no error margin is shown for this rate).

the difference with baseline was significant only for semi-aggressive interactions ($n = 17$, $t = 24$, $p < 0.01$, one-tailed). The magnitude of the contact increase, however was much smaller for aggressors than for recipients of aggression: 10 out of 13 individuals observed in both roles showed a higher rate of received contact during the first 2 minutes after having been the recipient than after having been the aggressor ($n = 13$, $t = 19$, $p < 0.05$, one-tailed). This confirms prediction 3. So, all three predictions derived from the consolation hypothesis were supported.

Behavior during these interactions resembled that reported for the Arnhem Zoo colony by de Waal & van Roosmalen (1979). Embraces and gentle touches were observed in 76.4% of all contacts between recipients of aggression and bystanders during the first 2 minutes following a conflict, which is about three times as high as the equivalent proportion in control contacts. Thus, the data on the Yerkes group supported previous findings. This, together with qualitative data from the wild (Goodall, 1986), leads us to conclude that chimpanzees console recipients of aggression by means of hugging and gently touching.

Although consolation in macaques has not been investigated nearly as extensively as reconciliation, none of the studies addressing this topic have produced evidence for consolation. These studies were carried out on four species living under different conditions. Despite various measures and statistical methods, the finding was the same:

Postconflict behavior and cognition 97

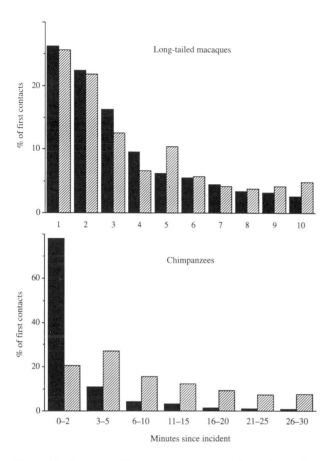

Figure 4.5. Percentage of first contacts between recipients of aggression and bystanders in the minutes following the incident (postconflict) and during control periods. Note that the time window is different for long-tailed macaques (data from Aureli & van Schaik, 1991a) as compared to chimpanzees. The graph differs from Figure 4.4 in that it is restricted to first contacts, both given and received, and in that it pools data for all individuals. The chimpanzee postconflict data are limited to periods following aggressive incidents. Black bars, postconflict; dashed bars, controls.

affiliative contact between recipients of aggression and bystanders did not occur more often following a conflict than during control periods (Aureli, 1992; Aureli & van Schaik, 1991a; Aureli et al., 1993b, 1994; Judge, 1991).

Figure 4.5 shows that in long-tailed macaques, for example, the proportion of first affiliative contacts between a recipient of aggression and a bystander following conflict did not differ from that during control periods. This contrasts with the distribution of first contacts for the Yerkes chimpanzees, in which almost 80% of all such contacts within 30 minutes of a conflict occurred within the first 2 minutes, which is at least three times higher than during control periods. (In interpreting this Figure, and comparing it with Figure 4.4, it should be noted that a distribution of *first* contacts, such as is shown in Figure 4.5, by definition results in a sloping graph because once a contact between two individuals has occurred, subsequent contacts are ignored. This applies also for the control graph. Figure 4.4 is different in that it presents *all* contacts following conflict, which does not result in a sloping graph in case contacts are randomly distributed. In addition, Figure 4.4 is limited to contacts received from bystanders, whereas Figure 4.5 includes both given and received contacts.)

Before concluding from these studies that consolation is totally absent in macaques, we need to rule out the possibility that the phenomenon is restricted to specific cases. Since in macaques, relatives of recipients of aggression are potentially the parties most interested in initiating calming contact (because of relationship value; pp. 87–9), they are prime candidates for consolation. Thus, instead of examining affiliative contact between recipients of aggression and all bystanders, some studies have limited the analyses to contacts between recipients of aggression and their own kin. However, in both long-tailed and pig-tailed macaques such contact did not increase in frequency following aggression (Aureli, 1992; Aureli & van Schaik, 1991a; Judge, 1991), whereas in Barbary and Japanese macaques such contact even decreased (Aureli *et al.*, 1993b, 1994; see also de Waal & Ren, 1988). Thus, not only do macaques show no evidence for consolation, they may actually avoid contact with recent recipients of aggression. In addition, no evidence for consolation was found when the initiative to postconflict contact was taken into account (initiated by the recipient of aggression or a bystander; Judge, 1991) or the kind of behavior shown (e.g. specific calming behavior such as allogrooming; Aureli & van Schaik, 1991a).

Nevertheless, we have witnessed incidents that suggest macaques may possess a consolatory disposition. For example, one day in a rhesus macaque group at the Wisconsin Primate Center several high-ranking adults attacked a juvenile, named Fawn. They viciously bit her from various sides. Fawn was completely overcome with fear when her attackers finally left the scene; she lay on her belly screaming, then jumped up and fled. Afterwards, she sat in a hunched position, looking exhausted. Within 2 minutes, her older sister approached to put an arm around her. The sister gently pushed and pulled at Fawn as if trying to activate her, then embraced her again, and remained with her for a while. Such cases suggests that macaques may, on rare occasions, engage in consolatory behavior. We also have regularly seen infant rhesus macaques being attracted to the screams of one amongst them (e.g. after a fall or punishment by an adult), approach the screaming infant, and seek contact with it (Figure 4.6). Occasionally, this response resulted in a pile of several infants clambering over each other.

In macaques, instances of consolation are uncommon, though, whereas one only needs to watch a group of chimpanzees for a short while to witness consolation. The contrast is not only in the frequency but also in the richness and conspicuousness of these acts. Chimpanzees have developed specific behavior to console and reassure one another, the need and expectation for which probably derive from maternal gestures to

Postconflict behavior and cognition 99

Figure 4.6. Two 3 month old infant rhesus macaques at the Wisconsin Primate Center following an aggressive incident in which the monkey in the foreground was seriously bitten by a dominant female. The distressed victim, a female, is embraced by a male peer. At this age, rhesus macaques do seem to show affiliative responses to distressed conspecifics, but these responses almost entirely disappear at a later age. Photograph by Frans de Waal.

comfort and calm infants (Mason, 1964). Chimpanzees also have special signals to "complain" and invite contact when distressed (such as pouting, whimpering, and yelping) along with a different set of signals used to invite agonistic support (de Waal & van Hooff, 1981). Macaques show the latter, but not the former kind of signals.

THE SOCIAL CONSTRAINTS HYPOTHESIS

In considering the social constraints hypothesis (see p. 84) of differences in postconflict behavior, probably the most important factor is *dominance style*, defined as the strictness of the hierarchy and the degree to which dominants tolerate or exploit subordinates (de Waal, 1989a). Whereas dominance styles vary considerably within the macaque genus (de Waal & Luttrell, 1989; Thierry, 1986), an even greater variation exists between macaques, and chimpanzees. For example, macaques compete over, and rarely share, food outside the mother–offspring relationship, whereas chimpanzees appear to virtually cancel dominance relationships in the feeding context resulting in begging and sharing on a large scale, both up and down the hierarchy (de Waal, 1989b; Goodall,1963; Nishida, 1970; Teleki, 1973). These differences in the expression of social dominance may be responsible for differences in consolation behavior, because in a strict hierarchy consolation may be rather risky.

What kind of risk is involved in consoling individuals living in strict hierarchies? It is known that recipients of aggression in macaque groups continue to attract aggression in the period immediately following the incident (Aureli, 1992; Aureli & van Schaik, 1991b; Cords, 1992; de Waal & Yoshihara, 1983). Furthermore, Cords (1992) has demonstrated experimentally that intolerance between opponents persists after unreconciled confrontation. If aggression causes a similar disturbance of relationships in the wild, this would be particularly serious as it would force the recipient of aggression to closely monitor the behavior of others, and hence sacrifice time devoted to important subsistence activities. Indeed, Aureli (1992) observed in the field that long-tailed macaques foraged less than usual after having received aggression. If the elevated chance of further aggression and intolerance towards recipients of aggression extends to bystanders that approach them, this is a risk bystanders will need to reckon with. Only if bystanders, alone or together with the recipient of aggression, were able to repel aggressors would this risk be eliminated.

A strict hierarchy and kin-based alliance system, however does not leave much room for a protective role for bystanders. Interestingly, recipients of aggression in macaque groups do occasionally engage in postconflict contact with individuals who supported them in the previous incident, particularly if these individuals dominate the original aggressor. Since dominants are able to forestall further attacks, these exceptional contacts perhaps carry little risk. Rather than serving a consolatory function, they may represent a reciprocal exchange of services, i.e. social grooming for received support (de Waal & Yoshihara, 1983).

In the typical macaque social system (see pp. 86–7), recipients of aggression rarely, if ever, counter-attack dominant individuals. In a more egalitarian system, such as that found in chimpanzees, recipients do retaliate directly or team up with protectors to chase off attackers (de Waal & van Hooff, 1981). Thus, recipients of aggression in the Yerkes chimpanzee colony were supported by others in 20% of the observed incidents, whereas equivalent figures for macaques range from only 4% to 8% (Aureli *et al.*, 1992; Bernstein & Ehardt, 1985; de Waal, 1977; Kaplan, 1977; Watanabe, 1979; F. Aureli, unpublished data). Moreover, a direct comparison of intervention data on two macaque species and chimpanzees demonstrated a much greater reluctance in macaques than chimpanzees to intervene in fights against dominant individuals (de Waal & Luttrell, 1988).

In summary, differences in dominance style between chimpanzees and macaques may be partly responsible for the observed difference in consolation behavior. In macaques, bystanders approaching a recent victim of aggression are generally unable to protect the victim or themselves in cases of further aggression from the same source. This may discourage association with this individual in the aftermath of aggression. In chimpanzees, this inhibition may be absent because of the higher level of social tolerance, and the more flexible alliances.

This line of reasoning rests on the assumption that macaques are reluctant to take risks. This is no doubt a valid assumption (e.g. Chapais (1988) demonstrated that macaques are "conservative" in the pursuit of dominance strategies), yet it should also be noted that these monkeys do not avoid risks altogether. For example, they sometimes attack opponents more powerful than themselves in defense of a relative (e.g. Bernstein & Ehardt, 1985; de Waal, 1977). And even though reconciliation is most common between individuals with close relationships, exceptions do occur, and there are definite dangers involved in seeking contact with an individual from whom aggression has recently been received. That macaques are prepared to face these risks attests to the strength of motivation to protect kin and reconcile after fights. If, as argued here, similar risks keep them from approaching recipients of aggression, this would mean that the motivation to alleviate distress in others – if present at all – is not of the same order of strength as these other motivations.

THE SOCIAL COGNITION HYPOTHESIS

On pp. 93–9 we demonstrated a strong tendency in chimpanzees to care for distressed individuals. This tendency is also reflected in this species' behavioral repertoire, which, apart from contact patterns that provide reassurance, includes a range of expressions to *seek* reassurance from others. When agitated or upset, chimpanzees pout, whimper, yelp, beg with outstretched hand, or impatiently shake both hands in order to invite the calming contact they so urgently need. If this fails, even adults may break into a so-called "temper tantrum" by rolling around with apparent loss of control, screaming pathetically. Such elaborate displays would, of course, never have evolved in a species without strong succorant tendencies, and it is noteworthy that macaques show fewer and much less dramatic displays of this kind. The contrast between chimpanzees and macaques in the occurrence of consolation appears to be part, therefore, of a general difference in communication about, and responsiveness to, the emotional states and needs of others.

Greater responsiveness to emotional states may reflect greater sensitivity, which in turn may be based on processes of emotional contagion or identification (see p. 83). If this is what makes bystanders respond to a distressed individual, we would expect their facial expressions, gestures, and vocalizations to resemble those of the distressed individual (because of the similarity in aroused emotions). Possibly, this is the case when macaque infants hurry over to mount or embrace distressed peers (see p. 98 and Figure 4.6). These early responses need to be studied in greater detail. Our impression is one of a "spreading" of distress from one infant to the next, resulting in mutual contact-seeking in which it soon becomes immaterial which party was originally, and which secondarily, distressed. Emotional contagion may also explain responses to distress in young children and young chimpanzees.

However, when adult chimpanzees console a recipient of aggression they often do so without any sign of distress themselves, or after some delay when the most intense displays of distress have subsided. Their response may rest on cognitive evaluations in addition to emotional contagion.

This brings us to the possibility that chimpanzees identify with the others' emotions without letting these emotions overwhelm their own internal state, i.e. that they maintain a distinction between their own and the other's situation. If so, their consolatory tendency, as well as other succorant behavior, might be guided by an ability to imagine themselves in another individual's position. Presence of this ability in the chimpanzee is supported by research on the attribution of knowledge or mental states in other domains. Menzel's (1974) pioneering experiments revealed that young chimpanzees adjust their behavior according to the knowledge of others. He described, for example, attempts of dominant individuals to exploit the knowledge of subordinates and the counter-strategies of subordinates to keep dominants in the dark about the location of hidden food. The behavior of both parties suggested knowledge of knowledge in others. By contrast, Cheney & Seyfarth (1990a) reported that macaque mothers do not seem to attend to the knowledge states of their offspring when vocalizing in response to food or danger.

This possible difference between chimpanzees and macaques in attributional capacity was further explored by Povinelli and co-workers in a series of experiments in which chimpanzees and rhesus macaques were presented with similar test paradigms. The results indicate that chimpanzees may discriminate between the presence and absence of knowledge in human experimenters, whereas macaques may not (Povinelli *et al.*, 1990, 1991; but see the debate between Heyes, 1994a, and Povinelli, 1994). Similarly, experiments on role-taking indicate that chimpanzees grasp another individual's role from watching it, whereas macaques do not (Povinelli *et al.*, 1992a, b). It is conceivable, therefore, that unlike macaques, chimpanzees have a capacity to attribute knowledge, feelings, and intentions to others, and can picture themselves in someone else's position (cf. Premack & Woodruff, 1978).

If chimpanzees give and receive reassurance on a regular basis, we might expect them to exploit the solicitude-eliciting qualities of distress signals to their own advantage. An illustrative example is the behavior of the oldest male, Yeroen, of the Arnhem Zoo colony after he had been injured in a fight with a rival. The wound was not deep and he walked normally when out of sight of his rival. However, for a week, Yeroen limped heavily each time he passed in front of his rival. If injuries inhibit further aggression, Yeroen's attempt to create a false image of pain and suffering may have served to delay renewed hostilities. This deception is part of a wider range of deceitful tactics by the

Arnhem Zoo chimpanzees documented by de Waal (1982, 1986a). Chimpanzees may be special in this regard. Among 253 anecdotal accounts of spontaneous deception collected by Byrne & Whiten (1990) from field workers and other experts of primate behavior the most striking and complex cases concerned chimpanzees. If confirmed by research under more controlled conditions, the chimpanzee's deception of others might be a further indication of its capacity to take another individual's perspective.

There is continuing debate about the scientific admissability of certain kinds of information (e.g. qualitative descriptions) as well as the conclusiveness of some of the experiments designed to document nonhuman primate social cognition and self-awareness (de Waal, 1991a; Heyes, 1993, 1994a, b; Kummer *et al.*, 1990; Mitchell, 1993; Povinelli, 1994). Despite valid criticism of specific studies, however, it is hard to deny the emergence of an overall trend of differences between on the one hand, monkeys and on the other humans and apes. These two classes of primate seem to differ in multiple yet interlinked cognitive domains, which hints at an underlying organizing principle. The presence of consolation in chimpanzees but not macaques appears to further confirm this picture, and hence may derive from the same principle. It is unclear, though, what the exact nature of this principle may be, and whether the differences it produces are gradual or absolute.

CONCLUSION

The reconciliation behavior of macaques and chimpanzees differs only in degree, with a higher level of sophistication in the chimpanzee (e.g. as expressed in mediated peacemaking), but not necessarily a greater tendency to reconcile or a fundamentally different motivation. In both macaques and chimpanzees, reconciliation seems to serve the restoration of valuable relationships and the reduction of tension.

Interactions with third parties following aggressive incidents do seem to differ substantially between macaques and chimpanzees. Whereas chimpanzee bystanders frequently make affiliative contact with conflict participants, macaques do not; they rather seem to stay away from the scene of a fight in which they are not directly involved. Insofar as the chimpanzee's consolation behavior reflects empathy, the difference with macaques might reflect a difference of empathic capacity. A total absence of empathy in macaques seems unlikely as consolation does occasionally occur. It rather seems that the cognitive level of empathy is different in chimpanzees, who may possess greater emotional sensitivity to others and more developed capacities to figure themselves into someone else's position.

A comparative developmental approach (Parker & Gibson, 1990) might produce new insight into this possible difference between taxa. Young chimpanzees and macaques, like young children (see p. 85), may become emotionally disturbed in the presence of distressed individuals and so tend to contact them without fully distinguishing self from other. During development the perception of emotions in others may not change much in macaques, whereas chimpanzees may add certain cognitive evaluations resting on a greater capacity for perspective-taking.

In addition, apart from developmental differences, chimpanzees may have a lower threshold than macaques of response to others' distress because of differences in emotional arousability between them. Further research should focus on bystanders' responses to different levels of distress. Macaques may console another only when this individual is highly distressed.

Before accepting the social cognition hypothesis we need to exclude the possibility that the difference in consolation is a product of general differences in social organization between chimpanzees and macaques. The chimpanzee's more flexible and open social organization may make it less risky to contact recent recipients of aggression. There exist several ways to distinguish between this possibility and the cognitive hypothesis. One is to extend research to include other great apes and other monkey species to see how generalizable these differences are. Do all apes differ from all monkeys in this regard? Also, if even the most tolerant macaque species, such as Tonkean and stump-tailed macaques, fail to exhibit consolation, this would weaken the social constraints hypothesis.

The second way to find an answer would be through experiments that eliminate the risks associated with approaching a distressed conspecific (e.g. absence of potential aggressors). Another way to look at the effects of absence of risks on bystanders' behavior is to investigate the responses of group members to individuals that are distressed for reasons not associated with aggressive incidents, e.g. distress after a fall. If monkeys still showed no inclination to contact distressed groupmates under these circumstances, this would also weaken the social constraints hypothesis.

Without having conducted these additional observations and experiments, our conclusion is that chimpanzees and macaques differ substantially in their response to individuals distressed by previous aggression, and that it is possible yet unclear at this point whether this difference is part of a suite of differences related to species-typical levels of cognition and empathic capacity.

ACKNOWLEDGEMENTS

Research on various macaque species was supported in part by the Dutch and Italian Ministries of Education, and conducted by F. Aureli with the help of the following students of the University of Utrecht: Marjolijn Das, Carel van Panthaleon van Eck, Hans Veenema, and Desirée Verleur. We thank Jan van Hooff and Carel van Schaik for support during this period, and for discussions of the social constraints hypothesis. We are grateful to Josep Call, Amy Jones, Dario Maestripieri, Gabriele Schino, Emanuela Cenami Spada, Liesbeth Sterck, Peter Verbeek, and this book's editors for comments on earlier versions of the manuscript. Research on the chimpanzees at the Yerkes Field Station was conducted by F.B.M. de Waal with the invaluable assistance of Michael Seres. It was supported by a grant from the National Institutes of Mental Health (R03-MH49475). This research, as well as the writing of this chapter, was further supported by grant RR-00165 of the National Institutes of Health to the Yerkes Regional Primate Research Center (YRPRC). The YRPRC facilities are fully accredited by the American Association for Accreditation of Laboratory Animal Care.

REFERENCES

Anderson, J.R. (1983). Responses to mirror image stimulation and assessment of self-recognition in mirror and peer-reared stump-tailed macaques. *Quarterly Journal of Experimental Psychology*, **35B**, 201–12.

Aureli, F. (1992). Postconflict behaviour among wild long-tailed macaques (*Macaca fascicularis*). *Behavioral Ecology and Sociobiology*, **31**, 329–37.

Aureli, F., Cozzolino, R., Cordischi, C. & Scucchi, S. (1992). Kin-oriented redirection among Japanese macaques: an expression of a revenge system? *Animal Behaviour*, **44**, 283–91.

Aureli, F., Das, M. & Veenema, H.C. (1993a). Interspecific differences in the effect of kinship on reconciliation frequency in macaques. Abstract of paper presented at the 30th annual meeting of the Animal Behavior Society, Davis, CA, 24–29 July. *Abstract Book*, p. 5.

Aureli, F., Das, M., Verleur, D. & van Hooff, J.A.R.A.M. (1994). Postconflict social interactions among Barbary macaques (*Macaca sylvanus*). *International Journal of Primatology*, **15**, 471–85.

Aureli, F. & van Schaik, C.P. (1991a). Postconflict behaviour in long-tailed macaques (*Macaca fascicularis*): I The social events. *Ethology*, **89**, 89–100.

Aureli, F. & van Schaik, C. P. (1991b). Postconflict behaviour in long-tailed macaques (*Macaca fascicularis*): II Coping with the uncertainty. *Ethology*, **89**, 101–14.

Aureli, F., van Schaik, C.P. & van Hooff, J.A.R.A.M. (1989). Functional aspects of reconciliation among captive long-tailed macaques (*Macaca fascicularis*). *American Journal of Primatology*, **19**, 39–51.

Aureli, F., Veenema, H.C., van Panthaleon van Eck, C.J. & van Hooff, J.A.R.A.M. (1993b). Reconciliation, consolation, and redirection in Japanese macaques (*Macaca fuscata*). *Behaviour*, **124**, 1–21.

Balcomb, S.R., Yeager, C.P. & Berard, J.D. (1993). Reconcilition in a group of free-ranging rhesus macaques on Cayo Santiago. *American Journal of Primatology*, **30**, 295. (abstract.)

Bernstein, I.S. & Ehardt, C.L. (1985). Agonistic aiding: Kinship, rank, age and sex influences. *American Journal of Primatology*, **8**, 37–52.

Bischof-Köhler, D. (1988). Über den Zusammenhang von Empathie und der Fähigkeit sich im Spiegel zu erkennen. *Schweizerische Zeitschrift für Psychologie*, **47**, 147–59.

Byrne, R.W. & Whiten, A. (1990). Tactical deception in primates: The 1990 database. *Primate Report*, **27**, 1–101.

Chapais, B. (1988). Rank maintenance in female Japanese macaques: Experimental evidence for social dependency. *Behaviour*, **104**, 41–59.

Cheney, D.L. & Seyfarth, R.M. (1989). Redirected aggression and reconciliation among vervet monkeys, *Cercopithecus aethiops*. *Behaviour*, **110**, 258–75.

Cheney, D.L. & Seyfarth, R.M. (1990a). Attending to behaviour versus attending to knowledge: Examining monkeys" attribution of mental states. *Animal Behaviour*, **40**, 742–53.

Cheney, D.L. & Seyfarth, R.M. (1990b). *How Monkeys See the World*. Chicago: University of Chicago Press.

Cords, M. (1988). Resolution of aggressive conflicts by immature long-tailed macaques *Macaca fascicularis*. *Animal Behaviour*, **36**, 1124–35.

Cords, M. (1992). Postconflict reunions and reconciliation in long-tailed macaques. *Animal Behaviour*, **44**, 57–61.

Cords, M. & Thurnheer, S. (1993). Reconciliation with valuable partners by long-tailed macaques. *Ethology*, **93**, 315–25.

Demaria, C. & Thierry, B. (1992). The ability to reconcile in Tonkean and rhesus macaques. Paper

presented at the XIVth Congress of the International Primatological Society, Strasbourg, France, 16–21 August.
de Waal, F.B.M. (1977). The organization of agonistic relations within two captive groups of Java-monkeys (*Macaca fascicularis*). *Zeitschrift für Tierpsychologie*, **44**, 225–82.
de Waal, F.B.M. (1978). Exploitative and familiarity-dependent support strategies in a colony of semi-free living chimpanzees. *Behaviour*, **66**, 268–312.
de Waal, F.B.M. (1982). *Chimpanzee Politics: Power and Sex Among Apes*. New York: Harper & Row.
de Waal, F.B.M. (1984). Sex differences in the formation of coalitions among chimpanzees. *Ethology and Sociobiology*, **5**, 239–55.
de Waal, F.B.M. (1986a). Deception in the natural communication of chimpanzees. In *Deception: Human and Nonhuman Deceit*, ed. R.W. Mitchell & N.S. Thompson, pp. 221–44. Albany, NY: State University of New York Press.
de Waal, F.B.M. (1986b). The integration of dominance and social bonding in primates. *Quarterly Review of Biology*, **61**, 459–79.
de Waal, F.B.M. (1989a). Dominance "style" and primate social organization. In *Comparative Socioecology*, ed. V. Standen & R. Foley, pp. 243–63. Oxford: Blackwell.
de Waal, F.B.M. (1989b). Food sharing and reciprocal obligations among chimpanzees. *Journal of Human Evolution*, **18**, 433–59.
de Waal, F.B.M. (1989c). *Peacemaking among Primates*. Cambridge, MA: Harvard University Press.
de Waal, F.B.M. (1991a). Complementary methods and convergent evidence in the study of primate social cognition. *Behaviour*, **118**, 297–320.
de Waal, F.B.M. (1991b). Rank distance as a central feature of rhesus monkey social organization: A sociometric analysis. *Animal Behaviour*, **41**, 383–98.
de Waal, F.B.M. (1993). Reconciliation among primates: A review of empirical evidence and unresolved issues. In *Primate Social Conflict*, ed. W.A. Mason & S.P. Mendoza, pp. 111–44. Albany, NY: State University of New York Press.
de Waal, F.B.M. & Johanowicz, D.L. (1993). Modification of reconciliation behavior through social experience: An experiment with two macaque species. *Child Development*, **64**, 897–908.
de Waal, F.B.M. & Luttrell, L. (1988). Mechanisms of social reciprocity in three primate species: Symetrical relationship characteristics or cognition? *Ethology and Sociobiology*, **9**, 101–18.
de Waal, F.B.M. & Luttrell, L.M. (1989). Toward a comparative socioecology of the genus *Macaca*: Different dominance styles in rhesus and stump-tailed macaques. *American Journal of Primatology*, **19**, 83–109.
de Waal, F.B.M. & Ren, R. (1988). Comparison of the reconciliation behavior of stump-tailed and rhesus macaques. *Ethology*, **78**, 129–42.
de Waal, F.B.M. & van Hooff, J.A.R.A.M. (1981). Side-directed communication and agonistic interactions in chimpanzees. *Behaviour*, **77**, 164–98.
de Waal, F.B.M. & van Roosmalen, A. (1979). Reconciliation and consolation among chimpanzees. *Behavioral Ecology and Sociobiology*, **5**, 55–66.
de Waal, F.B.M. & Yoshihara, D. (1983). Reconciliation and redirected affection in rhesus monkeys. *Behaviour*, **85**, 224–41.
Dunbar, R.I.M. (1988). *Primate Social Systems*. London: Croom Helm.
Eaton, G.G. (1976). The social order of Japanese macaques. *Scientific American*, **235**, 96–106.
Eibl-Eibesfeldt, I. (1970). *Liebe und Haß: Zur Naturgeschichte elementarer Verhaltensweisen*. Munich: Piper.
Gallup, G.G. Jr (1970). Chimpanzees: self-recognition. *Science*, **167**, 96–106.

Gallup, G.G. Jr (1982). Self-awareness and the emergence of mind in primates. *American Journal of Primatology*, 2, 237–48.

Gallup, G.G. Jr (1991). Toward a comparative psychology of self-awareness: Species limitations and cognitive consequences. In *The Self: An Interdisciplinary Approach*, ed. G. R. Goethals & J. Strauss, pp. 121–35. New York: Springer-Verlag.

Gibson, J.J. (1979). *The Ecological Approach to Visual Perception*. Boston, MA: Houghton Mifflin.

Goodall, J. (1963). My life among wild chimpanzees. *National Geographic Magazine*, 124, 272–308.

Goodall, J. (1971). *In the Shadow of Man*. Boston, MA: Houghton Mifflin.

Goodall, J. (1986). *The Chimpanzees of Gombe: Patterns of Behavior*. Cambridge, MA: The Belknap Press.

Gopnik, A. & Meltzoff, A.N. (1994). Minds, bodies, and persons: Young children's understanding of the self and theory of mind research. In *Self-awareness in Animals and Humans: Developmental Perspectives*, ed. S.T. Parker, R.W. Mitchell & M.L. Boccia, pp. 166–86. New York: Cambridge University Press.

Harcourt, A.H. (1992). Coalitions and alliances: Are primates more complex than nonprimates? In *Coalitions and Alliances in Human and Other Animals*, ed. A.H. Harcourt & F.B.M. de Waal, pp. 445–72. Oxford: Oxford University Press.

Harcourt, A.H. & de Waal, F.B.M. (eds.) (1992a). *Coalitions and Alliances in Human and Other Animals*. Oxford: Oxford University Press.

Harcourt, A.H. & de Waal, F.B.M. (1992b). Cooperation in conflict: From ants to anthropoids. In *Coalitions and Alliances in Human and Other Animals*, ed. A.H. Harcourt & F.B.M. de Waal, pp. 493–510. Oxford: Oxford University Press.

Hatfield, E., Cacioppo, J.T. & Rapson, L. (1993). Emotional contagion. *Current Directions in Psychological Science*, 2, 96–9.

Hemelrijk, C.K. & Ek, A. (1991). Reciprocity and interchange of grooming and "support" in captive chimpanzees. *Animal Behaviour*, 41, 923–35.

Heyes, C.M. (1993). Anecdotes, training, trapping and triangulating: do animals attribute mental states? *Animal Behaviour*, 46, 177–88.

Heyes, C.M. (1994a). Cues, convergence and curmudgeon: A reply to Povinelli. *Animal Behaviour*, 48, 242–4.

Heyes, C.M. (1994b). Reflections on self-recognition in primates. *Animal Behaviour*, 47, 909–19.

Hinde, R.A. (1976). Interactions, relationships and social structure. *Man*, 11, 1–17.

Hoffman, M.L. (1977). Sex differences in empathy and related behaviors. *Psychological Bulletin*, 84, 712–22.

Hoffman, M.L. (1981). Perspectives on the difference between understanding people and understanding things: The role of affect. In *Social Cognitive Development*, ed. J.H. Flavell & L. Ross, pp. 67–81. Cambridge: Cambridge University Press.

Hoffman, M.L. (1987). The contribution of empathy to justice and moral judgment. In *Empathy and its Development*, ed. N. Eisenberg & J. Strayer, pp. 47–80. Cambridge: Cambridge University Press.

Humphrey, N.K. (1976). The social function of intellect. In *Growing Points in Ethology*, ed. P.P.G. Bateson & R.A. Hinde, pp. 303–17. Cambridge: Cambridge University Press.

Hutchins, M. & Barash, D.P. (1976). Grooming in primates: Implications for its utilitarian functions. *Primates*, 17, 145–50.

Johnson, D.B. (1982). Altruistic behavior and the development of self in infants. *Merrill-Palmer Quarterly*, 28, 379–88.

Jolly, A. (1966). Lemur social behaviour and primate intelligence. *Science*, 153, 501–6.

Judge, P.G. (1991). Dyadic and triadic reconciliation in pigtail macaques (*Macaca nemestrina*). *American Journal of Primatology*, 23, 225–37.

Kaplan, J.R. (1977). Patterns of fight interference in free-ranging rhesus monkeys. *American Journal of Physical Anthropology*, 47, 279–87.

Kappeler, P.M. & van Schaik, C.P. (1992). Methodological and evolutionary aspects of reconciliation among primates. *Ethology*, 92, 51–69.

Keverne, E.B., Martensz, N.D. & Tuite, B. (1989). Beta-endorphin concentrations in cerebrospinal fluid of monkeys are influenced by grooming relationships. *Psychoneuroendocrinology*, 14, 155–61.

Kummer, H. (1967). Tripartite relations in hamadryas baboons. In *Social Communication among Primates*, ed. S.A. Altmann, pp. 63–72. Chicago: University of Chicago Press.

Kummer, H. (1971). *Primate Societies: Group Techniques of Ecological Adaptation*. Arlington Heights, IL: AHM Publishing Corporation.

Kummer, H. (1978). On the value of social relationships to nonhuman primates: A heuristic scheme. *Social Science Information*, 17, 687–705.

Kummer, H., Dasser, V. & Hoyningen-Huene, P. (1990). Exploring primate social cognition: Some critical remarks. *Behaviour*, 112, 84–98.

Kurland, J.A. (1977). Kin selection in the Japanese monkey. *Contributions to Primatology*, 12, 00–00.

Lindburg, D.G. (1991). Ecological requirements of macaques. *Laboratory Animal Science*, 41, 315–22.

Maestripieri, D. (1993). Vigilance costs of allogrooming in macaque mothers. *American Naturalist*, 141, 744–53.

Mason, W. (1964). Sociability and social organization in monkeys and apes. In *Advances in Experimental Social Psychology*, ed. L. Berkowitz, pp. 227–305. New York: Academic Press.

Melnick, D.J. & Pearl, M.C. (1987). Cercopithecines in multimale groups: Genetic diversity and population structure. In *Primate Societies*, ed. B.B. Smuts, D.L. Cheney, R.M. Seyfarth, R.W. Wrangham & T.T. Struhsaker, pp. 121–34. Chicago: University of Chicago Press.

Menzel, E.W. (1974). A group of young chimpanzees in a one-acre field: Leadership and communication. In *Behavior of Nonhuman Primates*, ed. A. Schrier & F. Stollnitz, pp. 83–153. New York: Academic Press.

Mercer, P. (1972). *Sympathy and Ethics: A Study of the Relationship between Sympathy and Morality with Special Reference to Hume's Treatise*. Oxford: Clarendon Press.

Mitchell, R.W. (1993). Mental models of mirror-self-recognition: Two theories. *New Ideas in Psychology*, 11, 295–325.

Morin, P.A., Moore, J.J., Chakraborty, R., Jin, L., Goodall, J. & Woodruff, D.S. (1994). Kin selection, social structure, gene flow, and the evolution of chimpanzees. *Science*, 265, 1193–201.

Nagell, K., Olguin, R.S. & Tomasello, M. (1993). Processes of social learning in the tool use of chimpanzees (*Pan troglodytes*) and human children (*Homo sapiens*). *Journal of Comparative Psychology*, 107, 174–86.

Neisser, U. (1988). Five kinds of self-knowledge. *Philosophical Psychology*, 1, 35–59.

Nishida, T. (1970). Social behavior and relationships among wild chimpanzees of the Mahale Mountains. *Primates*, 11, 47–87.

Nishida, T. (1979). The social structure of chimpanzees of the Mahale Mountains. In *The Great Apes*, ed. D.A. Hamburg & E.R. McCown, pp. 73–121. Menlo Park, CA: Benjamin Cummings.

Nishida, T. (1983). Alpha status and agonistic alliance in wild chimpanzees. *Primates*, 24, 318–36.

Nishida, T. & Hiraiwa-Hasegawa, M. (1987). Chimpanzees and bonobos: Cooperative relationships among males. In *Primate Societies*, ed. B.B. Smuts, D.L. Cheney, R.M. Seyfarth, R.W. Wrangham & T.T. Struhsaker, pp. 165–77. Chicago: University of Chicago Press.

Parker, S.T. & Gibson, K.R. (1990). *"Language" and Intelligence in Monkeys and Apes: Comparative Developmental Perspectives.* New York: Cambridge University Press.

Parker, S.T., Mitchell, R.W. & Boccia, M.L. (eds.) (1994). *Self-awareness in Animals and Humans: Developmental Perspectives.* New York: Cambridge University Press.

Petit, O. & Thierry, B. (1995). Reconciliation in a group of crested macaques. *Dodo,* (Jersey Wildlife Preservation Trusts). (in press).

Povinelli, D.J. (1987). Monkeys, apes, mirrors and minds: The evolution of self-awareness in primates. *Human Evolution,* **2**, 493–509.

Povinelli, D.J. (1993). Reconstructing the evolution of mind. *American Psychologist,* **48**, 493–508.

Povinelli, D.J. (1994). Comparative studies of animal mental state attribution: A reply to Heyes. *Animal Behaviour,* **48**, 239–41.

Povinelli, D.J., Nelson, K.E. & Boysen, S.T. (1990). Inferences about guessing and knowing by chimpanzees (*Pan troglodytes*). *Journal of Comparative Psychology,* **104**, 203–10.

Povinelli, D.J., Nelson, K.E. & Boysen, S.T. (1992a). Comprehension of social role reversal by chimpanzees: Evidence for empathy? *Animal Behaviour,* **43**, 633–40.

Povinelli, D.J., Parks, K.A. & Novak, M.A. (1991). Do rhesus monkeys (*Macaca mulatta*) attribute knowledge and ignorance to others? *Journal of Comparative Psychology,* **105**, 318–25.

Povinelli, D.J., Parks, K.A. & Novak, M.A. (1992b). Role reversal by rhesus monkeys, but no evidence for empathy. *Animal Behaviour,* **44**, 269–81.

Premack, D. & Woodruff, G. (1978). Does the chimpanzee have a theory of mind? *Behavioral and Brain Sciences,* **1**, 515–26.

Rumbaugh, D.M. & Pate, J.P. (1984). The evolution of cognition in primates: A comparative perspective. In *Animal Cognition,* ed. H.L. Roitblat, T.G. Bever & H.S. Terrace. Hillsdale, NJ: Lawrence Erlbaum.

Schino, G., Scucchi, S., Maestripieri, D. & Turillazzi, P.G. (1988). Allogrooming as a tension-reduction mechanism: A behavioral approach. *American Journal of Primatology,* **16**, 43–50.

Scott, J.P. (1972). *Animal Behavior,* 2nd edn. Chicago: University of Chicago Press.

Seyfarth, R.M. (1977). A model of social grooming among adult female monkeys. *Journal of Theoretical Biology,* **65**, 671–98.

Seyfarth, R.M. (1980). The distribution of grooming and related behaviours among adult female vervet monkeys. *Animal Behaviour,* **28**, 798–813.

Straumann, C. & Anderson, J.R. (1991). Mirror-induced social facilitation in stump-tailed macaques (*Macaca arctoides*). *American Journal of Primatology,* **25**, 125–32.

Teleki, G. (1973). *The Predatory Behavior of Wild Chimpanzees.* Lewisburg, PA: Bucknell University Press.

Thierry, B. (1984). Clasping behavior in *Macaca tonkeana*. *Behaviour,* **89**, 1–28.

Thierry, B. (1986). A comparative study of aggression and response to aggression in three species of macaque. In *Primate Ontogeny, Cognition and Social Behavior,* ed. J.G. Else & P.C. Lee, pp. 307–13. Cambridge: Cambridge University Press.

Thierry, B. (1990). Feedback loop between kinship and dominance: the macaque model. *Journal of Theoretical Biology,* **145**, 511–21.

Trivers, R.L. (1971). The evolution of reciprocal altruism. *Quarterly Review of Biology,* **46**, 35–57.

van Hooff, J.A.R.A.M. (1974). A structural analysis of the social behavior of a semi-captive group of chimpanzees. In *Social Communication and Movement,* ed. M. von Cranach & I. Vine, pp. 75–162. London: Academic Press.

Veenema, H.C., Das, M. & Aureli, F. (1994). Methodological improvements for the study of reconciliation. *Behavioural Processes,* **31**, 29–38.

Visalberghi, E. & Fragaszy, D. M. (1990). Do monkeys ape? In *"Language" and Intelligence in Monkeys and Apes: Comparative Developmental Perspectives*, ed. S.T. Parker & K.R. Gibson, pp. 247–73. New York: Cambridge University Press.

Watanabe, K. (1979). Alliance formation in a free-ranging troop of Japanese macaques. *Primates*, 20, 459–74.

Whiten, A. & Ham, R. (1992). On the nature and evolution of imitation in the animal kingdom: Reappraisal of a century of research. In *Advances in the Study of Behavior*, vol 21, ed. P.J.B. Slater, J.S. Rosenblatt, C. Beer & M. Milinski, pp. 239–83. New York: Academic Press.

Wispé, L. (1991). *The Psychology of Sympathy*. New York: Plenum Press.

Woodruff, G. & Premack, D. (1979). Intentional communication in the chimpanzee: the development of deception. *Cognition*, 7, 333–62.

Wrangham, R.W. (1986). Ecology and social relationships in two species of chimpanzee. In *Ecological Aspects of Social Evolution: Birds and Mammals*, ed. D.I. Rubenstein & R.W. Wrangham, pp. 352–78. Princeton: Princeton University Press.

Zahn-Waxler, C., Radke-Yarrow, M., Wagner, E. & Chapman, M. (1992). Development of concern for others. *Developmental Psychology*, 28, 126–36.

Zahn-Waxler, C. & Smith, K. D. (1992). The development of prosocial behavior. In *Handbook of Social Development*, ed. V.B. von Hasselt & M. Hersen, pp. 229–56. New York: Plenum Press.

5

The misunderstood ape: Cognitive skills of the gorilla

RICHARD W. BYRNE

Gorillas most probably diverged from the chimpanzee–human lineage at 7.5 ($\pm$ 1.0) million years ago, only 2.0 ($\pm$ 1.0) million years before the human and chimpanzee lines themselves separated (Bailey *et al.*, 1992; Horai *et al.*, 1992). Yet the gorilla has often been regarded as lacking those characteristics that make chimpanzees of such interest to anthropologists and psychologists: the technological intelligence that chimpanzees demonstrate in tool-making; the similarity of their diet to that of human hunter–gatherers in the use of insects and red meat; the subtlety of social comprehension they show in deception and other 'political' maneuvering; the complexity and fluidity of their social organization; and, often suggested to be crucial to these phenomena, the chimpanzees' relatively large brain. Even the orangutan, separated from humans by at least 13–14 million years of independent evolution (Pilbeam & Smith, 1981), is sometimes believed to have more in common with humans than has the gorilla (e.g. Suarez & Gallup, 1981). Gorillas' size and power attracts public interest and sympathy, but they are seldom portrayed as other than a bit dim, intellectually. Attenborough (1979) expressed it nicely: "The placid disposition of the gorilla is connected with its diet and what it has to do to get it. It lives entirely on vegetation of which there is an infinite supply growing immediately to hand.... there is no need for it to be particularly nimble in either body or mind. The other African ape, the chimpanzee, ... has to be both agile and inquisitive." Gorillas are in danger of being stereotyped as nice, but dull.

Like all stereotypes, this one may be an over-simplification, or even a libel. The aim of this chapter is to review recent findings that relate directly to the cognitive and social complexity of the gorilla, to find out. This exercise is important, for reasons of wider importance than simple fairness to the animal: it is crucial to inferring evolutionary history. Cladistic comparisons are the most reliable means of inferring a sequence of evolutionary changes, and almost the only way in which changes concerning intellectual or behavioral adaptations can be reconstructed (see Chapter 16). Given the inadequate fossil record of nonhuman great apes, cladistic analysis of the morphology and behaviour of the living species is essential in evolutionary reconstruction of the common ancestors we share with great apes.

In cladistics, species are grouped into "clades" if they possess shared derived characters (see Ridley, 1986). The "shared derived" characters of a group of species are those characters that first evolved *de novo* in their common ancestor, and are unique to the group. A group of species may also share primitive characters, retained from earlier ancestors and therefore not unique to the group. Derived characters are best distinguished from primitive characters by their consistent presence within the clade and absence outside the clade. The most important comparison in this identification process is that with the "outgroup," the species or group of species most closely related to those within the clade. Crucially, the gorilla is the appropriate outgroup to the clade containing just chimpanzees and humans, and so a proper characterization of gorillas is essential for interpretation of human–chimpanzee shared characteristics. Also, of course, gorilla data have equal importance to those from the two chimpanzees and the orangutan for reconstructing the (earlier) ancestor of all modern great apes and humans. Misrepresentation of gorilla characteristics, therefore, will lead to errors in reconstructing the origins of the human behavioral heritage. My intention in this chapter is to give an accurate portrayal of gorilla intellect, as far as current evidence will support, to inform cladistic comparisons.

The results of the exercise may also help to illuminate a mystery: to what do we owe our own large brains? Of course, the rapid encephalization seen in the hominid line may have nothing to do with factors that have influenced variations among nonhuman primates. But this recent increase certainly began from a relatively high baseline. Haplorhine primates the tarsiers, monkeys, and apes are relatively large-brained, as a group, among mammals (see Deacon, 1990; Passingham, 1982); they are also noted for intelligence. The two evolutionary increases in brain size may or may not share similar origins, but, whatever caused the recent brain size increase among hominids, for a full understanding of the evolution of human intelligence we must discover the origins of the intellectual capacities of the last ancestor we share with a living primate. With the extant primates as a database to examine, this exercise may be the easier one of the two. However, there is a current controversy about the origin of large brains and intelligence in haplorhines: is what we observe a consequence of selection for *environmental skill*, or for *social skill*? The "Machiavellian Intelligence" hypothesis, originally suggested in different versions by Jolly (1966) and Humphrey (1976), would suggest that primate intelligence is a result of the cognitive demands of social sophistication, originally caused by the need to live in large groups. Alternatively, what might be called the "Food for Thought" hypothesis argues that primate intelligence is shaped by the cognitive demands of complexity encountered in primate diet acquisition (e.g. Gibson, 1986; Mackinnon, 1978; Parker & Gibson, 1977), some versions suggesting positive feedback between feeding skill and the energetic needs of brain enlargement (Milton, 1981). The gorilla is a rather crucial species in this debate: portrayed as cognitively complex but needing little intelligence to obtain its apparently simple diet, the gorilla can be used to argue for a social origin to intellect (e.g. by Humphrey, 1976); portrayed as a small-brained (and consequently less intelligent) folivore compared to the frugivorous chimpanzee, the gorilla can be used to support a foraging hypothesis for the origin of intellect (e.g. by

Clutton-Brock & Harvey, 1980). To progress on this issue, we will evidently need to examine the diet of the gorilla, as well as its cognitive skills and social behaviour.

IS FOOD ACQUISITION SIMPLE FOR A GORILLA?

The three subspecies of gorilla[1] show wide ecological variation, especially in their diet. The well-studied mountain gorillas (*Gorilla gorilla beringei*) of Karisoke, Rwanda, are alone in eating largely herb leaves and pith. Here, there is an almost complete lack of seasonal or spatial variation in availability of mountain gorilla food plants (Watts, 1984); no obvious argument can thus be made for the need of great skill in the gorillas' food finding. Mountain gorilla diet choice is also relatively simple, because the herbaceous vegetation in which they largely range is low in secondary compounds (Waterman *et al.*, 1983), and food plants are relatively abundant among the total vegetation (Vedder, 1984). Grauer's gorilla (*Gorilla gorilla graueri*) and the Ugandan population of *beringei*, both living at lower altitude, rely on a wider range of foods, including fruit (Goodall, 1977; A. Plumptre, personal communication). This trend is continued in Western lowland gorillas (*Gorilla gorilla gorilla*). Though this subspecies is insufficiently known to be certain of any complexity involved in food-finding, their diet is certainly much closer to that of the chimpanzee, including much ripe fruit and insects (Tutin & Fernandez, 1993).

The challenge of finding scarce forest fruits, often with asynchronous fruiting cycles that lack a simple 12 month periodicity, has been argued to select for mental mapping skill (Mackinnon, 1978; Milton, 1981). This could apply to those gorilla populations that rely more on fruit, and those are just the populations generally presumed closest to the ancestral, primitive state for gorillas – and, indeed, to the primitive state for all great apes. (It must be admitted that at present the case that frugivorous gorillas are challenged by complexity in food distribution is weakened by the lack of detailed, nutritional studies on frugivorous populations.) In addition, even Karisoke *beringei* make dramatic excursions from their normal ranging routes to inspect bamboo zones for seasonal edible shoots (Watts, 1984; and my personal observations); *graueri* make longer diversions, sometimes digging shoots from below ground level (Goodall, 1977). Considerable mental mapping ability is shown by these actions.

IS GORILLA TECHNICAL SKILL LIMITED?

Chimpanzees display signs of technical skill in their tool use for insect-gathering and other purposes (McGrew, 1992). Because gorillas do not use tools in foraging, the special arguments that have related the origin of intelligence to techniques of tool-making are certainly irrelevant to modern gorillas. However, gorillas' readiness to tackle problems involving tool use in captivity has always seemed paradoxical (see McGrew, 1989), and they have sometimes been presumed to have descended from tool-using ancestors (e.g. Parker & Gibson, 1977). One possible answer to this puzzle may come from examination of the techniques of food preparation.

The job of feeding does not cease once food is located. Ripe fruit may be intellectually trivial to process, but some techniques used by gorillas in feeding appear to be intellectually demanding. Mountain gorillas eat nothing but leaves and pith, but gorillas are not specialized herbivores: they could not possibly digest the secondary compounds that protect the mature leaves of tropical moist forests. The herbs mountain gorillas eat lack these poisons, but are instead protected by *physical* defences, including powerful stings, strong spines or hard outer casings. Feeding techniques of mountain gorillas are efficient at circumventing these defences, and show considerable technical complexity (see Byrne & Byrne, 1991, 1993; regrettably, no comparable data are yet available for other subspecies of gorilla). In fact, mountain gorilla food processing constitutes the main grounds at present for a negative answer to the question asked above, in gorillas' natural behaviour.

To illustrate the complexity of mountain gorilla plant feeding, flow charts are helpful. For the technique shown in Figure 5.1(*a*), for instance, the flow chart implies that we would find some sequences where a single item is picked and eaten, others where the picking is repeated until a bundle is accumulated. In both cases there is a significant hand preference: one hand is preferred for pulling off items and putting them into the mouth, whereas either may be used to pull the whole plant into range, if needed. This technique is about the simplest feeding method used by primates; mountain gorillas use it for occasional undefended leaves, but most of their diet is very different.

Consider the method used by a typical individual to eat the nettle, *Laportea alatipes*, shown in Figure 5.1(*b*). Iteration can occur in two alternative ways to accumulate a bigger bundle in either the left or right hand, showing an ability to treat parts of the sequence as subroutines. Four stages require deft bimanual coordination, with the two hands used in different roles. Some actions are specific to just this food, whereas the delicate precision movements used in picking out small fragments of inedible leaves are used more generally. This nettle-eating technique minimizes the number of stings that contact the palm, fingers and especially the lips of a gorilla. Each type of plant leaf requires a different technique. For instance, the bedstraw, *Galium ruwenzoriense*, is not defended by stings like a nettle, but the whole plant is covered in tiny hooks, and gorillas' technique for this (see Figure 5.1(*c*)) enables a tight bundle to be eaten with shear bites, rather than popping a parcel through the lips, minimizing the tendency of hooks to attach to the mouth. The same delicate cleaning precision movements and iterative repetition of part of the sequence are employed, but in this case loose stems are tucked and folded in to the bundle, both one-handed and with bimanual coordination.

Behavioral laterality is a striking feature of gorilla plant processing. All mountain gorillas have strong hand preferences that are consistent across independent bouts of processing, and processing efficiency is identical in left- and right-handers (Byrne & Byrne, 1991). Efficiency is slightly greater with the preferred hand, as would be expected from the effects of practice. Hand preference is often linked to task complexity (Fagot & Vauclair, 1991), but complexity alone is insufficient to elicit asymmetry in these gorillas: pith from the stem of thistle, *Carduus nyassanus*, a food just as complex to process as bedstraw or celery, is usually eaten symmetrically. In thistle-eating, the hands reverse roles regularly within a bout of processing, apparently to avoid the difficulty of maneuvering a long spiny stem. In sharp contrast, even the simple technique for undefended leaves was found to elicit hand preferences.

Unlike the case of a human right-hander, where most skilled actions show right hand dominance, hand preferences are somewhat task specific in mountain gorillas. Hand preferences are general across some leaf-eating tasks, but for pith-eating tasks, such as eating bamboo shoot and especially celery, *Peucedanum linderi* (see Figure 5.1(*d*)), a second, wholly independent preference is found (Byrne & Byrne, 1991). The two groups of tasks, for which shared hand preferences are found, are not closely similar in structure or elements used (in Figure 5.1, compare (*d*) with (*a*) and (*b*)). However, the very different techniques applied to various plants in adulthood may develop from a common stage earlier in development – when all leaves are perhaps eaten in a more similar way, skill asymmetries may develop and cause adult preference patterns.

The particular direction of animals' preferences is not the result of injury – being forced to use the unhurt hand causing a lasting preference. Animals are highly resistant to changing hands even when they are hurt, but most telling are data from animals with permanent hand damage (Byrne & Byrne, 1991). On the hypothesis that injury causes preference, these individuals should show the strongest preference for the uninjured hand; in fact, they proved to have unusually *low* preference for either hand. Injury masks preference, rather than causing it. There is a significant lateral bias at the population level, towards right-handed fine precision manipulation for leaf techniques, but none whatsoever for pith-eating technique (Figure 5.2); the latter is perhaps phylogenetically more ancient, because western lowland gorillas also eat pith (Tutin & Fernandez, 1993). Population-level handedness may result from the same slight neural bias (in the gorilla–human common ancestor) as does human laterality: when there was a need for laterality to develop later in human evolution, it was in the direction of this bias.

All these skills are fully established by weaning at 3 years old, after which animals display adult-like behaviour and show no correlation between age and efficiency (Byrne & Byrne, 1991). Each technique is composed of many elements that vary idiosyncratically, animals having different repertoires of preferred ways of achieving each subgoal. In contrast, the techniques themselves, structured sequences of elements, are remarkably standardized across the population: subgoals are generally attempted in a fixed order by every animal (Byrne & Byrne, 1993). Another contrast between the "fine grain" of exactly how an item is held or moved, and the overall organization of the process, emerges in the changes with age. The repertoire of elementary actions used by an individual does not increase with age beyond 3 years, whereas the range of techniques does increase significantly with age among the juvenile and adult population (Byrne & Byrne, 1993). It seems that, having established one fully satisfactory way of processing each food by the time of weaning, variants of this technique are added with further experience, perhaps to allow efficient processing when limited in some way, such as when hanging in a tree or supporting an infant.

The feeding techniques shown in Figure 5.1(*b*) to (*d*) are all complex skills, in the sense that they consist of several component processes, each one precise and often

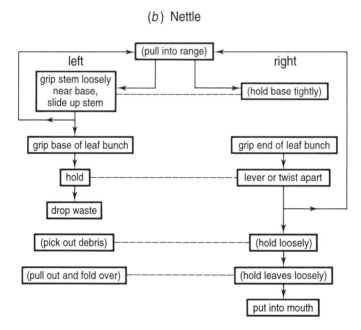

Figure 5.1. In each flow chart, the sequence of actions (which begins when an animal finds a food to eat) starts at the top and moves down. Rectangular boxes show actions, loosely described by the words in them, whereas lozenges represent branch points, with the apparent criteria for the decision indicated in words in the lozenge. A process may therefore repeat or iterate until some criterion for

Cognitive skills of the gorilla

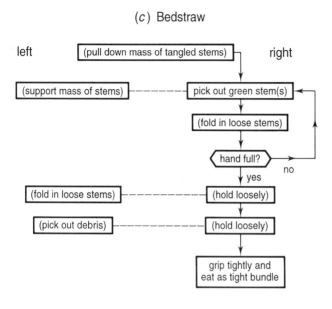

(c) Bedstraw

(d) Celery

a different decision is reached. Actions normally done by the left hand are put on the left, and the reverse for those done with the right hand; if there is no statistically significant preference, the box is put in the middle. Each sequence ends with a handful of processed food put in the mouth.

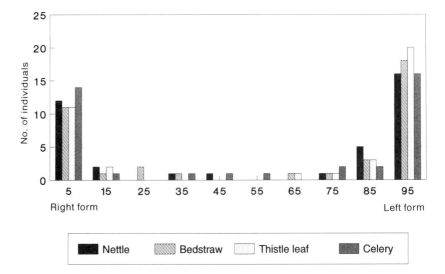

Figure 5.2. Distribution of individual gorillas' hand preferences for four feeding techniques. The index varies from 0, for the mirror form in which the left hand is used for most precision actions (coded "right form"), to 100, for the mirror form in which the right hand is used for most precision actions (coded "left form"). Note that the gorilla population is skewed towards right-hand precision usage (significant on Kolmogorov–Smirnov test for bedstraw $p = 0.01$, thistle $p = 0.01$, nettle $p = 0.05$).

involving bimanual coordination, structured together into the whole program, in which a subroutine of variable size may be used repeatedly. How are these complex programs of action acquired so reliably and with such great standardization across the population? The probability of every individual ending up with essentially the same logical programs for each food type by trial-and-error learning must be very low, and genetic encoding of behavioral sequences is implausible; some sort of imitation is suggested. (In a naturally occurring task, tackled in a similar way by all study animals, the possibility that there *is* only one way to do it cannot logically be discounted; but humans questioned about these tasks readily suggest *several* workable methods, seldom including the actual one used by gorillas, increasing our confidence in the likelihood of some sort of imitation by the gorillas.)

In animal behaviour, "imitation" is usually interpreted to mean copying of strings of particular motor acts from observing the behaviour of another (e.g. Galef, 1988; Tomasello *et al.*, 1987); this might be called *action-level imitation*. In mountain gorilla food preparation techniques, imitation in this sense can be ruled out by our data. If gorillas were imitating at the action-level, both repertoires of preferred elements and hand preferences would tend to "run in families." This is because the only adults who allow infants near them when they feed, the only potential models for imitation, are the infant's own mother, and the silverback leader – normally the father. In fact, element repertoires do not show any pattern among families, and not the slightest association exists between the hand preferences of offspring and their parents. We found that 11 out of 22 mother–infant pairs showed shared laterality (Byrne & Byrne, 1991), as did 13 out of 22 father–infant pairs (unpublished results), in both cases very much as would be expected by chance. The idiosyncratic variability in element repertoires is also a clear pointer to acquisition by trial-and-error (Byrne & Byrne, 1993).

However, because complex behaviour typically has hierarchical structure (Dawkins, 1976; Lashley, 1951), imitation at higher levels in the hierarchy is possible. Indeed, this sort of imitation might be of much greater utility than slavish duplication of details, where these could easily be learnt by trial-and-error. In the case of gorilla feeding techniques, we found that once we sorted the elements into categories defined by their function or result, almost every individual possessed an identical set of these (superordinate) actions (Byrne & Byrne, 1991). Furthermore, as noted already, the organization of these actions into complete programs was essentially identical between animals. For these reasons, I have suggested that complex food processing techniques may be understood hierarchically by the animals, and thus acquired partly by imitation – but imitation at the level of task structure, subroutines and subgoal steps: *program-level imitation* (Byrne, 1993a, 1994). This could account for the common currency of overall strategies for each task, whereas the fine details are apparently fleshed out by individual experience.

Unfortunately, comparing with other apes is not straightforward, because they apparently rely less on structured techniques of this kind. (Detailed data on the orangutan are not yet available, but it is possible that orangutans are closer to gorillas in the complexity of their feeding techniques: Galdikas & Vasey, 1992.) Laboratory tasks designed to test for imitation are usually simple to the point of lacking all potential for hierarchical structure (e.g. Whiten & Ham, 1992), so it is not yet clear whether the much-tested chimpanzee shows program-level imitation. However, common chimpanzees' reliable acquisition of tool-making techniques with population differences across Africa (McGrew, 1992; Chapter 18), and the complex behaviors that orangutans have been shown to imitate from humans (Russon & Galdikas, 1993), suggest that all great apes have the capability to understand and so copy the structural organization of behaviour.

Returning to the question of gorilla technical intelligence, it is too early to say whether the manual skills shown by mountain gorillas reflect a specific "module" of enhanced ability at motor skill learning and understanding, or a more general ability to represent objects and intended actions in mind. The latter interpretation is suggested by a study that directly compared a young gorilla with a Japanese monkey, *Macaca fuscata*, in their development (Natale *et al.*, 1986). They found the gorilla to show behaviour that in children would be treated as clear evidence of the ability to hold a mental representation, whereas the monkey did not.

IS GORILLA SOCIAL ORGANIZATION SIMPLE?

As typically represented in the literature, the social organizations of chimpanzees and gorillas are strikingly different in complexity (e.g. Jolly, 1985; Richard, 1985). Whereas chimpanzees live in large communities of up to 105 members, gorillas live in small groups, of median size nine in East and five in West African subspecies (or species, as noted above); further, gorilla groups are cohesive and stable, whereas chimpanzee communities fragment and coalesce, allowing members to range in a fluid, fission–fusion organization; and the gorilla group is typically one-male, unlike the multimale communities of chimpanzees (for full details, see Nishida & Hiraiwa-Hasegawa, 1987; Stewart & Harcourt, 1987). Chimpanzee social organization must lead to complexity of choice for individual males and females. A female chimpanzee has a real choice of mating partner, which she can exercise by leaving the core subgroup on "safari" with a nonalpha male of her choice over her estrous period (Tutin, 1979[2]); in one-male gorilla groups, a female's only choice is to change groups. A male chimpanzee consequently has the choice of two mating strategies: either devoting his energies towards attaining high dominance rank (with reproductive benefits in the core subgroup), or spending his time in building up relationships with females that may lead to their choice of him as mating partner (with reproductive benefits in safari liaisons). Also, because in chimpanzees male competitors live together and also cooperate, there is scope for benefits of political and Machiavellian behaviour, which has been described (e.g. de Waal, 1982; Nishida, 1983). By contrast, according to this view, gorilla males compete only by physical strength and fighting ability, and losers are excluded from breeding groups (Stewart & Harcourt, 1987).

The picture of simple social life presented above may be misleading, at least in the mountain gorilla, *G. g. beringei*. At the long-term study site of Karisoke in Rwanda, all regularly observed groups of mountain gorillas now have more than one silverback (my personal observation), whereas smaller one-male groups used to be found commonly – as they still are in Western lowland gorillas, *G. g. gorilla*, and Grauer's gorilla, *G. g. graueri* (Tutin & Fernandez, 1993; Yamagiwa, 1983). The virtual disappearance of small groups in the well-protected areas of *beringei*'s range raises the possibility that the small groups once observed there were an artifact of human interference, since silverback males tend to be killed in operations to obtain infant gorillas for zoos. In two-male groups, I have often watched the two silverbacks cooperate effectively to defend and keep females (and see Sicotte, 1993). One male herds females back, away from a rival lone male or group, while the other goes forward alone to display at the rival. This "team-work" may be unbeatable in contests with single male groups, so a multimale tendency may be spreading as a means of competing for females. Groups with three and four silverbacks at Karisoke have been observed to be stable for periods of years; the males in these groups are usually, but not always, father and son(s). Up to seven silverbacks in a group has been recorded in census data of Ugandan populations. The dynamics of groups with multiple silverbacks in intergroup encounters is as yet unknown.

The presence of more than one male immediately makes social choices more complex for both females and males, in just those ways found in chimpanzees. Female gorillas have never been recorded to show safari behaviour, but nevertheless I have seen secret matings frequently. As well as remaining behind with favored nonleader males when the group moves off, females actively solicit silverbacks to mate secretly, and both sexes successfully inhibit their normal copulation calls (my observations are cited as record nos 161, 163–5, 166–8 in Byrne & Whiten, 1990). These data were collected during other work, and as yet no systematic study has been done; however, it is clear that gorillas normally *do* give copulation calls when there is no need for secrecy. This can be illustrated by the changes in behaviour of one male, the silverback Beetsme. When he was the dominant animal, he was noted to give loud copulation calls on each of the 16 times he was recorded as mating (Karisoke Research Center records). However, by 1989 another silverback, Titus, was dominant. Beetsme then inhibited his copulation calls when mating with a young female favored by Titus (record no. 161).

The nature of social comprehension underlying hiding and call-suppression will be taken up in the next section, but the *option* of secret mating certainly allows female gorillas to exercise mate choice within groups as well as between. Similarly, males do have real options, between contest by power (by leaving the natal group and attempting to amass females by aggressive display at other breeding groups) or subtlety (by remaining in the natal group, cooperating with father or half-brother in defence, and fertilizing those females who permit secret matings). These options are comparable to those facing chimpanzees, and the cohesive ranging behaviour of gorillas only serves to make secrecy a more *demanding* intellectual challenge than it would be in the fission–fusion society of chimpanzees. However, many monkey societies are of comparable complexity, and similarly span cohesive and fission–fusion grouping patterns; any attempt to link complex social organization to higher intelligence depends crucially on the social understanding of the animals themselves, which is considered in the next section.

IS THE SOCIAL COMPREHENSION OF GORILLAS LIMITED?

Theory of mind (Premack & Woodruff, 1978) is the term used for any understanding of the mental states of other individuals, compared with one's own. Many of the recent claims of a "theory of mind" in the chimpanzee are based on single experiments. These have not yet been investigated over a range of conditions (some have only been carried out on a single individual chimpanzee), nor in the main have they been extended to the gorilla, a species which is seldom conveniently available to psychologists. Chimpanzees but not gorillas have shown:

1. Comprehension of a cooperating individual's role in a joint task, without explicit training (Povinelli, Nelson & Boysen, 1992).
2. Choice of another individual as a source of information, on the basis of the informant's knowledge – e.g. preferring the advice of a person who has been in a position to see which food-cup was baited, rather than one whose vision had been restricted (Povinelli *et al.*, 1990; Premack, 1988).
3. A preference for an individual whose past failures appeared to result from clumsiness rather than deliberate malice (Povinelli, 1991).
4. An ability to pick solutions appropriate to the problems of another individual viewed on videotape (Premack & Woodruff, 1978; but see Premack & Dasser, 1991).

All of these tasks have been criticized as failing to give clear evidence of theory of mind in chimpanzees; problems include the possibility of nonsocial comprehension, and cue-learning over repeated trials (see Heyes, 1993b; Tomasello & Call, 1994); however, several of the studies also tested monkeys, which uniformly failed.

When gorillas have been posed the same problems as those in which chimpanzees have distinguished themselves, the results are sometimes strikingly similar. For instance, Gómez chose to use a gorilla rather than a chimpanzee to investigate an observation made in passing by Köhler (1925): when chimpanzees want to reach an out-of-reach object, their first approach is to attempt to enlist the help of a human. Only by preventing this was Köhler able to elicit the famous box-piling actions of the chimpanzees. Gómez (1991) found that a young gorilla first used people as if they were boxes, pushing them bodily to a position in which they could be climbed. Later, however, she switched to gentle hand-pulling and repeated gaze alternation between the human's eyes and the desired goal, enlisting the human's help *as a causal agent*. This work implies that the two species of ape do not differ in their ability to use objects as tools to solve problems, nor in having some understanding of voluntary agency of other individuals. Redshaw (1978) argued that gorilla infants *lack* understanding of individuals as causal agents, whereas human infants show clear evidence of this from 11 months onwards. (Redshaw used Piagetian scales developed for human children, justifying the comparison by the fact that gorillas follow, in general, strikingly similar trajectories to human infants.) However, the infant gorilla studied by Gómez began to do so at just the age at which Redshaw's studies ceased (18 months), so the difference would seem to be only one of lag.

Other sources of evidence furnish data for all great apes. Gallup's (1970) famous experiment on mirror-self-recognition, for instance, has been repeatedly applied to different species of apes. At first, gorillas seemed to fail the test: when marked with colored paint under anesthetic, they did not show the characteristic reaction of surprise and investigation when a mirror later enabled them to see a mark on a hidden part of their anatomy (Suarez & Gallup, 1981). This result was puzzling, when orangutans and chimpanzees both succeeded, as it suggested the trait had evolved twice (independently), or been secondarily lost in gorillas. However, Patterson & Cohn (1994) have now tested the home-reared gorilla Koko, subject of a sign-language project, and unambiguously found evidence of self-recognition: she passed a mark test performed by stealth rather than anesthetic, explored parts of her body that she could not normally see, and so forth. Recently, Heyes (1994) has criticized the Gallup studies, arguing cogently that, although significant increases in face-touching were found in the presence of mirrors, this face-touching may reflect only an artifact of recovery from anesthesia. This leaves the experiment on Koko as the best evidence for mirror-self-recognition in any great ape. Whether this ability is related to possession of a concept of "self" is a moot point; however other, entirely "nonsocial," explanations are possible (see Heyes, 1994; Mitchell, 1993).

Deception is another sphere in which evidence of social comprehension has been obtained from a range of apes. Observational records of tactical deception in gorillas are frequent[3] (Byrne & Whiten, 1990; Mitchell, 1991), and span various techniques: concealment of actions, inhibition of vocalization, inhibition of visual attention, distraction of others' attention, and various actions more nebulously categorized as "creating an image." This repertoire is not specific to apes, but found also in monkeys, usually in circumstances that make learning by trial-and-error the most likely explanation (Byrne & Whiten, 1992). Occasionally, however, gorilla deception shows signs of real planning. For instance, at Twycross Zoo in Britain during a phase in which a young male increasingly tended to challenge an older male, he was once observed to creep up on the older male in order to effect a surprise attack: "about 5 m from [him], [the young male] changed the nature of his approach, putting his feet down carefully, with his whole body adopting a tense demeanor" (P. Hart, record no. 160 in Byrne & Whiten, 1990). Byrne & Whiten (1991) noted that it is possible – though not very plausible – to explain this tactic as a result of trial-and-error learning.

Deceptive communication is particularly useful in evaluating social comprehension. In the context of surprise initiation of play, a gorilla's deliberate hiding of the "play-face" expression by placing a hand over the mouth has been observed (Tanner & Byrne, 1993). This suggests that gorillas have awareness of their spontaneous facial expressions and the consequences they entail. How this might be acquired is more problematic. In the analogous vocal case, that of mountain gorillas' inhibition of their characteristic copulation calls during secret matings (noted above), learning apparently takes place, to judge from the following observations. A low-rank silverback gorilla, Shinda, had several times been seen to mate secretly without inhibiting his calls, but on 26 August 1989 several observers (including myself) recorded their distinct impression that he was trying unsuccessfully to prevent sounds emerging from his mouth – an impression based on his contorted facial expression (cited as record no. 163, Byrne & Whiten, 1990). During September, he continued to use the tactic of secret mating, giving "quiet" copulation calls (records no. 164–5). Then, on 14 November 1989, Shinda was first noted to *fully* inhibit his copulation calls (record no. 166). Trial-and-error is perhaps sufficient to account for such learning, as failures are punished by the dominant silverback, although usually only the female is in fact physically chastized.

A final possible source of evidence of social comprehension is imitation. Imitation, in the sense of copying motor actions by direct observation of another individual's behaviour, has been used as evidence for comprehension of others' mental states (e.g. Bruner, 1972; Tomasello & Call, 1994). Unfortunately, it has been hard to find experimental evidence of this "action-level imitation," as I have described it above, even in chimpanzees (see Custance & Bard, 1994; Tomasello *et al.*, 1993), and none exists for the gorilla. However, there are suggestive anecdotes of action-level imitation by gorillas. For instance, a zoo gorilla that had often watched its cage being mopped out by a caretaker, was once noticed to pick up the mop when it was inadvertently left in the cage and replicate the same action of mopping-up as used by the human (G. Thienes, personal communication). Also, a recent study of handedness in zoo gorillas may give evidence of copying human preference. The paper presents the data as evidence of a species tendency to right-handedness (Olson *et al.*, 1990), but splitting the sample into zoo-born versus wild-born show that only zoo-born gorillas are right-handed; wild-born subjects are equally right- and left-preferent. This pattern is readily explained if gorillas are capable of (and prone to) imitating their human keepers, although other explanations are possible.

The case that gorillas show "program-level imitation" has been discussed above; however, note that this sort of imitation need not require understanding of the mental states of the individual serving as model. Program-level imitation likely depends on an understanding of the logical structure of action sequences, but this does not seem to be a matter of "social" comprehension.

ARE GORILLA BRAINS SMALL?

The bodily substrate of intelligence is in the brain, and brain enlargement can be taken to index – in some way – a species' specialization in intelligence as an evolutionary strategy. In general in animal biology, allometric scaling is used to compare bodily dimensions "fairly" across species, to take account of changes of body form with growth in absolute size (see e.g. Jungers, 1985). Allometric methods compare the actual size of a species' organ with that expected for an average member of the taxonomic group in question. (The approximate straight line, obtained by plotting many species on a log–log graph of organ size against body size, is used to compute this expectation.) Gorillas are relatively small-brained among haplorhine primates by traditional allometry (Clutton-Brock & Harvey, 1980), and this is often used to bolster the case for a relative lack of intelligence in gorillas. (Though not always: as noted above, Parker & Gibson (1977) suggested that the gorilla–chimpanzee common ancestor was an intelligent, tool-using, extractive forager – and that gorillas *retain* enhanced intelligence from this ancestor despite their current lack of need for tool-use. See Chapter 16.)

There are reasons however, to be skeptical of standard allometric methods as applied to brain sizes. First, allometry implicitly assumes that body size is a fixed feature against which the relative size of component parts change with selective pressures. But it has been suggested that gorillas and chimpanzees have in fact only recently evolved their differences in body size and sexual dimorphism, with the more conservative brain retaining a basic similarity in size and shape (Deacon, 1990; Shea, 1983). If so, then scaling against body size seriously underrates gorilla brains. Secondly, relative brain size may *always* be an inappropriate measure for judging intelligence. If the brain were like a telephone switchboard, with more input–output traffic requiring proportionately more central machinery, then judging the effective size of this machinery – the brain – against some measure of the input–output traffic would make good sense (although spinal cord cross-sectional area would be better than body size for this purpose). Clearly this approach gives part of the truth, but any view of the brain as "intelligent", as a cognitive organ, assumes the brain also functions as a Turing machine – producing computational solutions to problems, not simply reflex responses (Byrne, 1995). A Turing machine is limited by the *absolute* number of elements, whereas its size relative to its routine input–output traffic (and so body size) is irrelevant to its power (see Byrne, 1993b, 1994). However, it is likely in practice that primate brains subserve both functions: switchboard and Turing machine. How might we best extract a useful estimate of the absolute amount of brain tissue available for cognitive computation? Several methods have been tried, but neocortex ratio (the ratio of neocortical to other brain tissue) is perhaps the best, because it can be determined reliably, and because variance in primate brain size is largely a function of differences in neocortex (see Byrne, 1994, 1995, and Dunbar, 1992).

Gorillas have a relatively large neocortex ratio among haplorhines, although it is still smaller than that of the other great apes and *Papio* baboons. Even this measure makes the implicit assumptions that only the neocortex is seriously involved in intelligence (a view consistent with the rapid change that has occurred specifically in this region across the primate radiation), and that the relative proportion of neocortex is more relevant than its absolute size. Ideally, the spinal cord cross-sectional area might be used to assess input–output traffic, and brain "spare capacity" computed from this; but until such data are available perhaps absolute neocortical (or brain) volume is as reliable a measure as any. Indeed, Gibson (1990) took absolute brain size to be crucial, and Parker (1990) argued for cortical surface area, which scales with absolute volume. In absolute brain volume, and in volume of neocortex, the gorilla is second only in size to *Homo* among the primates (at 500 grams, the brain is slightly greater than the 400 gram brains of chimpanzee and orangutan: Stephan *et al.*, 1981).

With this imperfect understanding of how to compare brain sizes in order to estimate intellectual potential, it would therefore be most unwise to judge the gorilla as "small-brained." Specific comparisons with chimpanzees and orangutans had best be delayed until we have a much sharper understanding of how the brain works, so that the contribution to brain efficiency from tissue volume can be sensibly measured.

DISCUSSION

The evidence sketched in this chapter should lay to rest the popular, but always rather strange idea, that gorillas are anomalous among the great apes in lacking the intelligence and large brains of all the others. Given the phylogenetic relationships of the species (forming a clade with the two chimpanzee species and humans, with the orangutan as the closest outgroup), this would have implied secondary loss of intelligence. A strong selection pressure for intelligence has apparently acted more than once during primate evolution, to judge from neocortical expansion – once producing the haplorhines, and once producing the later *Homo* species. A sudden lack of any such need would be very puzzling; but if gorillas are comparably intelligent to other great apes, the puzzle does not arise.

The "small brain" claim was always based on brain size relative to body size, now increasingly realized to be an inappropriate measure for judging intellectual efficacy (though entirely appropriate for assessing the metabolic cost of the organ). Most likely, rapid, recent selection for increased body size in the more folivorous-adapted gorilla left the brain relatively unchanged in size from the primitive state (i.e. the chimpanzee–gorilla common ancestor). In order to eat coarse food, such as the pith and leaves that gorillas rely on more than do chimpanzees, a primate with unspecialized digestion needs a large gut, and hence a large body. Given the metabolically high cost of gut tissue (as high as brain tissue: Aiello & Whitcombe, 1995), selection for reduced brain size might have been expected; but this has not happened, because the gorilla brain (and neocortex) is the largest of any nonhuman primate. Unusually, the gorilla cerebellum is particularly enlarged (Stephan et al., 1981), and the significance of this is unclear; however, note that the human cerebellum is also unusually large (Passingham, 1982). While controversy remains about which of the body-independent measures is most likely to assess computational potential, the gorilla brain is not particularly small on any of them.

With modern evidence, the feeding adaptation of gorillas emerges as more similar to that of chimpanzees than used to be realized. The widely distributed western lowland gorilla (*G. g. gorilla*) inhabits what is presumably the species' environment of evolutionary adaptedness, the primitive gorilla niche. Gorillas should therefore be seen as tropical forest frugivores, like chimpanzees, supplementing a nutritionally inadequate fruit diet with a range of leaf, shoot, bark, fungus, and animal matter. The ability of some populations of the two eastern subspecies (*G. g. beringei* and *graueri*) to colonize montane habitats is, I suggest, related to gorilla intellect. To acquire the skilled techniques of leaf-processing necessary to access nutritious, protein-rich herbs that are defended by physical barriers, gorillas must be able to organize and structure complex sequences of coordinated actions: program-level imitation seems likely to be the ontogenetic origin of these skills. The *origin* of gorilla technical skill, however, should not be sought in these unusual gorilla habitats; indeed, it may lie further back in evolutionary history, if orangutans prove to have similar skills, as suggested by Galdikas & Vasey (1992).

In social organization, too, gorillas are now realized to be more chimpanzee-like, with complex mating options for both sexes, and subtle tactics used to manipulate other individuals. However, note that many species of monkey show seemingly intelligent tactics that they almost certainly acquire by trial-and-error learning (Byrne & Whiten, 1992). In captive testing, gorillas show "theory of mind" abilities typical of other great apes: understanding of humans as causal agents, mirror-self-recognition, and action-level imitation. Whether any of these abilities necessarily require understanding of other individuals as having mental states is less clear (see Byrne, 1995; Gómez, 1991; Heyes, 1993a; Tomasello & Call, 1994).

Realization of the similarity of gorillas and chimpanzees makes their points of difference especially interesting. The lack of hunting and tool use in wild gorillas may perhaps prove to be ecologically driven (the pygmy chimpanzee also does not hunt and hardly uses tools in the wild; see Kano, 1982; McGrew, 1989); but the difference in the gorilla's behavioral style is very striking, though hard to document. Caroline Tutin (personal communication) noted a difference, when caged apes of both species were experimentally shown an unfamiliar insect: the gorillas watched and touched the insect gently, the chimpanzees killed it first, then examined the body! Chimps seem to be born experimentalists, testing anything they see to destruction, whereas gorillas rather observe, like ethologists. The famous film of the gorilla Digit taking, examining and returning Dian Fossey's ballpen and notebook is a striking example of the characteristic style (National Geographic Video Library, 1986); anyone working with mountain gorillas will have similar anecdotes to this. It is to be hoped that future research can begin to describe and quantify these intriguing differences.

NOTES

1 Recent mitochondrial DNA sequencing suggests that the two Eastern subspecies (*beringei* and *graueri*) may be better treated as belonging to a separate species to the Western lowland *gorilla* (Ruvolo, 1994).
2 Male chimpanzees also attempt to coerce unwilling females on safari with them; the screams of such a female will bring aid from other males, but if not, she may be forced to accompany a male not of her choice (Goodall, 1986, p. 481).
3 Comparison of frequency across species is not easy, because of systematic differences in number of studies, total observation times, observation conditions, and the focus of investigation.

REFERENCES

Aiello, L. & Whitcombe, E. (1995). Are there anatomical foundations for cognition? In *Modelling the Early Human Mind*, ed. P. Mellars & K.R. Gibson. Cambridge: McDonald Institute Monograph Series. (in press).

Attenborough, D. (1979). *Life on Earth*. London: Collins.

Bailey, W.J., Hayasaka, K., Skinner, C.G., Kehoe, S., Sieu, L.C., Slightom, J.L. & Goodman, M. (1992). Re-examination of the African hominoid trichotomy with additional sequences from the primate β-globin gene cluster. *Molecular Phylogenetics and Evolution*, 1, 97–135.

Bruner, J.S. (1972). Nature and uses of immaturity. *American Psychologist*, 27, 687–708.

Byrne, R.W. (1993a). Hierarchical levels of imitation. Commentary on M. Tomasello, A. Kruger & H.H. Ratner "Cultural learning". *Behavioral and Brain Sciences*, 16, 516–7.

Byrne, R.W. (1993b). Do larger brains mean greater intelligence? Commentary on R.M. Dunbar "Co-evolution of neocortex size, group size and language in humans". *Behavioral and Brain Sciences*, 16, 696–7.

Byrne, R.W. (1994). The evolution of intelligence. In *Behaviour and Evolution*, ed. P.J.B. Slater & T.R. Halliday, pp. 223–65. Cambridge: Cambridge University Press.

Byrne, R.W. (1995). *The Thinking Ape: Evolutionary Origins of Intelligence*. Oxford: Oxford University Press.

Byrne, R.W. & Byrne, J.M.E. (1991). Hand preferences in the skilled gathering tasks of mountain gorillas (*Gorilla g. beringei*). *Cortex*, 27, 521–46.

Byrne, R.W. & Byrne, J.M.E. (1993). Complex leaf-gathering skills of mountain gorillas (*Gorilla g. beringei*): Variability and standardization. *American Journal of Primatology*, 31, 241–261.

Byrne, R.W. & Whiten, A. (1990). Tactical deception in primates: The 1990 database. *Primate Report*, 27, 1–101.

Byrne, R.W. & Whiten, A. (1991). Computation and mindreading in primate tactical deception. In *Natural Theories of Mind: Evolution, Development and Simulation of Everyday Mindreading*, ed. A. Whiten, pp. 127–41. Oxford: Basil Blackwell Ltd.

Byrne, R. W. & Whiten, A. (1992). Cognitive evolution in primates: Evidence from tactical deception. *Man*, 27, 609–27.

Clutton-Brock, T.H. & Harvey, P.H. (1980). Primates, brains and ecology. *Journal of Zoology*, 190, 309–23.

Custance, D. & Bard, K.A. (1994). The comparative and developmental study of self-recognition and imitation: The importance of social factors. In *Self-awareness in Animals and Humans: Developmental Perspectives*, ed. S.T. Parker, R.W. Mitchell & M.L. Boccia, pp. 207–26. New York: Cambridge University Press.

Dawkins, R. (1976). Hierarchical organisation: A candidate principle for ethology. In *Growing Points in Ethology*, ed. P.P.G. Bateson & R.A. Hinde, pp. 7–54. Cambridge: Cambridge University Press.

Deacon, T.W. (1990). Fallacies of progression in theories of brain-size evolution. *International Journal of Primatology*, 11, 193–236.

de Waal, F. (1982). *Chimpanzee Politics*. London: Jonathon Cape.

Dunbar, R.I.M. (1992). Neocortex size as a constraint on group size in primates. *Journal of Human Evolution*, 20, 469–93.

Fagot, J. & Vauclair, J. (1991). Manual laterality in non-human primates: A distribution between handedness and manual specialization. *Psychological Bulletin*, 109, 76–89.

Galdikas, B.M.F. & Vasey, P. (1992). Why are orangutans so smart? Ecological and social hypotheses. In *Social Processes and Mental Abilities in Non-human Primates: Evidence from Long-term Field Studies*, ed. F.D. Burton, pp. 183–224. Lewiston, NY: The Edward Mellon Press.

Galef, B.G., Jr (1988). Imitation in animals: History, definition and interpretation of data from the psychological laboratory. In *Social Learning: Psychological and Biological Perspectives*, ed. T. Zentall & B.G. Galef Jr, pp. 3–28. Hillsdale, NJ: Lawrence Erlbaum.

Gallup, G.G., Jr (1970). Chimpanzees: Self-recognition. *Science*, 167, 86–87.

Gibson, K.R. (1986). Cognition, brain size and the evolution of embedded food resources. In *Primate Ontogeny, Cognition and Social Behaviour*, ed. J.G. Else & P.C. Lee, pp. 93–104. Cambridge: Cambridge University Press.

Gibson, K.R. (1990). New perspectives on instincts and intelligence: brain size and the emergence of hierarchical mental construction skills. In *"Language" and Intelligence in Monkeys and Apes: Comparative Developmental Perspectives*, ed. S.T. Parker & K.R. Gibson, pp. 97–128. New York: Cambridge University Press.

Gómez, J.C. (1991). Visual behaviour as a window for reading the mind of others in primates. In *Natural Theories of Mind: Evolution, Development and Simulation of Everyday Mindreading*, ed. A. Whiten, pp. 195–207. Oxford: Basil Blackwell Ltd.

Goodall, A.G. (1977). Feeding and ranging behaviour of a mountain gorilla group (*Gorilla gorilla graueri*) in the Tshibinda-Kahuzi region (Zaire). In *Primate Ecology*, ed. T.H. Clutton-Brock, pp. 450–79. New York: Academic Press.

Goodall, J. (1986). *The Chimpanzees of Gombe: Patterns of Behavior*. Cambridge, MA: The Belknap Press.

Heyes, C.M. (1993a). Imitation, culture and cognition. *Animal Behaviour*, 46, 999–1010.

Heyes, C.M. (1993b). Anecdotes, training, trapping and triangulating: Can animals attribute mental states?. *Animal Behaviour*, 46, 177–88.

Heyes, C.M. (1994). Reflections on self-recognition in primates. *Animal Behaviour*, 47, 909–19.

Horai, S., Satta, Y., Hayasaka, K., Kondo, R., Inoue, T., Ishida, T., Hayashi, S. & Takahata, N. (1992). Man's place in hominoidea revealed by mitochondrial DNA genealogy. *Journal of Molecular Evolution*, 35, 32–43.

Humphrey, N.K. (1976). The social function of intellect. In *Growing Points in Ethology*, ed. P.P.G. Bateson & R.A. Hinde, pp. 303–17. Cambridge: Cambridge University Press.

Jolly, A. (1966). Lemur social behavior and primate intelligence. *Science*, 153, 501–6.

Jolly, A. (1985). *The Evolution of Primate Behavior*, 2nd edn. New York: Macmillan.

Jungers, W.L. (ed.) (1985). *Size and Scaling in Primate Biology*. New York: Plenum Press.

Kano, T. (1982). The social group of pygmy chimpanzees (*Pan paniscus*) of Wamba. *Primates*, 23, 171–88.

Köhler, W. (1925). *The Mentality of Apes*. London: Routledge & Kegan Paul Ltd.

Lashley, K.S. (1951). The problem of serial order in behavior. In *Cerebral Mechanisms in Behavior: The Hixon Symposium*, ed. L.A. Jeffress, pp. 112–36. New York: Wiley.

Mackinnon, J. (1978). *The Ape Within Us*. London: Collins.

McGrew, W.C. (1989). Why is ape tool use so confusing? In *Comparative Socioecology: The Behavioral Ecology of Humans and Other Mammals*, ed. V. Standen & R.A. Foley, pp. 457–72. Oxford: Blackwell Scientific Publications Ltd.

McGrew, W.C. (1992). *Chimpanzee Material Culture*. Cambridge: Cambridge University Press.

Milton, K. (1981). Distribution patterns of tropical plant foods as a stimulus to primate mental development. *American Anthropologist*, 83, 534–548.

Mitchell, R.W. (1993). Mental models of mirror-self-recognition: Two theories. *New Ideas in Psychology*, 11, 295–325.

Natale, F., Antinucci, F., Spinozzi, G. & Poti, P. (1986). Stage 6 object concept in nonhuman primate cognition: A comparison between gorilla (*Gorilla gorilla gorilla*) and Japanese macaque (*Macaca fuscata*). *Journal of Comparative Psychology*, 100, 335–39.

National Geographic Video Library (1986). *Search for the Great Apes*.

Nishida, T. (1983). Alpha status and agonistic alliance in wild chimpanzees. *Primates*, 24, 318–36.

Nishida, T. & Hiraiwa-Hasegawa, M. (1987). Chimpanzees and bonobos: Cooperative relationships among males. In *Primate Societies*, ed. B.B. Smuts, D.L. Cheney, R.M. Seyfarth, R.W. Wrangham & T.T. Struhsaker, pp. 165–77. Chicago: University of Chicago Press.

Olson, D.A., Ellis, J.E. & Nadler, R.D. (1990). Hand preferences in captive gorillas, orang-utans and gibbons. *American Journal of Primatology*, 20, 83–94.

Parker, S.T. (1990). Why big brains are so rare: Energy costs of intelligence and brain size in anthropoid primates. In *"Language" and Intelligence in Monkeys and Apes: Comparative Developmental Perspectives*, ed. S.T. Parker & K.R. Gibson, pp. 129–54. New York: Cambridge University Press.

Parker, S.T. & Gibson, K.R. (1977). Object manipulation, tool use, sensorimotor intelligence as feeding adaptations in cebus monkeys and great apes. *Journal of Human Evolution*, 6, 623–41.

Passingham, R.E. (1982). *The Human Primate*. Oxford: Freeman.

Patterson, F. & Cohn, R. (1994). Self-recognition and self-awareness in lowland gorillas. In *Self-awareness in Animals and Humans: Developmental Perspectives*, ed. S.T. Parker, R.W. Mitchell & M.L. Boccia, pp. 273–90. New York: Cambridge University Press.

Pilbeam, D. & Smith, R. (1981). New skull remains of *Sivapithecus* from Pakistan. *Memoirs of the Geological Survey of Pakistan*, 11, 1–13.

Povinelli, D.J. (1991). *Social Intelligence in Monkeys and Apes*. Ph.D. thesis. Yale University, New Haven, CT.

Povinelli, D.J., Nelson, K.E. & Boysen, S.T. (1990). Inferences about guessing and knowing in chimpanzees (*Pan troglodytes*). *Journal of Comparative Psychology*, 104, 203–10.

Povinelli, D.J., Nelson, K.E. & Boysen, S.T. (1992). Comprehension of role reversal in chimpanzees: Evidence of empathy? *Animal Behaviour*, 43, 633–40.

Premack, D. (1988). "Does the chimpanzee have a theory of mind?" revisited. In *Machiavellian Intelligence: Social Expertise and the Evolution of Intellect in Monkeys, Apes and Humans*, ed. R.W. Byrne & A. Whiten, pp. 94–110. Oxford: Clarendon Press.

Premack, D. & Dasser, V. (1991). Perceptual origins and conceptual evidence for theory of mind in apes and children. In *Natural Theories of Mind: Evolution, Development and Simulation of Everyday Mindreading*, ed. A. Whiten, pp. 253–66. Oxford: Basil Blackwell Ltd.

Premack, D. & Woodruff, G. (1978). Does the chimpanzee have a theory of mind? *Behavioral and Brain Sciences*, 4, 515–26.

Redshaw, M. (1978). Cognitive development in human and gorilla infants. *Journal of Human Evolution*, 7, 133–41.

Richard, A.F. (1985). *Primates in Nature*. New York: Freeman.

Ridley, M. (1986). *Evolution and Classification: The Reformation of Cladism*. London: Longman.
Russon, A.E. & Galdikas, B.M.F. (1993). Imitation in free-ranging rehabilitant orangutans (*Pongo pygmaeus*). *Journal of Comparative Psychology*, 107, 147–61.
Ruvolo, M. (1994). Molecular evolutionary processes and conflicting gene trees: The hominoid case. *American Journal of Physical Anthropology*, 94, 94–113.
Shea, B.T. (1983). Phyletic size change and brain/body allometry: A consideration based on the African pongids and other primates. *International Journal of Primatology*, 4, 33–61.
Sicotte, P. (1993). Inter-group encounters and female transfer in mountain gorillas: Influence of group composition on male behaviour. *American Journal of Primatology*, 30, 21–36.
Stephan, H., Frahm, H. & Baron, G. (1981). New and revised data on volumes of brain structures in insectivores and primates. *Folia Primatologica*, 35, 1–29.
Stewart, K.J. & Harcourt, A.H. (1987). Gorillas: Variation in female relationships. In *Primate Societies*, ed. B.B. Smuts, D.L. Cheney, R.M. Seyfarth, R.W. Wrangham & T.T. Struhsaker, pp. 155–64. Chicago: University of Chicago Press.
Suarez, S. & Gallup, G.G. (1981). Self-recognition in chimpanzees and orangutans, but not gorillas. *Journal of Human Evolution*, 10, 175–88.
Tanner, J.E. & Byrne, R.W. (1993). Concealing facial evidence of mood: Perspective-taking in a captive gorilla? *Primates*, 34, 451–56.
Tomasello, M. & Call, J. (1994). Social cognition of monkeys and apes. *Yearbook of Physical Anthropology*, 37, 273–305.
Tomasello, M., Davis-Dasilva, M., Camak, L. & Bard, K. (1987). Observational learning of tool-use by young chimpanzees. *Human Evolution*, 2, 175–83.
Tomasello, M., Savage-Rumbaugh, E.S. & Kruger, A.C. (1993). Imitative learning of actions on objects by children, chimpanzees and encultured chimpanzees. *Child Development* 64, 1688–1705.
Tutin, C.E.G. (1979). Mating patterns and reproductive strategies in a community of wild chimpanzees. *Behavioral Ecology and Sociobiology*, 6, 39–48.
Tutin, C.E.G. & Fernandez, M. (1993). Composition of the diet of chimpanzees and comparisons with that of sympatric lowland gorillas in the Lope Reserve, Gabon. *American Journal of Primatology*, 30, 195–211.
Vedder, A.L. (1984). Movement patterns of free-ranging mountain gorillas (*Gorilla gorilla beringei*) and their relation to food availability. *American Journal of Primatology*, 7, 73–88.
Waterman, P.G., Choo, G.M., Vedder, A.L. & Watts, D. (1983). Digestibility, digestion-inhibitors and nutrients of herbaceous foliage and green stems from an African montane flora and comparison with other tropical flora. *Oecologia*, 60, 244–9.
Watts, D.P. (1984). Observations on the ontogeny of feeding behaviour in mountain gorillas (*Gorilla gorilla beringei*). *American Journal of Primatology*, 8, 1–10.
Whiten, A. & Ham, R. (1992). On the nature and evolution of imitation in the animal kingdom: Reappraisal of a century of research. In *Advances in the Study of Behavior*, vol. 21, ed. P.J.B. Slater, J.S. Rosenblatt, C. Beer & M. Milinski, pp. 239–83. New York: Academic Press.
Yamagiwa, J. (1983). Diachronic changes in two eastern lowland gorilla groups (*Gorilla gorilla graueri*) in the Mt. Kahuzi region, Zaire. *Primates*, 24, 174–83.

6

Ostensive behavior in great apes: The role of eye contact

JUAN CARLOS GÓMEZ

INTRODUCTION

This chapter is about the origins and significance of an apparently simple behavior – looking into the eyes of other people. Eye contact and related patterns such as gaze following are an important and conspicuous part of human interaction. There are few face-to-face interactions in which eye contact (or its avoidance) does not appear (Argyle & Cook, 1976; Fehr & Exline, 1987). Furthermore, eye contact seems to be one of the earliest face-to-face interaction patterns to appear in human ontogeny (Adamson & Bakeman, 1991).

In this chapter I suggest a theoretical framework for interpreting the significance of eye contact in humans, and then I explore its occurrence in nonhuman primates. I suggest that eye contact in humans is a case of "ostensive behavior" (Sperber & Wilson, 1986) – that is, a way to express and assess communicative intent – and, as such, an important sign of higher cognitive processes. I then explore a purported behavioral difference between monkeys and apes – their reactions to eye contact – and to what extent it reflects an important difference in cognition and communication. My suggestion is that in the great apes there is evidence that eye contact has evolved into an ostensive behavior associated with the expression and assessment of communicative intent. Let me begin with the theoretical significance of eye contact in human communication.

EYE CONTACT IN HUMAN COMMUNICATION

A typical distinction in human communication is that between intentional and accidental communication. Compare the following examples:

1. Beatrice sees John wearing a waistcoat and understands that he has bought a new one.
2. John addresses Beatrice and points to his new waistcoat.

In example (1) there is an involuntary transmission of information. John does not explicitly intend Beatrice to know that he has bought a new waistcoat. In example (2), however, John *shows* Beatrice his new waistcoat: this information is intentionally transmitted to Beatrice.

This could lead us to define a communicative intention as the intention to transmit certain information to another person. However, human communication is more complex than that. For example, imagine that for some reason John thinks that Beatrice does not want to be addressed by him, but he, nevertheless, wants her to see his new waistcoat:

3. John sees Beatrice standing in the hall, and he decides to walk past her with the hope that she will see him wearing his new waistcoat.

Note that here John has the intention of informing Beatrice that he has bought a new waistcoat, but he is not communicating it overtly. He manipulates the situation in such a way that Beatrice will notice the information but (hopefully) not his intention to show that information. This gives us an important insight into the nature of human intentional communication: in normal, overt communication people show not only the information, X, but also their intention to show that information. In example (2) when John addresses Beatrice and points to his new waistcoat, he is not only showing his waistcoat but also his intention to show it.

Since Paul Grice's insightful analysis of "meaning" (Grice, 1989), it has become apparent that the intention to communicate has a peculiar structure: it involves not only having an informative intention, but also showing it; it involves not only showing X, but also showing your intention to show X. The structure of communicative intentions, as analyzed by Grice, has been the subject of considerable debate (for recent analyses of the problem, see Avramides, 1989; Sperber & Wilson, 1986). Here, I want to concentrate on the issue of how we show our intention to show something – what Sperber & Wilson call the problem of *ostension*.

An overt, ostensive communicative act seems to consist of two parts:
Showing X + Showing the intent to show X.

As to the first part, we show X by directing somebody's attention onto X. The human pointing gesture seems to be an adaptation for this function. In a sense in linguistic communication we are also manipulating the attention of others, making their minds focus upon particular "points." However, I have concentrated my discussion of human communication mostly upon the case of nonverbal behavior, because this will facilitate later comparison with nonhuman primates. Directing the visual attention of someone usually involves making him or her look at particular things or, alternatively, placing things in front of that person's eyes. For example, in (2) John directs Beatrice's visual attention onto his waistcoat, whereas in (3) he tries to make his waistcoat fall under Beatrice's eyes. In each case, the point is to make the addressee's attention meet the target object. The mechanism involved in showing something to someone would seem, then, to be fairly simple.

But what about the second part of ostensive communication – how does one "show the intention to show?" Consider the above examples: the difference between covertly managing to make Beatrice notice the waistcoat and overtly showing it to her lies in the fact that, in the latter case, John *addresses* Beatrice. What is involved in *addressing* someone? A possibility is that addressing someone essentially consists of calling his or her attention upon oneself. But this definition is insufficient: in the above example, when John parades in front of Beatrice wearing his new waistcoat, he is trying to capture Beatrice's attention upon himself, but indeed he is precisely avoiding addressing her. A true address consists of calling someone's attention not just upon oneself, but *upon one's own attention*. Consider the following example:

4. John has been parading in front of Beatrice for a time but she seems to be absent-minded and apparently she has not noticed John and his waistcoat. John decides to do something to get her attention: he pretends to stumble on a dustbin with the hope that the noise will make Beatrice look in his direction, thereby noticing him and his new waistcoat.

If John's plot succeeds, he will have called Beatrice's attention upon himself but *not addressed* her. When you address someone, you call his or her attention upon your own attention: when the addressee looks at you, he or she finds you are attending to him. As suggested by Gómez (1991, 1992), communication is built upon attention contact – intentional communication is first of all a meeting of attentions.

One of the usual ways of indicating and recognizing attention contact is through eye contact. As argued by Gómez (1994a), looking into each other's eyes amounts to attending to each other's attention. Whatever you do after establishing eye contact is overtly addressed to the other person's attention. Eye contact not only gives you the information that the other person is attending to you, but also gives to the other person the information that you are attending to him or her; that is, through eye contact one simultaneously gets and transmits information about mutual engagement with others. The structure of eye contact has the mutuality features that characterize intentional communication in the Gricean sense (for further exploration of this point, see Gómez, 1994a). This feature of mutual visual attention might explain the important role played by eye contact in human interaction. Eye contact or looking into each other's eyes seems to be a privileged way to transmit communicative intent, to turn any behavior into an ostensive behavior.

Empirical studies show that eye contact starts very early in human life. It is an important part of the face-to-face interaction patterns displayed by babies. According to some authors (e.g. Trevarthen, 1979), early eye contact expresses a level of "primary intersubjectivity" between the human baby and the adult. However, more interesting for our purposes is that at around 9–12 months infants coordinate eye contact with the production of gestures like pointing, and the resulting patterns are interpreted as the beginnings of intentional communication (Bates *et al.*, 1975). Sarriá (1989) has shown that this eye contact is indeed interpreted as an index of communicative intentionality by judges asked to assess videotapes of infant interactions. This fits with the interpretation of eye contact as a signal of ostension.

Futhermore, eye contact has been extensively studied in human adults within the field of nonverbal communication research. A plethora of descriptive, quantitative studies have tried to analyze the relation between eye contact and different variables (for reviews, see Fehr & Exline, 1987; Kleinke, 1986). The results show that eye contact is pervasive in verbal and nonverbal interactions. For example, it has been found that two conversing people can look at each other's eyes for about 30% of the time and the listener looks twice as much at the eyes of the speaker as the speaker does at the listener. The occurrence of eye contact is also related to the grammatical structure of ongoing speech, the regulation of conversational turns, and the transmission of emotional information (Kleinke, 1986). However, a common finding of these studies is that the meaning of eye contact is highly sensitive to contextual variables (Argyle, 1987; Kleinke, 1986). Mutual gaze is not a signal with a definite meaning in humans. In Argyle's words:

> the basic effect of gaze is to show attention and to increase arousal, but the meaning can vary with the situation and facial expression – from threat to sexual attraction.
> *(Argyle, 1987, p.247)*

Although nonverbal communication studies have rarely analyzed eye contact in relation to cognitive theories of communication, conclusions like Argyle's seem to fit well into the ostension hypothesis outlined above. With eye contact, mutual attention is established, but then the contents of the interaction depend upon context and other expressive behaviors.

In summary, eye contact is a pervasive pattern in human interaction. I have argued that this is because it plays a fundamental role in expressing and assessing communicative intent. It is in this light that I will consider the problem of eye contact in nonhuman primates. Do nonhuman primates use eye contact in an ostensive way? Can they help us to understand the origins of ostensive communication?

EYE CONTACT IN NONHUMAN PRIMATES

The primary purpose of the eyes is to collect information. They collect information distally from the environment and have the property of orienting themselves to the source of the information they are collecting. This orientation is what we call "gaze." But, as Argyle & Cook (1976, p. 1) pointed out, "Whenever organisms use vision, the eyes become signals as well as channels."

The eyes have, then, the interesting property of advertising their information-processing activity and its targets, and this interesting property has been exploited in evolution. The eyes have become *social signals*, that is they have become stimuli capable of releasing reactions in other animals, and it is generally assumed that the most common reaction in the animal world to being gazed at is arousal plus flight and/or attack by an animal that is looked at by another (Argyle & Cook, 1976; Goodenough et al., 1993). A pair of eyes looking at an organism seem to act as a powerful danger signal for the latter. It has been speculated that the evolutionary advantage of this lies in the high likelihood that if an animal is being looked at it is as a prey. Among the signals that best announce one's imminent conversion into the dinner of someone else – so the speculation goes – is a pair of eyes gazing at you. This basic function – detecting the danger of predation – could explain what seems to be a widespread sensitivity to gaze in animals.

Whatever the evolutionary explanation, the point is that the eyes seem to be a stimulus to which animals of many species show a high sensitivity, usually expressed as avoidance or fear. This "relationship between eyes and threat" (as Goodenough et al. (1993) put it) has been experimentally demonstrated in some animal species, such as plovers (Ristau, 1991), iguanas (Burger et al., 1992), chickens (Suarez & Gallup, 1982) or snakes (Burghart & Green, 1988), which react defensively to organisms who are staring at them. Indeed some species seem to exploit this sensitivity to eyes as an antipredator tactic: for example, some insects develop false eyespots to scare their vertebrate predators away (Blest, 1957; Tinbergen, 1974).

Primates are no exception in this sensitivity to gaze. In early field and captivity studies of nonhuman primates many authors noted that direct staring served as a threat signal, whereas gaze avoidance seemed to act as appeasement or submission. For example, in a representative review of the field, Redican (1975, p. 115) stated that the direct stare "is no doubt the most universal component of threat for nonhuman primates and is a basis on which further elaboration of the threat displays takes place." Threat displays could be enhanced by other components such as facial expressions of menace and menacing behaviors, but, according to Redican (1975, pp. 115–6), "A direct stare even by itself is a mild form of threat and is specifically mentioned as such in a large number of studies for a wide range of genera," among them lemurs, macaques, baboons, chimpanzees, and gorillas.

Correspondingly, the complementary pattern of gaze-avoidance is also typical of primates. Again according to Redican (1975), gaze aversion is a response to another monkey's stare as universal among primates as the very pattern of staring in threat displays. It is interesting that this sensitivity to being looked at also seems to operate on an interspecific basis. For example, Schaller (1963) described characteristic gaze-avoidance responses by gorillas when they were looked at by human observers.

Thus, early studies of nonhuman primate communication seemed to indicate that prolonged eye contact or gazes directed at the eyes of a partner had an *intrinsic* threatening value in contrast to the arousing, but without any definite meaning, effects of eye contact in humans. Although not everybody agreed with this interpretation (for example, Oppenheimer (1977) suggested that "prolonged looking . . . should be regarded as an expression of interest that may have a number of different motivations," the particular motivation involved in a stare depending upon the presence of other signals such as facial expressions[1]), some experimental evidence seemed to support it. For example, Wada (1961) found electroencephalographic evidence of arousal in response to stares in rhesus monkeys. Keating & Keating (1982) examined the visual scanning patterns of two rhesus monkeys looking at monkey faces and found that the eyes received greater fixation times than other areas. This only confirms that the eyes are "special" for rhesus, but interestingly "gazing" faces without emotional expressions were equivalent to "gazing and threatening" faces in terms of the scanning time they elicited.

Exline & Yellin (1969, as cited in Exline, 1972) devised a situation in which human experimenters stared at four male rhesus monkeys until eye contact was established. Then, they either kept staring at them or looked down. The monkeys reacted in a threatening way upon establishing eye contact: they crouched and looked back fixedly at the experimenter while drawing their brows in a V shape. In the continued stare condition, the monkeys escalated their responses to clearer threats and eventually attacked the experimenter (safely guarded beyond the cage bars!) or fled. In the interrupted eye contact condition, however, monkeys produced a significantly lower number of threat displays. In a control situation it was found that the rhesus monkeys did not react aggressively if the experimenter performed the orientation pattern to the monkey but kept his eyes closed. This singled out eye contact as the relevant aversive stimulus for the monkeys. (However, because there was no reaction when the experimenter was wearing a hood with holes for his eyes, it seems appropriate to say that it is a pair of eyes *on a face* that is menacing, not just the eyes.)

According to Exline's descriptions, the very fact of establishing eye contact is distressing for rhesus monkeys, because before they experience its prolongation, they first react with the cautious, mildly threatening crouch with drawn-up brows. Eye contact per se seems to have an intrinsic threatening value, and its prolongation enhances this threatening effect.

Thomsen (1974) studied the reaction of a variety of primate species to a human staring at them. Unfortunately he did not provide any description of the nature of the expressive or behavioral reaction accompanying eye contact. However, he found that most subjects gave only brief looks at the observer and they were more likely to make eye contact when they were in their home cages than when they were in a new cage. Of special interest (given the bias of most other studies towards rhesus monkeys) is his finding of pronounced species differences in the readiness to make eye contact with the staring observer: adult talapoins, patas and crab-eating monkeys looked at the observers more frequently than adult stump-tailed, squirrel or rhesus monkeys. Age and sex also seemed to be important variables, young female rhesus monkeys being as good "lookers" as adult patas and talapoins.

Several authors working in the field of human nonverbal behavior (Ellsworth,1975; Exline, 1972; Hindmarch, 1973; Vine, 1970), drawing on the above findings, assumed that eye contact is phylogenetically linked to aggression. As an example, Diebold (1968) speaks of a "shared primate ethogram" (in which eye contact initially constitutes an aversive stimulus), and that there would be a dividing line between the functions of eye contact in humans (variable and dependent upon context) and in the other primates (inherently threatening).

If this story were true, eye contact could have no ostensive function in nonhuman primates because it would be a signal designed for a specific communicative function – threat. However, the study of the great apes reveals that this may be only part of the story about eye contact in primates.

Eye contact in the great apes

Many informal reports by experienced observers point to the existence of a generalized use of eye contact in the great apes. For example, Goodall (1986, p. 121) summarizes her extensive experience with chimpanzees pointing out that "in most nonhuman primates, including gorillas, prolonged eye contact between adults functions as a threat. However in chimpanzees, as in man, relatively long bouts of eye contact may also accompany friendly interactions." Similarly, de Waal (1982, 1989) and de Waal & van Roosmalen (1979) in a study with captive chimpanzees found that adult males, after an aggressive encounter, need to engage in "reconciliation" and "consolation" behaviors, involving specific behavior patterns, such as touching their hands, hugging, or grooming. It turned out that an essential component of these patterns is the establishment of eye contact by the former opponents. De Waal considers this to be a prerequisite for reconciliation in chimpanzees. "It is as if chimpanzees do not trust each other's intentions without a look into the eyes" (de Waal, 1989, p. 43). Other observers agree with the importance of eye contact both in friendly and unfriendly interactions of common chimpanzees (Goodall, 1968; Nishida, 1970) and bonobos (de Waal, 1989; Kano, 1980; Kuroda, 1980; White, 1992).

However, Goodall's suggestion that in gorillas eye contact is threatening (see quotation above and Schaller's (1963) report of dominant silverback stares as a threat signal) is challenged by Yamagiwa (1992). He found that staring at close range between wild gorilla males need not have aggressive implications at all. In the all-male mountain gorilla group he studied, "social staring" appeared in a variety of nonaggressive social contexts, among them postconflict approaches.

The pattern found in chimpanzee reconciliations contrasts strongly with that of rhesus monkeys where former adversaries do not look at each other's eyes when reconciling. They tend rather to look in all directions except in the face of their former opponents (de Waal, 1989; de Waal & Yoshihara, 1983). De Waal's conclusion is that the "rules" for eye contact seem to be different in chimpanzees and rhesus monkeys. According to him:

> Both humans and apes avoid eye contact during strained situations and seek it when ready to reconcile. Rhesus monkeys, in contrast, look each other straight in the eye during conflict; dominants intimidate subordinates by fixedly staring at them.
> *(de Waal, 1989, p. 114)*

De Waal & Yoshihara (1983) do not hesitate to put together humans and chimpanzees as contrasted to macaques in terms of their rules for eye contact.

These findings suggest that the dividing line between eye contact as a sign of threat and eye contact as a generalized component of communication could be drawn at a different point – not between humankind and the other primates, but between monkeys, on the one side, and apes and humankind on the other. The bias towards rhesus monkeys in the monkey findings about eye contact together with the interspecies variability suggested by Thomsen (1974) indicate that this hypothesis be treated with caution[2]. However, what seems to be clearly established is that eye contact is not necessarily a threat signal in some of the great apes. To what extent is it really a generalized component of communication in all the great apes? Is it indicative of an ostensive function that is, do apes use eye contact as an expression of their intention to communicate and to address other individuals?

EYE CONTACT AS OSTENSIVE BEHAVIOR IN APES AND MONKEYS

One prediction from the interpretation that apes's eye contact is ostensive is that they will make eye contact in any communicative situation in which they have to obtain and direct the attention of others to a particular object or event. A good test of this prediction would be to observe what happens in experimental situations where an individual has to make a request from another. Informal descriptions of natural behavior suggest that looks at the face and perhaps eye contact occur in such situations (for example, in food-sharing (Goodall, 1968; White, 1992)). What happens in research settings where the occurrence of eye contact can be systematically observed?

Gorillas

Let us start with my own studies on gorilla communication. In a longitudinal study of a female hand-reared gorilla named Muni (Gómez, 1990, 1992), I found that she would begin to coordinate requestive gestures with eye contact when she was about 1 year old. For example, if she wanted to be taken in arms by a human she would put her hands on the human's body and look at his or her eyes while waiting to be raised, whereas before that age she would simply climb up the person. When the gesture was related to an external object, e.g. taking the human's hand to the latch of a closed door, its coordination with eye contact did not appear until several months later (when the gorilla was about 20 months old). I interpreted the emergence of eye contact behaviors in this gorilla as a strategy to control the visual attention of the human addressee, comparable to that of human infants at the beginning of preverbal intentional communication (Bates, 1979; Bates *et al.*, 1975).

In a study of four juvenile hand-reared gorillas (among them, the same subject of the study outlined above), I found that the requests addressed to humans were structured in a pattern where attention-getting was differentiated from attention-redirection (Gómez, 1992). For example, a typical displacement request (where a gorilla would ask a person to move to a particular location) consisted of first calling the attention of the person (e.g. pulling clothes), then offering a hand, and then, if the person accepted by taking the hand offered by the gorilla, leading him or her to the desired location. After claiming the attention of a person the gorilla would wait until he or she looked down, whereby the person's and the animal's eyes would meet. Then, and only then, would the gorilla take the person's hand and carry out his or her request. Eye contact could be made at any point of the request, but most eye contact occurred during the attention-getting phase or when the destination was reached and the request was specified.

These hand-reared gorillas developed a varied repertoire of gestures specifically for calling the attention of people; that is, gestures whose aim was to establish eye contact before making a request. Among them were pulling the clothes, patting the leg, touching a hand, holding out the hand towards the person, turning the human's head towards their eyes, throwing straw to the human, and, at least in the case of a young female, using the pig-grunt vocalization – a typical gorilla vocalization described by Fossey (1972) in wild mountain gorillas. In all these cases when the humans looked at the gorilla, they would find the gorilla's eyes waiting for theirs.

In these examples eye contact seems to act as a prerequisite to communication: attention contact is sought by the gorilla before carrying out the gesture that will direct the attention of the addressee to an object or a place. This coincides with the scheme developed in the first section of this chapter, when I distinguished between these two aspects of ostension. There seems to be a neat distinction between the gestures specialized for directing the attention of the human to a particular target, and the gestures whose aim is to establish attention contact (as expressed by eye contact). At least in hand-reared, captive gorillas some form of ostensive function seems to be fulfilled through eye contact.

Chimpanzees

In a classic study of the ability of chimpanzees to communicate the location of hidden objects to their conspecifics, Menzel (1973, 1974) found that looking at the face of the addressee was an important component of his subjects' communicative strategies. Typically, the knowledgeable chimpanzee would start walking in the direction of the hidden object, then stop and turn, looking at the others' faces. Menzel (1973) remarked that, in his view, the communicative actions of his chimpanzees were not so much based upon a shared repertoire of codified behaviors (such as facial expressions and vocalizations) as upon their ability to watch and show each others' normal behaviors and interpret them in purposeful terms. This is highly reminiscent of the distinction between *coded* and *inferential* communication (Sperber & Wilson, 1986). Essential to Sperber & Wilson's notion of inferential communication is the idea of "attention-claiming" as a way to confer communicative value to otherwise ordinary pieces of behavior.

Woodruff & Premack (1979) found that chimpanzees confined in a cage were able to point out to a human which of two boxes contained food, provided the human would share afterwards the contents of the boxes. The chimpanzees' pointing behavior was frequently accompanied by gaze alternation between the target and the person's face. In a similar study with an adult female chimpanzee, I could determine that the chimpanzee's looks at the human's face that accompanied the gestures were specifically directed at the human's eyes (J.C. Gómez, P. Teixidor & M.V. Laá, unpublished results). (To differentiate between eye contact and simple looks at the face the experimenters were instructed to signal when the subject made eye contact with them.)

Tomasello et al. (1985) found that after 3 years of age young captive chimpanzees would combine eye contact with gestures and "expectant pauses" in their social interactions with conspecifics. Their eye contact behaviors were part of a broader pattern of gaze alternation between the addressee and the focus of communication. Carpenter et al. (1995) compared two chimpanzees with two human babies in terms of their looks at the face of an experimenter during an imitation experiment involving object manipulation. Chimpanzees would look at the face of the person less often and for shorter periods of time. Carpenter *et al.* also mention that their chimpanzee subjects produced imperative gestures and even a declarative one, but they don't clarify if these were accompanied by looks at the face. O'Connell (1994) found that eye contact and gaze alternation to objects was present in a captive group of chimpanzees in their own interactions, although no imperative or declarative gestures were observed. However, the chimpanzees would produce imperative gestures with eye contact and gaze alternation addressed to conspecifics in the possession of novel objects or to humans sitting by a desirable object.

All in all, there seem to be consistent data that chimpanzees use eye contact as a component of communication in situations where they have to direct the attention of someone upon something. This is highly reminiscent of the ostensive communication patterns found in humans.

Bonobos

In an informal description of bonobo development in hand-rearing conditions (Savage-Rumbaugh, 1984) eye contact was reported to emerge after 1 year of age, and with it caregivers' increasing attribution of deliberateness to the bonobo. For example, one caregiver commented that, when she touched the young bonobo on his shoulder, he no longer turned, startled, as if looking for the source of the disturbing stimulation; instead, he would turn and look at the eyes of the keeper. In an experimental study, O'Connell (1994) found eye contact, gaze alternation and imperative gestures in captive bonobos, with both conspecifics and humans who were holding a novel object. She found that in everyday interactions bonobos would use gaze as a means to regulate their dyadic playful interactions.

Orangutans

Gómez & Teixidor (1992), in an experiment about theory of mind in an orangutan, found that their subject (a captive, adult female) was capable of performing pointing gestures coordinated with eye contact addressed to humans in order to request food padlocked in a box. Moreover, her pointing or, more precisely, reaching gestures were not limited to request functions. She was able to perform something closer to "informative" pointing when the human was unable to find the keys necessary to open the boxes where the food was hidden. Although the final goal of the orangutan was clearly the possession of food or an object, her pointing to the keys (frequently accompanied by eye contact) was not a case of request, but something closer to what is known as "protodeclarative" pointing. Bard (1990) described the occurrence of eye contact coordinated with gestures among semi-wild orangutans in food-requesting situations. Babies were capable of begging with gaze alternation by at least 2 years of age. Some animals would also beg from humans; at least one of them would combine begging gestures with gaze alternation between food and the human's eyes accompanied with a vocalization. Call & Tomasello (1994) also found requestive pointing in orangutans accompanied by looks at the face of the experimenter.

In summary, combining this material on orangutans with that on the other great apes, we have systematic evidence that all the great apes use eye contact as part of different communicative patterns. In experimental situations where the aim of the apes is to get food or objects from humans, eye contact or looks at the face are frequently used in conjunction with the performance of gestures.

Monkeys

Data on monkeys in similar situations are extremely scarce and their results somewhat contradictory. Blaschke & Ettlinger (1987) trained four rhesus monkeys to "point" to a box containing food to signal to experimenters to give that food to them (the procedure being roughly the same as that described in Woodruff & Premack, 1979). An average of 428 trials were needed for the animals to learn this. The authors reported that two of the rhesus monkeys "looked at the face of the experimenter during pointing" (Blaschte & Eltlinger, 1987, p. 1521). Apparently the animals were not required to look at the experimenter's face as part of the training procedure. So, their looks at the human's face were spontaneous. Unfortunately the authors do not mention whether the looks were aggressive or ordinary.

In her comprehensive study of joint attention, O'Connell (1994) also tested seven spider monkeys. The monkeys never showed eye contact or gaze alternation in their spontaneous interactions. In an experimental, requestive situation without previous training, none of the monkeys showed pointing gestures of any kind. Only one regularly showed gaze alternation between the experimenter and the object and, according to O'Connell, these were threat stares. This contrasts with the behavior she observed in chimpanzees and bonobos in the same situations, reported above. Altogether the differences between the spider monkeys and the great apes were statistically significant.

Thus, whereas there is consistent evidence that great apes use eye contact with request gestures, the empirical evidence on monkeys' use of eye contact is too scarce to permit drawing any reliable conclusions. The fact that great apes use eye contact in their requests supports the hypothesis that eye contact is ostensive for them, but it is not conclusive evidence. Is their use of eye contact really ostensive? This I consider in the next section.

OSTENSION AND THE GREAT APES

As the reader will remember, we considered ostensive behavior to go beyond the mere act of showing something, to "showing the intention to show." I argued that establishing eye contact with the addressee was a straightforward way to make a gesture ostensive (indeed, to turn the other person into an "addressee").

As pointed out by Altmann (1967), nonhuman primates can "address" one another by means of body or head orientation. A baboon can address a threat or play gesture by means of its body or face orientation. Indeed, this system can be considered to be a primitive form of ostension (probably shared by many other vertebrates) in that it helps to show for whom a message is intended. Even in humans, physical address can be enough to make a gesture work: Beatrice's hand extended to John's pen may be enough for John to interpret that she is requesting his pen. But ostension is more than physical address. Consider, for example, that John decides to ignore Beatrice's request (perhaps because he doesn't like lending his personal belongings!). Having failed to make eye contact, she could not reproach John for ignoring her request, because she could not be sure that her request "reached" John's attention. However, if she had made eye contact with John while holding her hand to his pen, his lack of response would have amounted to a denial. With attention contact established, Beatrice's request is ostensive and it can no longer be simply ignored.

In other words, ostension through eye contact not only makes it overt that the sender is trying to communicate something, but also that the addressee has perceived this attempt. Does eye contact in apes fulfill these two functions of ostension?

Eye contact as a sign of ostensive intent

The extensive use of eye contact by apes in requestive situations cannot be considered conclusive evidence of this function, since some alternative explanations are possible. For example one reason why apes could look at the eyes of the receiver might be not to show him or her that he or she is being addressed, but to check whether attention is being directed correctly. Imagine that apes understand that for another organism to act upon an object it is necessary that its visual attention is directed to that object. Looking at the eyes of the other would then be a means to check whether this prerequisite for action is or is not fulfilled. Note that in this case eye contact would be accidental: the real aim of the sender would be to check the direction of the other's attention, and accidentally this would happen to be his or her own attention!

How can we distinguish between these two alternatives? If apes were trying to make eye contact per se, they should distinguish between the *addressing* (getting the attention of the other) and the *referential* (redirecting the other's attention to the goal) components of communication. As argued on p. 132, real ostension exists only if this distinction is made.

The data from Gómez (1992) about the existence of a repertoire of attention-claiming gestures in hand-reared gorillas (see above) can be taken to support the ostensive nature of communication in these animals. Establishing eye contact seems to be an important preliminary step in their requests; eye contact seems to be purposefully "looked for" and not a mere accidental consequence of looking into the eyes of the other.

In an experiment with six juvenile chimpanzees, J.C. Gómez, P. Teixidor & M.V. Laá (reported in Gómez, 1994b) tried to assess further their appreciation of the addressing component of communication. The chimpanzees were in a cage from which they could request food from a human sitting in front of the cage. The food was visible on a box. In normal trials the human was looking at the chimpanzee and would immediately answer any request (irrespective of whether the chimpanzee made eye contact or not). In experimental trials the human was not attending to the chimpanzee; instead she was in a variety of *inattentive* positions: body oriented to chimp, but eyes closed; back turned to chimp; eyes looking sideways; and so forth. The aim was to see if in the inattentive conditions the chimpanzees would call the attention of the human to themselves as a way of making their gestures operative. Overall the six chimpanzees claimed the human's attention before making or before repeating a gesture in about 50% of the "inattentive" trials. However, there were important individual differences. Three of the chimps called the attention of the inattentive human in only about 50% of the trials, whereas the other three did so in about 85% of the trials in which the human was inattentive. More importantly, when the experimenter turned her eyes to the chimpanzee in response to the latter's call, eye contact was established; that is, the chimpanzees seemed to be waiting for the human's gaze to be directed at them.

The individual differences found among the chimpanzees could be related to Tomasello *et al.*'s (1993; see also Chapter 17) concept of "enculturation." The three chimps who performed better were those who had a more extensive hand-rearing history with humans. The other three, though habituated to friendly contact with humans in a zoo setting, had not experienced "individualized" hand-rearing. (It is important to remember that the gorillas who developed attention-claiming procedures, as described above, were also hand-reared[3]).

Thus, there seems to be positive evidence that eye contact is really used by apes (at least, hand-reared apes) to fulfill the ostensive function of making the request overt to the addressee.

Eye contact as a sign of ostensive success

A second, complementary function of eye contact is recognizing that ostension has occurred; that is, that the addressee has noticed the ostensive effort of the sender. Observations like those of de Waal (1982) in which a chimpanzee rejects another's reconciliatory attempts by refusing to make eye contact are strongly suggestive of this.

I had the opportunity to study the reactions of one of my gorillas (Muni) to the failure of her requests (Gómez, 1992). A request may fail for two reasons: either because the addressee refuses to comply with it or because the request fails to reach the addressee. In the first case, the request has become ostensive, but it receives a negative answer; in the second, the problem is that the request has failed to become ostensive. A negative answer implies a recognition of the request (an indication that the request has reached the addressee's attention) even if it is not going to be answered. This recognition consists, first of all, of accepting attention contact with the sender, and this, once more, seems to be typically accomplished by means of eye contact paired with gestures and/or vocalizations. In this way, the sender knows that the request failed not because it didn't reach the addressee, but because the addressee did not want or was unable to answer. This information is crucial for any sender.

The gorilla I studied gave evidence of making this distinction between a failure to make attention contact and a request rejection. Before she began to use eye contact in her requests, the humans tended simply to ignore the demands they didn't want to comply with. During this period, only 35% of rejected requests were accompanied by a negative (that is, a recognition of the request plus a marking of the rejection). However, as soon as eye contact was included in the requestive strategy of the gorilla, over 95% of rejected requests were accompanied by some form of "negative." Humans felt compelled to give at least an answer to requests accompanied by eye contact. Did Muni benefit from it?

Apparently she did, because in 60% of the "negatives" she would not pursue the request further. For example:

> Muni approaches the human, takes his hand and leads him to a door, where she takes his hand towards the latch as a request to open the door. The human says "no" shaking his finger in the human way. Muni then protrudes her lips and starts emitting typical gorilla complaint vocalizations while looking at the eyes of the human. Muni does not repeat the request and goes away.

In other occasions Muni would repeat the request to a different person or would try a different strategy with the same person. What she did not do was use attention-claiming devices in those situations: she already had the attention of the human who was denying her request. Indeed, even if the final goal of a request is usually to get something done in the physical world, a rejection is something that happens in the realm of attentional and emotional exchanges – the realm of ostension. The gorilla has succeeded in her partial goal of carrying her request through to the human, but the human fails to comply. However, eye contact (and other expressive behaviors) make it clear that the request has not failed to become ostensive. The failure is elsewhere, in the human's will, and this cannot be solved through the use of attention-claiming devices.

In contrast, when requests failed because the human did not notice them, the repair procedures used by the gorillas consisted mainly of attention-claiming behaviors.

Thus, the scarce evidence available shows that at least young hand-reared gorillas and chimpanzees use eye contact to fulfill ostensive functions: they are able to separate attention-claiming from attention-directing in their requestive strategies, and show some appreciation of the difference between failing to be ostensive and failing to elicit a response.

DISCUSSION

I have argued that an apparently simple behavior – looking into the eyes of others to make eye contact – is an important index of a complex cognitive process in humans: ostension or expressing and assessing communicative intent. The ability to use specific signals for expressing communicative intent opens the possibility of engaging in inferential communication. Following the theories of Sperber & Wilson (1986), one can argue that there are two kinds of communicative system: *code* systems, based upon the use of signals with codified meanings (be they relatively fixed, as in animal display repertoires, or open and creative as in human language), and what these authors call *inferential* communication, based upon "producing and interpreting evidence." In one sense inferential communication is present everywhere in the animal world. Any behavior performed by animals (even their mere presence) is "evidence" that may affect the behavior of others, even in the absence of codified signals (Green & Marler (1979) called this kind of interaction "adventitious communication"). However, Sperber & Wilson refer to the *deliberate* use of inferential communication: almost any piece of otherwise ordinary behavior can be converted by humans into a communicative gesture. To do this, humans use a special function, ostension (or showing that one wants to show something). In my own analysis, ostension essentially consists of calling the other person's attention upon one's own attention before performing a gesture or a behavior. This will then function as a gesture. Eye contact is a privileged way to establish this attention contact necessary for inferential communication.

In this chapter I have tried to show that there is evidence that the great apes are "reaching into ostension"; that is, that they organize their interactions distinguishing between the behaviors that serve to call the attention of others and those that transmit particular information. There is also some suggestive (albeit as yet insufficient) evidence that in this the great apes may essentially differ from monkeys.

The ability to engage in ostensive behaviors is not banal. The core of ostension lies in the expression and attribution of intentions to others: making clear that one means something or inferring that others mean something. Even linguistic communication seems to be incorporated into this broader cognitive framework (it is very rare that we literally encode with words what we mean; rather we give cues and rely upon the ability of others to infer our meaning; for further discussion of this point, see Austin, 1962; Searle, 1969; Sperber & Wilson, 1986). In this sense, ostensive communication is closely related to the problem of "theory of mind" (Whiten, 1991) and most likely to the problem of language evolution. To engage in ostension one has to be capable of some degree of mindreading, including the attribution of mindreading abilities to the receiver. Thus, if great apes are capable of some forms of ostension, this would mean that they are reaching into one of the most complex aspects of human communication.

The results of our inquiry into great apes' ability for ostension are admittedly tentative. Although for many primatologists the difference between eye contact in monkeys and apes is a well-established fact of personal experience, consistent observational and experimental evidence is, as we have seen, very scarce and, in the case of monkeys, mainly limited to the rhesus macaque. An interesting point is the fact that the best evidence for ostension seems to be found in young gorillas and chimpanzees who have undergone some degree of hand-rearing. Tomasello *et al.* (1993; and see Chapter 17) suggested that hand-rearing may "enculturate" the young apes humanizing their joint-attention patterns. Do apes need the help of humans to reach into ostension? This possibility is somehow undermined, in my view, by the fact that in their natural interactions chimpanzees are reported to resort to apparently ostensive strategies. For example, the well-known deceptive behaviors observed by de Waal and others (for a catalog, see Byrne & Whiten, 1990) seem to imply a control of ostensive behaviors (usually to suppress them). However, most of these observations have been carried out on adult individuals, whereas the above experiments were carried out with juvenile subjects (4–6 year olds). This suggests that the beneficial effects of hand-rearing might consist only of an acceleration in the development of ostensive behavior which, otherwise, would eventually appear in the adult apes (see Chapter 17).

Another important point I have not discussed in this chapter is the extent to which the ostensive behaviors of apes are comparable to those of humans. It seems that even the ostensive abilities of hand-reared apes do not develop the full potentiality of ostension. A most important ostensive behavior – showing things for the sake of showing them, or what Bates et al. (1975) called "protodeclaratives" – seems to be absent in apes (Gómez et al., 1993). The ostensive abilities of great apes seem to be confined to the realm of request and control functions. Their eye contact behaviors are usually embedded in request procedures. Even the examples of "informative" pointing reported by Gómez & Teixidor (1992) in an orangutan can be interpreted as making the experimenter do what the ape wants. The absence of "protodeclarative" gestures even in highly "enculturated" apes is probably one of the main cues about the differential evolution of ostension in humankind and the great apes.

A possible speculation is that the sensitivity to gaze and eye contact apparent in many vertebrates is similar to the substrate on which selection operated in the past to build "mindreading" abilities. (These involve the adaptation to the mental processes of other individuals, and in their most sophisticated manifestation involve a "theory of mind" or the ability to represent mental states explicitly (see Whiten, 1991).) Monkeys seem to have developed an arousal reaction to eye contact that remains linked mainly to aggressive situations (although there are indications of some flexibility in this; see Gautier & Gautier, 1977; Oppenheimer, 1977). They have developed aggressive facial expressions and displays built around this sensitivity to gaze and eye contact[4]. However, it seems that at least in great apes the sensitivity to eye contact has been used by evolution to build or further develop the function of ostension. One could speculate that the richness of expressive information contained in great ape facial expressions probably promoted (or was promoted by) the liberation of eye contact from exclusively aggressive connotations. On the one hand, looking into each other's face became an important way of transmitting and gathering information. On the other hand, the rich information about ostension potentially available in eye contact was developed, and making (or avoiding) eye contact became a way to engage in or avoid interaction with others.

The ostensive use of eye contact has many implications that probably have remained "implicit" in the great ape lineage. However, the evolutionary path leading to humans seems to have developed these implications. For example, human infants produce "protodeclaratives", a behavior in which ostension is liberated from requestive goals, thereby allowing the exploration of ostension by itself (Gómez, 1994c). Humans have also developed sophisticated forms of ostension involving metacognitive processes (e.g. "Beatrice reckons that by doing gesture X John will think that she wants him to understand that she doesn't want to be with him"). It is an open question to what extent these forms of ostension are an extension of mechanisms based upon mutual attention or involve the addition of new cognitive mechanisms (for a proposal of different mechanisms, see Baron-Cohen (1995)). What is important is to bear in mind that the presence of ostensive behaviors in apes does not mean that great apes possess the underlying cognitive machinery of human ostension. A point I have defended elsewhere (Gómez, 1991, 1994a; Gómez et al., 1993) is that mutual attention, as expressed in eye contact and gaze following, allows mindreading and ostension without metarepresentational mechanisms (held by many to constitute the basis for theories of mind). The understanding of attention should not be confused with the understanding of epistemic mental states like believing or thinking, but not understanding these, in turn, should not be confused with lacking the ability for intentional, ostensive communication.

The aim of this chapter was to explore to what extent a purported behavioral difference between monkeys and apes – their reactions to eye contact – could reflect a difference in an important cognitive function, ostension. The result of this inquiry is that the great apes seem to be capable of some sort of ostensive function by means of eye contact. However, although the purported behavioral difference with the monkeys has turned out not to be sufficiently documented, there is some evidence that it actually exists (at least in some monkey species) and could be related to a lack of ostensive functions. The problem of ostension is closely related to that of mindreading or theory of mind. Ostension through eye contact seems to involve some mechanisms close to what has been described as theory of mind and it could give us a clue as to the origins of this complex cognitive ability.

ACKNOWLEDGEMENTS

The writing of this chapter and my research mentioned in it have benefited from a DGICYT grant (PB92-0143-C02-02). I am grateful to the three editors of the book for their clarifying comments on earlier versions of the manuscript. Thanks to them the final result has gained in substance and precision. Needless to say, any remaining shortcomings are my sole responsibility. S. Baron-Cohen and I have had the opportunity to discuss some of the topics treated in this chapter several times under the Anglo-Spanish Acciones Integradas Program. I am grateful to E. Sarriá, A. Brioso and J. Tamarit for sharing their ideas on the topics of our common research with me during the past few years.

NOTES

1 This interpretation is similar to Altmann's (1967) view of body orientation and gaze as "metasignals" announcing to whom a display is directed. However, the discrepancy may be due to the fact that both Oppenheimer and Altmann seem to be speaking of gaze direction, rather than eye contact.
2 Unfortunately, the role of eye contact in friendly and aggressive interactions has not been pursued by ethologists, despite the recent burgeoning of literature on the topic of primate conflict resolution (Harcourt & de Waal, 1992; Mason & Mendoza, 1993). Reconciliation (but usually not consolation) has been identified in several species of monkeys, but generally the authors don't specify whether the macaque versus chimpanzee pattern of eye contact obtains in these monkeys as well (Aureli et al., 1994; de Waal, 1993).
3 However, the amount of hand-rearing varied among them. Muni had the most extensive hand-rearing experience. I have proposed elsewhere that it is important to distinguish different degrees of "enculturation" (Gómez, 1993).
4 This part of the speculation seems to be supported by recent neurological studies showing that nonhuman primates have neurological systems specialized in detecting and processing faces. Neurons have been identified that respond to particular facial expressions or facial orientations. What is of interest to us is that other neurons have been identified whose sole specialized function seems to be the detection of gaze and visual contact. Some neurons seem to be specifically designed to detect eye contact. The origins of these systems seem to be phylogenetically old. Nonprimate mammals posses

some of these detectors (thus giving support to the idea that mammals have a special sensitivity to gaze), but it is in primates that certain cerebral systems have specialized to process the information provided by faces. Among this information, gaze and eye contact seem to be of special interest (for a review of these studies, see Grüser & Landis, 1991). Thus, neuropsychological evidence confirms that gaze information is singled out as a special kind of information in the primate brain.

REFERENCES

Adamson, L.B. & Bakeman, R. (1991). The development of shared attention during infancy. In *Annals of Child Development*, ed. R. Vasta, vol. 8, pp. 1–41. London: Jessica Kinsley Publishers.

Altmann, S.A. (1967). The structure of primate social communication. In *Social Communication Among Primates*, ed. S. Altmann, pp. 325–62. Chicago: University of Chicago Press.

Argyle, M. (1987). Eye contact. In *The Oxford Companion to the Mind*, ed. R.L. Gregory, pp. 247–8. Oxford: Oxford University Press.

Argyle, M. & Cook, M. (1976). *Gaze and Mutual Gaze*. Cambridge: Cambridge University Press.

Aureli, F., Das, M., Verleur, D. & van Hooff, J. (1994). Postconflict social interactions among barbary macaques (*Macaca sylvanus*). *International Journal of Primatology*, 15, 471–85.

Austin, J.L. (1962). *How to do Things with Words*. Oxford: Clarendon Press.

Avramides, A. (1989). *Meaning and Mind: An Examination of a Gricean Account of Language*. Cambridge, MA: MIT Press.

Bard, K.A. (1990). "Social tool use" by free-ranging orangutans: A Piagetian and developmental perspective on the manipulation of an animate object. In *"Language" and Intelligence in Monkeys and Apes: Comparative Developmental Perspectives*, ed. S.T. Parker & K.R. Gibson, pp. 356–78. New York: Cambridge University Press.

Baron Cohen, S. (1995). *Mindblindness and the Language of the Eyes*. Cambridge, MA: MIT Press.

Bates, E. (ed.) (1979). *The Emergence of Symbols: Cognition and Communication in Infancy*. New York: Academic Press.

Bates, E., Camaioni, L. & Volterra, V. (1975). The acquisition of performatives prior to speech. *Merrill-Palmer Quarterly*, 21, 205–26.

Blaschke, M. & Ettlinger, G. (1987). Pointing as an act of social communication by monkeys. *Animal Behaviour*, 35, 1520–3.

Blest, A. D. (1957). The function of eyespot patterns in the lepidoptera. *Behaviour*, 11, 209–56.

Burger, J., Gochfeld, M. & Murray, B.G., Jr (1992). Risk discrimination of eye contact and directness of approach in black iguanas (*Clerosauna similis*). *Journal of Comparative Psychology*, 106, 97–101.

Burghardt, G.M. & Greene, H.W. (1988). Predator simulation and duration of death feigning in neonate hognose snakes. *Animal Behaviour*, 36, 1842–4.

Byrne, R.W. & Whiten, A. (1990). Tactical deception in primates: The 1990 database. *Primate Report*, 27, 1–101.

Call, J. & Tomasello, M. (1995) Production and comprehension of referential pointing by orangutans (*Pongo pygmaeus*). *Journal of Comparative Psychology*, 108, 307–17.

Carpenter, M., Tomasello, M. & Savage-Rumbaugh, S. (1995). Joint attention and imitative learning in children, chimpanzees, and enculturated chimpanzees. *Social Development*. (in press).

de Waal, F. (1982). *Chimpanzee Politics: Power and Sex Among Apes*. London: Jonathan Cape.

de Waal, F. (1989). *Peacemaking Among Primates*. Cambridge, MA: Harvard University Press.

de Waal, F. (1993). Reconciliation among primates: A review of empirical evidence and theoretical issues. In *Primate Social Conflict*, ed. W.A. Mason & S.P. Mendoza, pp. 111–44. New York: State University of New York Press.

de Waal, F. & van Roosmalen, A. (1979). Reconciliation and consolation among chimpanzees. *Behavioral Ecology and Sociobiology*, **5**, 55–66.

de Waal, F. & Yoshihara, D. (1983). Reconciliation and redirected affection in rhesus monkeys. *Behaviour*, **85**, 224–41.

Diebold, A.R. (1968). Anthropological perspectives: Anthropology and the comparative psychology of communicative behavior. In *Animal Communication*, ed. T.A. Sebeok, pp. 525–71. Bloomington, IN: Indiana University Press.

Ellsworth, P.C. (1975). Direct gaze as a social stimulus: The example of aggression. In *Nonverbal Communication of Aggression*, ed. P. Pliner, L. Krames & T. Alloway, pp. 53–75. New York: Plenum Press.

Exline, R. V. (1972). Visual interaction: The glances of power and preference. In *Nebraska Symposium on Motivation 1971*, ed. J. Cole, pp. 163–206. Lincoln, NB: University of Nebraska Press. [Quoted from the 1974 reprint in *Nonverbal Communication: Readings with Commentary*, ed. S. Weitz, pp. 65–92. Oxford: Oxford University Press].

Fehr, B.J. & Exline, R.V. (1987). Social visual interaction: A conceptual and literature review. In *Nonverbal Behavior and Communication*, ed. A.W. Siegman & S. Feldstein, 2nd edn., pp. 225–326. Hillsdale, NJ: Lawrence Erlbaum.

Fossey, D. (1972). Vocalization of the mountain gorilla (*Gorilla gorilla beringei*). *Animal Behaviour*, **20**, 36–53.

Gautier, J.-P. & Gautier, A. (1977). Communication in old world monkeys. In *How Animals Communicate*, ed. T.A. Sebeok, pp. 890–964. Bloomington, IN: Indiana University Press.

Gómez, J.C. (1990). The emergence of intentional communication as a problem-solving strategy in the gorilla. In *"Language" and Intelligence in Monkeys and Apes: Comparative Developmental Perspectives*, ed. S.T. Parker & K.R. Gibson, pp. 333–55. New York: Cambridge University Press.

Gómez, J.C. (1991). Visual behavior as a window for reading the minds of others in primates. In *Natural Theories of Mind: Evolution, Development and Simulation of Everyday Mindreading*, ed. A. Whiten, pp. 195–207. Oxford: Basil Blackwell Ltd.

Gómez, J.C. (1992). *El Desarrollo de la Comunicación Intencional en el Gorila*. Ph.D. thesis, Universidad Autónoma de Madrid.

Gómez, J.C. (1993). Intentions, agents and enculturated apes. *Behavioral and Brain Sciences*, **16**, 520–1.

Gómez, J.C. (1994a). Mutual awareness in primate communication: A Gricean approach. In *Self-awareness in Animals and Humans: Developmental Perspectives*, ed. S.T. Parker, R.W. Mitchell & M.L. Boccia, pp. 61–80. New York: Cambridge University Press.

Gómez, J.C. (1994b). Some issues concerning theory of mind in development and evolution. Paper presented at the Hang Seng Center for Cognitive Studies conference on Theories of Theories of Mind. Sheffield, UK, 13–16 July.

Gómez, J.C. (1994c). Developing what is enveloped: Comments on *Beyond Modularity*. Paper presented at the Hang Seng Center for Cognitive Studies Ist Workshop on Language and Thought. Sheffield, UK, 19–20 November.

Gómez, J.C., Sarriá, E. & Tamarit, J. (1993). The comparative study of early communication and theories of mind: Ontogeny, phylogeny and pathology. In *Understanding Other Minds: Perspectives from Autism*, ed. S. Baron-Cohen, H. Tager-Flusberg & J.D. Cohen, pp. 397–426. Oxford: Oxford University Press.

Gómez, J.C. & Teixidor, P. (1992). Theory of mind in an orangutan: A nonverbal test of false-belief appreciation? Paper presented at XIVth Congress of the International Primatological Society, Strasbourg, France, 16–21 August.

Goodall, J. (1968). The behaviour of free-living chimpanzees in the Gombe Stream area. *Animal Behaviour Monographs*, 1, 161–311.

Goodall, J. (1986). *The Chimpanzees of Gombe: Patterns of Behaviour*. Cambridge, MA: The Belknap Press.

Goodenough, J., McGuire, B. & Wallace, R. (1993). *Perspectives on Animal Behavior*. New York: Wiley.

Green, S. & Marler, P. (1979). The analysis of animal communication. In *Handbook of Behavioral Neurobiology*, vol. 3: Social Behavior and Communication, ed. P. Marler & J. G. Vandenberg, pp. 73–158. New York: Plenum Press.

Grice, H.P. (1989). Meaning. In *Studies in the Way of Words*, ed. P. Grice, pp. 213–23. Cambridge, MA: Harvard University Press. (First published in 1957 in *Philosophical Review*).

Grüsser, O.J. & Landis, T. (1991). *Visual Agnosias and Other Disturbances of Visual Perception and Cognition*. London: MacMillan.

Harcourt, A.H. & de Waal, F. (eds.) (1992). *Coalitions and Alliances in Humans and Other Animals*. Oxford: Oxford University Press.

Hindmarch, I. (1973). Eyes, eye-spots and pupil dilation in nonverbal communication. In *Social Communication and Movement*, ed. M. van Cranach & I. Vine, pp. 299–321. London: Academic Press.

Kano, T. (1980). Social behavior of wild pygmy chimpanzees (*Pan paniscus*) of Wamba: A preliminary report. *Journal of Human Evolution*, 9, 243–60.

Keating, C.F. & Keating, E.G. (1982). Visual scan patterns of rhesus monkeys viewing faces. *Perception*, 11, 211–9.

Kleinke, C.L. (1986). Gaze and eye contact: A research review. *Psychological Bulletin*, 100, 78–100.

Kuroda, S. (1980). Social behavior of the pygmy chimpanzees. *Primates*, 21, 181–97.

Mason, W.A. & Mendoza, S.P. (eds.) (1993). *Primate Social Conflict*. New York: State University of New York Press.

Menzel, E.W. (1973). Leadership and communication in young chimpanzees. In *Precultural Primate Behavior*, ed. E.W. Menzel, vol. 1, pp. 192–225. Fourth International Primatology Congress Symposia Proceedings. Basel: Karger.

Menzel, E.W. (1974). A group of young chimpanzees in a one-acre field. In *Behavior of Nonhuman Primates*, ed. M. Schrier & F. Stollnitz, vol. 5, pp. 83–153. New York: Academic Press.

Nishida, T. (1970). Social behavior and relationships among wild chimpanzees of the Mahale mountains. *Primates*, 11, 47–87.

O'Connell, S. (1994). Joint attention in chimpanzees, bonobos and squirrel monkeys. Paper presented at the Hang Seng Center for Cognitive Studies conference on Theories of Theories of Mind. Sheffield, UK, 13–16 July.

Oppenheimer, J.R. (1977). Communication in new world monkeys. In *How Animals Communicate*, ed. T. A. Sebeok, pp. 851–89. Bloomington, IN: Indiana University Press.

Redican, W.K. (1975). Facial expressions in nonhuman primates. In *Primate Behavior*, ed. L.A. Rosenblum, vol. 4, pp. 103–94. New York: Academic Press.

Ristau, C. (1991). Before mindreading: Attention, purposes and deception in birds? In *Natural Theories of Mind: Evolution, Development and Simulation of Everyday Mindreading*, ed. A. Whiten, pp. 209–22. Oxford: Basil Blackwell Ltd.

Sarriá, E. (1989). La intención comunicativa preverbal: Observación y aspectos explicativos. Ph.D. thesis. Universidad Nacional de Educación a Distancia.

Savage-Rumbaugh, E.S. (1984). *Pan paniscus* and *Pan troglodytes*: Contrasts in preverbal communicative competence. In *The Pygmy Chimpanzee: Evolutionary Biology and Behavior*, ed. R.L. Susman, pp. 395–419. New York: Plenum Press.

Schaller, G.B. (1963). *The Mountain Gorilla: Ecology and Behavior*. Chicago: University of Chicago Press.

Searle, J. (1969). *Speech Acts: An Essay in the Philosophy of Language*. Cambridge: Cambridge University Press.

Sperber, D. & Wilson, D. (1986). *Relevance: Communication and Cognition*. Cambridge, MA: Harvard University Press.

Suarez, S.D. & Gallup, G.G. Jr (1982). Open-field behaviour in chickens: The experimenter is a predator. *Journal of Comparative and Physiological Psychology*, 96, 432–9.

Thomsen, C.E. (1974). Eye contact by nonhuman primates toward a human observer. *Animal Behaviour*, 22, 144–9.

Tinbergen, N. (1974). *Curious Naturalists*, 2nd edn. Harmondsworth: Penguin.

Tomasello, M., George, B., Kruger, A., Farrar, J. & Evans, E. (1985). The development of gestural communication in young chimpanzees. *Journal of Human Evolution*, 14, 175–86.

Tomasello, M., Kruger, A.C. & Ratner, H.H. (1993). Cultural learning. *Behavioral and Brain Sciences*, 16, 495–552.

Trevarthen, C. (1979). Communication and cooperation in early infancy. In *Before Speech: The Beginnings of Human Communication*, ed. M. Bullowa, pp. 321–47. Cambridge: Cambridge University Press.

Vine, I. (1970). Communication by facial-visual signals. In *Social Behaviour in Birds and Mammals*, ed. J.H. Crook, pp. 279–354. London: Academic Press.

Wada, J.A. (1961). Modification of cortically induced responses in brain stem of shift of attention in monkeys. *Science*, 133, 40–2.

White, F.J. (1992). Meat and fruit sharing in *Pan paniscus*. Paper presented at XIVth Congress of the International Primatological Society, Strasbourg, France, 16–21 August.

Whiten, A. (ed.) (1991). *Natural Theories of Mind: Evolution, Development and Simulation of Everyday Mindreading*. Oxford: Basil Blackwell Ltd.

Woodruff, G. & Premack, D. (1979). Intentional communication in the chimpanzee: the development of deception. *Cognition*, 7, 333–62.

Yamagiwa, J. (1992). Functional analysis of social staring behavior in an all-male group of mountain gorillas. *Primates*, 33, 523–44.

7

Imitation in everyday use: Matching and rehearsal in the spontaneous imitation of rehabilitant orangutans (*Pongo pygmaeus*)

ANNE E. RUSSON

INTRODUCTION

True imitation, which generates new behavior on the basis of observation rather than experience, may be the most powerful and efficient of the social learning processes. It is only one of several psychological processes that can generate imitative behavior, or reproductions, the more general product of behavior inspired by demonstrated actions. It has special status because it is viewed as involving symbolic mental processes, often deemed the exclusive province of humanity, and as having played a key role in the evolution of human intelligence (Galef, 1988; Piaget, 1952; Whiten & Ham, 1992).

These views have spotlighted great apes because, as humans' closest living genealogical relatives, they are among the most likely candidates to share this capacity. Indeed, they are the most prominent of the nonhuman species who have recently shown the capacity for true imitation (Custance & Bard, 1994; Heyes, 1993; Moore, 1992; Russon & Galdikas, 1993; Tomasello *et al.*, 1993b). These views have also favored "acid test" approaches to empirical research, primarily experimental attempts to elicit unambiguous, conclusive cases of true imitation in great apes. Such approaches, although important, have their limits: experimentally induced true imitation may bear little resemblance to spontaneous, self-selected true imitation (Lock, 1993; Speidel & Herreshoff, 1989) and unambiguous cases are rarely typical ones. A balanced understanding will require, in addition, consideration of true imitation as it is used spontaneously in everyday life.

My approach to understanding true imitation comparatively has been to identify true imitation in great apes' spontaneous behavior then compare it directly with human true imitation. For these purposes, I developed a methodology for identifying true imitation in spontaneous behavior, applied it to free-ranging rehabilitant orangutans (Russon & Galdikas, 1993), then compared the orangutan with human true imitation on several properties, including how behavior generated by true imitation matches its demonstration and when, how, and why true imitation is enacted in the normal stream of behavior. To develop methods for identifying true imitation in spontaneous behavior, I worked from experimental research on nonhuman species, where researchers have meticulously articulated many of the processes that can generate reproductions and developed methods for identifying their use (for reviews, see Galef, 1988; Moore, 1992). To identify appropriate bases for comparative analyses of the properties of great apes' true imitation, I looked to published research on humans, where properties of true imitation have been explored. The experimental research could not be used here, because it has operated from a narrow, experimentally driven view of true imitation that has not generated the empirical basis to reliably reflect these properties.

This chapter presents the comparative analyses of two facets of orangutan true imitation, match and rehearsal, and their implications for great apes' uses of true imitation. I proceed by discussing the process and properties of spontaneous true imitation, the methodology developed for identifying spontaneous true imitation, analyses of the properties of orangutan as compared with human spontaneous true imitation, and implications of the findings.

True imitation: Definitions and terms

What I call "true imitation" has also been called observational or imitative learning (Galef, 1988), level 4 imitation (Mitchell, 1987), or simply imitation (e.g. Heyes, 1993; Tomasello, 1990; Whiten & Ham, 1992; Chapters 14 and 17). True imitation is defined as one individual learning new behaviors demonstrated by another, by observation of the demonstration. It is generally agreed that the attempt to learn is purposeful and goal-directed, that observation of the demonstration is sufficient instigation for the observer's reproduction, and that learning about the form of the demonstrated behavior is involved, not merely about the demonstrator's goals or about affordances of the environment (Galef, 1988; Heyes, 1993; Moore, 1992; Thorpe, 1956; Tomasello, 1990; Whiten & Ham, 1992).

Properties of spontaneous true imitation in humans

I determined critical properties of spontaneous human true imitation from published research on imitation in young children's language acquisition. Language is not outwardly comparable to nonverbal behavior, the substance of great ape imitation, but many agree with Piaget (1952, 1954, 1962) that, at early levels, language and nonverbal behavior do not constitute modular domains but rather developmentally constructed products of a suite of relatively generalized cognitive processes such as symbolic thought, combinatorial abilities, and true imitation (e.g. Case, 1985; Gibson & Ingold, 1993; Meltzoff & Gopnik, 1989). Initially, then, language and nonverbal behavior may reflect the constraints and properties of the processes that construct them to a greater degree than they reflect the constraints and properties of their own domains. Data concur: in children under 3 or 4 years of age, language, nonverbal action, and cognitive skills are closely linked in content and organization (e.g. Bates et al., 1979; Case, 1985; Goodnow & Levine, 1973; Greenfield, 1991; Ingold, 1986; Lock, 1993; Meltzoff & Gopnik, 1989).

Properties of a process are normally identified when it is enacted in behavior. Unfortunately, imitators do not always enact true imitation, so the properties of true imitation have remained elusive. Hints as to how true imitation is likely to be enacted can be found through a consideration of the problems that face observers attempting to learn by true imitation, and some evidence on the circumstances under which imitative behavior is enacted (these circumstances include observers' efforts to further their own learning (Meltzoff, 1990; Moerk, 1989; Parker, 1993; Speidel & Nelson, 1989b) so they may apply to true imitation, one of the processes that generates imitative behavior).

Learning by true imitation is likely to require multiple observational probes and enactments because demonstrations always comprise complex arrays of behavioral elements, only some of which are essential to effective reproduction. Of the essential components alone, some are novel by definition when true imitation is involved and some may be opaque to observation (e.g. what a probe does after disappearing inside a container). Any one observation and its associated enactment are then likely to act on some subset of the demonstrated components, and the complete ensemble demonstrated will likely be only partly understood or apprehended in one encounter. Consequently, the match between any enactment and its demonstration is likely to be inexact, and rehearsals, or enactment patterns, are likely to involve multiple rather than single occurrences; others concerned with these issues agree (e.g. Moerk, 1989; Snow, 1981; Whiten & Ham, 1992). Matching and rehearsal patterns thus figure as integral parts of and cues to the workings of true imitation; they are discussed using a conceptual framework largely devised by Moerk (1989).

Match

A match, some type of similarity between observer's and demonstrator's behavior, is a defining criterion for all imitative behavior. Matching is commonly inexact so it is measured by degree – exact, reduced, or expanded (Snow, 1981). Exact matches occur when imitative behavior reproduces all demonstrated elements with no additions or omissions; reduced, when a subset is reproduced; expanded, when a subset is reproduced and independent elements are added.

The degree of match expressed can vary with observers' preferences and abilities. In true imitation, imitators actively prefer to reproduce novel behaviors performed by favored demonstrators and just beyond their own current capacities (Moerk, 1989; Snow, 1981, 1989; Speidel & Nelson, 1989b; Whitehurst & DeBaryshe, 1989; Yando et al., 1978). Imitators' abilities also constrain the degree of match achieved. All imitative behavior is generated from coded and stored input, and both encoding and storage are affected by imitators' capacities to understand the demonstrated behavior they observed. Reduced matches suggest difficulty in handling all of the information demonstrated, while expanded matches suggest advanced abilities (Moerk, 1989). Since processing abilities vary with development, the degree of match achieved may follow suit (Speidel & Nelson, 1989b). Reduced matches may be the most common at base level cognitive skills and expanded ones at more advanced levels; exact ones are rare and not necessarily the most advanced (Moerk, 1989).

Matching can vary in the level at which correspondence is achieved, ranging from molecular actions to relations between units and overall behavioral strategies (Moerk, 1989; Speidel & Nelson, 1989b). Since observers selectively imitate demonstrations that challenge their understanding, matching levels will reflect development levels (see e.g. Yando et al., 1978). At the early symbolic levels of understanding found in young children, matches would be expected beyond the level of individual actions, at the level of relational and organizational features (Meltzoff & Gopnik, 1989; Moerk, 1989). True imitation could shift the level matched across enactments, depending on which facets of a demonstration are novel or challenging to the imitator.

Finally, the time delay between demonstration and reproduction can affect the match, in part owing to the intervening memory processes. Short-term memory processes influence behavior performed up to 4–5 minutes after input ceases; they have echoic qualities. Long-term memory processes underlie behavior produced after longer delays; they code and store input in semantic terms, after interpretation. Correspondingly, immediate imitation, where reproduction and demonstration co-occur, tends to show rote-like, detailed matches with demonstrations; whereas deferred imitation, where the two are temporally separate, shows more derived, less exact matches (Moerk, 1989).

Rehearsal
Any enactment of imitative behavior constitutes rehearsal, including imitators' reproductions of demonstrations as well as of their own prior performances (self-repetition). Rehearsal entails experiential learning because changes to later enactments may be guided by experiential feedback from earlier ones; however, it can also entail true imitation when observation provides "feed forward" to later enactments. Observational material can derive from demonstrations intervening between enactments or from memory stores. Some findings bear this out: children's efforts to reproduce novel actions commonly involve several attempts, changing in the direction of what was observed (Moerk, 1989).

Rehearsal can serve learning purposes in several ways. It serves self-teaching functions by assisting retention of new information in short-term memory, allowing further analysis, and assisting transfer of new learning from short- to long-term memory, helping to solidify new learning (Meltzoff, 1990; Parker, 1993). Within interaction, it communicates an observer's understanding of a demonstration; if understanding was incomplete, this could elicit teaching. The purpose served may vary with an enactment's temporal position within the learning process: initial ones may serve retention or communication, later ones to solidify refinements or elaborations. Enactments of true imitation could be expected when the imitator aims to analyze what was newly observed more fully, to solidify new learning, or to communicate understanding of novel demonstrated actions. Amount of rehearsal per demonstration, the contexts of rehearsal, and the direction of changes over enactments should serve as indices of the operation of true imitation.

Implications

These properties support portrayals of true imitation as an active process of abstraction and reconstruction, one which may be used in conjunction with other processes and repeatedly in a learning process that can be complex and lengthy. They do not support the accepted behavioral prototype of true imitation, a rather passive and faithful echoing of an arbitrary demonstration achieved in a single, immediate attempt (Moerk, 1989).

Evidence suggests these human properties may likewise apply to great apes. Great apes' cognitive development follows the human course: they can achieve symbolic and combinatorial skills to levels similar to those of 3–4 year old human children; they are capable of true imitation; and they can attain elementary linguistic skills (Custance & Bard, 1994; Dumas & Dore, 1986; Matsuzawa, 1991; Meador *et al.*, 1987; Parker & Gibson, 1979; Premack, 1988; Russon & Galdikas, 1992, 1993; Tomasello *et al.*, 1993b; Chapter 13). Great apes match demonstrations partially and at various levels; they show similar demonstrator and action preferences; and their efforts at true imitation have involved multiple enactments changing directly towards the demonstration (Byrne & Byrne, 1993; Custance & Bard, 1994; Custance *et al.*, 1994; Meador *et al.*, 1987; Russon & Galdikas, 1995; Tomasello *et al.*, 1993a; Visalberghi & Fragaszy, 1990; Chapter 13).

Finally, these human patterns suggest that methods aimed at detecting true imitation within reproductions generated by a mix of processes may be more realistic than methods aimed at inducing or detecting reproductions based on true imitation alone. The procedures described below represent one attempt to develop such methods.

Detecting true imitation in spontaneous reproductions

In comparative psychology it has been considered that true imitation can be identified empirically only by excluding all the simpler processes that can also generate imitative behavior. True imitation is an inaccessible psychological process, the most cognitively complex of several that can generate imitative behavior in nonhuman species: it produces no accepted unique behavioral markers; it acts in isolation only in the first enactment of reproduction because all later attempts are confounded by experiential learning; and first enactments are rarely witnessed (Galef, 1988; Mitchell, 1987; Moore, 1992; Speidel & Nelson, 1989b; Tomasello, 1990). Many simpler processes that mimic true imitation have been identified (e.g. local and stimulus enhancement, social facilitation, Pavlovian conditioning, and instrumental conditioning such as matched dependent learning – see Galef, 1988; Moore, 1992), all of which generate the form of new imitative behavior by experiential learning. Experimental criteria do exist for excluding experiential learning as being responsible for imitative behavior.

In the methodology developed here, these experimental criteria were adapted for observational use by shifting the focus of exclusion and establishing appropriate research conditions. The focus for exclusion was shifted to the *sufficiency* of experiential learning for acquiring imitative behaviors rather than its *involvement*, because in everyday life imitative behavior is probably derived from a mix of processes including both true imitation and experiential learning. Appropriate research conditions are those where extensive data are available on subjects' backgrounds, because empirically establishing these criteria requires historical data. If imitative behavior satisfies any of these criteria, it could not have been acquired by experiential processes alone; then, because nonhuman imitative behavior is believed to be acquired only through experiential learning and true imitation, the acquisition processes must have included true imitation.

Caveats are, first, that this modified approach may be possible only for complex reproductions whose essence is a novel arrangement or application of already-known elements, not for the simple actions demonstrated in studies of social learning in nonhuman species or human infants (Galef, 1988; Meltzoff, 1988). This is not a serious constraint because one of the hallmarks of true imitation, from the perspective of cognitive development, is coordination among elements (Case, 1985; Fischer, 1987; Meltzoff & Gopnik, 1989; Mitchell, 1987). At complex levels, novelty is in reorganizing rules rather than individual elements (Bandura, 1986; Byrne & Byrne, 1993; Visalberghi & Fragaszy, 1990). Secondly, these criteria can not identify all cases of true imitation: they allow identification of some cases under favorable conditions. Thirdly, this approach has, like all others, a methodological bias against conspecific-demonstrated true imitation: true imitation between conspecifics favors species-appropriate behavior, so excluding experiential processes is more difficult than it is for cross-species imitation.

Excluding experiential learning
Experiential learning can be either direct or indirect that is, guided by stimuli causally relevant to the behavior acquired or by stimuli not so related. The two are treated separately.

Criteria excluding direct experiential learning are features in a reproduction that are imitative and at the same time "misfits:" they match a demonstration and they are inconsistent with direct experiential learning. I used four misfit features: arbitrariness, exceptionality, rapid acquisition of productively novel acts, and atypicality (see Table 7.1). Finding such misfit features in a reproduction suggests that processes beyond direct experiential learning were involved in acquiring them; these misfit features are imitative, so true imitation is a plausible contributing process. So, however, is indirect experiential learning. Concluding that true imitation was used requires exclusion of indirect experiential learning as well.

Indirect experiential learning, by definition, involves stimuli that lack causal relevance to the behavior acquired – here, the four misfit features. Experiential learning is known to favor conditions where stimuli and behavior co-occur and are saliently causally related or causally relevant (e.g. taste–illness associations are more readily induced than buzzer–illness ones; Dickinson, 1989). Causal relevance can be cued by contingency, contextual relevance, referentiality, modality, and domain (social–nonsocial); frequency of co-occurrence likely plays a role, but a complex interactive one (Dickinson, 1989; Pepperberg, 1988; Whitehurst & DeBaryshe, 1989). By definition, stimuli underlying indirect experiential learning lack such favorable qualities with respect to misfit features and so likely constitute relatively weak forces in their acquisition.

With this in mind, the sufficiency of indirect experiential learning for acquiring each misfit feature is assessed. For rapid acquisition, it is immediately excluded because this feature is incompatible with all experiential learning. For the other features, nonsocial

Table 7.1. *Misfit features of reproductions*

Feature	Definition
Arbitrariness	Features inconsistent with stimulus contingencies relevant to the reproduction (Visalberghi & Fragaszy, 1990; Zentall, 1988), and so not plausibly shaped by direct experiential contingencies *Ex.*: using a demonstrated technique to solve a problem when that technique is ineffective (Ladygina-Kots, 1935/1982) or when other equally effective techniques are more common (Cheney & Seyfarth, 1990; Galef, 1988; Lefebvre & Palameta, 1988; Visalberghi & Fragaszy, 1990)
Exceptionality	Features not acquired in the normal course of experiential learning in the target species and for which there is no species predisposition (Beck, 1980; Moore, 1992; Pepperberg, 1988; Thorpe, 1956; Visalberghi & Fragaszy, 1990, Walker, 1983; Wright, 1972; Zentall, 1988) *Ex.*: great apes' acquiring language or making fire
Rapid acquisition of productively novel actions	Features novel to the imitator that are acquired rapidly after observation. Speed is typical of true imitation but not experiential learning, where the typical acquisition pattern is slow and gradual (Lefebvre & Palameta, 1988; Masur, 1988; Meltzoff, 1988; Thorpe, 1956; Visalberghi & Fragaszy, 1990; Whiten, 1989). The distinction is not absolute: experiential learning can occur in one trial and imitative learning is not always fast (Moore, 1992; Zentall, 1988) *Ex.*: great apes' washing windows the first time they are offered the use of washing tools
Atypicality	Features of behavior inconsistent with an imitator's normal spontaneous behavior and independent experiential learning *Ex.*: delayed echolalia in autism and other routinized utterances that children copy as fixed units from adult speech and which show lexical, grammatical and/or stylistic levels beyond those normal for the child (these derive from deferred imitation: Ratner, 1989; Snow, 1981)

and social indirect stimuli (NIS, SIS) are treated separately. Concerning NIS, exceptional and atypical features are behaviors outside species and individual norms, so stimulus contingencies should be particularly strong to overcome whatever species or individual constraints inhibit their acquisition; and generating arbitrary features requires stimuli sufficiently salient and strong to overshadow causally related stimuli, such as task-intrinsic contingencies, and social reinforcers (Dickinson, 1989). Since NIS are intrinsically weak in relation to misfit features, they therefore seem implausible as sufficient for acquisition. Food may be an exception, although attempts to train two-way linguistic communication in avian mimids and chimpanzees using food have been less successful than attempts using social reinforcers, which derive from a more relevant domain; this has been interpreted as reflecting the relative weakness of indirect stimuli as compared with direct ones (Andrew, 1962 cited in Tharp & Burns, 1989; Pepperberg, 1988). What success was achieved with indirect food stimuli may stem from its highly systematic, social administration; both support qualities may be lacking in free-living conditions. Overall, NIS seem unlikely to generate misfit features under free-living conditions; without compelling positive evidence (e.g. involvement of food) they were therefore excluded. Excluding SIS requires showing that exceptional, atypical, or arbitrary features were not tutored, guided or shaped by social stimuli. Two conditions can show this. First, all social stimuli discouraged reproduction. This is likely when reproductions are unacceptable (e.g. orangutans starting motor boats or torching buildings) because social stimuli aim to quash learning and performance; although they may render actions or tasks more salient, they shape the form of actions with difficulty. Second, shaping by SIS is considered unlikely in nonauditory modalities (Galef, 1988).

In summary, the presence of any such misfit features in imitative behavior, along with evidence that the misfit features were not generated by social forces, excludes experiential learning as being sufficient for acquiring the imitative behavior; it implies, as the only plausible alternative, that true imitation was also involved.

The rehabilitant orangutan study
To investigate spontaneous true imitation in great apes, I chose to study a community of free-ranging rehabilitant orangutans. These rehabilitants were not behaviorally or ecologically normal but their behavior occurred freely, with few situational constraints, and in accordance with their own choices and interests. Imitative behavior produced by sophisticated orangutans should tend to replicate relational or structural features of demonstrations rather than individual actions, so the specific behaviors imitated were relatively unimportant and these extraordinary conditions were in fact valuable. This chapter's focus is the matching and rehearsal patterns in the true imitation identified in these rehabilitant orangutans.

METHODS

Subject selection and data collection procedures were identical to those reported by Russon & Galdikas (1993, 1995). These common features are sketched briefly here; more attention is devoted to methodological issues not covered in detail in these other papers.

Subjects and setting
Subjects were rehabilitant orangutans at Camp Leakey, Tanjung Puting National Park, Central Indonesian Borneo. The rehabilitation programme supports ex-captive orangutans

returning to forest life with supplementary provisioning, caregiving, and forest training. Other than those needing protective care, all rehabilitants and their offspring are free-ranging. Virtually all rehabilitants had been captured from the wild and orphaned. Their captive experiences had varied considerably: some had been freed from human captivity as early as 2–3 months of age, some as late as subadults after years of captivity; most had arrived when they were older infants or juveniles. In addition to rehabilitation to forest life, some remained at ease with camp settings; also, some of the rehabilitants' offspring, through their mothers, were at ease in the camp. I focused on adult female rehabilitants with offspring, but collected some data on peer-raised immature orphans.

Subjects were observed in Camp Leakey and in the forest. The camp comprises several buildings, occupying about 1 ha of park land on the Sekonyer Kanan river. During this study, the camp employed 10–20 Indonesian staff who handled maintenance, rehabilitants' care, data collection on wild orangutans, etc. The forest around the camp is mostly lowland dipterocarp and peat swamp forest interrupted by a few abandoned rice fields; a trail system transects a formal 35 km^2 study area (Galdikas, 1988).

Data collection

Behavior was sampled using the focal subject method in the 1989 and 1990 dry seasons (June–August). Focal subjects were followed for half-days in 1989 (waking to leaving camp; entering camp to nesting) and full days in 1990 (waking to nesting; over three consecutive days where possible). I conducted most follows assisted by Indonesian field workers and, in 1990, by students and Earthwatch research volunteers. Raw data were collected by continuously recording the focal orangutan's behavior during the follow, concentrating on reproductions and indices of true imitation.

I took systematic video samples, which together with ad lib reports from B. Galdikas, knowledgeable staff, and myself, supplemented data from the follows. Reports confirmed as reliable provided additional reproductions and information on subjects' histories (Russon & Galdikas, 1993, 1995). The full database encompassed all reproductions where the focal orangutan was either demonstrator (one who independently performs behavior observed by another) or imitator (one who observes a demonstrator's behavior then reproduces it). Data for each reproduction included imitator and demonstrator identification (name, age, sex, species, demonstrator's relationship to the imitator) and descriptions of both demonstrated and reproduced behaviors. After a reproduction was identified, we sometimes collected additional information about it through further observation, published literature, or staff reports.

Data coding

The empirical criteria used to identify reproductions and true imitation and to assess match and rehearsal are given below.

Reproductions

Imitative behavior, or a reproduction, occurs when an observer's behavior matches a demonstrator's behavior and it is contingent on having observed the demonstrated behavior. Empirically establishing a match requires showing that an observer's behavior replicates some features of the demonstrated behavior (Hall, 1968; Nadel, 1986). Establishing contingency requires showing that the observer's behavior did not occur spontaneously, independent of the demonstration: observational criteria are appropriate temporal contiguity between the observer's and the demonstrator's behavior and several measures of the observer's behavioral and social history prior to the demonstration (for details, see Russon & Galdikas, 1993, 1995).

Rehearsal and match considerations predict multiple enactments of demonstrated actions. Multiple enactments did occur, often in a concentrated bout, suggesting that the imitator was making an integrated, sustained effort at reproduction. I therefore segmented each interconnected set of enactments, not each single enactment, as one reproduction. I considered that a reproduction had ended when the imitator shifted the focus of behavior for more than 10 minutes, either changing the content of the imitative behavior or changing to nonimitative behavior. Time delay affects the match produced so I distinguished immediate from deferred reproductions as those enacted within 15 minutes of demonstration and within the same social and physical context versus those enacted after longer delays and under altered social and physical contexts.

True imitation

Empirical criteria covered the four misfit features (see Table 7.1) and the imitator's learning history.

Arbitrary features included standardized or stylized but nonessential techniques, model-consistent problem-solving errors, and "empty" or pretend performances that lacked functionality. Exceptional features included actions outside the range of species-typical repertoires, based on published reports (i.e. the action occurred not at all or rarely in normal orangutan behavior, or it was inconsistent with normal orangutan behavior) and actions appearing only after demonstration despite prior opportunities for independent learning. Novel features were actions that no knowledgeable humans had observed the imitator perform prior to demonstration (assessment required extensive knowledge of the imitator's history); rapid acquisition was identified by direct observation. Atypical features were facets of a reproduction that were unlike the imitator's normal behavior, based on reports by knowledgeable humans about the imitator's typical behavior.

The learning history data used were those needed to assess the role of social stimuli in acquiring reproductions, namely the occurrence and valence of current and prior social stimuli associated with a reproduction. These data came from observations and reliable reports by knowledgeable humans.

Matching

I assessed the degree of match by comparing the reproduction to an "essential" version of the demonstration, i.e. those actions and structures judged essential to achieving the demonstration's ostensible goal. For deferred imitations, this was possible when the demonstration involved camp traditions, i.e. standardized techniques and behavioral routines characteristic of, and peculiar to, the camp. According to the views of Moerk

(1989) and others, the degree of match was *exact* when all essentials were duplicated; *reduced* when some essentials were omitted, or present but ineffective; or *expanded* when some essentials were duplicated and new elements were independently contributed by the imitator. Expanded imitations were divided into *substitutions* (most elements duplicated, few elements independent) or *chaining* (most elements independent, few elements duplicated).

I examined the level of match by reinterpreting analyses of competence indices in the actions orangutans selected to imitate (Russon & Galdikas, 1995) and analyzing patterns occurring in substituting, chaining, and errors.

Rehearsal
I defined rehearsal as the enactment of imitative behavior derived from one demonstration, including reenactments. I treated intrinsically repetitive activities, such as sweeping or sawing, as one enactment unless individual replications were systematically and individually varied. I noted time lapse, intervening demonstrations, and three distinct rehearsal patterns with implications for the learning purposes served (isolated rehearsal, self-repetition, and recurrent rehearsal). I noted: isolated rehearsal when I observed a single enactment; self-repetition when an imitator replicated its own behavior, enactments were explicitly varied, and at least some enactments occurred without intervening demonstration; and recurrent rehearsal for repeated enactments of a demonstration, either across or within reproductions.

Learning purposes of enacting true imitation
I examined the ensemble of matching and rehearsal patterns in each reproduction that showed an influence of true imitation for the possible goals of orangutan true imitation. I considered each of the purposes suggested earlier, retaining new material in memory, communicating understanding, and solidifying refinements or elaborations within ongoing learning.

RESULTS AND DISCUSSION

A total of 395 hours of direct observation plus supplementary material was collected, documenting 354 reproductions by 26 individual orangutans. Using the above criteria, 54 of 354, plus 5 incidents identified from data collected in 1991–93, were identified as involving true imitation (TI) (Russon & Galdikas, 1993).

Table 7.2 summarizes the assessments of degree of match and rehearsal for these 59 TI cases. In the discussions which follow, case numbers correspond to numbering in Table 7.2. It should be kept in mind that these TI cases do not necessarily show true imitation in action. Probably some do, but most are products of mixed attempts at imitative, experiential and insightful learning in the past. Also, quantitative measures are biased: sample size per orangutan varied with accessibility, and there is an inherent methodological bias against true imitation demonstrated by orangutans.

Degree of match

Exact matches

Only one TI case was classified as an exact match, a case of immediate imitation (no. 1).

No. 1. Volunteers and I were sitting on the bunk house porch with Siswi, the adult daughter of a rehabilitant. Siswi stole a plastic bottle of insect repellant from a volunteer (JS) and squirted it on the porch floor. Several volunteers dabbed fingers in the spilled repellent and rubbed it on themselves. JS, who was sitting on the porch railing beside Siswi, did so too, rubbing it onto her own knee. Siswi watched, then dabbed her own fingers into the spilled repellant and likewise rubbed it onto JS's knee.

Reduced matches

Thirty-five TI cases were classified as reduced matches. Of these cases, 16 were very close to being exact matches. The techniques and structures replicated were relatively simple in all but two of these. Their content ranged from simple coordinations such as sawing wood or shading eyes against the sun to lengthy routines such as putting on a T-shirt or brushing teeth. All 16 cases matched their demonstrations virtually exactly in their superficial behavioral forms but only partially achieved demonstrated goals (e.g. an adult female, Supinah, hammered nails, but not hard enough to embed them in wood; she swept paths, but did not clean them of debris; an adolescent female, Davida, reproduced the camp routine for brushing teeth in all visible details, but probably did not effectively clean them).

Three patterns of reduction were identified: poor differentiation of essential elements from a demonstration (20 cases e.g.: no. 2, rubbed a sharpening stone anywhere on a blade versus along its edge; no. 21, swung a hoe so that it chopped into versus scraped the earth); omissions of elements essential to complete the demonstrated routine (23 cases e.g.: no. 2, did not wet the sharpening stone before rubbing a blade with it; no. 21, did not target weeds when hoeing); and poor organization (1 case: no. 30, Siswoyo tried to insert a siphon into a fuel drum before uncapping the drum; she corrected her error in a later attempt).

All 35 reductions suggested poor comprehension or poor skills, but other explanations were sometimes possible. Rehabilitants may have omitted elements because materials were not available at the time (e.g. no. 27; no shampoo was available when Pegi washed her hair) or because their goals differed from those of the demonstrators (e.g. no. 31, Supinah locked her infant daughter in a cage like human demonstrators did, but probably for different reasons).

Expanded matches: Substituting and chaining

Substituting expansions were found in 16 TI cases. All were more complex than the cases classified as exact or reduced. Commonly, orangutans substituted either the tools or the techniques they used to achieve demonstrated goals (e.g. a stick instead of a hoe, a piece of bark instead of a plate, and boards instead of canoe paddles; they removed weeds by pulling instead of chopping them, and fastened rope by winding and threading instead

Table 7.2. *Reproductions suggesting true imitation*

No.	Imitator[a]	Content sketch	Match[b]	Rehearsal[c]	Delay[d]
1	SW	Apply insect repellant to volunteer's knee	X	I	I1
2	SP	Sharpen axe blade (wet stone, rub it on blade)	RX	R S	I1
3	UN	Saw or scrape wood, cut hair, as with a knife	RX	I S	I1
4	SS	Saw wood	RX	I s	I2
5	SP	Sand blowgun darts	RX	R s	I2
6	SP	Saw board	RX	R S	I2
7	DA	Brush teeth (two sides), spit foam over porch edge	RX	I	D
8	PS	Sweep path with twig broom, realign bristles	RX	I s	D
9	SW	Shade eyes with hand when looking into sun	RX	R	D
10	??	Ring dinner bell, using conventional stick	RX	R	D
11	SP	Sweep porch with household broom	RX	R s	D
12	SP	Sweep floor with household broom	RX	R s	D
13	SP	Sweep path with twig broom, realign bristles	RX	R s	D
14	SW	Shade eyes with hand, when looking into sun	RX	R S	D
15	SP	Nail two boards together, hammer with third board	RX	R S	D
16	SP	Chop wood with a hatchet	RX	R S	D
17	DA	Put on a T-shirt	RX	I S	D
18	??	Hoe ground (with hoe or shovel)	R	R s	I1
19	SP	"Examine" a sick infant, like a nurse	R	I	I2
20	HB	Tamp and wiggle a grave marker	R	I s	I2
21	PS	Remove weeds from path with hoe	R	I S	I2
22	SP	Sand blowgun darts	R	R s	I2
23	SP	Hold burning stick to cigarette butt, twice	R	I	D
24	PS	Hold Rambo toy gun like a machine gun	R	I	D
25	PS	Wipe whole surface of ladder treads	R	I	D
26	SP	Hammer a grave marker into the ground	R	I	D
27	PG	"Shampoo" hair (splash water on, then rub wet hair)	R	I	D
28	PS	Apply insect repellant on foot and shin	R	I S	D
29	UN	Hang up hammock (wrap, fasten), insecurely	R	I S	D
30	SS	"Siphon" fuel from capped fuel drum	R	I S	D
31	SP	Put daughter in cage, lock it, wipe it with towel	R	I S	D
32	??	Start motor boats (unsuccessfully)	R	R	D
33	??	Wipe perineal area with leaves	R	R	D
34	TT	Hammer newly built wooden platform with a stick	R	R s	D
35	TT	Hammer one piece of wood with another	R	R s	D
36	SP	Pour fuel on burning stick	R	R S	D
37	PR	Bend sugarcane to break and squeeze out juice	ES	I	I1
38	SS	Chop weeds at path edge, collect them in a row	ES	R s	I1
39	SS	Chop weeds at path edge, move them aside	ES	R s	I1
40	PS	Saw and scrape wood, file nail, as with a knife	ES	I S	I1
41	SP	Hammer nails into boards	ES	R S	I1
42	SP	Saw boards	ES	R S	I1

Table 7.2. (cont.)

No.	Imitator[a]	Content sketch	Match[b]	Rehearsal[c]	Delay[d]
43	SW	Rehang a hammock (hook), ride in it	ES	R S	I2
44	??	Build log bridges, cross the river on them	ES	R S	I2
45	??	Put rice on a bark plate, offer to Galdikas	ES	I	D
46	SP	Apply insect repellant on head, arms, legs	ES	I S	D
47	PS	Unlock door bar-lock, after making a probe-key	ES	R	D
48	??	Take canoes for rides	ES	R	D
49	??	Pull up cassava-like plants	ES	R	D
50	SP	Wipe dry paint on porch and wall (recently painted)	ES	R S	D
51	SP	Wipe dry paint on porch (recently painted)	ES	R S	D
52	SP	Saw a beam with a two-handle saw	ES	R S	D
53	SP	Paint porch and shelves, wipe spilled paint	EC	R S	I1
54	SP	Pour drink into a bottle, shake it, then drink it	EC	I	I2
55	SP	Rehang a hammock (wrap and fasten), then ride in it	EC	I S	I2
56	SP	Bale water from dugout by rocking it, then ride it	EC	I S	D
57	SW	Rehang hammock (hooking), then ride in it	EC	R S	D
58	SP	"Siphon" from an empty fuel drum to a gerry can	EC	R S	D
59	SP	Fire-related activities	EC	R S	D

[a]Imitator: DA, Davida; HB, Herbie; PG, Pegi; PR, Prince; PS, Princess; SS, Siswoyo; SW, Siswi; SP, Supinah; TT, Tutut; UN, Unyuk; ??, unidentified.
[b]Match: X, exact; RX, reduced, almost exact; R, reduced; ES/EC, expanded substitutive/chaining.
[c]Rehearsal: I, isolated; S/s, self-repetition with/without variation; R, recurrent.
[d]Delay: I1, immediate within 4–5 min.; I2, immediate within 5–15 mins.; D, delay.

of knotting it). Some substitutions ($n = 8$), like reductions, suggested poor comprehension of, or skill with, demonstrated techniques (e.g. rehabilitants seemed unable to tie knots, so they substituted other fastening techniques). Other substitutions ($n = 9$) showed considerable innovation (boards for canoe paddles, hoes, or hammers) that was sometimes unavoidable because the demonstrated tool or technique was inaccessible (e.g. in weeding, the human demonstrator had the only hoe). All substitutions were nevertheless appropriate and so suggest considerable cognitive sophistication: to make these substitutions, imitators had to analyze goal and subgoal structures of demonstrations, identify problematic components, and generate effective alternatives.

Chaining expansions were found in seven cases. Supinah performed six of them. The case showing the greatest proportion of independent over imitative contribution was no. 56, Supinah's use of a canoe (for a description see Russon & Galdikas, 1993). Her goal and behavioral strategy seemed completely independent. Her goal appeared to be stealing soap and laundry from workers washing clothes on a raft moored off the camp dock. Her strategy involved preparing and riding a canoe alongside the dock, bypassing the guard stationed on the dock near the raft (against the very possibility of her harassing workers), until the canoe reached the raft. She then jumped onto the raft, scared the workers away, took their soap and laundry, and began washing clothes herself. What she duplicated were standardized camp techniques for two subgoals: removing water from the canoe by rocking it from side to side, and washing laundry by dunking clothes into water, rubbing soap on them then scrubbing and wringing them. She expanded even these (e.g. she stopped rocking the canoe, looked at the water left in it, then resumed rocking it until more water was sloshed out). These chaining cases suggest even greater cognitive sophistication than the substitutions do: imitators not only reproduced demonstrated techniques but also identified them as organized units, reintegrated them into independently structured strategies, and sometimes modified them to suit immediate circumstances.

Level of match

Earlier analyses relevant to the levels at which matches were achieved (Russon & Galdikas, 1995) indicated that TI cases were constructed on the basis of a limited repertoire of basic elements, many of which were reused. These elements reflected two cognitive levels, actions on objects (e.g. bend, collect, scrape) and object–object relations (e.g. pour, thread, hook over, scrape, hammer or pry with a tool). The limited repertoire and the reuse showed that the elements themselves were not the new behaviors acquired imitatively. What was unique to each case, and so probably was acquired imitatively, was the particular arrangement of these elements or their application. For example, many TI cases involved routines centered on rubbing or scraping one object against another (nos. 53, 2, 40, 3, 22, 11–13; see Russon & Galdikas, 1995). By varying the pairs of objects used and adding refinements, very complex routines were generated (e.g. no. 7).

No. 7. Davida, an adolescent female rehabilitant, came to the bunk house porch one morning about the time when people came out to wash. This was a favorite haunt of hers, probably because of the attention. A visitor gave her a toothbrush with toothpaste on it. Davida and others had in the past stolen toothbrushes or toothpaste but rarely both together, especially combined this way. She put the brush in her mouth, nibbled the paste, then brushed: with the brush in her fist, she inserted its bristle end in one side of her mouth, closed her lips around it and scrubbed the bristles back and forth, horizontally and audibly, over her teeth and/or tongue. She then twisted the brush to the other side of her mouth and likewise scrubbed. When she finished brushing, she climbed onto the porch railing, spit the used toothpaste over the railing onto the ground, then moved off. Her technique of brushing, including spitting used toothpaste over the rail onto the ground, was identical to the standardized technique used by camp visitors staying at that house.

The level at which matching occurred was also indicated by patterns of substituting, chaining, and errors. Substitutions and chaining involved choice of tool within an established technique or choice of technique within a complex routine; this suggests that neither basic elements themselves nor the surface forms of techniques were imitated, but rather their instrumental and structural parameters. Errors, where a reproduction did not achieve the demonstrated goal, commonly comprised poorly differentiated but correct techniques or poorly skilled performance (e.g. no. 47, confusing which direction to slide a bar to unlock a door; no. 29, wrapping and fastening tie ropes did hang a hammock but not securely enough for riding). Poor differentiation reflects conceptual limitations and suggests matching at the level of overall strategies or structures.

These patterns converge in pointing to matching aimed at the level of structures or of strategies for combining elements more than at the level of molecular elements like the detailed form of actions.

Rehearsal

Isolated rehearsal

Isolated rehearsal characterized 14 TI cases. This is probably an overestimate owing to sampling errors and/or poorly detailed data. Of these 14 cases, five were immediate imitation that occurred during interaction or parallel activity with the demonstrator, or immediately after a commotion surrounding the demonstration. The nine deferred cases ranged from sweeping with a twig broom (nos. 8 and 13) or hammering a marker (no. 26) to lighting a cigarette butt with a burning stick (no. 23); the most complex was Davida's toothbrushing routine (no. 7). Isolated rehearsal cases were characterized by (a) enactment triggered by reencountering stimuli relevant to the routine, not initiated spontaneously and (b) no evident purpose beyond replicating the routine (e.g. in brushing her teeth, Davida did not likely aim to clean them).

Self-repetition

Self-repetitions occurred in 28 cases; 11 of 28 occurred within single reproductions, 17 of 28 recurred across reproductions.

Six cases involved immediate imitation, during interaction with the demonstrator and within 5 minutes of demonstration; all involved intervening demonstration (e.g. no. 3).

No. 3. Russon and a field worker, Ucing, were following an adult female Unyuk. Ucing began modifying a walking stick with a Swiss army knife: he sawed off its end and scraped off its bark. Unyuk watched intently then imitated his sawing and scraping, using one stick moved against a second. Probably due to her interest, Ucing opened scissors and snipped them near her forehead, pretending to cut her hair. She winced as if afraid but did not leave. The second he withdrew the scissors, she looked at him, grabbed her own head hair in her right fist and both her sticks in her left fist, and made a cutting motion with the sticks across and against the fistful of hair. In response Ucing resumed snipping near her forehead; she winced but held her head near the scissors while he snipped. When Ucing paused again, she again grabbed her head hair and made a cutting motion with a stick against it, looking intently at the ground; Ucing resumed play-snipping, Unyuk held her head to the scissors and held still while he snipped. Twice more, when Ucing paused, Unyuk grabbed a fistful of hair on her arm and leaned it towards him; Ucing resumed snipping near the spot she indicated, whereupon she sat still and watched the snipping intently.

In five of six of these interactive cases, later repetitions more closely matched the demonstrated actions than initial ones (e.g. no. 2). In no. 2 and one other case, improvement was consistent with observation of intervening demonstrations, but not practice. This suggests that subsequent attempts were directed on the basis of observation and not experience, a key criterion in identifying symbolic level imitation (Meltzoff & Gopnik, 1989).

No. 2. Supinah watched a worker sharpen an axe blade: he dipped a stone in a pail of water then rubbed the stone back and forth along the blade's edge. After a minute he offered her the blade and stone and gestured to her to do the same. She accepted them and rubbed the stone against the blade's face. The worker retrieved his tools and resumed work; Supinah resumed watching. Five minutes later she motioned for the blade and stone, and the worker gave them to her. She took them, but this time retrieved the pail of water and dipped the stone into it before rubbing the stone against the blade's face.

The clearest example of self-repetition in deferred imitation was Davida's reproduction bout with 12 attempts at putting on a T-shirt (no. 17). Each attempt, (1) to (12), was separated by removing the shirt.

No. 17. Davida had a T-shirt, probably pilfered, and carried it up a tree. Hanging by one arm, she (1) put the other arm through arm and neck holes; (2) pulled the shirt over her head via the body hole and put that arm and her head through the head hole; (3) put that arm through both armholes, like a pole; (4) pulled the shirt over her head through the body hole, poked that arm and her head through the head hole, poked the other arm through one armhole, and pulled the shirt down over her midriff; (5) to (9) repeated (4); (10) pulled the shirt around her midriff through body and head holes, shirt upside down; (11) repeated (4); (12) climbed into the shirt though body then head holes with the shirt upside down, then pulled it all around her waist.

Although the variations between attempts could stem from trial-and-error, they follow a systematic time-related pattern in their match with the demonstration: first reduced, then closer to exact, finally more expansive and creative.

Recurrent rehearsal
In 34 cases, attempts at replication recurred across reproductions: 17 also showed imitative self-repetition within reproductions (in most others it was likely but poorly documented). The greatest contributor was Supinah (18 cases). Her most elaborate series involved six attempts to hammer nails into wood over 10 days (no. 41). She was not followed on 5 of the 10 days so some attempts may have been missed.

No. 41. 11.07.90, 14:25: Staff were making a bench near staff quarters using saws, hammers, boards, and nails. Supinah joined in, hammering nails with a saw. She took a saw, nails, and a board under the building, then positioned a nail properly on the board and hammered it with the saw. She pushed the nail in with her mouth, then her hand, then hammered it with the saw. She fled to the dining hall with her nail when Siswoyo, a dominant female, approached. There she found a new saw, reassembled her materials, moved under the dining hall with them, and resumed hammering nails into boards.

12.07.90, 8:08: Supinah went under the staff building, retrieved boards, picked up a nail a few meters away, then used one board to hammer the nail into two others she placed together. She shifted boards around, separated them, and hammered the nail again. She removed the nail with her mouth, repositioned it and hammered it again, but this knocked it over. She pulled the nail out, fitted it carefully back into its small indentation and hit it with the biggest piece of wood, fitted it again into its indentation then into a new hole. She finally left with the nail in her mouth.

12.07.90, 9:08: Supinah went under the dining hall, retrieved a hatchet, began chopping wood, then found a nail. She put the nail into the hatchet's handle very close to the blade, then removed it. She put the nail into an old stump, pushed it hard with both hands, and hammered it in with the axe blade. She poked the nail into the ground and pushed it in, removed it, poked it into the stump then the hatchet handle then the stump again, and hammered it with the flat of the axe blade. This bent the nail. She inspected it closely, put its point into her mouth, then holding it upright, hammered it gently with the edge of the axe. She finally abandoned her assembly.

13.07.90, 10:09: Supinah was near the staff quarters, found three boards, and hammered apart a rotten piece of wood with her largest board.

13.07.90, 11:15: Supinah went under the staff building, got five pieces of wood, placed two pieces together and hammered them with a third.

20.07.90: Supinah found a board with a nail in it, pulled the nail out, put it in a groove in a second board and hammered it with a third board.

In cases of recurrent rehearsal, enactments matched demonstrated techniques (e.g. hammer nails into wood with a tool) and demonstrated goals (e.g. embed nails, by whatever technique). Rehabilitants often persisted in attempting demonstrated techniques even in the face of failure to achieve their ostensible aims, apparently discounting much experiential feedback: for instance, Supinah repeatedly attempted to hammer a nail into wood despite persistent failure; she similarly repeated failing attempts at reproducing removing paint, making fire, and siphoning fuel. In some cases reenactment was triggered by reencountering relevant stimuli, but in others the imitator independently recreated the essential stimulus complex apparently for the express purpose of reenactment: for instance, Supinah reassembled a carpentry kit, collecting essential items from various places and transporting them to a convenient work spot, in order to resume hammering nails into wood. The rehearsals appeared more strongly guided by internal images than by external circumstances; although apparently serious, they thus resembled pretend play (Mitchell, 1994).

Properties of recurrent rehearsal were: (a) bouts recurring over hours or days, many without intervening demonstration; (b) no obvious purpose beyond replicating the observed strategy and its immediate outcome (e.g. no. 59 make a fire, or no. 41 hammer nails using camp techniques, with no sign that either would serve any further purpose); and (c) extensive imitative self-repetition, with explicit variation between attempts.

Purposes of orangutan true imitation

The TI cases probably constituted later rather than initial enactments, so their features were probably products of imitative learning that had occurred earlier. The possibility remains that some of these cases themselves further learning, so I explored their possible role in the learning process.

Cases showing isolated rehearsal probably did not serve learning. They tended to be triggered by encountering relevant stimuli and they lacked apparent functional value for the imitator. They probably acted to express existing understanding or recognition, in line with Piaget's (1952) and Werner's (1957) notions of knowing through action (Piaget's "recognitory assimilation"), that perceptual events may not be fully known without the actions they afford: when events are encountered, acting on them is part of understanding or recognizing them (see Meltzoff & Moore, 1992). When understanding is derived from observation rather than direct experience, actions may be imitative. This use of imitative behavior occurs in children under 4 years old (Meltzoff & Moore, 1992; Speidel & Nelson, 1989a), so it is plausible at the cognitive levels found in great apes.

True imitation enacted for the purpose of retaining new information in short-term memory would involve reproductions enacted within 4–5 minutes of demonstration, reduced matches (since new information is unlikely to be completely understood), and attention focused on the actions duplicated. Such attempts were found in no. 3, in Unyuk's several attempts at sawing then scraping a stick and her second attempt at snipping hair. The field worker demonstrated each of these repeatedly and within seconds of his pausing the demonstrations, Unyuk attempted to reproduce them; her matches were reduced, and her attention focused closely on her actions (her expression was intent and slightly puzzled, and her mouth dropped half open).

True imitation enacted for communicative purposes would involve cases of reduced immediate imitation, during interaction with the demonstrator, and with attention to the demonstrator. Evidence was found in no. 3 again, in Unyuk's first and several later enactments of snipping hair. In the first, as soon as the field worker stopped snipping near her forehead, she simulated a portion of the demonstration while looking intently at him; he immediately resumed play-snipping and she monitored intently. Later, when he paused snipping, she grabbed a fistful of the hair on her arm, held it towards him while squeaking quietly at him, then watched intently as he resumed play-snipping.

Learning can be refined or elaborated through rehearsal if enactments are explicitly varied; if variation is derived from observation versus experience, rehearsals involve true imitation. Later versus first enactments probably involve learning if rehearsal involves multiple attempts that are massed versus spaced; they probably also involve true imitation if variations between enactments add new actions that are attributable to observation and not experience. Ten TI cases showed all these qualities. In addition, in some the variation between attempts seemed to actually discount experiential feedback. A simple example is no. 1: Supinah's second attempt at sharpening an axe blade differed from her first one, 4 minutes earlier; she added wetting the sharpening stone before rubbing it on the blade. Her change better matched the demonstration; it had no evident basis in experience.

CONCLUSION

This methodology for identifying true imitation differs from traditional comparative psychological approaches. Its aim is to identify any effects of true imitation in spontaneous reproductions rather than "pure" expressions of true imitation. As in other approaches, it identifies true imitation by excluding simpler processes; it differs in that exclusion concerns whether simpler processes were sufficient for acquisition not whether they were used at all, and exclusion rests on several converging criteria, not on a single one. Use of this approach allowed effects of true imitation in the reproductions performed by rehabilitant orangutans to be identified.

In these orangutans, true imitation was found in reproductions wherein the degree of match was more commonly reduced or expanded than exact. The nature of the match could vary systematically with rehearsal. As in young human children, true imitation in rehabilitant orangutans seemed to operate on fragments of demonstrations rather than their totality. The orangutans' fragmenting showed they effectively segmented the demonstration into organized routines that represented techniques or subgoals. Two types of expansion considered, substitutions and chaining, suggested orangutans analyzed and reconstructed these fragments at sophisticated cognitive levels. The level matched more commonly reflected structural than surface features of demonstrations, supporting suggestions that true imitation may more commonly target relational or "program" levels than behavioral details (Byrne & Byrne, 1993; Visalberghi & Fragaszy, 1990).

Rehabilitant orangutans may have been using true imitation repeatedly as they acquired demonstrated routines. Repeated rehearsal was common in the TI cases and some of the later enactments appeared to advance learning on the basis of observation rather than experience. Improvements in imitative matching as a result of imitative rehearsal have also been documented in other orangutans and chimpanzees. Chantek's attempts to reproduce stomping on a stair (see Chapter 13) and 4 year old chimpanzees' efforts to imitate novel gestures (Custance et al., 1994) show patterns very similar to those found in the rehabilitants and reported for human children.

Analyses of the potential learning purposes of TI cases suggested that true imitation in these rehabilitant orangutans may serve learning in a variety of ways. They have implications for current discussions on the significance of one form of true imitation, impersonation. Impersonation involves reproducing demonstrated techniques but not corresponding demonstrated goals; it is also termed "mimicking," carrying connotations of rote, inferior understanding like that of mimids who slavishly copy speech (e.g. see Tomasello, 1990 after Wood, 1989; Tomasello et al., 1993b; Chapter 17). Impersonation was found in orangutan TI cases, but it suggested different interpretations. When orangutans impersonated, reproducing demonstrations of no obvious use such as placing grave markers, locking their infants in cages, or sharpening blowgun darts, they selectively imitated demonstrators who were preferred social partners. By impersonating, they achieved participation in the demonstrator's social circle and increased their own resemblance with the demonstrator. Orangutan impersonators may simply have had goals different from those of their demonstrators, such as these interpersonal ones. Interpersonal goals have been proposed to represent a fundamental function of human imitation, of equal importance to its learning function (Užgiris, 1981; Yando et al., 1978). Piaget (1962) reported human toddlers "becoming" the demonstrator by imitating his or her actions as early as 2 years of age (e.g. crawling on all fours saying "meow," strutting around in imitation of a favored cousin while describing the imitative actions out loud as the *cousin's* actions). Alternatively, some orangutan impersonations may have constituted active attempts to understand demonstrated actions (e.g. when they enacted reproductions as though analyzing or consolidating new information) or requests for teaching (e.g. when they addressed imitative behavior to the demonstrator during interaction). Such impersonation does show limited understanding of demonstrated actions but it suggests active attempts to further understanding rather than slavish mimicking. Both latter uses of impersonation are consistent with suggestions that learning in great apes may be importantly handled by apprenticeship, and that imitation serves self-teaching functions (e.g. Parker, 1993; Chapters 16 and 18).

It is important to revisit the notion of "first" attempts at reproduction, because first attempts are considered critical in identifying the presence of true imitation. The arguments and findings here suggest that true imitation tends to target fragments of demonstrations, serially. For a naive observer, any demonstration constitutes an ensemble of fragments and its acquisition by true imitation seems to require multiple observational probes and enactments. The question of what constitutes the first attempt is thus complicated, because the first attempt at a given fragment may not occur until after many attempts at reproducing other fragments of the ensemble. The first attempt at reproducing any aspect of the ensemble is the one that has been traditionally used, but findings from this study are that it may actually represent a minor and rather atypical component of the process of learning by true imitation.

My sense is that this spontaneous imitation in rehabilitant orangutans offers a rich basis for exploring true imitation, and that noninterventionist studies of spontaneous imitation occurring in everyday practice are essential to a balanced understanding of true imitation. These observationally based findings on how imitation plays a role in orangutans' learning should also be considered in the light of their striking resemblance to patterns reported for human children functioning at comparable cognitive levels. The advantages gained by unshackling behavior from controls and constraints far outweigh their costs when the issues under question concern how behavior reflects cognition as it operates in everyday life.

ACKNOWLEDGEMENTS

Research for this study was funded by York University Research Funds through Glendon College and the National Sciences and Engineering Research Council (NSERC) of Canada, and by individual NSERC operating grants. Sponsorship by the Perlindinguan Hutan dan Pelestarian Alam Directorate of the Forestry Department, The Indonesian Sciences Institute (Lembaga Ilmu Pengetahuan Indonesia), and Fakultas Biologi, Universitas Nasional, made the work possible in Indonesia. Assistance from many individuals contributed to realizing the empirical work underlying this chapter. First and foremost, thanks to Dr B.M.F. Galdikas, who was an invaluable collaborator on other aspects of this work, developed the rehabilitation programme, provided the facilities for the study, and provided information and support in innumerable ways. Thanks also to others who assisted the study in Indonesia, including the Orangutan Research and Conservation Project staff, W. Russon, D. Wahyudi, V. Gobeil, J. Fitchen, C. Russell, Earthwatch volunteers, and Pak Doran and Pak Bidap of the Pangkalan Bun Police Department. For comments on earlier versions of the manuscript, thanks to S.T. Parker, K. Bard, M. Tomasello, B. Moore, R. Mitchell, and P. Vasey. Finally, thanks to the many agencies and individuals who have provided the funds and the many other contributions that supported the rehabilitation project for over 20 years; without this support, the study could not have been conceived or carried out.

REFERENCES

Bandura, A. (1986). *Social Foundations of Thought and Action: A Social Cognitive Theory*. Englewood Cliffs, NJ: Prentice Hall.

Bates, E., Benigni, L., Bretherton, I., Camaioni, L. & Volterra, V. (1979). *The Emergence of Symbols: Cognition and Communication in Infancy*. New York: Academic Press.

Beck, B.B. (1980). *Animal Tool Behavior: The Use and Manufacture of Tools by Animals*. New York: Garland STPM Press.

Byrne, R.W. & Byrne, J.M.E. (1993). Complex leaf-gathering skills of mountain gorillas (*Gorilla g. berengei*): Variability and standardization. *American Journal of Primatology*, 31, 241–261.

Case, R. (1985). *Intellectual Development: Birth to Adulthood*. New York: Academic Press.

Cheney, D.L. & Seyfarth, R.M. (1990). *How Monkeys See the World: Inside the Mind of Another Species*. Chicago: University of Chicago Press.

Custance, D. & Bard, K.A. (1994). The comparative and developmental study of self-recognition and imitation: The importance of social factors. In *Self-awareness in Animals and Humans: Developmental Perspectives*, ed. S.T. Parker, R.W. Mitchell & M.L. Boccia, pp. 207–26. New York: Cambridge University Press.

Custance, D.M., Whiten, A. & Bard, K.A. (1994). The development of gestural imitation and self-recognition in chimpanzees (*Pan troglodytes*) and children. In *Current Primatology. Selected Proceedings of the XIVth Congress of the International Primatological Society*, vol. 2: *Social Development, Learning and Behaviour*, ed. J.J. Roeder, B. Thierry, J.R. Anderson & N. Herrenschmidt, pp. 381–7. Strasbourg: Université Louis Pasteur.

Dickinson, A. (1989). *Contemporary Animal Learning Theory*. Cambridge: Cambridge University Press.

Dumas, C. & Doré, F.-Y. (1986). *Intelligence Animale: Recherches Piagetiennes* [Animal intelligence: Piagetian studies]. Sillery, Québec, Canada: Presses de l'Université du Québec.

Fischer, K.W. (1987). Relations between brain and cognitive development. *Child Development*, 58, 623–32.

Galdikas, B.M.F. (1988). Orangutan diet, range, and activity at Tanjung Puting, Central Borneo. *International Journal of Primatology*, **9**, 101–19.

Galef, B.G., Jr (1988). Imitation in animals: History, definition, and interpretation of data from the psychological laboratory. In *Social Learning: Psychological and Biological Perspectives*, ed. T.R. Zentall & B.G. Galef Jr, pp. 3–28. Hillsdale, NJ: Lawrence Erlbaum.

Gibson, K.R. & Ingold, T. (eds.) (1993). *Tools, Language, and Cognition in Human Evolution*. Cambridge: Cambridge University Press.

Goodnow, J. & Levine, R. (1973). "The grammar of action": Sequence and syntax in children's copying. *Cognitive Psychology*, **4**, 82–98.

Greenfield, P.M. (1991). Language, tools and the brain: The ontogeny and phylogeny of hierarchically organized sequential behavior. *Behavioral and Brain Sciences*, **14**, 531–95.

Hall, K.R.L. (1968). Social learning in monkeys. In *Primates: Studies in Adaptation and Variability*, ed. P. Jay, pp. 383–97. New York: Holt, Rinehart & Winston.

Heyes, C.M. (1993). Imitation, culture and cognition. *Animal Behaviour*, **46**, 999–1010.

Ingold, T. (1986). Tools and *Homo faber*: Construction and the authorship of design. In *The Appropriation of Nature: Essays on Human Ecology and Social Relations*, ed. T. Ingold, pp. 40–78. Manchester: Manchester University Press.

Ladygina-Kots, Nadezda N. (1935/1982). Infant ape and human child. *Storia e Critical della Psicologia*, **3**, 122–89. (First published in Russian in 1935).

Lefebvre, L. & Palameta, B. (1988). Mechanisms, ecology, and population diffusion of socially learned, food finding behavior in feral pigeons. In *Social Learning: Psychological and Biological Perspectives*, ed. T.R. Zentall & B.G. Galef Jr, pp. 141–64. Hillsdale, NJ: Lawrence Erlbaum.

Lock, A. (1993). Human language development and object manipulation: Their relation in ontogeny and its possible relevance for phylogenetic questions. In *Tools, Language and Cognition in Human Evolution*, ed. K.R. Gibson & T. Ingold, pp. 279–99. Cambridge: Cambridge University Press.

Masur, E.F. (1988). Infants' imitation of novel and familiar behaviors. In *Social Learning: Psychological and Biological Perspectives*, ed. T.R. Zentall & B.G. Galef Jr, pp. 301–18. Hillsdale, NJ: Lawrence Erlbaum.

Matsuzawa, T. (1991). Nesting cups and metatools in chimpanzees. *Behavioral and Brain Sciences*, **14**, 570–1.

Meador, D.M., Rumbaugh, D.A., Pate, J.L. & Bard, K.A. (1987). Learning, problem solving, cognition, and intelligence. In *Comparative Primate Biology*, vol. 2B, *Behavior, Cognition, and Motivation*, ed. G. Mitchell & J. Erwin, pp. 17–84. New York: Alan R. Liss.

Meltzoff, A.N. (1988). The human infant as "homo imitans". In *Social Learning: Psychological and Biological Perspectives*, ed. T.R. Zentall & B.G. Galef Jr, pp. 319–42. Hillsdale, NJ: Lawrence Erlbaum.

Meltzoff, A.N. (1990). Foundations for developing a concept of self: The role of imitation in relating self to other and the value of social mirroring, social modeling, and self practice in infancy. In *The Self in Transition: Infancy to Childhood*, ed. D. Cicchetti & M. Beeghly, pp. 129–64. Chicago: University of Chicago Press.

Meltzoff, A.N. & Gopnik, A. (1989). On linking nonverbal imitation, representation and language learning in the first two years of life. In *The Many Faces of Imitation in Language Learning*, ed. G.E. Speidel & K.E. Nelson, pp. 23–51. New York: Springer-Verlag.

Meltzoff, A.N. & Moore, M.K. (1992). Early imitation within a functional framework: The importance of person identity, movement, and development. *Infant Behavior and Development*, **15**, 479–505.

Mitchell, R.W. (1987). A comparative-developmental approach to understanding imitation. In *Perspectives in Ethology*, ed. P.P.G. Bateson & P.H. Klopfer, vol. 7, pp. 183–215. New York: Plenum Press.

Mitchell, R.W. (1994). The evolution of primate cognition: Simulation, self- knowledge, and knowledge of other minds. In *Hominid Culture in Primate Perspective*, ed. D. Quiatt & J. Itani, pp. 177–232. Boulder, CO: University of Colorado Press.

Moerk, E.L. (1989). The fuzzy set called "imitations". In *The Many Faces of Imitation in Language Learning*, ed. G.E. Speidel & K.E. Nelson, pp. 277–303. New York: Springer-Verlag.

Moore, B.R. (1992). Avian movement imitation and a new form of mimicry: Tracing the evolution of a complex form of learning. *Behaviour*, 122, 231–63.

Nadel, J. (1986). *Imitation et Communication entre Jeunes Enfants*. Paris: PUF.

Parker, S.T. (1992). Imitation as an adaptation for apprenticeship in foraging and feeding. Paper presented at the XIVth Congress of the International Primatological Society, Strasbourg, France, 16–21 August.

Parker, S.T. (1993). Imitation and circular reactions as evolved mechanisms for cognitive construction. *Human Development*, 36, 309–23.

Parker, S.T. & Gibson, K.R. (1979). A developmental model for the evolution of language and intelligence in early hominids. *Behavioral and Brain Sciences*, 2, 367–408.

Pepperberg, I.M. (1988). The importance of social interaction and observation in the acquisition of communicative competence: Possible parallels between avian and human learning. In *Social Learning: Psychological and Biological Perspectives*, ed. T.R. Zentall & B.G.Galef Jr, pp. 279–300. Hillsdale, NJ: Lawrence Erlbaum.

Piaget, J. (1952). *The Origins of Intelligence in Children*. New York: Norton.

Piaget, J. (1954) *The Construction of Reality in the Child*. New York: Basic Books.

Piaget, J. (1962). *Play, Dreams, and Imitation in Children*. New York: Norton.

Premack, D. (1988). "Does the chimpanzee have a theory of mind?" revisited. In *Machiavellian Intelligence: Social Expertise and the Evolution of Intellect in Monkeys, Apes, and Humans*, ed. R.W. Byrne & A. Whiten, pp. 160–79. Oxford: Clarendon Press.

Ratner, N.B. (1989). Atypical language development. In *The Development of Language*, 2nd edn, ed. J. Berko-Gleason. Columbus, OH: Merrill.

Russon, A.E. & Galdikas, B.M.F. (1992). The complexity of object-related intelligence in rehabilitant orangutans. Paper presented at the XIVth Congress of the International Primatological Society, Strasbourg, France, 16–21 August.

Russon, A.E. & Galdikas, B.M.F. (1993). Imitation in free-ranging rehabilitant orangutans (*Pongo pygmaeus*). *Journal of Comparative Psychology*, 107, 147–61.

Russon, A.E. & Galdikas, B.M.F. (1995). Constraints on great ape imitation: Model and action selectivity in rehabilitant orangutan (*Pongo pygmaeus*) imitation. *Journal of Comparative Psychology*, 109, 5–17.

Snow, C.E. (1981). Saying it again: The role of expanded and deferred imitations in language acquisition. In *Children's Language*, vol. 4, ed. K.E. Nelson. Hillsdale, NJ: Lawrence Erlbaum.

Snow, C.E. (1989). Imitativeness: A trait or a skill? In *The Many Faces of Imitation in Language Learning*, ed. G.E. Speidel & K.E. Nelson, pp. 73–90. New York: Springer-Verlag.

Speidel, G.E. & Herreschoff, M.J. (1989). Imitation and the construction of long utterances. In *The Many Faces of Imitation in Language Learning*, ed. G.E. Speidel & K.E. Nelson, pp. 181–97. New York: Springer-Verlag.

Speidel, G.E. & Nelson, K.E. (eds.). (1989a). *The Many Faces of Imitation in Language Learning*. New York: Springer-Verlag.

Speidel, G.E. & Nelson, K.E. (1989b). A fresh look at imitation in language learning. In *The Many Faces of Imitation in Language Learning*, ed. G.E. Speidel & K.E. Nelson, pp. 1–22. New York: Springer-Verlag.

Tharp, R.G. & Burns, C.E.S. (1989). Phylogenetic processes in verbal language imitation. In *The Many Faces of Imitation in Language Learning*, ed. G.E. Speidel & K.E. Nelson, pp. 231–50. New York: Springer-Verlag.

Thorpe, W.H. (1956). *Learning and Instinct in Animals*. London: Methuen.

Tomasello, M. (1990). Cultural transmission in the tool use and communicatory signaling of chimpanzees? In *"Language" and Intelligence in Monkeys and Apes: Comparative Developmental Perspectives*, ed. S.T. Parker & K.R. Gibson, pp. 274–311. New York: Cambridge University Press.

Tomasello, M., Kruger, A.C. & Ratner, H.H. (1993a). Cultural learning. *Behavioral and Brain Sciences*, 16, 495–552.

Tomasello, M., Savage-Rumbaugh, E.S. & Kruger, A.C. (1993b). Imitative learning of actions on objects by children, chimpanzees, and enculturated chimpanzees. *Child Development*, 64, 1688–705.

Užgiris, I.C. (1981). Two functions of imitation during infancy. *International Journal of Behavioral Development*, 4, 1–12.

Visalberghi, E. & Fragaszy, D. (1990). Do monkeys ape? In *"Language" and Intelligence in Monkeys and Apes: Comparative Developmental Perspectives*, ed. S.T. Parker & K.R. Gibson, pp. 247–73. New York: Cambridge University Press.

Walker, S. (1983). *Animal Thought*. London: Routledge & Kegan Paul.

Werner, H. (1957). The concept of development from a comparative and organismic point of view. In *The Concept of Development*, ed. D.B. Harris, pp. 125–48. Minneapolis, MN: University Minnesota Press.

Whiten, A. (1989). Transmission mechanisms in primate cultural evolution. *Trends in Ecology and Evolution*, 4, 61–2.

Whiten, A. & Ham, R. (1992). On the nature and evolution of imitation in the animal kingdom: Reappraisal of a century of research. In *Advances in the Study of Behavior*, ed. P.J.B. Slater, J.S. Rosenblatt & C. Beer, vol. 21, pp. 239–83. San Diego: Academic Press.

Whitehurst, G.J. & DeBaryshe, B.D. (1989). Observational learning and language acquisition: Principles of learning, systems, and tasks. In *The Many Faces of Imitation in Language Learning*, ed. G.E. Speidel & K.E. Nelson, pp. 251–76. New York: Springer-Verlag.

Wood, D. (1989). Social interaction as tutoring. In *Interaction in Human Development*, ed. M. Bornstein & J. Bruner, pp. 59–80. London: Basil Blackwell Ltd.

Wright, R.V.S. (1972). Imitative learning of a flaked tool technology: The case of an orangutan. *Mankind*, 8, 296–306.

Yando, R., Seitz, V. & Zigler, E. (1978). *Imitation: A Developmental Perspective*. Hillsdale, NJ: Lawrence Erlbaum.

Zentall, T.R. (1988). Experimentally manipulated imitative behavior in rats and pigeons. In *Social Learning: Psychological and Biological Perspectives*, ed. T.R. Zentall & B.G. Galef Jr, pp. 191–206. Hillsdale, NJ: Lawrence Erlbaum.

Zentall, T.R. & Galef, B.G., Jr (eds.) (1988). *Social Learning: Psychological and Biological Perspectives*. Hillsdale, NJ: Lawrence Erlbaum.

8

"More is less": The elicitation of rule-governed resource distribution in chimpanzees

SARAH T. BOYSEN

Since the classic studies of Yerkes & Yerkes (1929), Kohts (1921, 1928, cited in Yerkes & Yerkes, 1929), Köhler (1970), and other pioneers in primatology who explored the cognitive–behavioral capacities of chimpanzees (*Pan troglodytes*) (e.g. Hayes & Hayes, 1951; Hayes & Nissen, 1971; Kellogg & Kellogg, 1933), the chimpanzee has continued to fascinate the scientific community as well as the lay public. More recent studies of captive apes have probed further into the range of abilities and social flexibility demonstrated by chimpanzees (e.g. Boysen, 1994; de Waal, 1982, 1989; Limongelli *et al.*, 1995; Matsuzawa, 1985; Premack, 1976, 1986), and the chimpanzee continues to emerge as the ultimate challenge to traditional definitions of human uniqueness. While significant morphological and behavioral differences are readily apparent between chimpanzees and humans, remarkable cognitive abilities continue to be demonstrated by chimpanzees as their human tutors endeavor to tease apart the gray zone of intellectual overlap between the two species. Such is the challenge in our own laboratory, as we further explore the range of skills that chimpanzees appear capable of demonstrating in the area of numerical competence (Boysen, 1993; Boysen *et al.*, 1993).

WHY STUDY COUNTING IN CHIMPANZEES?

Chimpanzees have been the subject of study with respect to their intellectual and cognitive abilities for many years, but perhaps the most thought-provoking and controversial work has occurred in the last 30 years. This has included the extraordinary findings from field studies of wild chimpanzees from which we have learned that chimpanzees have remarkably dynamic social systems, modify and use several types of material as tools, including sticks for termite-fishing and ant-dipping (McGrew, 1992; McGrew *et al.*, 1979), and leaves for sponges, and also hunt cooperatively for meat which is then ritualistically shared (Boesch, 1994; Boesch & Boesch, 1989; Goodall, 1986; Teleki, 1973).

Studies of captive animals have been no less startling, as numerous attempts to examine the acquisition of symbolic language-like skills were investigated in chimpanzees across several projects, using different symbol modalities (Gardner et al., 1989; Premack, 1976; Rumbaugh, 1977; Savage-Rumbaugh, 1986). Their remarkable work was soon followed by that of others, including Patterson (1978) with the gorilla Koko, and Miles (1994) with the orangutan Chantek. While the topic of ape language no longer provokes discussion and controversy with the same level of intensity as a decade ago (e.g. Seidenberg & Pettito, 1979), the notion that another species might view the world in ways similar to our own, and might share that view with us through some acquired symbol system, continues to be of interest. However, many of the questions about great ape cognitive abilities that no doubt figured significantly in these investigators' eagerness to pursue such studies remain unanswered. Thus, the chimpanzee remains a species with a special intrigue, and continues to remind us of its unique relationship to the genus *Homo*, and its potential for providing new insights toward understanding both our shared, and distinct, evolutionary pasts.

It was with this frame of reference that we first undertook the study of numerical skills in young chimpanzees. First, chimpanzees had previously demonstrated remarkable potential in the area of symbol manipulation and representational abilities, though the mechanisms and processes that subserve these skills were as elusive in chimpanzees as those that likely support complex cognitive skills in humans. Secondly, chimpanzees were capable of interacting in a highly social environment in the company of humans such that a true teaching context was possible between chimpanzee pupil and human teacher (Boysen, 1992). In addition, sensitivity to quantity had been previously demonstrated in a considerable range of species (e.g. birds, Koehler, 1950; raccoons, Davis, 1984; rats, Capaldi & Miller, 1988; monkeys, Hicks, 1965, including chimpanzees, Ferster, 1964; Matsuzawa, 1985), and number may be a type of natural category for many species, given the significance of space, time, and number to the survival of many animals (Gallistel, 1989). Thus, quantity-based abilities may represent a continuum of capacities that could be fruitfully explored across species, including *Homo sapiens*.

WHAT COUNTS TO A CHIMPANZEE

Over the past several years, we have documented the ability of chimpanzees to acquire considerable expertise in counting, to demonstrate an understanding of ordinality, to utilize the identity principle, to invoke the apparent use of addition algorithms, and to comprehend symbols representing fractions (Boysen, 1992, 1993; Boysen & Berntson, 1989, 1990). With each new task we devised, the animals ultimately emerged successful in demonstrating a more complex understanding of number concepts. We sometimes greeted these new accomplishments with disbelief and chagrin, particularly when new, untrained capabilities emerged *de novo* (Boysen, 1993). Since the details of our training methods have been reported in considerable detail elsewhere (Boysen, 1993; Boysen & Berntson, 1989; Boysen et al., 1995), they will not be further outlined. However, one detail of our procedure warrants further discussion, as it likely impacted on studies to be described shortly.

A basic strategy that we employed during all number-related tasks was to allow the animals to learn about counting and number concepts through the use of edible stimuli. Being hardy pragmatists and certainly opportunistic when it comes to food, the chimpanzees readily paid attention to the initial counting task in which arrays of gumdrops were presented in groups of one to three items to be matched with placards bearing a corresponding number of markers (one to three small magnets that were glued to the surface of plastic placards). Arabic numerals were then introduced to replace the placards, and the chimpanzees' task was to pick the number symbol that represented the array of gumdrops presented on a given trial. The introduction of all subsequent numbers to the chimpanzees' counting repertoire was made directly as Arabic numerals, without the matching task described with the placards. In all cases, whenever the chimpanzee was correct in its choice of a number, he or she was permitted to literally eat the items in the array. Though the animals were never food-deprived and were also accustomed to receiving high-incentive foods on a regular basis (e.g. for compliant behavior or simply as treats), the candy arrays were nevertheless attractive and encouraged continued motivation on tasks that required a great deal of attention and were cognitively complex.

Our success in teaching complex enumeration skills to nonverbal, nonhuman apes was, without question, significantly enhanced by the use of this procedure, as it also contributed to a multifaceted perspective on individual quantities in some unique ways. Not only did the chimpanzees see the individual units that comprised arrays of varying set sizes, they also heard the English word associated with it, they watched the teacher indicate, through pointing, how many items comprised the array, and in a sense, they assimilated individual numerosities as they consumed the stimuli at the end of each trial. It is likely that seeing five gumdrops, touching five gumdrops, hearing the word that represents five gumdrops, and then finally eating those five gumdrops (which includes how long it takes to eat that many candies, and how satiated you feel after you eat them, etc.) provides a very different perspective on "fiveness" to a chimpanzee than would have been provided by a less dynamic teaching context. Thus, our teaching approach may have contributed to bridging the abstractness of the counting process and its symbol system for the animals because of their immediate experience with the stimuli during counting.

To date, we have explored a variety of number-related concepts with our chimpanzees, including counting of edibles and objects, summation of items presented in separate arrays, summation of Arabic numerals, subtraction, ordinality and transitivity, fractions, as well as the introduction of zero. All tasks and approaches we have attempted have been successful, and demonstrations of the acquisition or spontaneous emergence of expertise with number concepts seemed limited only by our creativity and persistence as teachers and experimenters. It was precisely at this point that the chimpanzees came through with a tantalizing peek into that gray area, between great ape and hominid mind, and cognitive potential, to alert us to some very real and enigmatic differences between apes and humans.

In light of the animals' success and demonstrated flexibility with number concepts, we sought to explore the possible use of deception by chimpanzees who had previous training in counting and a broad familiarly with number concepts. Our two female chimpanzees, who at that time were 9 year old Sheba and 32 year old Sarah, served as subjects for the proposed number deception study. Both animals, particularly Sarah (Gillan et al., 1981; Premack, 1976, 1986; Premack & Woodruff, 1978), had extensive experience on a wide range of cognitive tasks (e.g. Boysen & Berntson, 1989b), and they were also concurrently immersed in continuing training with counting throughout the study period. Since Sarah had joined the Comparative Cognition Project more recently, her counting repertoire was more limited (use of numbers 1–4) than Sheba's, who had been involved in number-related training in one form or another since 1984. Sheba's counting repertoire included the use of numbers 0–8.

In the original design, we had planned to teach the two animals a fairly simple discrimination task, and then impose a condition whereby the choices were hidden from view from one chimpanzee. The initial idea was for the chimpanzees to work cooperatively. Two food dishes would be made available, each containing different amounts of candy. One chimpanzee would be designated as the "selector" chimpanzee, and the other, the "receiver." The selector chimpanzee would be permitted to choose one of the two food dishes, and then the experimenter would intervene and provide the contents of that dish to the receiver chimpanzee. The contents of the second, remaining dish would then be provided to the selector chimpanzee. Since two different amounts of candy were available in the dishes, in ratios of $1:2$, $1:4$ or $1:6$ gumdrops, it was to the selector chimpanzee's advantage to always choose first the dish that contained the smaller array. That way, the selector would garner the larger remaining array from the second dish (since the experimenter would have provided the receiver chimpanzee with the contents of the chosen dish).

Once the animals had learned the discrimination task, we planned to introduce a partition so that only the selector chimpanzee would be able to see the available candy in the two dishes. We were interested in whether or not the selector chimpanzee would consistently select the smaller array, particularly given that the receiver chimpanzee would have no knowledge of the actual number of candies available. That way, the selector chimpanzee could always receive the greater number of reinforcements. Such a setting would provide an interesting opportunity for the selector chimpanzee to take advantage of the other animal's lack of knowledge, and thereby attempt to "deceive" the receiver chimpanzee about the true state of the available resources.

LEARNING OR MOTIVATIONAL CONSTRAINTS?

We did not, however, get past the initial discrimination task. The animals' performance was nevertheless remarkable, but for a change because of their *inability* to learn (Boysen & Berntson, 1995). When the task was first introduced, Sarah was chosen as the selector chimpanzee, and over 32 training sessions during which she was given the opportunity to choose one of the two dishes containing increasingly disparate ratios of candy

(1 versus 2 gumdrops for 8 sessions; 1 versus 4 for 8 sessions; 1 versus 6 in the last 8 sessions; and a final 8 sessions of mixed ratios), Sarah's performance never exceeded chance. She showed no evidence of recognizing the rules of enforced food-sharing we had devised, and continued to persist at selecting the larger array, despite its contingent effects of limiting her share of the spoils. We were surprised, but not too dismayed. Other seemingly idiosyncratic individual differences in learning some tasks had been demonstrated previously by other chimpanzees in the group. Kermit, for example, had struggled with minimal skills at counting for a number of years, despite the same type of training and the same teacher, under which the other chimpanzees had readily acquired counting and other number-related abilities (Boysen, 1992, 1994). Perhaps Sarah was just not attuned to the task, and at this point in the study, we had no reason to suspect that a more significant issue might be at hand.

In an attempt to clarify Sarah's difficulty in maximizing the task to her advantage, we had the two chimpanzees change roles. Now Sheba was to serve as the selector animal, and given her considerable experience as the passive receiver in the first phase of training, we anticipated that acquisition of the basic discrimination might be somewhat easier for her. But once again, Sheba's performance indicated that our hypothesis was wrong.

Following sessions with each of the individual ratios (1:2, 1:4, and 1:6) as we had done with Sarah, Sheba was also unable to recognize the strategy of choice that would reap her the larger amount of candy. Her overall performance for each ratio was similarly poor and was, in fact, significantly *below* chance. In addition, Sheba performed more poorly when the disparity between the two amounts of candy increased. That is, she was more likely to choose the larger amount as the difference between the two arrays increased, a situation in which she had the potential to benefit the most if she chose the smaller amount (so that she would subsequently obtain the larger, remaining amount).

These results did not reconcile with the significant record of cognitive achievements for chimpanzees in general, and certainly not for these two animals, in particular. Between the two of them, Sarah and Sheba had a record of contributing at the top end in the field of animal cognition, and Sarah as an individual had been the subject of a series of unprecedented demonstrations of new heights in chimpanzee cognitive abilities (e.g. Premack, 1986). Both animals' inability to acquire what could be framed as a fairly straight forward discrimination problem nearly defied credulity. It was particularly striking to watch the animals interact during the task, as they both exhibited considerable behavioral distress (e.g. vocalizations and striking the apparatus) over the inequitable distribution of candy. Sheba's performance in the selector role clearly did not indicate any appreciable enhancement of performance relative to her previous partnership with Sarah; there was simply no evidence of transfer of training. If anything, Sheba's performance on the task actually deteriorated over the course of testing.

The overall results called for a serious rethinking of any notion we had of studying deception, and instead we were now faced with explaining the chimpanzees' *inability* to learn something – a novel circumstance indeed. In an effort to explore whether the animals' difficulties with the task were related to the rules of food-sharing or the particular stimuli we were using (candy), we began by substituting with symbols. These included the Arabic numerals 1, 2, 4, and 6. Sheba, who had considerably more experience in all aspects of counting and number concepts, was elected to continue in her role as the selector. She was now permitted to choose between number symbols presented in pairs on each trial, including the numerals 1 versus 2; 1 versus 4; or 1 versus 6, placed in the two otherwise empty food dishes. As for the food-sharing rules devised in the first experiment, in this phase of the study, the number symbol selected by Sheba resulted in a corresponding number of candies being given to her partner Sarah. Sheba then received the number of candies that corresponded to the remaining nonselected numeral in the second dish. Three sessions were completed using numerical stimuli only, followed by six additional test sessions of an **ABBA** design, with A sessions employing Arabic numerals as stimuli, and B sessions using candy arrays.

When numerals served as test stimuli, Sheba readily chose the smaller of the two numerals, thereby receiving the larger amount of reinforcement a remarkable 74% of the time. Moreover, her significant performance was observed during the very first session that number symbols were made available. With this change, Sheba responded correctly on the very first trial, and she continued to respond significantly above chance over the next three sessions. During the fourth session, candy arrays were reintroduced, and as seen in Table 8.1, Sheba's performance deteriorated immediately. Candies and numerals were then alternated as stimuli over the next six sessions, and Sheba's performance rose and fell according to which stimuli were used in a given session (mean performance with candies = 17%; performance with symbols = 72%). These differences in performance using the two types of stimuli were consistent and also highly significant. Thus, Sheba was consistently unable to select the food array that was smaller if candy stimuli were presented, while her performance when numerals were first introduced suggested that she understood the food-division rules that had been enforced since the start of the experiment. To explore her understanding of the conceptual rules of food-division, a third phase of the study was conducted in which all possible combinations of number symbols between 0 and 6 were employed as stimuli. Since Sheba may have fortuitously learned to simply select the number "1" in the previous task, and thus not really have grasped the rule of selecting the smaller array (in order to receive the larger remainder), a novel test using all possible number pairs would permit us to examine her conceptual knowledge of the food-sharing rules.

Sheba continued to serve in the selector role for the third experiment, in which all possible arrays composed of numerals 1–6, and candy arrays composed of one to six items were used as stimuli. Test sessions were again alternated as to which arrays of candies or Arabic numerals were used as stimuli. Sheba's performance with these novel combinations of quantities and numerals were such that she reliably selected the smaller array when given a choice between novel combinations of Arabic numerals, but continued to choose the larger array (and thus received the smaller food reward) if food arrays were used. During sessions in which Arabic symbols were used, her performance was significantly above chance, but with food arrays as stimuli, her performance dramatically deteriorated to below chance. This was strong support for the hypothesis

Table 8.1. *Performance (Sheba only) as selector in the enforced food-sharing task with novel symbol combinations of arabic numerals 1–6 and candy arrays of 1–6 items*

Session	Stimuli	Correct/total trials	% correct
1	Numerals	11/15	73
2	Candy	4/15	26
3	Candy	2/15	13
4	Numerals	10/15	67
5	Numerals	11/15	73
6	Candy	2/15	13
7	Candy	5/15	33
8	Numerals	10/15	67

that Sheba was able to use a "smaller/larger" criterion that would reap her the larger reward when selecting between Arabic numerals. These data suggested that Sheba had an understanding of the task and the rules associated with food-division; if symbols were used, she was able to correctly choose between the two numbers so as to maximize the number of reinforcers she received. However, if she had to choose between two differing quantities of real candy, Sheba consistently selected the array that ultimately yielded her smaller rewards. And, as in the previous tasks, her performance with food arrays significantly deteriorated when the disparity between the two food amounts was larger.

The disposition to respond to the larger food arrays did not appear to have prevented learning of the rules and contingencies of the task. Instead, a powerful interference effect, in the form of a competing evaluative disposition, was apparent in Sheba's behavioral response. As noted, this interference increased with increasing arithmetic disparity between the choices in the candy arrays, and may have reflected an evaluative gradient associated with differences in the incentive properties of the task stimuli. In contrast to the results with the edible arrays, her performance improved as a function of the arithmetic disparity when number symbols were used as stimuli. Clearly, symbols imbued the animal with a powerful tool for moving beyond what may be an innate predisposition or evaluative preference to respond, when given a choice, to a larger and/or more concentrated high incentive food source.

RELATED FINDINGS IN CHILDREN'S REPRESENTATIONAL STRATEGIES

While the same task has not yet been conducted with children, a number of findings from studies exploring children's cognitive strategies suggest some similarities to the chimpanzees' performance. One of the most significant and readily comparable comes from a study of strategic deception in young children (Russell *et al.*, 1995). Tests of strategic deception are assumed to measure a child's understanding of mental representation, typically demonstrable in 4 year old children, but not 3 year olds (Russell *et al.*, 1995). Accordingly, such deceptive behavior in young children necessitates an understanding that false beliefs can be provoked in another person, and that one must suppress behavior(s) which indicate that one actually knows the true state of affairs in some situation, while at the same time indicating some false information or state. The former cognitive ability is known as metarepresentational insight (Leslie, 1987; Perner, 1991). The second cognitive skill, the active suppression of one knowledge state and expression of a false one, requires what Russell *et al.* (1995) refer to as "executive control," that is, the execution of a strategy.

Russell and his colleagues were interested to see just when children could pass formal tests of deception, since the literature suggested that 4 year olds could engage in such strategies, while 3 year olds could not. Interestingly, our original intent in teaching the chimpanzees the quantity judgement task, as noted earlier, was to explore the very same phenomenon by employing a task that serendipitously was highly similar to that designed by Russell *et al.* (1991). That is, if given a choice between two amounts of candies when the location of both were known only to one animal, would (or could?) that chimpanzee exhibit strategic deception? Because the chimpanzees failed to learn the initial discrimination, we were unable to answer this intriguing question.

In their study, however, Russell *et al.* (1991) explored the question of strategic deception with normal 3 and 4 year olds, a group of autistic children, and a fourth group of nonautistic, mentally handicapped children. Both of the compromised groups had mental ages of around 4 years, and had been included in the study to address the proposal that autistic children, lacking some capacities that are critical for a theory of mind (Baron-Cohen *et al.*, 1985), would not demonstrate strategic deception. In their "windows task," children worked with a partner, and two boxes, one of which was baited with a chocolate. The subject was to point to one of the two boxes to indicate which box the other child (opponent) should open. If the box contained the chocolate, the opponent was permitted to keep it. If the box was empty, then the subject got the chocolate. It is important to note that in this phase of the experiment, neither child knew which of the boxes was baited; they merely knew the rules of division, once a box was selected. During the 15 trials of the training phase, both children came to learn which outcome would be to their advantage.

In the test phase, however, the rules were changed. The boxes themselves now had windows facing one of the children, which enabled that child (who was also doing the pointing) to see what was inside each box. Thus, the subject could choose to employ strategic deception by pointing to the empty box, knowing that their opponent would select that box, and therefore relinquish the chocolate in the other, unselected box. In this study, both the 3 year old and the autistic groups typically pointed to the baited box on the first experimental trial, while the 4 year olds and nonautistic mentally handicapped children did not. However, the two failing groups also persisted in selecting the baited box despite repeated observations that the contents went to their opponents. This suggested more of a failure in executive functioning to the authors, although other alternative explanations were possible.

In a second study, Russell et al. (1995) attempted to remove one requirement of the task by testing the children without the presence of opponents. They also included a condition during which the children indicated their choice verbally rather than gesturally, to control against impulsive motor responses. Both 3 and 4 year olds were tested, and assigned to one of the four possible conditions: (a) with opponent, using a pointing response; (b) no opponent, using a pointing response; (c) with opponent, and using a verbal response; or (d) no opponent, verbal response. The results indicated that more children failed to grasp the rules of the game in the no-opponent condition, with more 3 year olds (11) and two 4 year olds failing to meet criteria for continuing to the test phase of the study. Thus, the principal findings of this study were twofold: (a) younger children (3 years) were significantly more likely to fail the test; and (b) removing the opponent in the task did nothing to change the difficulty of the task presented to the younger children. These findings were consistent with the perspective that the windows task was difficult in terms of its executive demands, which require disengagement from the object and reference to an empty box, rather than its requirement to consider the informational state of an opponent (Russell et al., 1995). Even though half of the younger subjects had no opponent to be actively deceived, they nevertheless indicated the location of the chocolate on the first test trial, and often continued to do so throughout the test phase.

These findings are strikingly similar to those observed in our chimpanzee subjects during the initial training phases of the quantity judgement task, but did not rule out the alternative hypothesis that some visual/perceptual features of the task took precedence. To address the possibility that perceptual features were capturing the children's attention, Russell et al. (1995) ran an additional experiment during which the children were verbally informed of the chocolate's location during the test phase, rather than viewing the location of the chocolate. In this case, only the older children in the opponent condition outperformed the younger subjects; performance of the two different age groups were similar in the no-opponent condition. Four year olds performed significantly better in the opponent condition, when the location of the chocolate was whispered to them instead. The authors suggested that perhaps stating the location of the chocolate, as in the no-opponent condition, was suggestive of an instruction, while the whispered statement (in the opponent condition) set the stage for using that information strategically. Despite these differences with 4 year olds, however, twice as many 3 year olds were wrong on the first test trial, and thus the critical difference observed between these two age groups in previous windows tasks was again demonstrated under these testing conditions (Russell et al., 1995).

Comparisons with the chimpanzees' performance on the nearly identical task are striking. While Russell et al. (1995) raised a number of possible hypotheses to account for the varying performance of younger versus older children, they also highlighted the perseverative response in the younger children. Recall that the chimpanzees also exhibited such perseverance, session after session, and Russell et al. (1995) reminded us that such responses are "a quintessentially executive function." Like the children, the chimpanzees were required to make a determination between two quantities of edibles employing a rule that was socially counter-intuitive – a behavioral act quite the opposite of a more natural choice between two amounts (i.e. "more is better"). If the quantity judgement task is, like Russell's "windows task," dependent upon cognitive capacities that are a reflection of a "theory of mind," one might speculate that species other than humans or great apes would be unable to solve the task, regardless of the stimuli employed.

POSSIBLE EVOLUTIONARY IMPLICATIONS

What do failures to inhibit the impulse to take the larger food array imply regarding limitations on cognitive processing in chimpanzees? What does it promise to add to our understanding of human evolution? Humans may readily recognize how one might optimize receipt of the larger number of food items in the task, but may elect to behave altruistically toward their partner. In a broader evolutionary context, however, the behavior of chimpanzees under these task conditions hints at more significant differences between the two species perhaps related, in part, to their respective history of food-sharing, as well as further elaboration of some cortical areas, such as the frontal lobe in humans, which likely subserve "executive" functions.

Observations of extant apes have stimulated speculation on the contributions that food-sharing has made to the hominization process (Goodall & Hamburg, 1974; McGrew, 1992; McGrew & Feistner, 1992; McGrew et al., 1979; Wrangham, 1987), as have paleontological and archeological finds and ethnographic data on modern hunter–gatherers (Isaac, 1978; Lovejoy, 1981; Parker, 1987; Parker & Gibson, 1979; Tanner, 1981). When Isaac's (1978) roster of food-sharing and related activities for *Homo* were compared with current observations of *Pan*, some qualitative differences did emerge. However, the similarities were more striking. Of the ten differences representing major differences in food sharing behaviors between chimpanzees and humans (Isaac, 1978), the most recent evidence from chimpanzee field work indicates that only two of the original ten remain as real differences between the two species (McGrew & Feistner, 1992). These include first, the fact that chimpanzees eat food when and where it is located, as opposed to humans, who can postpone eating of gathered foods, and, secondly, that chimpanzees have no permanent home base, compared with a home base that serves as a site from which to focus foraging and ranging for humans. The remaining eight differences in food-sharing between apes and humans have been dispelled on the basis of additional observations from the wild (e.g. Boesch & Boesch, 1983; McGrew, 1987; McGrew & Feistner, 1992; Teleki, 1973). Clearly, some of the differences between apes and humans are a matter of degree, and as McGrew & Feistner (1992) noted, the gulf between great ape and hominid, in terms of a reconstruction of their common hominoid ancestry, appears to have been significantly narrowed.

How do we reconcile the limitations evidenced by our chimpanzees during the food-sharing tasks described earlier with McGrew & Feistner's (1992) roster of similarities in food-sharing behaviors in chimpanzees and humans? In many respects, our findings, in which chimpanzees failed to acquire an "optimal" discrimination strategy, but succeeded at the task if symbols were employed, may dovetail quite nicely with field observations of chimpanzee behavior and hypotheses concerning ancestral humans. The numerical representational system we provided in our experiment allowed our subjects to overcome a highly prepared preferential response that would maximize an individual's food intake under natural conditions. The unique contribution of symbols, embedded within a socio-cultural framework, was to free the chimpanzees from the limiting features of their own biological and perceptual frame of reference. It requires minimal creative speculation to consider how an ancestral hominid group, poised to capitalize upon the behavioral plasticity and cognitive potential of its increasingly complex brain, might also immediately capitalize upon the capacity to better regulate social interactions and other relationships, including food-sharing, via some type of representational system.

These studies of extant chimpanzees offer us a tantalizing hint at what might have occurred at the evolutionary crossroads that separated chimpanzees and humans. The rapidity with which cultural innovations such as abstract symbols can overcome behavioral limitations, as shown by our animals, provide strong support for the explosive impact that such phenomena likely had on early hominids, easily catapulting them into a protocultural domain with enormous potential for further cultural development, including tool use, language, cooperation, altruism and other socially based behaviors.

REFERENCES

Baron-Cohen, S., Leslie, A.M. & Frith, U. (1985). Does the autistic child have a 'theory of mind'? *Cognition*, 21, 37–46.

Boesch, C. (1994). Chimpanzee–red colobus monkeys: A predator–prey system. *Animal Behaviour*, 47, 1135–48.

Boesch, C. & Boesch, H. (1983). Optimisation of nut-cracking with natural hammers by wild chimpanzees. *Behaviour*, 83, 265–85.

Boesch, C. & Boesch, H. (1989). Hunting behavior of wild chimpanzees in the Taï National Park. *American Journal of Physical Anthropology*, 78, 547–73.

Boysen, S.T. (1992). Pongid pedagogy: The contribution of human/chimpanzee interaction to the study of ape cognition. In *The Inevitable Bond: Examining Scientist-Animal Interaction*, ed. H. Davis & D. Balfour, pp. 205–17. New York: Cambridge University Press.

Boysen, S.T. (1993). Counting in chimpanzees: Nonhuman principles and emergent properties of number. In *The Development of Numerical Competence: Animal and Human Models*, ed. S.T. Boysen & E.J. Capaldi, pp. 39–59. Hillsdale, NJ: Lawrence Erlbaum.

Boysen, S.T. (1994). Individual differences in cognitive abilities in chimpanzees (*Pan troglodytes*). In *Chimpanzee Cultures*, ed. R.W. Wrangham, W.C. McGrew, F.B.M. de Waal & P. Heltne, pp. 335–50. Cambridge, MA: Harvard University Press.

Boysen, S.T. & Berntson, G.G. (1989). Numerical competence in a chimpanzee (*Pan troglodytes*). *Journal of Comparative Psychology*, 103, 23–31.

Boysen, S.T. & Berntson, G.G. (1990). The emergence of numerical competence in the chimpanzee (*Pan troglodytes*). In: *"Language" and Intelligence in Monkeys and Apes: Developmental Perspectives*, ed. S.T. Parker & K.R. Gibson, pp. 435–50. New York: Cambridge University Press.

Boysen, S.T. & Berntson, G.G. (1995). Quantity judgments: Perceptual vs. cognitive mechanisms in chimpanzees (*Pan troglodytes*). *Journal of Experimental Psychology: Animal Behavior Processes*, **21**, 83–6.

Boysen, S.T., Berntson, G.G., Shreyer, T.A. & Hannan, M.B. (1995). Use of indicating acts during counting by a chimpanzee (*Pan troglodytes*). *Journal of Comparative Psychology*, **109**, 47–51.

Boysen, S.T., Berntson, G.G., Shreyer, T.A. & Quigley, K.S. (1993). Processing of ordinality and transitivity by chimpanzees (*Pan troglodytes*). *Journal of Comparative Psychology*, **108**, 208–15.

Capaldi, E.J. & Miller, D.J. (1988). Counting in rats: Its functional significance and the independent cognitive processes which constitute it. *Journal of Experimental Psychology: Animal Behavior Processes*, **14**, 3–17.

Davis, H. (1984). Discrimination of the number "three" by a raccoon (*Procyon lotor*). *Animal Learning and Behavior*, **12**, 409–13.

de Waal, F.B.M. (1982). *Chimpanzee Politics*. New York: Harper & Row.

de Waal, F.B.M. (1989). *Peacemaking Among Primates*. Cambridge, MA: Harvard University Press.

Ferster, C.B. (1964). Arithmetic behavior in chimpanzees. *Scientific American*, **210**, 98–106.

Gallistel, C.R. (1989). Animal cognition: The representation of space, time and number. *Annual Review of Psychology*, **40**, 155–89.

Gardner, R.A., Gardner, B. & van Cantfort, T.E. (1989). *Teaching Sign Language to Chimpanzees*. Albany, NY: State University of New York Press.

Gillan, D.J., Premack, D. & Woodruff, G. (1981). Reasoning in the chimpanzee: I. Analogical reasoning. *Journal of Experimental Psychology: Animal Behavior Processes*, **7**, 1–17.

Goodall, J. (1986). *The Chimpanzees of Gombe: Patterns of Behavior*. Cambridge, MA: The Belknap Press.

Goodall, J. & Hamburg, D. (1974). Chimpanzee behavior as a model for the behavior of early man. New evidence on possible origins of human behavior. *American Handbook of Psychiatry*, **6**, 14–43.

Hayes, K.J. & Hayes, C. (1951). The intellectual development of a home-reared chimpanzee. *Proceedings of the American Philosophical Society*, **95**, 105–9.

Hayes, K.H. & Nissen, C. (1971). Higher mental functions of a home-reared chimpanzee. In *Behavior of Non-Human Primates*, ed. A.M. Schrier & F. Stollnitz, vol. 4, pp. 59–115. New York: Academic Press.

Hicks, L. H. (1956). An analysis of number-concept formation in the rhesus monkey. *Journal of Comparative and Physiological Psychology*, **49**, 212–8.

Isaac, G. (1978). The food sharing behavior of protohuman hominids. *Scientific American*, **238**, 90–108.

Kellogg, W.N. & Kellogg, L.A. (1933). *The Ape and the Child*. New York: McGraw-Hill.

Koehler, O. (1950). The ability of birds to "count". *Bulletin of Animal Behaviour*, **9**, 41–5.

Köhler, W. (1970). *The Mentality of Apes*. New York: Liveright. (First published in German in 1925).

Leslie, A.M. (1987). Pretense and representation. *Psychological Review*, **94**, 412–26.

Limongelli, L., Boysen, S.T. & Visalberghi, E. (1995). Comprehension of cause–effect relationships in a tool-using task by chimpanzees (*Pan troglodytes*). *Journal of Comparative Psychology*, **109**, 18–26.

Lovejoy, C.O. (1981). The origins of man. *Science*, **211**, 341–50.

Matsuzawa, T. (1985). Use of numbers by a chimpanzee. *Nature*, **315**, 57–9.

McGrew, W.C. (1987). Tools to get food: The subsistants of Tanzanian chimpanzees and Tasmanian aborigines compared. *Journal of Anthropological Research*, **43**, 247–58.

McGrew, W.C. (1992). *Chimpanzee Material Culture: Implications for Human Evolution*. Cambridge: Cambridge University Press.

McGrew, W.C. & Feistner, A.T.C. (1992). Two nonhuman primate models for the evolution of human food sharing: Chimpanzees and callitrichids. In *The Adapted Mind*, ed. J.H. Barkow, L. Cosmides & J. Tooby, pp. 229–43. Oxford: Oxford University Press.

McGrew, W.C., Tutin, C.E.G. & Baldwin, P.J. (1979). Chimpanzees, tools, and termites: Cross-cultural comparisons of Senegal, Tanzania, and Rio Muni. *Man*, 14, 185–214.

Miles, H. L. (1994). ME CHANTEK: The development of self-awareness in a signing orangutan. In *Self-awareness in Animals and Humans: Developmental Perspectives*, ed. S.T. Parker, R.W. Mitchell & M.L. Boccia, pp. 254–72. New York: Cambridge University Press.

Parker, S.T. (1987). A sexual selection model for hominid evolution. *Human Evolution*, 2, 235–53.

Parker, S.T. & Gibson, K.R. (1979). A developmental model for the evolution of language and intelligence in early hominids. *Behavioral and Brain Sciences*, 2, 367–408.

Patterson, F.G. (1978). The gestures of a gorilla: Language acquisition in another pongid. *Brain and Language*, 3, 72–97.

Perner, J. (1991). *Understanding the Representational Mind*. Chicago: University of Chicago Press.

Premack, D. (1976). *Intelligence in Ape and Man*. Hillsdale, NJ: Lawrence Erlbaum.

Premack, D. (1986). *Gavagai*. Cambridge: Cambridge University Press.

Premack, D. & Woodruff, G. (1978). Does the chimpanzee have a theory of mind? *Brain and Behavioral Sciences*, 1, 515–26.

Rumbaugh, D.M. (1977). *Language Learning by a Chimpanzee: The Lana Project*. New York: Academic Press.

Russell, J., Jarrold, C. & Potel, D. (1995). What makes strategic deception difficult for children – the deception or the strategy? *British Journal of Developmental Psychology*, 12, 301–14.

Russell, J., Mauthner, N., Sharpe, S. & Tidswell, T. (1991). The 'windows task' as a measure of strategic deception in preschoolers and autistic subjects. *British Journal of Developmental Psychology*, 9, 331–49.

Savage-Rumbaugh, E.S. (1986). *Ape Language: From Conditioned Response to Symbol*. New York: Columbia University Press.

Seidenberg, M.S. & Pettito, L.A. (1979). Signing behavior in apes: A critical review. *Cognition*, 7, 177–215.

Tanner, N.M. (1981). *On Becoming Human*. Cambridge: Cambridge University Press.

Teleki, G. (1973). *The Predatory Behavior of Wild Chimpanzees*. East Brunswick, NJ: Bucknell University Press.

Wrangham, R.W. (1987). The significance of African apes for reconstructing human social evolution. In *The Evolution of Human Behavior: Primate Models*, ed. W.G. Kinsey, pp. 51–71. Albany, NY: State University of New York Press.

Yerkes, R.M. & Yerkes, A.W. (1929). *The Great Apes*. New Haven, CT: Yale University Press.

9

Tool-using behavior in wild *Pan paniscus*: Social and ecological considerations

ELLEN. J. INGMANSON

INTRODUCTION

Pan paniscus have demonstrated impressive abilities under captive conditions in the understanding and acquisition of language, tool use, and social skills (Greenfield & Savage-Rumbaugh, 1990; Jordan, 1982; Ingmanson, 1987, 1988; Savage-Rumbaugh, 1988; Savage-Rumbaugh & Rumbaugh, 1993). The results of these studies indicate complex cognitive capabilities. In order to explore the possible adaptive significance of these abilities, I have been observing the ways in which *P. paniscus* express these skills in their natural habitat. One focus of my research is identifying the extent and nature of object manipulation and tool-using behavior by *P. paniscus*. These behaviors are clearly associated with intelligence, as seen through their use in problem-solving and goal-directed activities (Gibson, 1993; Ingold, 1993).

In spite of incomplete agreement on what constitutes a tool, the use of a detached object in order to mediate the attainment of a desired goal is the core of the classic definition of tool use (Beck, 1980). This strict definition, though, does not adequately reflect the cognitive abilities driving the behavior (Parker & Gibson, 1977). For example, species as diverse as chimpanzees and otters meet the criteria for using tools, whereas they are generally considered to be vastly different in their cognitive capacities. Also, evidence for intelligence based on tool use verges on nonexistent for wild orangutans, but Bard (1990, 1993) demonstrated that cognitive abilities equally complex to those involved in tool use are exhibited by wild orangutans in behaviors such as communicative gestures and movement through the trees. It becomes clear, therefore, that if we are to understand what intelligence is and how it has evolved, we must not restrict ourselves to narrowly conceived or narrowly interpreted definitions, but also examine other closely related behaviors that are indicative of similar levels of competency. I have therefore collected evidence on behavior related to tool use, including instances that do not meet all the definitional criteria and instances where objects "mediate" attaining a goal through communicative rather than physical causes. In this chapter, I describe both the types of tool use and some tool-related behaviors that I have observed among *P. paniscus* in the wild. I also discuss the contexts in which they occur, as well as some of the similarities and differences with other great ape species, in order to further explore the breadth of the great ape mind.

SPECIES CHARACTERISTICS

Since *P. paniscus* differ in important ways from the better known *P. troglodytes*, I begin with a brief sketch of important features of their lives. *Pan paniscus* are extant only in the Equateur Province of the Republic of Zaire, within the southern basin formed by the Zaire river. The research site where I studied, Wamba, is located in the eastern portion of this area, on the equator. This region is dense tropical rainforest, intersected by many rivers that form tributaries to the larger Zaire river. The area around Wamba is composed of primary forest, swamp forest, secondary forest, and cultivated areas. All of these vegetation types are utilized by *P. paniscus* at different times.

Research at Wamba was first initiated by Takayoshi Kano in 1973, and has continued under his direction through Kyoto University in Japan. Three separate unit-groups have been habituated to human observation. In addition, the ranges of several other groups overlap with those of the habituated groups, and these individuals are occasionally encountered by observers. The habituated groups have been provisioned with locally grown sugar cane, but the amount provided has varied considerably over time, with the season, and for the different groups.

The social organization of *P. paniscus* is a fission–fusion type (Kano, 1982, 1992a; Kuroda, 1980). The unit-groups consist of between 30 and 70 individuals, with relatively equal male–female ratios (Kano, 1987). The group will divide into parties of various sizes for foraging, especially during the dry season, but remains a cohesive unit with most members meeting on a daily basis. Feeding and travel often involves the entire unit-group. Group membership is based on male residency in the natal group, with strong ties between mothers and sons lasting throughout their lives (Kano, 1992b). Females transfer at adolescence. Unlike *P. troglodytes*, though, *P. paniscus* females form strong bonds with each other once they join a group and have not been observed to transfer again (Kano, 1992a). Bonds between adult males appear to be weaker in *P. paniscus* than *P. troglodytes*, possibly related to a lack of cooperative patrolling and territory defense in *paniscus* (Kano, 1987).

Few reports have been published describing the object manipulation and tool using behaviors of *P. paniscus*. Information from the wild is particularly rare. Field research was only begun in the mid-1970s, and only two sites, Wamba and Lomako, have been examined in any depth. Kano (1982) briefly described the use of rain coverings by a few individuals in one group at Wamba, and the inclusion of objects into play (Kano, 1986). Neither behavior however, was systematically studied. One tantalizing observation occurred at Lomako, where a stick was found thrust into a termite mound (Badrian *et al.*, 1981). It was not possible to definitely attribute this to *P. paniscus*. Two studies conducted in zoo settings (Ingmanson, 1988; Jordan, 1982) indicate that this species does use a variety of objects as tools under captive conditions. The types of tool included cups, sponges, probes for extracting food, ladders, and signalling devices. The lack of extensive tool-using activities, especially during food acquisition, is one of the behavioral characteristics that sets *P. paniscus* apart from many groups of *P. troglodytes* (Boesch & Boesch, 1990; Goodall, 1986; Kano, 1992a; McGrew, 1993a,b).

METHODS

My research at Wamba was conducted from September 1987 to January 1988, and September 1990 to February 1991. Altogether, I collected close to 900 hours of direct observation, over 178 days. My primary study group was the habituated E2 group, though many observations were also made of the E1 group, especially during 1987–88. In February 1991, the E2 group consisted of 53 members. Based on age classes defined by Goodall (1986) and Kano (1992a), there were 12 adult males, 14 adult females, 4 adolescent females, 2 adolescent males, 8 juveniles, and 13 infants. Their range covered approximately 46 km^2. The E1 group had 29 members: 8 adult males, 9 adult females, 2 adolescent males, 3 juvenile males, 3 juvenile females, and 5 infants.

In collecting data on tool-using behavior, I utilized a combination of observation techniques, including *ad lib* recording for all observed occurrences of tool use and tool-related behaviors, focal animal follows, and behavioral sampling – where I focused on a particular behavior (Altmann, 1974; Martin & Bateson, 1986). Although some of these observation methods can present difficulties for quantitative analyses, they can also provide an opportunity to collect a wide range of data in a relatively short period. As this study focused on discovering the types and extent of tool-using behavior, the potential quantitative difficulties were less important than they might otherwise be. I used *ad lib* observations and recordings of all observed occurrences primarily in the initial stages of research during 1987–88, and when encountering unfamiliar groups. This permitted me to determine the presence or absence of particular behaviors, the relative frequencies of some behaviors, especially if they were very rare, to identify situations in which tools were used, and to sample groups whose individuals I was unable to identify.

During 1990–91, my observations were concentrated on the E2 group (89 out of 101 observation days) and consisted of both focal animal and focal behavior sampling (Martin & Bateson, 1986). All adults and juveniles in the E2 group were targeted for focal individual follows of 30 minutes duration each, with behaviors and context recorded at 1 minute intervals, and changes in behaviors during the intervals noted. Few focal follows were conducted on adolescents as they tended to be less well habituated and travelled on the periphery of the group. Any instance of observed tool-using was recorded, regardless of whether the individual involved was the focal target at that moment. Certain behaviors were also targeted for specific observation, such as branch dragging, play with objects, and use of rain hats in order to clarify the manipulative and social characteristics of each. In addition, activities that might be expected to involve tool use, such as digging in the ground for mushrooms, were targeted for specific focal behavior sampling.

PAN PANISCUS TOOL USE AND RELATED BEHAVIOR

The observed forms of tool use and related behavior at Wamba can be divided into two different categories based on how well the behavior meets all of the defined parameters of tool use. These categories are (a) tool-related behavior, and (b) tool use. The behaviors observed for each of these categories are described in this section.

Tool-related behavior

The behaviors that have been included in the category of tool-related behaviors are tree drumming, bridging, object dropping, branch waving, and animate "step ladders." These behaviors are labeled tool-related instead of tool use because they do not meet all of the stringent criteria for tool use now commonly accepted (Beck, 1980; Goodall, 1986; McGrew, 1992). For instance, the manipulated objects lack the characteristic of having been altered, and usually remain *in situ* as opposed to being transported to a new location. Nevertheless, they indicate that there has been an intentional incorporation of an object into a goal-directed activity, and that the object is used to mediate the attainment of the goal. As such, I consider they can provide considerable information relevant to *P. paniscus* tool-using activities, and deserve attention.

Tree drumming

Tree drumming generally occurred at the end of a communicative act, such as a charging display or branch dragging sequence (see below), and was usually performed by a male. This behavior seems to serve as added emphasis. Typically, an individual ran towards a tree, slapping the soles of his feet against the trunk. The sound produced was very distinct and carried clearly through the forest. In some instances, only one foot was slapped against the tree, though frequently both were used, creating a sequential bang bang. On a few occasions, one particular older male was observed to simply walk toward a tree, brace himself with one hand, then repeatedly strike one foot against the tree.

Bridging

Bridging behavior consisted of individuals using their own body weight on a small tree to carry them between two larger trees. This behavior differed somewhat from other movement through the trees using branches or lianas. First, the distance between the larger trees is usually too great to allow direct transfer. Second, the individual must combine several different pieces of information in order to successfully complete the task (Bard, 1993; Chevalier-Skolnikoff *et al.*, 1982). The height of the intermediate tree is crucial – it must be tall enough to bridge the required distance. It must also be strong enough at the appropriate height to sustain the weight of an adult individual, while remaining supple enough to actually bend. These bits of information must then be put together. Each bridging event appears to be dealt with as an instance of a type of problem for which solutions and knowledge have been developed progressively through extensive experience and learning. Although mistakes did occur, most individuals seemed to quickly assess the various properties of available trees, select and position themselves appropriately, and proceed with their travel. Infants of 2 and 3 years of age were often observed testing the movement and strength of various branches and trees, slowly moving further up or along them, giving small bounces as they went – usually with one hand firmly anchored to another branch. This activity appears to provide learning opportunities for development of the skills necessary for later successful bridging behavior. I have also observed juveniles to inaccurately judge one component of the task, such as the height of the intermediate tree, and therefore fail to make a successful arboreal crossing. Similar differences in skill associated with age have been reported for orangutans (Bard, 1993).

Object dropping
Many field researchers report the uncanny skill primates seem to show in dropping a variety of objects on the researcher (Jolly, 1985). This is also true for *P. paniscus*. In many cases these instances can be dismissed as coincidental. The observer is after all trying to obtain the closest view possible of the subject, which is often directly underneath. However, there are occasions when the situation may be more deliberate. Individuals will seem to carefully position themselves, then watch as the object is dropped. I have observed 3 and 4 year old infant *P. paniscus* retrieve an object after watching it drop, then proceed to drop it again. The sequence will be repeated three or four times, and can often be considered to be play (see below). While a target, such as an observer, may not always be apparent, it cannot be ruled out either. Similar behavior has been reported for orangutans (Rijksen, 1978).

Branch waving
Branch waving occurred frequently, and was observed almost daily. The branch being waved was not removed, but remained attached to the bush, tree, or vine. It can be either a calm movement of the attached branch with the hand, or a more intense, sharp movement. It was directed towards another individual, as when a mother waved a branch at her infant, or an adult male waved a branch at an adult female, in an apparent attempt to get the other's attention. The branch wave was then followed by a more precise gesture – the mother reaching out an arm to her infant, or a male soliciting copulation with a female. Branch waving was also used as a mild threat gesture.

Animate step ladders
A problem situation occurs when an individual is unable to reach the lower branches of a tree and a form of step ladder is needed in order to climb into the tree from the ground. In this case, the young individual uses objects to mediate achieving a goal by including another individual as the object (Gómez, 1990; Köhler, 1925). Infant and juvenile *P. paniscus* will often use a convenient stump or log as a first step when climbing large trees with few lower branches. Sometimes, though, one is not readily available. On four

such occasions, 4 and 5 year old individuals were observed slightly manipulating another individual into the appropriate position, then using their back or shoulders as the step stool.

Whether any of these activities should technically be considered tool use is of relatively minor importance. More critical is what these behaviors tell us concerning *P. paniscus* cognitive skills. All demonstrate a grasp of object–object or object–individual relations, the effect one object can have on another. Some require the combination of more than one piece of information to achieve a goal, and both learning and development are necessary for success. These are all skills that are also part of tool use and tool making (Chevalier-Skolnikoff *et al.*, 1982).

Tool use

This category includes behaviors I observed at Wamba that are clearly examples of tool use – detached objects being directly manipulated to mediate obtaining a goal. A variety of forms of tool use were observed, some of which occurred more frequently than others. I observed some forms of tool use only two or three times each, during both of my research periods, and some of them had never been observed by other researchers at Wamba (T. Kano, personal communication). In addition, I observed some in only one individual. Other forms of tool use were observed several times a day. All of these examples provide information on the extent and variety of tool-using activities by *P. paniscus* at Wamba. Some may constitute tool-using traditions within the group, in that they were observed on many occasions and performed by many different individuals (Nishida, 1987).

Wipers/napkins

Leaves were observed being used as wipers or napkins on three occasions. Two adult males used then to remove feces that had fallen on them or that they had stepped into. One adult female wiped her infant's urine off her thigh.

Toothpick

A single adult male, Jes, was twice observed to use a small twig as a toothpick after eating sugar cane. On one occasion he picked up a twig that was on the ground next to him and immediately began to use it. On another occasion, he used his teeth to shape the end prior to using it, adding the element of tool-making to his activity.

Fly swatter

On one occasion, an older adult female, Kame, was observed using a long stick (approximately 30 cm) with a few leaves attached to swat at bees around her shoulders. The bees were apparently attracted to rotting fruit lying on the ground in the sun where the group had been feeding during the morning. Kame kept the stick for almost 20 minutes, periodically swatting at the bees. She dropped the stick as she moved away.

Scratchers

Two individuals, one adult male and one adult female, were each observed once using a stick to scratch their own backs. In both cases the behavior appeared casual, as the individual used a stick within reach while resting on the ground.

Rain hats

In comparison to the above behaviors, the use of rain hats was observed on many occasions. An obvious, though often overlooked, prerequisite when searching for tool use, is whether or not the need for a particular pattern exists. For example, in contrast to some *P. troglodytes* habitats (Goodall, 1986; Nishida, 1980), there is a surplus of water in *P. paniscus* environments. Being a rainforest, it does rain frequently. This removes the problem of finding drinking water that some *P. troglodytes* may face, but it may also create a different problem for *P. paniscus* – getting wet. The vegetation in the forest is dense, mitigating the impact of rain at the lower forest levels. *Pan paniscus*, though, do spend time in more open areas, such as in natural clearings, or in day nests. Infants invariably stop playing when it begins to rain, returning to huddle with their mothers, suggesting a sense of discomfort.

In such situations, avoiding rain is obviously a problem for *P. paniscus* to solve. They do so by making rain hats. Basically, an individual places small, leafy branches on his or her head and shoulders, providing some relief from the rain. The use of rain hats may seem like a fairly simple, mundane sort of activity. In examining this behavior in *P. paniscus*, I have found that several steps are involved in the production and use of rain hats. I have placed them into four groups: (a) having the concept of a rain covering; (b) selecting the materials to use; (c) effectively arranging the materials; (d) behaving appropriately in conjunction with the rain hat. All four components are required.

The first concept needed to address the problem of finding protection from the rain is that of a rain covering. While orangutans also use rain coverings (Rijksen, 1978), mountain gorillas and *P. troglodytes* do not. Goodall (1971) has described how miserable the Gombe chimpanzees appear when it rains. So, since the problem apparently exists, discomfort is experienced, and materials are available for solving the problem, they may not have discovered the concept of a rain covering. At Wamba, not all of the *P. paniscus* use rain hats, either. In the two habituated communities I observed most frequently, approximately half of the adult individuals, both male and female, were recorded using rain hats. The pattern of users and nonusers was not random. Where relationships were known, the pattern of use appeared to follow maternal kin lines – if a mother was observed using rain hats, all of her offspring in the group were observed using them as well. In four cases where a mother was never observed using a rain hat, none of her offspring were observed using them either, even though other individuals were present who could have served as models for the behavior. I observed the use of rain hats by some individuals in four different unit-groups – E1, E2, P, and S groups – indicating the pattern is fairly widespread among groups at Wamba. Since female *P. paniscus* transfer to a new group at adolescence, while males remain in their natal group, it is likely that females carry this tradition with them and then serve as the source for transmitting the knowledge and skills to the next generation. This would account for some of the individual variation in use that I observed. Although a genetic component to the use of

Tool-using behavior in wild *Pan paniscus* 197

Figure 9.1. Materials used as rain hats by *Pan paniscus*.

rain hats would also explain the individual variation, this seems less likely considering the learning apparently involved in the process. At this stage of research and analysis, it is impossible to speculate why individuals seem to learn this behavior only from their mothers, and not from other companions. I never observed any indication of active teaching on the part of mothers, though this cannot be ruled out.

Once the idea of a rain covering exists, the next step is acquiring construction materials. A variety of materials was used, ranging from leafy branches to small twigs or sticks that had been frayed by chewing before use (see Figure 9.1). Usually, the individual would collect materials from whatever was within reach when the need arose. The quality of materials was not consistent, however, either across individuals, or within a single individual. I was not able to determine the characteristics affecting which materials were selected, though they may have been related to the force of the rain. As the rain fell harder, one adult female abandoned her poor quality covering and gathered new materials. One of the best examples I observed was when a young adult male (Tawashi) moved about 10 m, pulled up a large leafy plant, broke off the bottom of the stem, carried the main part of the plant back to where he had been sitting, and held it over his head in exactly the same manner as we would hold an umbrella. In fact, many of the leaves *P. paniscus* utilized as rain hats closely resemble those used by the village women and children during rain.

After selecting materials, they must be arranged in a manner that will provide protection. Considerable variation occurred in this. Some individuals would hold them as umbrellas. More common was to drape them over the head and shoulders, and maybe the back if lying down. Still others would not completely detach their branches, but break them until they bent, forming a tent-like structure over themselves. This pattern was especially common if the individual was already in a nest.

The last major component that needs to be learned involves the manner in which one should behave while using a rain hat. The typical response to rain by an unprotected individual is to move about and frequently shake the water off. This, though, is inappropriate behavior with a rain hat, as the carefully selected and arranged materials are easily dislodged by movement. The rain hat user must sit very still in order for the rain hat to be effective. I observed individual rain hat sessions that lasted from less than 10 minutes, to over 1 hour. Even during the longest sessions, the rain hat users remained very still, while nonusers kept shifting positions and shaking. If the covering materials were lost due to movement, they were usually immediately retrieved and rearranged. The difference in behavior between users and nonusers was quite striking when both types of individuals were sitting near each other.

Successfully coordinating all of these components, behavioral strategies, and skills is necessary in order to effectively manufacture and use rain hats. As with nut-cracking and termite-fishing by *P. troglodytes* (Boesch, 1993; Boesch & Boesch, 1984a,b; Goodall, 1986), several years appear to be required to achieve a high level of success.

A good example of the difficult and lengthy acquisition process necessary for successful rain hat making comes from my observations of one 7 year old male, Senta. He clearly demonstrated his incomplete mastery of the total rain covering technique. He sat in a small tree about 3 m off the ground. As the rain began, he started to bend branches over, as though to make a shelter. The branches he selected were all leafy and should have formed a good barrier against the rain. After almost 30 minutes of breaking and bending branches, Senta had not succeeded in protecting himself, and appeared somewhat perplexed. In fact, he was now more exposed than at the beginning of the process. Although having acquired the appropriate concept of a covering, the types of material needed, and some of the technique, he clearly had an incomplete knowledge of the relationship between actions and objects. Rather than bending his branches up high, where they would have covered his head and shoulders, he had bent all of them at waist level. Thus at most, Senta was managing to protect his knees. Senta's work and effort, though, was not a complete waste. His mother and four other adult females were sitting directly beneath him, now well sheltered from the rain. Three years later, when Senta was 10, I observed him constructing fully effective rain coverings. He had achieved full mastery in the intervening time. While it is impossible to determine from this single example the nature of the change that had occurred – practice, observation of the technique, cognitive development or increased manual dexterity – it was clear that sometime during the period between the two sets of observations, Senta had mastered the behavior. I did not observe any individuals younger than 10 years old make a rain hat successfully.

At first sight, the use of rain hats may not appear to be much more complicated than nest-building – both require a similar understanding of the quality of objects selected, and how one's actions can influence an object. Rain hats, though, may be more difficult. As noted above, 2 year old infants were often seen testing the effect of their actions on various objects, such as carefully bouncing on a branch while moving further and further along it, suggesting they do understand object–force relations. While 4 or 5 year old chimpanzees are observed building quite adequate nests (Goodall, 1986; McGrew, 1977; my personal observations), I only observed adolescents and adults effectively using rain-hats.

Tool use in play
At Wamba, objects are often the focus of play activity (Kano, 1992a). The objects I observed being played with most frequently were sticks and small leafy branches, approximately 20–45 cm in length. These are highly abundant and easy to obtain within the forest environment. Stones, which are the focus of object play in some species (Huffman, 1984), are not readily available. Small pebbles occur in the stream beds, but I never observed these being manipulated. Occasionally, a piece of fruit or a flower was used as a target of play.

Both solitary and social play with objects was observed. Solitary play with an object included both casual play with the object and play directly focused on the object itself. Social play was defined when two or more individuals' play incorporated an object into the activity. For this study, a "bout" of social play with objects was defined as beginning when two or more individuals were first observed playing with an object, and ending when play stopped, the object was left behind, and another activity begun. Data from a 3 month sampling period in 1990, which yielded 270 hours of observation during which I particularly focused on the use of objects in play, were examined for various characteristics. During that time period, I recorded a total of 47 bouts of object play, utilizing the technique of all observed occurrences (Altmann, 1974). When I first noted play occurring with an object, I recorded the time, who and how many individuals were involved, and the nature of the activity. I continued to observe and record the behavior until the play bout ended. Of the observed play bouts with objects, 26 (55%) were solitary play, while 21 (45%) were social play with objects. The total number of solitary play bouts with objects was greater than the total number of social play bouts with objects. However, the proportion of solitary bouts to social play bouts with an object differed with age, suggesting developmental changes may occur.

The youngest age at which infants were observed to play with objects was at around 3 months of age, the same time they would have developed sufficient hand–eye coordination to allow them to accurately reach for and grasp a particular object. For infants younger than 2 years of age, the primary form of object play observed was solitary. Infants of more than 3 years of age, however, were rarely observed in solitary object play at all, but instead engaged in social play with objects. These differences closely follow the general development in infant motor skills and social behavior, where in the first year of life, the infant is rarely out of contact with its mother (Ingmanson, 1985, and unpublished results; Nicolson, 1977). When infants do begin to leave their mother, it is initially at a very small distance. By age 3, however, infants are becoming very active in peer play and independent locomotion – though they will quickly return to mother when the group begins to travel.

Solitary object play in these youngsters included a variety of actions with the object, such as waving, dangling, slapping it against the ground, and play nest building. Infants of 2 and 3 years old were often observed engaged in play characteristic of Piaget's stage 5 object–force relations (Piaget, 1952, 1954), repeatedly dropping an object, letting a stick roll down the body, or pulling a branch and watching it bounce back. Though clearly not tool use, an understanding of this type of relationship is crucial to eventual tool-using activities. The age at which they are first observed is indicative of the developmental and learning processes involved.

I have divided the occurrence of social object play into two categories: play directly with an object, and play where the object functions as an intermediary. Direct object play included activities such as wrestling with an object, tug-of-war, and games of keep-away. Play where the object served as an intermediary included play initiations, play chases, and cases where the object was carried during play. Social play with objects as intermediaries is the more interesting of the two with regard to tool use since the object becomes a tool to achieve a social goal, rather than the direct focus of attention.

Play initiations generally consisted of breaking off a small branch and waving it at another individual. This is a fairly common pattern that is seen in some monkeys as well as apes (Goodall, 1986; Ingmanson, 1988; Simonds, 1974). Once play began, the small branch became superfluous and was left behind.

Play chases, unlike keep-away, did not appear to be attempts to acquire the object. Instead, the stick seemed to indicate the leader, identifying the individual who was "it." If caught, the individuals might wrestle or tickle, but did not contend over possession of the stick. The chase resumed when someone picked up the stick and ran off. During the entire sequence, the stick was present and appeared to serve the function of identifying the leader.

In addition to these forms of social play with objects, I also observed many occurrences where one individual, in a play group of two to four individuals, simply carried an object. This behavior caught my attention very early in my research, as I would often see an infant carrying a stick while playing with others, but apparently not playing with the stick. This activity reminded me of a behavior I had often observed in the captive *P. paniscus* group at the San Diego zoo, where a stick would be placed in the mouth and shaken back and forth in a consistent form of play invitation. I then began to examine the behavior at Wamba more consistently. It differed from the play chase in some key ways. Unlike in the play chase, the individual in possession of the stick did not appear to be identified for a particular role, and several forms of play might actually occur, such as group locomotor play or wrestling, though not with the stick. Interestingly, if the stick was dropped during the course of play, the play group often stopped while the same stick was retrieved, or replaced with a new one. Only then did play resume. If the entire group began to travel, the stick was immediately dropped as the infants or juveniles scattered to join their respective mothers. It appeared that the presence of the stick was an important indicator that play was occurring, and that without the stick, it would not continue. This behavior could generally be observed daily, and did not seem to simply be a coincidence, such as the stick being leftover from previous play. My impression was that bouts that included sticks were longer and more frequent than play bouts without sticks. An empirical analysis is not possible because I did not collect data systematically on bouts without sticks.

The use of an object as an intermediary during play was more frequent than play directly with an object. During this sample period, I recorded only 6 cases of direct social play with an object, but 27 cases of play where the object was used as an intermediary. What these three types of play have in common is that while the object is present during play, the focus is not on the object per se, but on its use as a social facilitator. Players are not playing with the stick, but the stick enhances the play, signalling to other players information and focusing attention on the activity itself. Thus, the use of an object as an intermediary during play can be considered tool use because it functions as a communicative signal that mediates coordination of social activity.

Branch dragging

The last form of tool use I observed at Wamba, and one of the ones I have found most interesting, was branch dragging. This behavior was performed primarily by adult males, with adult females and adolescents exhibiting it only rarely. Branch dragging consisted first of obtaining a "branch," often a small tree, approximately 2 m in length. These often appeared to be selected carefully, with several possibilities sometimes being rejected prior to finding the "right" one. Length and leafiness of the top appeared to be the principle characteristics chosen. Once selection was made, a delay of up to 30 minutes might occur before use. During this time the individual held onto his branch as he sat, possibly eating or grooming someone (see Figure 9.2). Finally, the individual ran through the forest, dragging the branch behind him (see Figure 9.3). This resulted in considerable noise as the leafy end of the branch moved through the underbrush. Both quadrupedal and bipedal running were employed, often within the same sequence.

Branch dragging occurred in a variety of contexts. The main ones were excitement, dominance or mild threat interactions, and group movement. During one 2 month sample period from 1990, I recorded a total of 604 occurrences of branch dragging, representing every case I observed. Of these, approximately 23% were during excitement, such as when entering a feeding site, 3% during dominance/threat interactions, and 64% were in association with group movement. I was unable to identify a context for the remaining examples. While branch dragging during excitement and dominance interactions may include aggressive components, these were absent in the latter context. Pilo-erection and pant-hooting were not exhibited. And while individuals were moved out of a branch dragger's path to avoid being run over, they did not flee, exhibit fear or perform submissive gestures.

In all of these contexts, branch dragging was used as a communicative gesture, enhancing an individual's other behavior by adding emphasis, or providing additional information. *Pan troglodytes* also perform branch dragging during excitement and dominance/threat exchanges (Goodall, 1986). However, only *P. paniscus* appear to use

Figure 9.2. *Pan paniscus* sitting with a branch selected for branch dragging.

Figure 9.3. *Pan paniscus* branch dragging.

branch dragging in association with group movement, promoting social cohesiveness rather than accenting an aggressive display. Within this context, branch dragging is used to initiate group movement, to indicate the direction of movement, to signal directional changes once movement is underway, and to keep straggling group members together.

In order to initiate group movement, one of the central, high-ranking males would begin branch dragging from the current location of the group, off into the forest for 20–30 m. He would often then return to the group along the same route, continuing to branch drag. This process would be repeated several times. The rest of the group would slowly begin to move off, following the same direction the male had used while branch dragging. Frequently, more than one male would branch drag simultaneously, sometimes in conflicting directions. It was then necessary for the group to make a choice as to whom they would follow. The basis of this decision is still unclear, but may be related to the rank of the branch dragger, the intensity of the activity, or his persistence.

Branch dragging was clearly predictive of the subsequent direction of movement of the group, and was not being performed randomly, similar to the tree drumming described by Boesch (1991) for *P. troglodytes* in the Taï Forest. Of a total of 349 individual occurrences of branch dragging during which I was able to record the direction, nearly 81% accurately predicted the course of movement. Further supporting the view that the individual had a specific location in mind while branch dragging was my ability to identify a single, discrete, goal at the end of movement. As branch dragging prior to movement usually consisted of several series of individual branch drags, I was able to group them into clusters of activity. Thus, 386 occurrences of branch dragging became 74 clusters. I recognized an apparent destination subsequent to the movement following 28 (37.84%) of these clusters. These destinations included a large fruit tree, a road that bisected this group's range, a stream where they foraged, the location of a neighboring group or a small party of the same group, and a provisioning site. The lack of a higher percentage probably has more to do with my inability to consistently follow a group and identify their target, than their lack of planning. Longer distance movement was usually fairly rapid and on the ground, allowing the group to outdistance me easily.

Within the context of group movement, most of the branch dragging did occur prior to, and at the beginning of, movement. Once movement was underway, the frequency of branch dragging decreased almost immediately and often stopped completely. In particular, branch dragging back towards the group ceased. Branch dragging is therefore principally a behavior used to initiate movement in a specific direction, although there were times when branch dragging would occur later during movement.

Occasionally, branch dragging could be heard coming from the front of the group. Sometimes this was continuing in the same direction of travel. At other times, a major change of direction in movement was initiated. The males leading the group would again branch drag, this time moving from a point on the original heading, off in a new direction. In these situations, the entire group would gradually shift to the new orientation.

In addition to indicating the direction of movement from the front of the group, one or more of the adult males would sometimes drop back to the rear of the group and begin branch dragging in a semi-circular pattern behind the last group of individuals. This behavior seemed to be directed primarily towards fully adult females, who frequently carried infants. Adult females were often the last to leave a feeding spot when the group began to travel, and might continue to forage as they moved. Thus, they were often slower than the males and tended to lag behind. This "herding" behavior did not appear to be directed toward adolescents or juveniles, who were also often scattered at the rear of the group. The action of the males in all of these situations served to keep the group oriented in the proper direction and to prevent slow individuals from becoming separated from the group.

In addition to these situations involving group movement, I also clearly observed one occasion of branch dragging being used to achieve desired movement from a specific individual. One morning in October 1987, I arrived at the nesting site prior to the awakening of the party. After about 10 minutes, one adult male, Mon, climbed out of his nest and descended to the ground. For several minutes he sat on the ground beneath the nesting tree, alternately gazing up at another nest and towards a fruit tree that was about 8 m off to the side. After a short while, Mon selected a sapling, broke it off, and began branch dragging between the nesting tree and the fruit tree. After 10 minutes of this, a head appeared over the edge of one of the remaining nests. It was Ika, another adult male, older than Mon. Mon stopped branch dragging, gave a few excited squeaks and bounces, and looked up at Ika. After only a few moments, Ika disappeared back into his nest. Mon once again began branch dragging with his previous implement, back and forth. Five minutes later, Ika again looked down, then began a slow descent towards the ground. As soon as Mon saw Ika begin his descent, he stopped branch dragging and became very excited. As Ika reached the ground, Mon ran to the fruit tree, and for the first time entered it. Ika followed more slowly. Mon had succeeded in getting a friend to join him for breakfast.

In the rainforest habitat where visibility is limited to short distances, the noise produced by branch dragging during all of these uses serves several functions: to focus attention on group leaders, to get the group moving in a desired direction, and to keep everyone together during movement. Even if not directed at particular individuals, group members can still utilize the information presented and stay with the rest of the group. It can also be used to communicate that an action is desired from a specific individual, as in the case of Mon and Ika. During branch dragging, very specific information is communicated concerning the directionality of the desired action, i.e. action is requested in the direction of branch dragging. It thus functions in part as a statement regarding intent concerning group activity. This is similar to the behavior reported for Kanzi at the Yerkes Language Center, where he uses symbols to communicate with others the planning of complex sequences of social activity (Greenfield & Savage-Rumbaugh, 1990).

DISCUSSION

Many of the kinds of tool-using behavior observed in the *P. paniscus* at Wamba are also exhibited by other great apes – *P. troglodytes*, *Gorilla gorilla*, and *Pongo pygmaeus*. For example, the use of another individual to achieve a goal, as with animate step ladders, was described in *P. troglodytes* very early by Köhler (1925), and more recently for a gorilla infant by Gómez (1990). Bard describes a similar manipulation of other individuals to obtain goals by infant orangutans – the goal consisting of food (Bard, 1990), or transport between trees (Bard, 1995). Orangutans also use bridging behavior to move between large trees in the same manner that I have described for *P. paniscus* (Bard, 1993; Chevalier-Skolnikoff *et al.*, 1982), as well as rain coverings. These behavioral similarities with orangutans are clearly associated with similarities in the environment – the dense rainforest presenting some of the same problems to be solved whether in Indonesia or central Africa.

Not all of the social play behaviors described above are unique to *P. paniscus*, either. *Pan troglodytes* also have tug-of-wars, play chases, and wave branches in play invitation (Goodall, 1986). What does seem to differ is the amount of social play with objects, particularly where the object is not the focus of the activity. On the basis of the published literature and discussions with other researchers (Goodall, 1986; Hiraiwa-Hasagawa, 1986; T. Kano, S. Kuroda and T. Nishida, personal communication), it appears to me that *P. troglodytes* infants may be engaging in more solitary object play and exploration than *P. paniscus* infants – playing at termite-fishing, ant-dipping, or hammering with stones or branches. These behaviors are important for adult *P. troglodytes*. But they are not present in the adult repertoire of *P. paniscus* at Wamba, and so should not be expected in immature individuals. Rather, *P. paniscus* use objects and tools in social contexts. The play behavior of infants and juveniles reflects this, with the use of objects in play serving to assist in the coordination of activity among individuals.

The use of branch dragging by *P. paniscus* to help coordinate group movement has not been reported for any other ape species. However, it is suggestive of communicative gestures described for *P. troglodytes* in the Côte d'Ivoire (Boesch, 1991). As at Wamba, the Taï Forest is dense, with poor visibility. These chimpanzees seem to use drumming on buttressed trees to convey information concerning the timing and direction of group movement. Boesch speculated that predation by leopards may be the factor that has selected for this behavior. At Wamba, while leopard predation does not currently appear to be a major threat, leopards do exist and have been known to follow the various groups. It is likely that prior to extensive human activity in the area, leopards posed a greater threat. The noise produced by the branch dragging may confuse or frighten a leopard, but can be decoded by other *P. paniscus* (and humans).

From my observations of the tool use and tool-related behaviors by *P. paniscus* at Wamba, the following preliminary conclusions can be made. First, *P. paniscus* at Wamba do not appear to use tools to assist in the acquisition of food. High quality food is easily available, and sophisticated methods of extraction are apparently not required. One context in which a tool may be useful is that in which the *P. paniscus* dig in the ground for mushrooms. This digging activity has been observed on many occasions, yet neither my colleagues nor I have ever observed a tool to be used in this context. For humans, using a stick to dig is so natural that I found it hard to suppress the desire to attempt to hand the digging individuals one, or to demonstrate their use. Digging with sticks has also been observed frequently in *P. troglodytes* (McGrew, 1992). Why is it then, that *paniscus* do not use tools in this context? Lack of materials is not the problem because there were adequate sticks available in the habitat. The ground where they search for mushrooms at Wamba is actually quite soft, and a stick may not add to the efficiency of the endeavor. It may be, though, that the looseness of the soil is a determining factor in deciding where they search.

Secondly, one of the things that is particularly interesting in *P. paniscus* is that they do not use objects as weapons, unlike *troglodytes*. I never observed objects being used to strike or attack another individual. Rather, *P. paniscus* use objects in cooperative social communication to facilitate social cohesion. Because of the highly cohesive nature of *P. paniscus* social structure (Kano, 1992a; Kuroda, 1980), the most frequent problems they face are in the social domain. Accurate communication is especially important in coordinating group activity. These communicative behaviors function to decrease inter-individual distance, and assist in maintaining group cohesion, as with both branch dragging and tools used in play. In contrast, branch dragging and waving in *P. troglodytes* often serve to increase the distance between individuals, being used in aggressive and threatening contexts (Goodall, 1986). At Gombe, there is greater competition for food, resulting in a selective advantage for greater spacing and looser social ties among individual.

At Wamba, the use of tools in social signalling by *P. paniscus* appears to constitute part of a suite of behaviors that control social activity, ease tensions among individuals, and promote friendly relations. Prominent among these are the various sexual behaviors that occur in nonreproductive contexts. The frequent and extensive genital–genital (g-g) rubbing between adult females has been interpreted as serving as a means of allowing genetically unrelated individuals to coexist with low levels of competition (Furuichi, 1987; Hashimoto & Furuichi, 1994; Kano, 1987, 1992a; Thompson-Handler *et al.*, 1984). It is a means of forming social bonds and indicating an acceptance of the other individual. When a new adolescent female enters a group for the first time, within a few hours she will repeatedly g-g rub with most of the resident females in the group, as well as copulate with the males (my personal observation). In contrast to *P. troglodytes*, infants and juveniles are more frequent participants in sexual activity (Hashimoto & Furuichi, 1994), affirming their membership within the group. Correspondingly, one of the more complex communicative gestures used by *P. paniscus* involves a basic rocking pattern that indicates a desire for closer contact with another individual (Ingmanson, 1992; Kuroda 1984). This basic pattern can then be modified with other gestures and postures to specify the type of contact requested – sexual, grooming, sitting together, or passing by. All of this suggests that *P. paniscus* intelligence, including tool use, may be focused in a different direction from that of *P. troglodytes*, with selection having favored the social dimensions.

Tool-using behavior has often featured significantly in some models of the evolution of human intelligence (Gibson & Ingold, 1993; Parker & Gibson, 1979, 1990). Attempts to elucidate the origins of human tool-using by looking at the tool-using behavior of the great apes, though, have been difficult due to the variation that has been observed among species, among different groups of a single species, and between captive and natural habitats (McGrew, 1993a,b). In nonhuman primates, tools have been traditionally categorized as feeding aids, weapons, or for bodily care; so consideration has been primarily restricted to their role in ecological adaptation. Occasionally, the use of an object to amplify a communicative gesture will be included (McGrew, 1992), generally in the context of intimidation. However, these are not the only activities in which tools may be utilized. Considering the importance of sociality to primates (Chance & Mead, 1988; Humphrey, 1976; Jolly, 1966, 1985), we should expect to see some coalescing of their social and manipulative behaviors. As suggested by Ingold (1993), the division in so much of the literature between technological and social factors in driving the evolution of intelligence, may in fact be unnecessary. The complexities of great ape lives in both domains provide many opportunities for selection to occur, favoring solutions to a variety of problems. In *P. paniscus*, the technological and social do come together. They manipulate objects and use tools in order to manipulate other individuals. Similar underlying intellectual abilities among the apes are expressed differentially depending on the requirements of the particular ecological and social environment.

ACKNOWLEDGEMENTS

The research and analysis described in this chapter were conducted with grants from the Wenner-Gren Foundation, NSF, and the Japan Society for the Promotion of Science. I wish to thank the Research Center for Natural Sciences of Zaire (CRSNZ) for assistance and permission to conduct research at Wamba. I am very grateful to Takayoshi Kano, Suehisa Kuroda, members of the Wamba Team, the people of Wamba, and Evelyn O. Vineberg, for their assistance, advice, and encouragement at various stages of my research. I also thank Anne Russon, Kim Bard, and Sue Taylor Parker for allowing me to contribute to this volume, and the advice they gave me in writing.

REFERENCES

Altmann, J. (1974). Observational study of behavior: Sampling methods. *Behaviour*, **49**, 227–67.

Badrian, N., Badrian, A. & Susman, R.L. (1981). Preliminary observations on the feeding behavior of *Pan paniscus* in the Lomako forest, Zaire. In *The Pygmy Chimpanzee: Evolutionary Biology and Behavior*, ed. R.L. Susman, pp. 275–99. New York: Plenum Press.

Bard, K.A. (1990). "Social tool use" by free-ranging orangutans: A Piagetian and developmental perspective on the manipulation of an animate object. In *"Language" and Intelligence in Monkeys and Apes: Comparative Developmental Perspectives*, ed. S.T. Parker & K.R. Gibson, pp. 356–78. New York: Cambridge University Press.

Bard, K.A. (1993). Cognitive competence underlying tool use in free-ranging orangutans. In *The Use of Tools by Human and Non-Human Primates*, ed. A. Berthelet & J. Chavaillon, pp. 103–17. Oxford: Oxford University Press.

Bard, K.A. (1995). Sensorimotor cognition in young feral orangutans (*Pongo pygmaeus*). *Primates*. (in press).

Beck, B.B. (1980). *Animal Tool Behavior: The Use and Manufacture of Tools by Animals*. New York: Garland STPM Press.

Boesch, C. (1991). Symbolic communication in wild chimpanzees? *Human Evolution*, 6, 81–90.
Boesch, C. (1993). Aspects of transmission of tool-use in wild chimpanzees. In *Tools, Language and Cognition in Human Evolution*, ed. K.R. Gibson & T. Ingold, pp. 171–83. Cambridge: Cambridge University Press.
Boesch, C. & Boesch, H. (1984a). Mental map in wild chimpanzees: An analysis of hammer transports for nut cracking. *Primates*, 25, 160–70.
Boesch, C. & Boesch, H. (1984b). Possible causes of sex differences in wild chimpanzees. *Journal of Human Evolution*, 13, 415–40.
Boesch, C. & Boesch, H. (1990). Tool-use and tool-making in wild chimpanzees. *Folia Primatologica*, 54, 86–99.
Chance, M.R.A. & Mead, A.P. (1988). Social behaviour and primate evolution. In *Machiavellian Intelligence: Social Expertise and the Evolution of Intellect in Monkeys, Apes, and Humans*, ed. R. W. Byrne & A. Whiten, pp. 34–49. Oxford: Clarendon Press.
Chevalier-Skolnikoff, S., Galdikas, B.M.F. & Skolnikoff, A. (1982). The adaptive significance of higher intelligence in wild orangutans: A preliminary report. *Journal of Human Evolution*, 11, 639–52.
Furuichi, T. (1987). Sexual swelling, receptivity, and grouping of wild pygmy chimpanzee females at Wamba, Zaire. *Primates*, 28, 309–18.
Gibson, K.R. (1993). Animal minds, human minds. In *Tools, Language and Cognition in Human Evolution*, ed. K.R. Gibson & T. Ingold, pp. 3–19. Cambridge: Cambridge University Press.
Gibson, K.R. & Ingold, T. (eds). (1993). *Tools, Language and Cognition in Human Evolution*. Cambridge: Cambridge University Press.
Gómez, J.C. (1990). The emergence of intentional communication as a problem-solving strategy in the gorilla. In *"Language" and Intelligence in Monkeys and Apes: Comparative Developmental Perspectives*, ed. S.T. Parker & K.R. Gibson, pp. 333–55. New York: Cambridge University Press.
Goodall, J. (1971). *In the Shadow of Man*. Boston: Houghton Mifflin.
Goodall, J. (1986). *The Chimpanzees of Gombe: Patterns of Behavior*. Cambridge, MA: The Belknap Press.
Greenfield, P.M. & Savage-Rumbaugh, S.E. (1990). Grammatical combination in *Pan paniscus*: Processes of learning and invention in the evolution and development of language. In *"Language" and Intelligence in Monkeys and Apes: Comparative Developmental Perspectives*, ed. S.T. Parker & K. R. Gibson, pp. 540–78. New York: Cambridge University Press.
Hashimoto, C. & Furuichi, T. (1994). Social role and development of noncopulatory sexual behavior of wild bonobos. In *Chimpanzee Cultures*, ed. R.W. Wrangham, W.C. McGrew, F.B.M. de Waal & P.G. Heltne, pp. 155–68. Cambridge, MA: Harvard University Press.
Hiraiwa-Hasegawa, M. (1986). Mother-infant Relationships in the Free-ranging Chimpanzees of the Mahale National Park, Tanzania. Ph.D. thesis, University of Tokyo.
Huffman, M.A. (1984). Stone-play of *Macaca fuscata* in Arashiyama B troop: Transmission of a non-adaptive behavior. *Journal of Human Evolution*, 13, 725–35.
Humphrey, N.K. (1976). The social function of intellect. In *Growing Points in Ethology*, ed. P.P.G. Bateson & R.A. Hinde, pp. 303–17. Cambridge: Cambridge University Press.
Ingmanson, E.J. (1985). The Relationship of Infant Independence to Length and Severity of Behavioral Stress Exhibited at Maternal Separation by Captive Infant Chimpanzees (*Pan troglodytes*). Ph.D. thesis. Ann Arbor: University Microfilms International.
Ingmanson, E.J. (1987). Clapping behavior: Non-verbal communication during grooming in a group of captive pygmy chimpanzees (*Pan paniscus*). *American Journal of Physical Anthropology*, 72, 173–4.

Ingmanson, E.J. (1988). The context of object manipulation by captive pygmy chimpanzees (*Pan paniscus*). *American Journal of Physical Anthropology*, 76, 224.

Ingmanson, E.J. (1992). *Pan paniscus* and the social context of complex communication. Paper presented at the VIIIth Annual meeting of the Language Origins Society, Cambridge, U.K.

Ingold, T. (1993). Technology, language, intelligence: A reconsideration of basic concepts. In *Tools, Language and Cognition in Human Evolution*, ed. K.R. Gibson & T. Ingold, pp. 449–72. Cambridge: Cambridge University Press.

Jolly, A. (1966). Lemur social behaviour and primate intelligence. *Science*, 153, 501–6.

Jolly, A. (1985). *The Evolution of Primate Behavior*, 2nd edn. New York: Macmillan.

Jordan, C. (1982). Object manipulation and tool-use in captive pygmy chimpanzees (*Pan paniscus*). *Journal of Human Evolution*, 11, 35–9.

Kano, T. (1982). The use of leafy twigs for rain cover by the pygmy chimpanzees of Wamba. *Primates*, 23, 453–7.

Kano, T. (1987). A population study of a unit group of pygmy chimpanzees of Wamba-with special reference to the possible lack of intraspecific killing. In *Animal Societies: Theories and Facts*, ed. Y. Ito, J.L. Brown & J. Kikawa, pp. 159–72. Tokyo: Japan Scientific Societies Press.

Kano, T. (1992a). *The Last Ape: Pygmy Chimpanzee Behavior and Ecology*. Stanford: Stanford University Press. (First published in Japanese in 1986).

Kano, T. (1992b). Effect of a mother's death on her offspring's social relationships with group members: Bonobos at Wamba, Zaire. *Bulletin of the Chicago Academy of Sciences*, 15, 16. (Abstract)

Köhler, W. (1925). *The Mentality of Apes*. London: Routledge & Kegan Paul Ltd.

Kuroda, S. (1980). Social behavior of the pygmy chimpanzees. *Primates*, 21, 181–97.

Kuroda, S. (1984). Rocking gesture as communicative behavior in the wild pygmy chimpanzees in Wamba, central Zaire. *Japanese Ethology*, 2, 127–37.

Martin, P. & Bateson, P. (1986). *Measuring Behaviour: An Introductory Guide*. Cambridge: Cambridge University Press.

McGrew, W.C. (1977). Socialization and object manipulation of wild chimpanzees. In *Primate Biosocial Development: Biological, Social, and Ecological Determinants*, ed. S. Chevalier-Skolnikoff & F.E. Poirier, pp. 261–88. New York: Garland STPM Press.

McGrew, W.C. (1992). *Chimpanzee Material Culture: Implications for Human Evolution*. Cambridge: Cambridge University Press.

McGrew, W.C. (1993a). The intelligent use of tools: Twenty propositions. In *Tools, Language and Cognition in Human Evolution*, ed. K.R. Gibson & T. Ingold, pp. 151–70. Cambridge: Cambridge University Press.

McGrew, W.C. (1993b). Brains, hands, and minds: Puzzling incongruities in ape tool use. In *The Use of Tools by Non-human Primates*, ed. A. Berthelet & J. Chavaillon, pp. 141–57. Oxford: Clarendon Press.

Nicolson, N.A. (1977). A comparison of early behavioral development in wild and captive chimpanzees. In *Primate Biosocial Development: Biological, Social, and Ecological Determinants*, ed. S. Chevalier-Skolnikoff & F.E. Poirier, pp. 529–60. New York: Garland STPM Press.

Nishida, T. (1980). Local differences in response to water among wild chimpanzees. *Folia Primatologica*, 33, 189–209.

Nishida, T. (1987). Local traditions and cultural transmission. In *Primate Societies*, ed. B.B. Smuts, D.L. Cheney, R.M. Seyfarth, R.W. Wrangham & T.T. Struhsaker, pp. 462–74. Chicago: University of Chicago Press.

Parker, S.T. & Gibson, K.R. (1977). Object manipulation, tool use and sensorimotor intelligence as feeding adaptations in cebus monkeys and great apes. *Journal of Human Evolution*, 6, 623–41.

Parker, S.T. & Gibson, K.R. (1979). A developmental model for the evolution of language and intelligence in early hominids. *Behavioral and Brain Sciences*, 2, 267–308.

Parker, S.T. & Gibson, K.R. (eds.) (1990). *"Language" and Intelligence in Monkeys and Apes: Comparative Developmental Perspectives.* New York: Cambridge University Press.

Piaget, J. (1952). *The Origins of Intelligence in Children.* New York: Norton.

Piaget, J. (1954). *The Construction of Reality in the Child.* New York: Basic Books.

Rijksen, H.D. (1978). *A Field Study on Sumatran Orangutans (*Pongo pygmaeus abellii, *Lesson 1827): Ecology, Behaviour and Conservation.* Mededelingen Landbouwhogeschool Wageningen. Wageningen, BV: H. Veenman and Zonen.

Savage-Rumbaugh, S.E. (1988). A new look at ape language: Comprehension of vocal speech. In *Nebraska Symposium on Motivation*, vol. 35, *Comparative Perspectives in Modern Psychology*, ed. D.W. Leger. pp. 201–56. Lincoln, NB: University of Nebraska Press.

Savage-Rumbaugh, S.E. & Rumbaugh, D.M. (1993). The emergence of language. In *Tools, Language and Cognition in Human Evolution*, ed. K.R. Gibson & T. Ingold, pp. 86–108. Cambridge: Cambridge University Press.

Simonds, P.E. (1974). *The Social Primates.* New York: Harper & Row.

Thompson-Handler, N., Malenky, R.K. & Badrian, N. (1984). Sexual behavior of *Pan paniscus* under natural conditions in the Lomako Forest, Equateur, Zaire. In *The Pygmy Chimpanzee: Evolutionary Biology and Behavior*, ed. R.L. Susman, pp. 347–68. New York: Plenum Press.

10

Comparison of chimpanzee material culture between Bossou and Nimba, West Africa

TETSURO MATSUZAWA AND GEN YAMAKOSHI

INTRODUCTION

Recent studies of chimpanzee material culture have revealed that individual communities of chimpanzees have their own cultural traditions (e.g. McGrew, 1992). Researchers who have compared findings from two or more chimpanzee communities have found many behavioral differences that appear to be culturally based (Nishida, 1987; Chapter 18). The most remarkable case involved two Tanzanian communities, Gombe and Mahale, about 150 km apart. Many differences between the Gombe and Mahale communities have been confirmed on the basis of long-term studies, including food repertoires (Nishida *et al.*, 1983), responses to water (Nishida, 1980), grooming postures (McGrew & Tutin, 1978) and tool use (Nishida & Hiraiwa, 1982).

With the progress achieved over many years of field research on wild chimpanzees, more exhaustive comparisons covering various communities across tropical Africa are becoming possible, especially comparisons of tool-using behavior (McGrew, 1992; Sugiyama, 1993). Nevertheless, we still know little about the case of neighboring communities where individual interchange is possible and genetic differences should be negligible. In such a case, an observed behavioral difference between communities can be attributed to cultural factors if environmental difference is insignificant.

We and our colleagues have studied a community of wild chimpanzees in Bossou, Guinea, West Africa, for about two decades (e.g. Sugiyama & Koman, 1979a). We also carried out an extensive survey of chimpanzees in the Nimba Mountains, the Côte d'Ivoire, only 10 km away from Bossou (Yamakoshi & Matsuzawa, 1993a). Between these communities, individual exchange is presumed to be possible, or at least to have been possible in former times (Sugiyama *et al.*, 1993b; Yamakoshi & Matsuzawa, 1993b).

The present chapter has the following aims. First, we describe the findings newly obtained from our preliminary study on the chimpanzees in Nimba, which were poorly known until now. New findings on tool use, nest-building on the ground, and so on, are reported. Secondly, we examine the behavioral similarities and differences between the two wild chimpanzee communities of Bossou and Nimba, West Africa. We report on stone tools for cracking open nuts and other forms of tool use, feeding habits, and medicinal leaves, and a field experiment on cultural transmission between the neighboring communities. Finally, we discuss the relationships between factors affecting the regional variation of traditional behaviors, such as environmental or cultural ones.

RESEARCH SITES AND OBSERVATIONS

Bossou

A small group of chimpanzees (*Pan troglodytes verus*) lives at Bossou, Republic of Guinea, West Africa. Bossou is located at 7° 39′ N and 8° 30′ W. The altitude is 550–700 m. The forest area at Bossou has been isolated from the nearby forest of the Nimba Mountains by a 3–4 km stretch of savanna vegetation. Habitat vegetation is very complex. Five small hills covered by primary forest constitute the chimpanzees' core area. Their home range covers a mosaic of primary, secondary, riverine and scrub forest, and farmland.

A preliminary study of this group was made in the 1960s (Albrecht & Dunnett, 1971). Its behavior and ecology have been investigated intensively since 1976 by Y. Sugiyama and his colleagues, who have studied demographics (Sugiyama, 1984, 1989, 1994b; Sugiyama & Koman, 1979a), social behavior (Sugiyama, 1988), ecology (Sakura, 1994; Sugiyama & Koman, 1992), and tool use (Kortlandt, 1986; Kortlandt & Holzhaus, 1987; Sugiyama, 1981; Sugiyama & Koman, 1979b). Intensive research on tool use has been carried out by T. Matsuzawa since 1986. We carried out a cooperative research study at Bossou from December 1992 to March 1993 and September 1993 to January, 1994. The present chapter is based primarily on data obtained during these periods.

Group size of the Bossou community has fluctuated between 16 and 23 since 1967 (Kortlandt, 1986; Sugiyama, 1989). Over the course of this cooperative study, the Bossou group consisted of about 18 chimpanzees including infants. All the Bossou chimpanzees are recognized individually.

Nimba

The Nimba Mountains are located to the southeast of Bossou on the borders of Guinea, Côte d'Ivoire, and Liberia (see Figure 10.1). Nimba is located to the northwest of the Taï Forest, Côte d'Ivoire, 230 km away, where a community of chimpanzees has been intensively studied since 1976 (Boesch, 1978). In Guinea and Côte d'Ivoire, the Nimba Mountains are designated as a national reserve, the Réserve Naturelle Intégrale du Mont-Nimba. The reserve contains relatively well-protected forest, approximately 170 km² in size. The Nimba Mountains' highest point is 1752 m. The region below 800 m is covered by tropical forest; above 800 m the vegetation changes gradually to savanna and the slope of the mountain becomes suddenly steeper.

Comparison of chimpanzee material culture

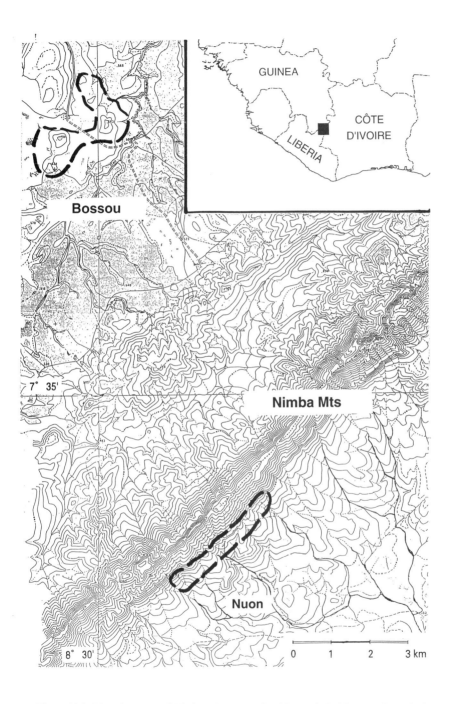

Figure 10.1. Map of Bossou and Nimba. The areas enclosed by the dashed lines are the study sites.

To date, little information is available about the population of chimpanzees in the Nimba Mountains region. Recently, F. Joulian conducted a preliminary survey on the Côte d'Ivoire side (about 50 km²) and estimated the chimpanzee population at about 50 individuals, with a density of 0.5 chimpanzees / km² (Boesch *et al.*, 1994; Hoppe-Dominik, 1991) from indirect evidence. Matsuzawa reached the summit of Mount Nimba from the Guinea side for reconnaissance in March 1986, then carried out a first survey of the Nimba Mountains in Côte d'Ivoire from January to March 1993 with Yamakoshi (Yamakoshi & Matsuzawa, 1993a).

Through the exploratory investigation of walking around this region in addition to the information from local people, we estimated that there were three groups of chimpanzees along the mountains' three rivers: the Goue, Yan, and Nuon. In January 1994, we intensively investigated a group of Nimba chimpanzees in the area of the Nuon river. We succeeded in directly observing chimpanzees of this Nuon community. This is the first report on Nimba chimpanzees involving direct observation as far as we know. The size and composition of the community were, however, still unknown. They were found living on the steep southeast side of the Nimba Mountains at an altitude of 700–1000 m. The center of this Nuon community's ranging area was located at 7° 32′ 50″ N and 8° 28′ 15″ W as measured by a Global Positioning System (SONY, PYXIS) using satellites to identify the locations.

Nut-cracking with stone tools and metatools
Bossou. A field experiment on stone tool use
Wild chimpanzees' nut-cracking behavior, cracking hard nuts with stone or wooden tools, has been discovered at several research sites (Anderson *et al.*, 1983; Beatty, 1951; Boesch & Boesch, 1981; Sugiyama & Koman, 1979b; Whitesides, 1985) within a very restricted area of West Africa (Boesch *et al.*, 1994). However, there are only two sites where direct observation of nut-cracking behavior is possible because of the difficulty of habituating chimpanzees. The two sites are the Taï Forest in Côte d'Ivoire (Boesch, 1991a; Boesch & Boesch, 1983) and Bossou in Guinea.

At Bossou, after Sugiyama's first discovery and preliminary work on the use of stone tools (Sugiyama, 1981; Sugiyama & Koman, 1979b), subsequent studies did not progress substantially because of difficulties with direct, close observation of the nut-cracking behavior in its natural habitat. The reasons are twofold. First, the bush beneath the palm trees providing the oil-palm nuts (*Elaeis guineensis*) is so thick that the nut-cracking behavior was not clearly visible. Secondly, the chimpanzees are extremely shy of humans under the palms, probably because these trees grow in secondary and scrub forest close to a village.

In order to better study the details of nut-cracking behavior at Bossou, we created an outdoor laboratory for field experiments (Matsuzawa, 1994). The outdoor laboratory was set at the top of a small hill named Gban, in the core area of the chimpanzees' home range. We transported stones and oil-palm nuts to the laboratory site at the top of the hill. Oil-palm nuts weigh 7.2 g on average (S.D. = 2.2 g); inside their hard shell is an edible kernel weighing 2.0 g on average (S.D. = 0.4 g). The nuts are oblong and round in shape, like rugby balls. The nutritional energy of the kernel is 663 kcal / 100 g (1 kcal = 4.184 kJ), close to that of walnuts, and rich in fat. The fatty acid in the kernel contains much lauryl acid in comparison to other kinds of nut. A chimpanzee cracks open a nut to obtain its kernel by placing the nut on an anvil stone then cracking the nut with a hammer stone held in the other hand (Figure 10.2).

The observer hid behind a screen made of grass (about 4 m long and 2 m high) about 20 m away from the experimental cracking site. The observer monitored activity at the cracking site continuously from 7 a.m. until 6 p.m. All episodes of nut-cracking behavior at this outdoor laboratory were observed directly and video-recorded as they occurred.

We studied nut-cracking at this site in four intermittent periods between 1990 and 1993. Chimpanzees came to the outdoor laboratory on 114 of 133 days of the field experiments (340 parties and 6564 minutes of total observation time). Our field experiments revealed a number of interesting characteristics of Bossou chimpanzees' nut-cracking behavior (for a review, see Matsuzawa, 1994). These are outlined below.

Flexibility of the nut-cracking behavior shows chimpanzees' cognitive ability to understand the relationships between tools and their functions (Sakura & Matsuzawa, 1991). Stones were used flexibly, as hammers and/or anvils, according to their shape, size, etc. Chimpanzees selected nuts on the basis of relevant characteristics, with the "best" stage being neither too old nor too fresh. Chimpanzees were able to use not only stones but also a tree trunk as an anvil, just as Taï chimpanzees do (Boesch & Boesch, 1983), when the availability of stones was limited by the experimenter.

Hand preference in nut-cracking behavior has been established (Fushimi *et al.*, 1991; Matsuzawa, 1994; Sugiyama *et al.*, 1993a). Each individual chimpanzee always uses the same hand, either right or left, as the hammer-holding hand; we emphasize that the adult chimpanzees hand preference is perfect for hammering. However, no left or right bias was found at the community level in hand preference. Longitudinal observations on hand preference demonstrate that individuals were ambidextrous and then showed strong preferences for using the same hand for hammering across development (Matsuzawa, 1994). Hand preference was not always congruent between mothers and their offspring. These findings are generally consistent with those reported from laboratory studies on chimpanzees' hand preference in reaching tasks (Tonooka & Matsuzawa, 1995) and from other studies in the wild (Boesch, 1991b; McGrew & Marchant, 1992).

"Metatool" use, the most complicated form of chimpanzee tool use found in the wild, was first observed at this outdoor laboratory. Three chimpanzees (an adult female, a $6\frac{1}{2}$ year old male, and a $10\frac{1}{2}$ year old male) were observed to perform this metatool use in the same situation. We prepared many stones varying in their shape and size. There was one stone, weighing 4.1 kg, whose upper surface was slanted. Although many chimpanzees tried to use the stone as an anvil, their attempts were unsuccessful because nuts rolled down quickly from the slanted upper surface. The three chimpanzees who solved this problem used another stone as a wedge to level and stabilize the upper surface of the anvil stone. This tool was called a metatool, that is a tool that serves as a tool for another tool (Matsuzawa, 1991, 1994).

One of our main interests in studying stone tool use at Bossou has been the

Figure 10.2. A Bossou chimpanzee cracking open an oil-palm nut with stone hammer and anvil.

development of nut-cracking ability (Inoue et al., 1994; Matsuzawa, 1994). Our longitudinal observations indicate that young chimpanzees at Bossou first succeed in using stone tools at between 3 and 5 years of age. On the basis of these data, we consider this age range to represent the critical period for acquiring the skill (Matsuzawa, 1994). The developmental changes of object–object manipulation shown in the acquisition of stone tool use illuminate the connection between tool use and cognition. The hierarchical nature of metatool use and its relation to symbol use shown in the laboratory (Matsuzawa, 1985) is discussed elsewhere (Matsuzawa, 1995).

Nimba: New discovery

At Nimba, we have directly observed chimpanzees but we have not yet witnessed stone tool use. However, we have found clear evidence from the remains left at the cracking sites that chimpanzees at Nimba have cracking techniques for some species of nut or fruit.

In areas along the Nuon river we found a site that had been used for cracking nuts by the chimpanzees. The target nut at Nimba was the coula nut (*Coula edulis* (Olacaceae)), and a pair of stones was used to crack the nuts. *C. edulis* is not available at Bossou but it is cracked frequently at Taï (Boesch & Boesch, 1983).

We also found three sites where chimpanzees had used a pair of stones to crack open the fruit of *Carapa procera* (Meliaceae). At one of the three sites, the chimpanzees had used the vertical plane of the tree trunk as an anvil (Figure 10.3). Bossou chimpanzees eat the young leaves, flowers, and the gum on the trunk of *C. procera* trees, but they do not eat the nut in the fruit (Sugiyama & Koman, 1992). *C. procera* is among the most predominant tree species at the Nimba Mountains.

Strychnos was also cracked by the Nuon chimpanzees. The hard shells of the tennis ball shaped *Strychnos* sp. are brought down to the ground and smashed against a rock without a hammering tool. This cracking technique was the same as that found in Gombe, Tanzania (Goodall, 1986). *Strychnos* is not available at Bossou.

We also found a site where the chimpanzees had smashed the fruit of an unknown species against the trunk of a tree in order to obtain its large seed (Figure 10.4). The outer tissue of this fruit is not particularly hard but it is sticky and bitter. Therefore, biting away the fruit leaves a bitter taste and tearing away the tissue manually leaves sticky hands, so neither of these techniques offers a desirable way to access the seed. Smashing the fruit against a tree trunk may offer the best technique.

We have found no evidence of oil-palm nut-cracking at Nimba simply because there are no oil-palm nuts in the Nuon chimpanzees' ranging area.

Other forms of tool use
Ant-dipping

Bossou chimpanzees use a variety of tools in addition to the stone tools. They use twigs to obtain Safari ants (*Dorylus sp.*: Sugiyama et al., 1988). We directly observed and video-recorded two examples of ant-dipping behavior, one by an adult female and one by a 6 year old male (see Figure 10.5(*a*) and (*b*)). In both cases, the chimpanzee broke off

(a)

(b)

Figure 10.3. *Carapa procera* was cracked with a stone hammer and a trunk anvil in Nimba.

Figure 10.4. Fruits of unknown sp. were smashed against the trunk of a tree to get the seeds in Nimba. For the illustration of the seed, one fruit was cut in half.

a twig or a grass stem then removed its leaves by biting them off. The processed stick or stem was then used as a wand for catching Safari ants. The wands were about 35 cm long in both cases (Figure 10.5(c); T. Matsuzawa & O. Sakura, unpublished data), comparable to the 25–55 cm long wands found in Sugiyama's 1988 study period. To date we have found no evidence of ant-dipping by the Nuon chimpanzees in Nimba although Safari ants are available. Sugiyama (1995) reported indirect evidence of ant-dipping in the Goue community.

Termite-fishing
East African chimpanzees fish for *Macrotermes* termites with twigs (Goodall, 1964). However, Bossou chimpanzees do not fish for *Macrotermes* termites from their nests but instead eat them by hand directly (Sugiyama & Koman, 1979b). We have found no evidence of termite-fishing at Nimba, where few *Macrotermes* termite mounds were found.

Use of leaves for drinking water
Bossou chimpanzees use leaves for drinking water from holes in trees (Figure 10.6; Sugiyama & Koman, 1979b; Tonooka *et al.*, 1995). To drink water, Bossou chimpanzees use a technique involving leaf-folding. They break off a leaf, put it into their mouth with one hand, fold it with the roof of their mouth, take it out, insert it into the hole in trees, then put back the folded leaf with water to the mouth. They showed a marked preference for selecting the leaves of *Hybophrynium braunianum* as tools for drinking

Figure 10.5. (*a*) and (*b*). A 6 year old male chimpanzee at Bossou ant-dipping with a wand. Print-out from a video taken by Noriko Inoue. (*c*) A selection of wands used for ant-dipping.

Comparison of chimpanzee material culture

(c)

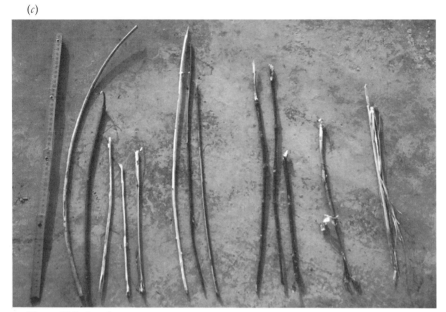

Figure 10.5. (*cont.*)

Figure 10.6. An adult female chimpanzee at Bossou drinking water with a leaf. Photograph taken by Kiyonori Kumazaki.

water. Interestingly, this leaf is also used by local people when they drink from small streams, although humans make a cup from the leaf instead of folding it.

Pestle-pounding

Recently Bossou chimpanzees were newly observed to pound then excavate the centers of oil-palm crowns with "pestles", making a hole from which they then obtained the soft and juicy pulp inside (Figure 10.7: Yamakoshi, 1994; Yamakoshi & Sugiyama, 1995). The part they eat is the apical meristem at the base of the young shoots, which may correspond to the "palm cabbage" or "palm heart" of other palm species used by humans as foods (Hartley, 1988). The chimpanzees used a petiole of the oil-palm as a pestle. In one case, a chimpanzee used a "fiber sponge" to extract sap from the hole after pounding (Sugiyama, 1994a). This type of tool use, "pestle-pounding," has never been reported from any other study site.

Feeding habits and medicinal leaves

The feeding habits of the chimpanzees at Bossou has been investigated exhaustively, generating a list of 200 species of plant food (Sugiyama & Koman, 1992) and various animal foods such as bee larvae, a species of termite (*Macrotermes bellicosus*), shrimp, fish, owl, and pangolin (Sugiyama & Koman, 1987). We have not as yet obtained corresponding data on the food repertoire of the chimpanzees at Nimba. However, we have discovered that snails (species unknown), which are not eaten by chimpanzees at Bossou, may be eaten by chimpanzees at Nimba. Further studies of the differences in flora and fauna between the two sites will extend knowledge of the differences between those two neighboring communities.

Chimpanzees are known to use medicinal plants (Rodriguez *et al.*, 1985; Wrangham & Nishida, 1983). Thus far, 13 plant species from 4 families and 7 genera have been identified as possibly used for their medicinal value by Mahale, Gombe, and Kibale chimpanzees (Huffman & Wrangham, 1994). The behavioral index of using medicinal leaves is swallowing the leaf whole, without chewing it (Wrangham & Nishida, 1983).

Chimpanzees at Bossou also use leaves of two species known to have medicinal properties, *Ficus mucuso* and *Polycephalium capitatum*. Their leaves, whole, have been found in Bossou chimpanzees' feces. Swallowing of whole *Polycephalium* leaves has also been observed directly. *Polycephalium* is used as a treatment for diarrhea by local people. At Nimba, we have also found whole *Polycephalium* leaves in chimpanzee feces (Figure 10.8). This is the only species of medicinal plant we have found this chimpanzee community using so far.

Nest building on the ground

For many years researchers have believed that wild chimpanzees almost always built their night nests in trees (Baldwin *et al.*, 1981; Fruth & Hohman, 1994; Izawa & Itani, 1966; Yerkes & Yerkes, 1929). Although some ground nests have been discovered at several study sites of both chimpanzees (Izawa & Itani, 1966; Reynolds & Reynolds, 1965) and bonobos (Horn, 1980; Kano, 1983), the proportion of ground nests to tree

Comparison of chimpanzee material culture

Figure 10.7. An adult female chimpanzee at Bossou pounding the hole of the top of a palm tree with a petiole as the "pestle".

Figure 10.8. Medicinal leaves found in the feces at Nimba.

nests was extremely small. At Gombe, the ground nests were very rare except for those made by sick individuals (Goodall, 1968). However, at Nimba we found an extraordinarily large number of chimpanzee nests on the ground as well as in the trees (Figure 10.9).

Of the total of 464 chimpanzee nests that we counted at Nimba, 164 (35.4%) were built on the ground. This percentage is extremely high. Chimpanzees do have a common habit of resting on the ground during the day and sometimes constructing nests (Goodall, 1962; Izawa & Itani, 1966). Because we did not directly observe the chimpanzees in the process of building ground nests, we are not yet certain whether the ground nests we found were used as night or day nests. Judgeing from the fact that we simultaneously found fresh ground nests in the same place as fresh tree nests, a certain proportion of the observed ground nests appear to have been used at night. Even if a part of the 35.4% is attributed to ground nests used in the day, it can be concluded that the chimpanzees at Nimba frequently build night nests on the ground, as frequently as ever known.

The ground nests could be divided into two categories. The first type was nests made of two to five small trees that were bent, formed into a circle, and/or broken. This technique of nest building and the appearance of the resulting nests were identical to regular chimpanzee nests built in trees: there was no fundamental difference between this first type of ground nest and regular tree nests, except for location. These ground nests were built on small rocks or, on some occasions, near tree trunks. The second type of ground nest was composed mainly of grass and had a cushion-like appearance. These two types of ground nest correspond to the two types of ground nest made by bonobos at Yalosidi (Kano, 1983).

Ground nests were found in high altitude areas. We located a main chimpanzee track moving from the northeast to the southwest at an altitude of 800 m. We found no ground nests at the lower altitudes below this track. These lower areas were not very steep and were covered by dense canopies of huge trees. The ground nests we found were concentrated in the higher altitude areas above this main track, on steep mountainside covered by zones of small trees, bushes, and finally grassland.

Some ecological factors may be involved in this unusual habit of building ground nests in the Nuon chimpanzees at Nimba: (a) geographical features, such as the high and steep slopes of the Nimba Mountains, do not provide good materials for tree nests; (b) it is extremely windy at these high altitudes, especially in the dry season, so tree nests may be dangerous; and (c) predation pressure by some carnivores and/or human hunters may be very low in high altitude areas.

Although further research is necessary, ours is the first reported finding of remarkable behavioral variation in chimpanzee "housing."

A field experiment reveals cultural transmission
In this final section we consider the possibility of cultural interaction between the two neighboring chimpanzee communities of Bossou and Nimba.

To explain the patterns of regional variation and the geographical distribution of behaviors between different communities of chimpanzees, some researchers (e.g. McGrew, 1992; Tomasello, 1990) have advanced the following hypothesis: some behaviors or

Comparison of chimpanzee material culture 225

(a)

(b)

Figure 10.9. Ground nests made by chimpanzees at Nimba.

traditions are transferred by individuals migrating from one community to another, so that a "cultural region" larger than the original communities is formed. In the case of chimpanzees, migrating individuals are likely to be females (Goodall, 1986; Morin et al., 1994). Through our field experiments on nut cracking we obtained interesting findings that support this hypothesis about cultural transmission.

In January 1993, we provided the Bossou chimpanzees with coula nuts (Yamakoshi & Matsuzawa, 1993b). The Bossou chimpanzees were unfamiliar with coula nuts because there are no coula trees in their home range. At Bossou, only oil-palm nuts had been cracked by the chimpanzees. Three coula nuts at a time were scattered in the outdoor laboratory area (Figure 10.10). Observers recorded all the reactions of the chimpanzees to the coula nuts directly and with video-recording, hiding themselves behind the grass screen. In the chimpanzees' first encounter with these coula nuts, 8 of the 14 community members over 3 years old sniffed the nuts, picked them up and tried to bite them, but did not attempt to crack them; 5 more simply ignored them. One adult female, Yo, estimated to be 31 years old, immediately placed a coula nut on her stone anvil, cracked it, and ate its kernel. In no time at all, she then ate the other two.

Coula nuts are green, round, and about 3 cm in diameter. Their edible kernel is embedded within a thick exocarp, so, from the outer appearance of the nut alone, it would be difficult to know that there is an edible kernel inside this nut and that the inner shell is too hard to crack without tools. Yo's immediate interest in the coula nuts and her skilled cracking suggest she was already familiar with them and very experienced in eating them.

When Yo first cracked and ate the coula nuts, a group of juvenile chimpanzees gathered around her and peered at her with great interest. However, they did not try to take the coula nuts. The next day, a 6 year old male named Vui, unrelated to Yo, successfully cracked open a coula nut without practice. Four days later, a 5 year old female named Pili followed suit. Both these juveniles cracked a nut open, sniffed its kernel, chewed it, then spat it out. Although these juveniles did not seem to swallow the kernels, they continued to crack the coula nuts during the following stage of the experiment. These observations suggest that the transmission of the behavioral skills can occur without any nutritional benefits, at least in the case of the younger individuals. They also suggest that the sense of taste is relatively conservative and must play some role in preventing individuals from swallowing unknown foods.

Although we provided coula nuts continuously for another 2 weeks, these two juveniles were the only group members who learned to crack them. No adults except Yo ever attempted to crack coula nuts, although they had a number of opportunities to observe Yo cracking the nuts. This means that all proficient tool-users in the Bossou group, except Yo, did not extend their cognitive proficiency to access this new food resource. This result strongly demonstrates the conservativeness or the "neophobia" tendency of adult chimpanzees toward unknown foods. The data are congruent with previous reports on the mechanisms of food-preference and food-aversion learning by Japanese monkeys, shown in the laboratory and in the wild (Hasegawa & Matsuzawa, 1981; Hikami et al., 1990; Matsuzawa & Hasegawa, 1982, 1983; Matsuzawa et al., 1983).

In February 1994, we conducted a further test of coula nut knowledge in the Bossou chimpanzees (Tonooka et al., 1994). We provided the chimpanzees with wooden balls

Figure 10.10. An outdoor experiment for cultural transmission of coula nut cracking at Bossou. Three nuts were provided (see arrow) but the adult chimpanzees simply neglected the unfamiliar nuts.

(3 cm in diameter) the same shape and size as coula nuts. Yo simply ignored the wooden balls. Other adult chimpanzees similarly ignored the balls or picked them up, sniffed them, then dropped them. Three young chimpanzees, Vui, Pili, and Na (an 8 year old) tried to crack the wooden balls as soon as they first appeared. This supplementary experiment demonstrates that Yo was not the sort of individual to try to crack any unfamiliar object with stone tools. The youngsters, however, appeared to try to crack any object resembling edible nuts even if the objects were unfamiliar. Their attempts to crack open wooden balls may then represent an existing tendency to try to crack open unfamiliar nut-like objects that was facilitated by their observations of Yo cracking new nuts.

Our interpretation of these results is as follows. The coula-cracking female, Yo, was born in another chimpanzee community where a tradition of cracking coula nuts already existed. She grew up and learned to crack coula nuts there, before her immigration to Bossou. Once at Bossou she would have had no further opportunity to crack coula nuts because no coula trees are found there. Our experimental manipulation reintroduced her to coula nuts; as a result, she functioned as an innovator by introducing a new kind of nut-cracking to the Bossou community.

This interpretation is congruent with the hypothesis described at the beginning of this section. New behavior was transmitted from an immigrant female to other members of her receiving community. Moreover, the field experiments demonstrated the transmission of knowledge from one generation to another. Although all the Bossou chimpanzees had the opportunity to access coula nuts, only younger chimpanzees learned to crack them by observing the informant, an adult female.

What community did Yo originally come from? It would have to be nearby with coula trees in its home range, so the Nimba Mountains are the most plausible place.

GENERAL DISCUSSION

We began our preliminary survey of the unknown chimpanzees of the Nimba mountains for the purpose of comparing neighboring communities. To date, our data from Nimba are admittedly very preliminary and sparse because direct observation of behavior is still difficult. However, we have found many provocative behavioral differences between the Bossou and Nimba chimpanzees, who live only 10 km apart, in several different types of behavior including tool use, feeding habits, and housing (see Table 10.1).

Table 10.1 *Comparison of chimpanzee behaviors between Bossou and Nimba*

Behaviors	Bossou	Nimba
Building nests on the ground	No	Yes
Medicinal uses of leaves	Yes	Yes
Polycephalium capitatum	Yes	Yes
Ficus mucuso	Yes	No
Eating snails	No	Yes
Stone tool use	Yes	Yes
Elaeis guineensis	Yes	N/A
Coula edulis	N/A	Yes
Carapa procera	No	Yes
Cracking *Strychnos* with stone	N/A	Yes
Ant-dipping	Yes	Yes
Macrotermes termite-fishing	No	No
Use of leaves for drinking water	Yes	No
Pestle pounding of palm trees	Yes	N/A

Yes, the behavior was observed; No, the behavior was not observed; N/A, the behavior was not observed because the target was not available.

In explaining these differences, we consider the possibility of genetic factors to be minimal because there has been direct genetic interchange between the two communities. What should be emphasized as potentially more influential are the environmental differences between the Bossou and Nimba regions. The former consists of small hills, mosaic vegetation, farm crops, and oil-palms. The latter is characterized by large mountains, steep slopes, great tropical forest, rich flora and fauna, and strong winds. These two environments differ so substantially from one another that the behavioral differences between their respective chimpanzee communities could be attributable to independent solutions to very different ecological pressures.

However, we hypothesize that along with independent adjustments to ecological conditions, some cultural factors must play an important role. For example, the reason why the Bossou chimpanzees except Yo were not able to crack coula nuts at first can be explained only by ecological factors because there have been no coula trees at Bossou. However, such a explanation citing only ecological factors is inadequate in the case of

Carapa procera nuts. There should be some cultural factors to explain why the Bossou chimpanzees do not crack *Carapa* nuts, although there are plenty of these nuts at Bossou.

In addition, there can be a dynamic cultural interchange between the Bossou and Nimba communities having different traditions, as our coula nut experiments suggest. We suggest that behavioral differences between communities are formed and maintained by balancing adjustments to local environments with dynamic cultural interchange. The behavior that is specific to one community can spread to neighboring communities through migrating individuals within the limitations of each community's environment. Moreover, these traditional behaviors must be maintained or changed across generations through the social transmission of traditional behaviors (Goodall, 1986; Matsuzawa, 1994). These explanations are applicable to the evolution of traditional behaviors distributed in geographically restricted areas, such as nut-cracking (Boesch *et al.*, 1994). Our future studies will focus on uncovering the mechanisms forming these behavioral differences between chimpanzee communities.

ACKNOWLEDGEMENTS

The present study was carried out with the collaboration of the people in Bossou and Direction National de la Recherche Scientifique et Technique, République de Guineé. Field research in the Nimba Mountains area was supported by the people in Yeale and the government of Côte d'Ivoire. We wish to thank Dr Yukimaru Sugiyama, who has conducted the field research at Bossou since 1976 and gave us the opportunity for research. Special thanks are due to the following colleagues during the research period: Jeremy Koman, Guano Gumi, Tino Camara, Osamu Sakura, Takao Fushimi, Hisato Ohno, Miho Nakamura, Kiyonori Kumazaki, Norikatsu Miwa, Naoto Yokota, Rikako Tonooka, and Noriko Inoue. The nutritional analysis of oil-palm nuts was done by Dr Kazuko Namiki. We thank Drs Anne Russon, Kim Bard, and Sue Parker for their helpful comments on the earlier draft. Financial support was given by a grant (no. 01041058) from the International Scientific Research Program of the Ministry of Education, Science, and Culture, Japan.

REFERENCES

Albrecht, H. & Dunnett, S.C. (1971). *Chimpanzees in Western Africa*. Munich: Piper.

Anderson, J.R., Williamson, E.A. & Carter, J. (1983). Chimpanzees of Sapo Forest, Liberia: Density, nests, tools and meet-eating. *Primates*, 24, 594–601.

Baldwin, P.J., Sabater Pi, J., McGrew, W.C. & Tutin, C.E.G. (1981). Comparisons of nests made by different populations of chimpanzees (*Pan troglodytes*). *Primates*, 22, 474–86.

Beatty, H. (1951). A note on the behavior of the chimpanzee. *Journal of Mammalogy*, 32, 118.

Boesch, C. (1978). Nouvelles observations sur les chimpanzés de la forêt de Taï (Côte d'Ivoire). *Terre et Vie*, 32, 195–201.

Boesch, C. (1991a). Teaching among wild chimpanzees. *Animal Behaviour*, 41, 530–2.

Boesch, C. (1991b). Handedness in wild chimpanzees. *International Journal of Primatology*, 12, 541–58.

Boesch, C. & Boesch, H. (1981). Sex differences in the use of natural hammers by wild chimpanzees: A preliminary report. *Journal of Human Evolution*, 10, 585–93.

Boesch, C. & Boesch, H. (1983). Optimization of nut-cracking with natural hammers by wild chimpanzees. *Behavior*, 83, 265–86.

Boesch, C., Marchesi, P., Marchesi, N., Fruth, B. & Joulian, F. (1994). Is nut cracking in wild chimpanzees a cultural behavior? *Journal of Human Evolution*, **26**, 325–38.

Fruth, B. & Hohmann, G. (1994). Comparative analyses of nest-building behavior in bonobos and chimpanzees. In *Chimpanzee Cultures*, ed. R.W. Wrangham, W.C. McGrew, F.B.M. de Waal & P.G. Heltne, pp. 109–28. Cambridge, MA: Harvard University Press.

Fushimi, T., Sakura, O., Matsuzawa, T., Ohno, H. & Sugiyama, Y. (1991). Nut-cracking behavior of wild chimpanzees (*Pan troglodytes*) in Bossou, Guinea, (West Africa). In *Primatology Today*, ed. A. Ehara, T. Kimura, O. Takenaka & M. Iwamoto, pp. 695–6. Amsterdam: Elsevier.

Goodall, J. (1962). Nest building behavior in the free ranging chimpanzee. *Annals of the New York Academy of Sciences*, **102**, 455–67.

Goodall, J. (1964). Tool-using and aimed throwing in a community of free-living chimpanzees. *Nature*, **201**, 1264–6.

Goodall, J. (1968). The behavior of free-living chimpanzees in the Gombe Stream Reserve. *Animal Behaviour Monographs*, **1**, 161–311.

Goodall, J. (1986). *The Chimpanzees of Gombe: Patterns of Behavior*. Cambridge, MA: The Belknap Press.

Hartley, C.W.S. (1988). *The Oil Palm*, 3rd edn. London: Longman.

Hasegawa, Y. & Matsuzawa, T. (1981). Food-aversion conditioning in Japanese monkeys (*Macaca fuscata*): A dissociation of feeding in two separate situations. *Behavioral and Neural Biology*, **33**, 237–42.

Hikami, K., Hasegawa, Y. & Matsuzawa, T. (1990). Social transmission of food preferences in Japanese monkeys (*Macaca fuscata*) after mere exposure or aversion learning. *Journal of Comparative Psychology*, **104**, 233–7.

Hoppe-Dominik, B. (1991). Distribution and status of chimpanzees (*Pan troglodytes verus*) on the Ivory Coast. *Primate Report*, **31**, 45–75.

Horn, A.D. (1980). Some observations on the ecology of the bonobo chimpanzee (*Pan paniscus*, Schwarz 1929) near Lake Tumba, Zaire. *Folia Primatologica*, **34**, 145–69.

Huffman, M.A. & Wrangham, R.W. (1994). Diversity of medicinal plant use by chimpanzees in the wild. In *Chimpanzee Cultures*, ed. R.W. Wrangham, W.C. McGrew, F.B.M. de Waal & P.G. Heltne, pp. 129–48. Cambridge, MA: Harvard University Press.

Inoue, N., Tonooka, R. & Matsuzawa, T. (1994). Development of use of stone tools by chimpanzees in the wild. Paper presented at the annual meeting of the Primate Society of Japan, Tokyo, Japan, 18 June.

Izawa, K. & Itani, J. (1966). Chimpanzees in Kasakati Basin, Tanganyika. 1. Ecological study in the rainy season 1963–1964. *Kyoto University African Studies*, **1**, 73–156.

Kano, T. (1983). An ecological study of the pygmy chimpanzees (*Pan paniscus*) of Yalosidi, Republic of Zaire. *International Journal of Primatology*, **4**, 1–31.

Kortlandt, A. (1986). The use of stone tools by wild-living chimpanzees and early hominids. *Journal of Human Evolution*, **15**, 77–132.

Kortlandt, A. & Holzhaus, E. (1987). New data on the use of stone tools by chimpanzees in Guinea and Liberia. *Primates*, **28**, 473–96.

Matsuzawa, T. (1985). Use of numbers by a chimpanzee. *Nature*, **315**, 57–9.

Matsuzawa, T. (1991). Nesting cups and metatools in chimpanzees. *Behavioral and Brain Sciences*, **14**, 570–1.

Matsuzawa, T. (1994). Field experiments on use of stone tools by chimpanzees in the wild. In *Chimpanzee Cultures*, ed. R.W. Wrangham, W.C. McGrew, F.B.M. de Waal & P.G. Heltne, pp. 351–70. Cambridge, MA: Harvard University Press.

Matsuzawa, T. (1995). Chimpanzee intelligence in nature and in captivity: Isomorphism of symbol use and tool use. In *Great Ape Societies*, ed. W.H. McGrew, T. Nishida & L. Marchant. Cambridge: Cambridge University Press. (in press.)

Matsuzawa, T. & Hasegawa, Y. (1982). Food-aversion conditioning in Japanese monkeys (*Macaca fuscata*): Suppression of key pressing. *Behavioral and Neural Biology*, **36**, 298–303.

Matsuzawa, T. & Hasegawa, Y. (1983). Food-aversion learning in Japanese monkeys (*Macaca fuscata*): A strategy to avoid a noxious food. *Folia Primatologica*, **40**, 247–55.

Matsuzawa, T., Hasegawa, Y., Gotoh, S. & Wada, K. (1983). One-trial long-lasting food-aversion learning in wild Japanese monkeys (*Macaca fuscata*). *Behavioral and Neural Biology*, **39**, 155–9.

McGrew, W.C. (1992). *Chimpanzee Material Culture: Implications for Human Evolution*. Cambridge: Cambridge University Press.

McGrew, W C. & Marchant, L.F. (1992). Chimpanzees, tools, and termites: Hand preference or handedness? *Current Anthropology*, **32**, 114–9.

McGrew, W.C. & Tutin, C.E.G. (1978). Evidence for a social custom in wild chimpanzees? *Man* (n.s.), **13**, 234–51.

Morin, P.A., Moore, J.J., Chakraborty, R., Jin, L., Goodall, J. & Woodruff, D.S. (1994). Kin selection, social structure, gene flow, and the evolution of chimpanzees. *Science*, **265**, 1193–201.

Nishida, T. (1980). Local differences in responses to water among wild chimpanzees. *Folia Primatologica*, **33**, 189–209.

Nishida, T. (1987). Local traditions and cultural transmission. In *Primate Societies*, ed. B.B. Smuts, D.L. Cheney, R.M. Seyfarth, R.W. Wrangham & T.T. Struhsaker, pp. 462–74. Chicago: University of Chicago Press.

Nishida, T. & Hiraiwa, M. (1982). Natural history of a tool-using behavior by wild chimpanzees in feeding upon wood-boring ants. *Journal of Human Evolution*, **11**, 73–99.

Nishida, T., Wrangham, R.W., Goodall, J. & Uehara, S. (1983). Local differences in plant-feeding habits of chimpanzees between the Mahale Mountains and Gombe National Park, Tanzania. *Journal of Human Evolution*, **12**, 467–80.

Reynolds, V. & Reynolds, F. (1965). Chimpanzees of the Budongo Forest. In *Primate Behavior*, ed. I. DeVore, pp. 368–424. New York: Holt, Rinehart & Winston.

Rodriguez, E., Aregullin, M., Nishida, T., Uehara, S., Wrangham, R.W., Abramowski, Z., Finlayson, A. & Towers, G.H.N. (1985). Thiarubrine-A, a bioactive constituent of *Aspilia* (*Asteraceae*) consumed by wild chimpanzees. *Experientia*, **41**, 419–20.

Sakura, O. (1994). Factors affecting party size and composition of chimpanzees (*Pan troglodytes verus*) at Bossou, Guinea. *International Journal of Primatology*, **15**, 167–83.

Sakura, O. & Matsuzawa, T. (1991). Flexibility of wild chimpanzee nut-cracking behavior using stone hammers and anvils: An experimental analysis. *Ethology*, **87**, 237–48.

Sugiyama, Y. (1981). Observation on the population dynamics and behavior of wild chimpanzees at Bossou, Guinea, 1979–1980. *Primates*, **22**, 435–44.

Sugiyama, Y. (1984). Population dynamics of wild chimpanzees at Bossou, Guinea, between 1976–1983. *Primates*, **25**, 391–400.

Sugiyama, Y. (1988). Grooming interactions among adult chimpanzees at Bossou, Guinea, with special reference to social structure. *International Journal of Primatology*, **9**, 393–407.

Sugiyama, Y. (1989). Population dynamics of chimpanzees at Bossou, Guinea. In *Understanding Chimpanzees*, ed. P.G. Heltne & L.G. Marquardt, pp. 134–45. Cambridge, MA: Harvard University Press.

Sugiyama, Y. (1993). Local variation of tools and tool use among wild chimpanzee populations. In

The Use of Tools by Human and Non-human Primates, ed. A. Berthelet & J. Chavaillon, pp. 175–87. Oxford: Clarendon Press.

Sugiyama, Y. (1994a). Tool use by wild chimpanzees. *Nature*, 367, 327.

Sugiyama, Y. (1994b). Age-specific birth rate and lifetime reproductive success of chimpanzees at Bossou, Guinea. *American Journal of Primatology*, 32, 311–8.

Sugiyama, Y. (1995). Tool-use for catching ants by chimpanzees at Bossou and Monts Nimba, West Africa. *Primates*, 36, 193–205.

Sugiyama, Y., Fushimi, T., Sakura, O. & Matsuzawa, T. (1993a). Hand preference and tool use in wild chimpanzees. *Primates*, 34, 151–9.

Sugiyama, Y., Kawamoto, S., Takenaka, O., Kumazaki, K. & Miwa, N. (1993b). Paternity discrimination and inter-group relationships of chimpanzees at Bossou. *Primates*, 34, 545–52.

Sugiyama, Y. & Koman, J. (1979a). Social structure and dynamics of wild chimpanzees at Bossou, Guinea. *Primates*, 20, 323–39.

Sugiyama, Y. & Koman, J. (1979b). Tool-using and making behavior in wild chimpanzees at Bossou, Guinea. *Primates*, 20, 513–24.

Sugiyama, Y. & Koman, J. (1987). A preliminary list of chimpanzees' alimentation at Bossou, Guinea. *Primates*, 28, 133–47.

Sugiyama, Y. & Koman, J. (1992). The flora of Bossou: Its utilization by chimpanzees and humans. *African Study Monographs*, 13, 127–69.

Sugiyama, Y., Koman, J. & Sow, M.B. (1988). Ant-catching wands of wild chimpanzees at Bossou, Guinea. *Folia Primatologica*, 51, 56–60.

Tomasello, M. (1990). Cultural transmission in the tool use and communicatory signaling of chimpanzees? In *"Language" and Intelligence in Monkeys and Apes: Comparative Developmental Perspectives*, ed. S.T. Parker & K.R. Gibson, pp. 274–311. New York: Cambridge University Press.

Tonooka, R., Inoue, N. & Matsuzawa, T. (1995). Leaf-folding behaviour for drinking water by wild chimpanzees at Bossou, Guinea: A field experiment and leaf selectivity. (In Japanese with English summary.) *Primate Research*, 10, 307–13.

Tonooka, R. & Matsuzawa, T. (1995). Hand preferences of captive chimpanzees (*Pan troglodytes*) in simple reaching for food. *International Journal of Primatology*, 16, 17–35.

Tonooka, R., Yamakoshi, G. & Matsuzawa, T. (1994). Field experiment on stone tool use by wild chimpanzees at Bossou, Guinea: Presentation of wooden balls similar to *Coula edulis*. Paper presented at the annual meeting of the Primate Society of Japan, Tokyo, Japan, 18 June.

Whitesides, G.H. (1985). Nut cracking by wild chimpanzees in Sierra Leone, West Africa. *Primates*, 26, 91–4.

Wrangham, R.W. & Nishida, T. (1983). *Aspilia* spp. leaves: A puzzle in the feeding behavior of wild chimpanzees. *Primates*, 24, 276–82.

Yamakoshi, G. (1994). Pestle-pounding behavior of wild chimpanzees at Bossou, Guinea: A newly observed tool-using behavior. Paper presented at the XVth Congress of the International Primatological Society, Kuta-Bali, Indonesia, 3–8 August.

Yamakoshi, G. & Matsuzawa, T. (1993a). Preliminary surveys of the chimpanzees in the Nimba Reserve, Côte d'Ivoire. (In Japanese with English summary.) *Primate Research*, 9, 13–7.

Yamakoshi, G. & Matsuzawa, T. (1993b). A field experiment in cultural transmission between groups of wild chimpanzees at Bossou, Guinea. Paper presented at the 23rd International Ethological Conference, Torremolinos, Spain, 2 September.

Yamakoshi, G. & Sugiyama, Y. (1995). Pestle-pounding behaviour of wild chimpanzees at Bossou, Guinea: A newly observed tool-using behaviour. *Primates*, 36, 489–500.

Yerkes, R.M. & Yerkes, A.W. (1929). *The Great Apes*. New Haven, CT: Yale University Press.

PART TWO

Organization of great ape intelligence: Development, culture, and evolution

11

Influences on development in infant chimpanzees: Enculturation, temperament, and cognition

KIM A. BARD AND KATHRYN H. GARDNER

INTRODUCTION

For decades, there has been discussion about whether chimpanzees possess cultural traditions (e.g. Goodall, 1973; Kawai, 1965; Nishida, 1987), and whether there are differences between groups in behavioral communicative expressions (i.e. grooming solicitations; McGrew & Tutin, 1978) or in tool use traditions (e.g. McGrew *et al.*, 1979). More recently, the topic of culture has reappeared in a narrower focus as a type of learning mechanism proposed to exist only in humans (i.e. Tomasello *et al.*, 1993). Discussions in this frame have served as a way to maintain the distinction between humans and other primates. In this chapter, we suggest an alternative approach: we discuss the process by which environmental influences impinge on a developing individual (enculturation) but we do not discuss culture in its anthropological sense of symbolically mediated knowledge transmitted across generations.

We decided to tackle this topic under the rubric of culture because of recent discussions suggesting that human "enculturation" creates new cognitive abilities in apes that otherwise would not exist (e.g. Tomasello *et al.*, 1993; Chapter 17). Indeed, results from our studies of newborn chimpanzees suggest that environmental influences can differentially affect behavioral organization as early as the first 30 days of life (Bard, 1994c). It is our plan in this chapter to illuminate some of the variables that underlie enculturation or socialization processes in order to better understand the manner in which behavioral performances are being changed.

The chimpanzees discussed in this chapter have been raised in human structured contexts. Briefly, there was a standard nursery environment in which chimpanzees who required hand-rearing were raised (for details of maternal care and lack thereof in chimpanzees, see Bard, 1994a, 1995a). The standard nursery conditions of the Yerkes Regional Primate Research Center of Emony University consist of chimpanzee groups of same-aged peers receiving regular daily interactions with adult human caregivers especially surrounding feeding times. There are two variations from these standard conditions: one is early exposure to maternal-rearing and the other is extra exposure to nurturant support of chimpanzee species-typical development. This latter chimpanzee group receives responsive nursery care while in groups of different-aged infants, 4 hours per day, 5 days per week with constant physical contact by two to five adult humans. Responsive care is designed specifically to nurture and support the development of chimpanzee species-typical behavior and competencies. The first group, the third chimpanzee "culture," consists of infants who were raised by their biological mother, for up to 25 days after birth, but then placed under conditions of standard nursery care due to insufficient and inadequate maternal care.

Effects of "cultural" differences on neonatal chimpanzees

We investigated the influence of human structured context on cognitive development, in part because of differences we found in chimpanzee neonates who were raised in the first 30 days of life under different environmental conditions. A summary of these findings is presented in this section.

Neonatal chimpanzees share many emotional expressions with human neonates. They exhibit smiles, irritable fussiness, crying, and anger (Bard & Fort, 1993). Early rearing environments, however, influenced expressions of emotional responsiveness in chimpanzees. Under standard-rearing conditions, chimpanzee infants when tested on a neonatal examination (Brazelton, 1984) responded to many items with fussiness. With responsive care, however, chimpanzee infants responded to some of these same items with anger (Bard & Fort, 1993; K.A. Bard, P. Herbert & K. McDonald, unpublished results). In fact, a significantly greater number of responsive care infants expressed anger than did standard care infants, and responsive care infants were found to be angry at an earlier age than standard care infants. Early rearing environments influenced the number of individuals who expressed some emotions and influenced the age at which some emotions were expressed (Bard, 1995b).

Preliminary results comparing the standard to responsive care groups on the neonatal tests that have been conducted have been reported (Bard et al., 1993). Most of the measures indicate that the neuro-behavioral integrity of chimpanzee neonates does not differ according to the rearing condition. A few measures do differentiate infants in each of the rearing conditions, as early as the first days of life. These measures indicate that some organizational aspects of neonatal behavior are influenced by early environmental variables. For instance, alertness differs between groups and across days: the responsive care and standard care groups are born exhibiting approximately the same level of alertness, but after 1 week of differential rearing the responsive care infants are significantly more alert compared with the standard care infants through the rest of the neonatal period. The quality of alertness reflects the neonate's ability to maintain a calm and alert behavioral state. The organization of behavioral state in neonates appears to be heavily influenced by early environment (Bard, 1994b; Nugent et al., 1989).

Responsive care infants exhibit significantly less ability to self-quiet compared with standard care infants. Both standard care and responsive care infants are typically able to quiet by themselves at least once during the Neonatal Behavioral Assessment Scale (NBAS) testing. The standard care infants, however, are able to quiet themselves briefly a few additional times. The more general ability to regulate behavioral state is measured by the State Regulation cluster score, reflecting the ability to cope when aroused (Brazelton, 1984). A significant difference is found between the groups in general coping ability. Thus, we again appear compelled to conclude that early environment strongly influences the developing neuro-behavioral organization of chimpanzees.

Responsive care appears to affect female and male neonatal chimpanzees differently. Under standard care, males have more muscle tone than females, both at birth and at 30 days of age. With responsive care, however, this difference is accentuated at 2 days of age. Males have more muscle tone than females at 2 days of age. At 30 days, however, responsive care females have greater muscle tone than responsive care males. This same pattern is evident in another motor item, activity, so it seems not to be a spurious effect. Muscle tone is thought to be an inborn characteristic that may differ between males and females in human neonates (e.g. Hopkins & Bard, 1993; Korner, 1973; Maccoby, 1966). We were surprised to find that apparently inborn characteristics in chimpanzees are malleable to environmental influences in the first 30 days of life.

Finally, under standard conditions, females exhibited more smiles. The number of smiles reflects social responsiveness. With responsive care, males smiled relatively more frequently and the females smiled relatively less frequently at 30 days of age compared with standard care. Responsive care appears to negate the sex differences in smiling.

Chimpanzee and human comparisons

One might argue that the findings of contextual influences on chimpanzee neonatal behavior do not pertain to similar influences on human neonates. Prior to conducting the comparisons between chimpanzee groups, however, comparisons were made between chimpanzee neonates and human neonates. These findings suggest that standard care chimpanzees are similar to human neonates in orientation performance and autonomic nervous system stability (Bard, 1994b; Bard et al., 1992). Differences were found between standard care chimpanzees and human infants in measures of neonatal coping. When additional comparisons were made with chimpanzee infants who remain with their mothers (Bard, 1994c; Hallock et al., 1989), however, it was apparent that few "pure" species effects exist. Moreover, few differences could be attributed solely to early contact with humans. Rather, it appeared to be the case that early rearing experiences and species membership jointly influenced the majority of behaviors and neuro-behavioral integrity in newborns (K.A. Bard, K.A. Platzman, S.J. Suomi, & B.M. Lester, unpublished results).

Effects on "cultural" differences in infant chimpanzees

Studies documenting attentional, organizational, and temperamental differences in neonatal chimpanzees (described above) set the stage for those that focus on chimpanzee infants up to 1 year of age (Bard, 1991). The study described in this chapter investigated the effect of early environmental influences on chimpanzees' performance on a standardized test of cognitive development. Cognitive development was measured with a human-based standardized test (Bayley Scales of Infant Development: Bayley, 1969) administered monthly from 3 to 12 months of age to all groups of chimpanzees. An important aspect of this cognitive test is the accompanying assessment of temperamental responses, as we expected to find increasing effects on cognition and behavioral expressions of temperament.

METHOD

Subjects and settings

Chimpanzees raised in the Yerkes Regional Primate Research Center nursery from 1988 through 1994 were the subjects. Fifteen infants were given standard care; nine infants were given responsive care; and five infants were given standard care after spending up to 25 days under the marginally adequate care of their biological mothers (Bard, 1994a, 1995a).

Standard care

It is important to begin with a brief description of standard nursery care (SC) practices at Yerkes, because infants raised in these standard conditions are the baseline group. In general, standard care for chimpanzees at the Yerkes Center consists of peer groups, formed as early as 6 weeks of age, who are given human contact during regular caregiving activities. Neonatal chimpanzees remain in an incubator for temperature regulation for the first 30 days of life, or until they reach 2000 grams in weight. Caregivers change the chimpanzees' diapers and feed bottles on a 4-hour schedule, and researchers administer behavioral tests (e.g. Bard *et al.*, 1992). Then the infants are placed in padded metal cages with towels and social companions (as many as are available). Around 5 months of age, the standard-reared chimpanzees are moved during the day, from their cage to a crib that is placed inside a new room. Around 7–8 months, the crib is removed in the day room and climbing structures are added. Around 1 year of age, bottles are replaced by cups, diapers are removed, and the chimpanzees no longer receive towels, primarily because the latter present a health risk to the animals when eaten. Weather permitting, at 1 year of age, standard care chimpanzees spend the day in an outdoor play yard.

Responsive care

The second chimpanzee group is responsive nursery care (RC) group. This rearing environment was designed to enhance behavioral well-being by providing care that more closely approximates species-typical care compared with standard care. Responsive care was designed to provide for both the physical needs and the emotional needs of the chimpanzee infants (in a manner similar to that proposed by Fritz & Fritz (1985)). In general, for a 4-hour period each afternoon, the chimpanzee infants are provided with an environment as much like a species-typical one as human caregivers can provide (Veira & Bard, 1994). Continuous physical contact with human adults is provided. Bottles are provided on demand, and the infants are fed with body contact appropriate for their age. Events that occur abruptly in standard care, are eased transitions in responsive care. For instance, in responsive care, solid food is presented as teeth emerge, often serving as a natural teething device. The transition from bottles to cups is made easier with a spill-proof cup, which the chimpanzees can learn to handle on their own, with success. Responsive care infants experience many new events at an early age. For instance, 6 month old responsive care infants sit, walk, and climb in the outdoor enclosures.

Responsive care enhances both physical development and social development of nursery-reared chimpanzees. Physical development is nurtured, for example, by the requirement that they learn to support their own weight rather than be carried; social development is nurtured, for example, with exposure to the sights, sounds, and smells of infant, juvenile, and adult chimpanzees. Responsive care infants learn to ride dorsally, around 6 months of age, as they would if they remained with their biological mother. Their development is nurtured with the researchers' physical and emotional support. Responsive care infants are given fleece in addition to towels. Fleece doesn't string like towels do. So, if the chimpanzees eat it, there is no health risk. Therefore, fleece can be provided for emotional comfort past 1 year of age, the time at which towels have to be taken away from standard care infants.

Late arrivals
The third chimpanzee group experienced a mixture of environments: mother-rearing for days or weeks, and then standard nursery-rearing. We have labeled these subjects as late arrivals (LA). A small group of chimpanzee infants initially remain with their biological mothers, but these mothers possess marginal maternal competence, which means that, although they exhibit sufficient maternal behaviors for the infant to survive initially (that is, cradling the infant and allowing the infant to nurse), they do not have sufficient maternal behaviors for the infant to remain past the first weeks of life. Typically, marginal behaviors include an unresponsiveness to the infant's distress, overgrooming, and putting the newborn on the ground (Bard, 1994a). When these behaviors occur with sufficient frequency to put the infant at risk for survival, then the infant is placed in the standard nursery. Every effort is made at the Yerkes Center to have every infant chimpanzee remain with the mother, so these infants can be quite stressed by the time they are removed from her marginally competent care.

Procedure
The Bayley Scales of Infant Development (BSID) were administered to a total of 29 chimpanzees. The BSID is one of the most widely used assessments for human infants (Francis *et al.*, 1987). We use the BSID, in particular, to assess behavioral, cognitive, and manipulative ability, to assess temperamental responsiveness, and to compare chimpanzee performance to that of human infants. This is a standardized test consisting of the presentation of objects to look at, to manipulate, to imitate actions upon, and to combine with other objects. Manipulative ability is age-graded for young infants and a Mental Development Index (MDI) is calculated which is used as a measure of cognitive

ability (Bayley, 1969). The MDI scores (like intelligence quotient (IQ) measures) are statistically normalized, which means that the score for a typical human infant is 100 and scores that deviate by more than three standard deviations (1 S.D. is 16 points) are considered exceptional; higher scores indicate more advanced cognitive ability.

The BSID was administered to each nursery-reared chimpanzee each month from 3–12 months of age. Some monthly tests were not conducted with some infants due to scheduling conflicts and occasional illness. Therefore, monthly sessions were averaged into 2 or 3 month bins, based on *a priori* considerations. The MDI obtained for each chimpanzee's manipulatory performance was a "generous" score. Chimpanzees were not penalized for failing language items that are not part of their natural repertoire. Preliminary analyses found no differences when "generous" versus "conservative" (i.e. no credit for language) scores were used (Bard *et al.*, 1991).

The Infant Behavior Record (IBR) is the second aspect of the Bayley examination and documents each individual's emotional responsiveness or personality (Bayley, 1969; Matheny, 1980). Each of the 30 items of the IBR measures a different aspect of emotionality, sociability, test-taking skills, or objective orientation. The third part of the Bayley examination measures motor development, which was not assessed in this study.

The 30 items of the IBR were subjected to a factor analysis. Twenty-three items of the IBR were loaded onto five factors: task, affect, activity, coordination, and audio-visual reactivity. Similar factors have been identified for human infants (e.g. Matheny, 1983; van der Meulen & Smrkovsky, 1985).

Repeated measures of analysis of variance (ANOVA) were used to compare group performance across the first year of life on MDI, the IBR factors, and IBR items. When significant developmental changes were found, then ANOVAs were conducted separately at each age. Pearson product moment correlations were conducted to assess the stability of each factor across age. This assessment is important to evaluate whether the variables are measuring stable individual characteristics.

RESULTS

Cognitive performance
Three month old chimpanzees typically oriented to sights and sounds and by 5 months were grasping and mouthing objects. In contrast, by 12 months, the chimpanzees were not combining objects or using objects in functional ways, including as tools. Chimpanzees' MDI scores were within the range of human infants. Nine subjects were classified as exceptional in the 3–5 month age group (4 of 15 standard care (SC); 5 of 9 responsive care (RC); and 0 of 5 late arrivals (LA)), scoring more than 3 S.D. above the human norm. In the 6–7 month age range, three chimpanzees had MDI scores more than 3 S.D. above the mean (one subject in each group). In the 8–9 month age range, one subject continued to score exceptionally highly (SC) but 3 subjects scored 3 S.D. below the human mean (2 of 15 SC; 1 of 5 LA). In the 10–12 month age range, only one chimpanzee (SC) had an exceptional score and it was below the mean.

There was no statistical difference among the groups on the MDI scores, the

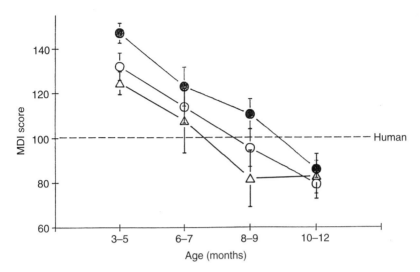

Figure 11.1. MDI scores for (●) responsive care, (○) standard care, and (△) late arrival (having stayed with mother < 25 days) chimpanzee subjects.

cognitive performance measure of the BSID (Figure 11.1). Additionally, there were no sex differences in MDI scores nor any significant interaction of sex with group or age. There was a trend for the responsive care group to perform better than the late arrival group ($F(1,12) = 4.06$, $p = 0.067$). Performance for all groups significantly declined from month 3–5 to month 10–12 ($F(3,78) = 42.89$, $p < 0.001$). Due to this significant developmental change, an additional analysis was run to compare groups on MDI performance on a month-by-month basis. Responsive care infants scored significantly higher than late arrival infants in the month 3–5 tests ($F(1,12) = 9.96$, $p < 0.01$) and in the month 8–9 tests ($F(1,12) = 4.76$, $p < 0.05$) but no other groups differed at any other months.

The MDI is normalized for human infants, so comparisons were conducted between the chimpanzee groups and human infants (Bayley, 1969). All groups of 3–5 month old chimpanzees performed at significantly higher levels compared with 3–5 month old human infants. At 6–7 months, standard care ($t = 2.88$ $p < 0.05$) and responsive care ($t = 3.8$ $p < 0.001$) chimpanzees performed significantly better than human infants, but late arrivals were no different from human infants. All groups of chimpanzees at 8–9 months performed at cognitive levels similar to that of human infants at 8–9 months of age. At 10–12 months, human infants begin to exhibit turn taking, tool use, and other object–object combinations, but nursery-reared chimpanzees do not. Human infants, 10–12 months of age, performed significantly better than standard care ($t = 4.4$, $p < 0.001$) and responsive care ($t = 2.39$, $p = 0.05$) chimpanzees, but only showed a tendency to perform better than late arrivals ($t = 2.2$, $p = < 0.1$). The small number of subjects in the late arrival group, however, contributes to some difficulty in obtaining statistically significant results. Preliminary analyses also found that chimpanzees performed

Table 11.1. *Percentage of subjects in each rearing group with a temperamentally reactive Bayley test*

	Month				
	3–5	6–7	8–9	10–12	At any time
SC ($n = 15$)	13	0	13	20	36
RC ($n = 9$)	22	0	0	0	22
LA ($n = 5$)	0	40	0	0	40

at lower levels, compared to humans at 11, 12, and 13 months of age (Bard et al., 1991).

Individual chimpanzees exhibited high stability in MDI scores in the first year of life. Correlations between consecutive months were high and statistically significant (month 3–5 with 6–7, $r = 0.47$, $p < 0.01$; 6–7 with 8–9, $r = 0.77$, $p < 0.001$; 8–9 with 10–12, $r = 0.67$, $p < 0.001$) whereas correlations remained statistically significant but were moderate between early and late tests (month 3–5 with 8–9, $r = 0.52$, $p < 0.01$; 6–7 with 10–12, $r = 0.44$, $p < 0.02$; 3–5 with 10–12, $r = 0.36$, $p < 0.058$). In other words, MDI scores in chimpanzees at 3–5 months were predictive of their cognitive performance throughout the first year of life.

Gross responsiveness

Some infants found the structure of the testing conditions distressing (Table 11.1). In the responsive care group, two infants were distressed at the 4 month test; in the late arrival group, two infants were distressed in the 7 month test. In standard care, one infant was distressed in both the 5 and 8 month tests, and one infant was distressed in both the 9 and 11 month tests; one infant was distressed in both the 4 and 12 month tests; and one infant was distressed in the 11 month test. A χ^2 test revealed that there was no group difference in the relative number of infants who found the standardized conditions distressful, even though three of the four standard care infants and one of the other groups reacted to more than one test with distress.

Personality Factors of the IBR

Factor analysis revealed that the 23 items of the IBR could be loaded onto five personality factors, listed and defined, with loading factors, in Table 11.2. There were no differences between male and female chimpanzees on any of the five factors nor any interactions between gender and group. Therefore, all subsequent analyses consisted of group comparisons. Table 11.3 contains the group means and standard deviation of these personality factors and items of the IBR. In the paragraphs that follow, we present the results of statistical analysis of each factor separately. Although these analyses compare groups on explicitly noncognitive measures, it is easy to extrapolate the

Table 11.2. *Results of the factor analysis: Loading of each IBR item onto factors*

Factor 1: Task (TASK)
  0.75 Responsiveness to objects (RsO)   1 = No interest, 9 = wants to keep object
  0.71 Mouth or sucks on objects (MoO)   1 = Never, 9 = always
  0.71 Attention span (ASp)   1 = Fleeting attention, 9 = absorbed
  0.69 Listen to sounds (Hear)   1 = Never, 9 = excessive
  0.68 Manipulate objects with hands (ManO) 1 = Never, 9 = excessive
  0.59 Goal-directedness (Goal)   1 = None, 9 = compulsive completion
  0.55 Look at sights (Look)   1 = Never, 9 = excessive

Factor 2: Affect – extraversion (AFFECT)
  0.80 Degree of happiness (Happ)   1 = Unhappy, 9 = radiates happiness
  0.76 Sociability to examiner (ScE)   1 = Avoidance, 9 = inviting
  0.72 Sociability to Mother (ScM)   1 = Avoidance, 9 = inviting
  −0.69 Fearfulness (Fear)   1 = Accepting, 9 = strong fear
  0.67 Cooperativeness (Coop)   1 = Resist, 9 = enthusiastic joining
  0.61 Endurance to demands of test (Endu) 1 = Tires, 9 = interested even if difficult

Factor 3: Activity (ACTIV)
  0.89 Body motion (Move)   1 = None, 9 = excessive
  0.86 Activity (Act)   1 = Quiet, 9 = hyperactive
  0.81 Energy and coordination for age (Ener) 1 = Low, 5 = high
  0.57 Produce sounds by banging (Bang) 1 = None, 9 = excessive

Factor 4: Coordination (COOR)
  −0.72 Fine muscle coordination (Fine) 1 = Smooth, 5 = poor
  −0.66 Gross muscle coordination (Gross) 1 = Smooth, 5 = poor
  0.59 Attachment to object (AtO)   1 = Yes, 2 = no

Factor 5: Audio-visual reactivity (AVAW)
  0.64 Tension of the body (Tens)   1 = Flaccid, 9 = rigid
  0.57 Reactivity or sensitivity (Reac) 1 = Unreactive, 9 = responds quickly
  0.56 Vocalize sounds (Voc)   1 = None, 9 = excessive

Loadings of 0.55 and higher were included in each factor. These are considered good to excellent by Tabachnick & Fidell (1983).

influence of personality variables on cognitive test performance. Increasingly, with age, these temperamental factors will impact all assessments of behavioral performance.

Task

The task factor measures general responsiveness to the demands of the BSID, such as: interest in objects; persistence in attaining goals; attention span; and sensory responsiveness to the sights, sounds, feel, and tastes of the test. Group performance on the task factor is

Table 11.3. *Values (means + standard deviation) on the items of the Infant Behavior Record for three groups of nursery-reared chimpanzees*

ITEM	3–5 month			6–7 month			8–9 month			10–12 month		
	SC	RC	LA	SC	RC	LA	SC	RC	LA	SC	RC	LA
TASK[a]												
RsO	4.3 ± 1.2	4.6 ± 0.9	3.9 ± 0.4	5.2 ± 1.3	5.4 ± 0.8	3.7 ± 1.2	5.4 ± 1.6	5.8 ± 0.5	4.8 ± 1.6	5.8 ± 1.4	6.4 ± 0.6	5.4 ± 1.6
MoO	4.2 ± 1.7	4.8 ± 1.2	3.6 ± 0.7	5.3 ± 2.0	5.7 ± 1.5	3.6 ± 1.6	5.6 ± 2.1	6.3 ± 0.6	4.5 ± 2.0	5.8 ± 2.4	6.8 ± 1.3	6.1 ± 1.7
ASp[a]	4.3 ± 2.1	4.4 ± 1.4	3.8 ± 1.3	6.4 ± 1.9	5.3 ± 1.0	5.0 ± 1.6	7.0 ± 1.7	6.4 ± 1.5	5.5 ± 1.9	6.2 ± 2.3	7.4 ± 1.3	5.6 ± 2.3
Hear[b]	4.2 ± 1.4	3.8 ± 1.2	3.1 ± 0.5	5.1 ± 1.5	5.0 ± 1.3	3.3 ± 1.8	5.5 ± 1.5	5.4 ± 0.8	4.1 ± 1.9	5.2 ± 1.6	5.4 ± 1.0	4.9 ± 1.8
ManO[a,b]	5.6 ± 0.9	6.4 ± 0.7	6.0 ± 0.8	6.0 ± 0.9	6.3 ± 1.2	5.2 ± 1.1	6.3 ± 1.2	6.7 ± 0.7	6.5 ± 2.0	6.2 ± 1.3	7.1 ± 0.8	6.3 ± 1.6
Goal[a]	3.6 ± 2.2	3.6 ± 1.1	2.8 ± 1.2	3.8 ± 1.1	4.3 ± 1.5	2.8 ± 1.3	3.7 ± 1.9	4.8 ± 1.1	3.2 ± 1.6	4.0 ± 1.1	5.8 ± 0.9	4.1 ± 1.5
Look	2.7 ± 1.6	3.3 ± 1.2	2.1 ± 0.8	3.8 ± 1.2	3.8 ± 1.3	2.2 ± 1.2	4.3 ± 1.5	4.6 ± 1.1	3.2 ± 1.0	5.3 ± 1.9	5.2 ± 1.1	3.8 ± 1.1
	6.9 ± 1.0	6.6 ± 0.9	6.8 ± 1.0	6.9 ± 1.0	7.3 ± 0.6	5.7 ± 1.4	7.0 ± 1.3	6.7 ± 0.7	6.7 ± 1.8	6.5 ± 1.3	6.8 ± 0.5	6.7 ± 1.6
AFFECT[a]												
Happ[a]	2.8 ± 1.0	2.5 ± 0.7	2.4 ± 0.4	2.6 ± 1.3	3.2 ± 1.2	1.6 ± 1.2	2.7 ± 1.5	3.8 ± 1.0	2.1 ± 1.4	2.8 ± 1.7	4.2 ± 0.6	3.1 ± 1.4
ScE	5.3 ± 1.3	4.5 ± 1.0	4.7 ± 0.4	4.7 ± 1.7	5.7 ± 1.8	3.4 ± 2.0	5.1 ± 2.2	6.7 ± 1.5	4.6 ± 2.4	5.3 ± 2.2	7.3 ± 1.1	5.5 ± 1.9
ScM	3.0 ± 0.8	2.7 ± 0.8	2.7 ± 0.5	2.9 ± 1.1	3.4 ± 1.0	2.0 ± 0.9	3.0 ± 1.3	3.6 ± 1.2	2.6 ± 1.4	3.0 ± 1.4	3.7 ± 0.7	2.8 ± 1.3
Fear[a,b]	3.8 ± 0.7	3.5 ± 0.4	3.9 ± 0.6	4.0 ± 0.8	4.2 ± 0.7	3.9 ± 0.8	3.9 ± 1.2	4.0 ± 0.6	3.9 ± 0.6	3.8 ± 1.2	4.4 ± 0.5	4.1 ± 0.6
Coop[a]	5.0 ± 1.3	4.8 ± 1.3	5.2 ± 1.7	5.2 ± 1.9	3.7 ± 2.5	7.1 ± 1.5	5.6 ± 1.8	2.9 ± 1.9	6.5 ± 1.6	4.6 ± 2.2	2.2 ± 0.9	4.5 ± 2.4
Endu	4.4 ± 1.2	4.4 ± 0.8	4.2 ± 0.6	4.5 ± 1.9	4.5 ± 1.8	2.8 ± 1.4	4.5 ± 1.8	5.6 ± 1.1	3.9 ± 1.5	4.2 ± 2.2	5.7 ± 1.7	4.7 ± 1.7
	5.5 ± 1.4	4.7 ± 0.8	4.4 ± 0.5	5.1 ± 1.4	5.4 ± 1.0	4.3 ± 1.6	5.8 ± 1.4	5.7 ± 1.2	4.7 ± 1.2	5.7 ± 1.6	5.9 ± 1.2	5.8 ± 1.8
ACTIV[a]												
Move[a]	3.8 ± 1.0	4.6 ± 1.0	3.4 ± 0.8	4.8 ± 1.9	5.2 ± 1.1	2.9 ± 1.7	4.7 ± 1.6	5.5 ± 1.1	4.1 ± 2.0	5.5 ± 1.6	5.9 ± 0.8	5.2 ± 1.6
Act[a]	5.2 ± 1.6	5.8 ± 1.7	3.4 ± 0.9	5.1 ± 2.4	6.4 ± 1.4	3.9 ± 2.4	5.6 ± 2.3	6.4 ± 1.5	4.8 ± 2.1	6.4 ± 2.3	7.0 ± 1.2	3.2 ± 2.3
Ener[a]	4.4 ± 1.5	5.5 ± 1.4	4.2 ± 0.9	4.9 ± 2.1	5.7 ± 1.5	2.9 ± 2.4	5.7 ± 2.1	6.3 ± 1.1	4.5 ± 2.1	6.2 ± 1.8	6.8 ± 0.8	5.9 ± 1.6
Bang	3.8 ± 0.9	3.9 ± 0.6	3.2 ± 0.5	3.5 ± 1.1	4.2 ± 0.9	2.5 ± 1.1	3.6 ± 1.2	4.4 ± 0.4	3.1 ± 1.4	4.0 ± 0.9	4.6 ± 0.5	3.8 ± 0.9
	1.8 ± 0.9	3.1 ± 1.4	1.5 ± 0.8	4.2 ± 2.4	4.6 ± 1.2	2.7 ± 1.0	3.5 ± 1.9	4.7 ± 2.4	3.8 ± 2.9	4.1 ± 1.9	5.3 ± 2.2	5.7 ± 2.2
COOR												
Fine	−1.0 ± 1.2	−0.8 ± 0.8	−1.2 ± 1.2	−0.9 ± 1.2	−1.1 ± 0.5	−0.4 ± 0.6	−1.4 ± 0.5	−0.8 ± 0.5	1.0 ± 1.1	−0.7 ± 0.7	−0.5 ± 0.4	−0.9 ± 0.6
Gross[b]	3.2 ± 0.8	3.0 ± 0.7	3.8 ± 0.8	3.0 ± 0.8	3.0 ± 0.7	2.8 ± 0.4	3.0 ± 0.5	2.6 ± 0.9	3.0 ± 1.3	2.3 ± 0.8	1.9 ± 0.6	2.8 ± 0.4
AtO[c]	3.1 ± 0.8	2.4 ± 0.9	3.5 ± 1.1	2.7 ± 0.8	2.3 ± 0.9	2.3 ± 1.0	2.5 ± 0.9	1.6 ± 0.6	2.4 ± 1.2	1.9 ± 0.8	1.5 ± 0.5	2.1 ± 0.8
	1.6 ± 0.3	2.0 ± 0.2	1.3 ± 0.3	1.4 ± 0.4	1.9 ± 0.2	1.6 ± 0.5	1.4 ± 0.4	1.8 ± 0.4	1.4 ± 0.2	1.7 ± 0.3	1.9 ± 0.2	1.7 ± 0.3
AV/AW												
Tens[a,b]	4.6 ± 1.2	4.5 ± 0.8	5.5 ± 0.5	5.0 ± 1.4	4.6 ± 0.9	5.0 ± 1.3	5.2 ± 1.3	4.1 ± 0.6	5.1 ± 1.3	4.9 ± 1.2	4.1 ± 0.6	4.5 ± 1.0
Reac	5.1 ± 1.0	4.4 ± 0.7	5.5 ± 0.8	5.3 ± 1.5	4.0 ± 0.7	5.3 ± 2.1	5.1 ± 1.5	4.0 ± 0.9	5.4 ± 1.7	4.7 ± 2.0	3.8 ± 0.3	4.5 ± 1.4
Voc	5.1 ± 2.0	5.6 ± 1.3	6.5 ± 1.1	5.9 ± 1.3	6.2 ± 1.5	4.9 ± 0.8	6.6 ± 1.1	6.0 ± 1.1	5.4 ± 1.8	6.8 ± 1.7	6.4 ± 1.3	6.7 ± 1.3
	3.8 ± 2.2	3.8 ± 1.6	4.4 ± 1.2	3.5 ± 2.0	3.6 ± 2.2	3.8 ± 2.1	3.8 ± 1.5	2.4 ± 1.2	4.5 ± 2.3	3.0 ± 1.3	2.3 ± 1.1	2.5 ± 0.9

SC, standard care (n = 15); RC, responsive care (n = 9); and LA, late arrivals (n = 5).
For item abbreviations see Table 11.2. [a]Significant difference between RC and LA. [b]Significant difference between SC and RC. [c]Significant difference between groups.

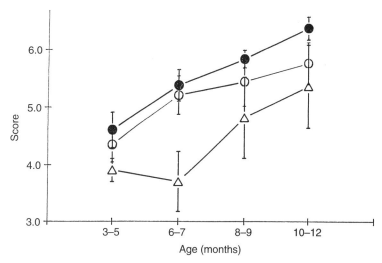

Figure 11.2. Task factor scores for responsive care, standard care, and late arrival chimpanzee subjects. Symbols as for Figure 11.1.

illustrated in Figure 11.2. Higher scores indicate better ability to handle the demands of the task. Task scores improved across the first year of life ($F(3,78) = 13.39, p < 0.001$). There was no statistical difference in task scores between standard care and responsive care chimpanzees. The responsive care and late arrival groups differed in their level of general responsiveness to the demands of the task ($F(1,12) = 7.40, p < 0.02$).

Consideration of the items included in the task factor revealed that responsive care chimpanzees compared with late arrival chimpanzees were more responsive to objects ($F(1,12) = 7.05, p < 0.05$), responsive care infants had a greater attention span ($F(1,12) = 4.69, p = 0.051$), responsive care infants exhibited more manipulation of objects with the hands ($F(1,12) = 8.85, p < 0.05$) and were more persistent in attaining goals ($F(1,12) = 8.17, p < 0.05$). Differences were also found between the responsive care and standard care groups on individual items. Responsive care infants exhibited greater amounts of listening to sounds than standard care infants ($F(1,20) = 5.32, p < 0.05$) and greater amounts of manipulating objects with the hands than standard care infants ($F(1,20) = 4.53, p < 0.05$). The groups did not differ in levels of mouthing objects or looking at sights.

Chimpanzees exhibited high and significant stability in task scores from test to test (month 3–5 with 6–7, $r = 0.50, p < 0.01$; 6–7 with 8–9, $r = 0.76, p < 0.001$; 8–9 with 10–12, $r = 0.58, p < 0.001$). There was moderate stability, however, in nonconsecutive months (month 3–5 with 8–9, $r = 0.36, p = 0.055$; 6–7 with 10–12, $r = 0.45, p < 0.05$). Task scores in the early test did not predict scores in the late tests (month 3–5 with 10–12, $r = 0.17, p = 0.38$).

Affect-extraversion

The affect scores measured the emotional responsiveness of infants during the BSID. Group differences on the affect-extraversion factor are illustrated in Figure 11.3. This factor included social orientation and responsiveness to the examiner and the mother, cooperation with the examiner, general emotional tone, fearfulness (with a negative loading), and endurance. The fearfulness scores are inverted prior to averaging so that higher affect scores indicate more extraversion with a more positive emotional tone.

A significant main effect of age ($F(3,78) = 5.74, p < 0.001$) and a significant interaction of age with group ($F(6,78) = 3.14, p < 0.01$) required further analysis. All groups exhibited equal levels of emotional responsiveness on the 3–5 month tests but standard care infants were the only group that remained consistently at the same level across the first year of life. Responsive care infants consistently obtained higher scores over the subsequent months, whereas late arrival infants obtained lower scores in the middle months, then returned to standard care levels at 10–12 month (group × month interaction: $F(3,66) = 4.14, p < 0.01$). A significant group difference was found between responsive care and late arrivals ($F(1,12) = 6.41, p < 0.05$). Consideration of the individual items clarifies these effects (see Table 11.3).

Responsive care infants were happier than standard care infants during the tests from 6 months to 12 months (group × month interaction $F(3,66) = 5.55, p < 0.01$) and happier than late arrival infants throughout testing ($F(1,12) = 5.73, p < 0.05$). Responsive care infants exhibited significantly less fear (more acceptance) than both standard care infants ($F(1,22) = 7.84, p < 0.01$) and late arrival infants ($F(1,12) = 9.38, p < 0.01$). Responsive care infants exhibited significantly more cooperation with the examiner than late arrival infants ($F(1,12) = 4.92, p < 0.05$). It is important to note that the groups did not differ in their endurance levels, responsiveness to the examiner, or responsiveness to their favorite caregiver ("mother").

Affect-extraversion was a stable characteristic of chimpanzees in consecutive tests (month 3–5 with 6–7, $r = 0.49, p < 0.01$; 6–7 with 8–9, $r = 0.77, p < 0.001$; 8–9 with 10–12, $r = 0.75, p < 0.001$). Moderate and significant stability of affect was also found in nonconsecutive months (month 3–5 with 8–9, $r = 0.43, p < 0.05$; 6–7 with 10–12, $r = 0.58, p < 0.001$). Affect-extraversion in chimpanzees at 10–12 months, however, could not be predicted by their scores at 3–5 months ($r = 0.25$, ns).

Activity

The third personality factor of the IBR section of the Bayley test is activity, a general measure of body movement and energy (Figure 11.4). Scores of 8 and 9 are associated with excessive hyperactivity. In this factor, optimal scores are those close to 5. All groups exhibited an increase in activity during the course of the year ($F(3,78) = 15.92, p < 0.001$). In the activity factor, no differences were found between responsive care and standard care chimpanzees. Responsive care infants, however, were more optimally energetic than late arrival infants ($F(1,12) = 6.23, p < 0.05$).

Analyses of individual items revealed differences between responsive care and late arrival groups (see Table 11.3). Responsive care infants exhibited greater freedom of

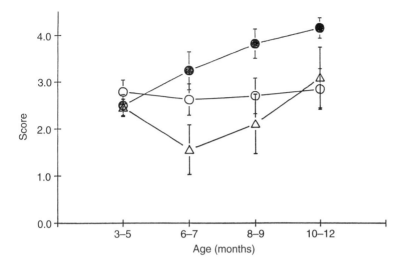

Figure 11.3. Affect-extraversion factor scores for responsive care, standard care, and late arrival chimpanzee subjects. Symbols as for Figure 11.1.

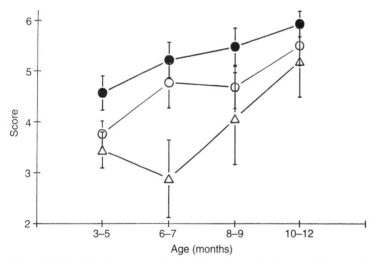

Figure 11.4. Activity factor scores for responsive care, standard care, and late arrival chimpanzee subjects. Symbols as for Figure 11.1.

movement than late arrival infants ($F(1,12) = 6.23, p < 0.05$). Both groups also exhibited increased levels of general activity with greater age ($F(3,36) = 10.6, p < 0.001$). Responsive care infants exhibited significantly higher levels of coordination and energy for their age compared with late arrival infants ($F(1,12) = 8.23, p < 0.05$). Responsive care infants exhibited higher levels of gross body movement than late arrival infants ($F(1,12) = 7.62$, $p < 0.05$) and more body motion than late arrival infants ($F(1,12) = 7.32, p < 0.05$). Activity was a highly and significantly stable individual characteristic throughout the BSID tests, evident both in consecutive tests (month 3–5 with 6–7, $r = 0.61, p < 0.001$; 6–7 with 8–9, $r = 0.78, p < 0.001$; 8–9 with 10–12, $r = 0.76, p < 0.001$) and among all tests (month 3–5 with 8–9, $r = 0.72, p < 0.001$; 3–5 with 10–12, $r = 0.49, p < 0.01$; 6–7 with 10–12, $r = 0.64, p < 0.001$).

Coordination

The fourth personality factor was coordination. Coordination was assessed for fine muscle movements and gross muscle movements. Fine and gross coordination were loaded negatively with the item attachment to objects, suggesting that strong object attachments limited coordinated activity. In this factor, the optimal score was 0. There were no significant differences between standard care and responsive care or between responsive care and late arrival infants on the coordination factor (Figure 11.5).

There was a significant difference between the groups as regards attachment to objects (either a piece of blanket or a towel). The responsive care infants exhibited no attachment to objects whereas both other groups were significantly more likely to exhibit attachment to objects (RC versus SC, $F(1,22) = 17.57, p < 0.01$; RC versus LA, $F(1,12) = 9.60, p < 0.01$). Attachment to a towel was especially evident in the early months of testing. The responsive care infants were rated as smoother in coordination of gross muscle compared with standard care infants ($F(1,19) = 4.90, p < 0.05$). The groups did not differ in coordination of fine muscle movements.

There was some stability in the coordination factor between the first test and subsequent tests (month 3–5 with 6–7, $r = 0.51, p < 0.01$; 3–5 with 8–9, $r = 0.47$, $p < 0.01$; 3–5 with 10–12, $r = 0.55, p < 0.01$) and between the last tests (months 8–9 to 10–12, $r = 0.59, p < 0.001$). Coordination in months 6–7 did not predict later coordination (month 6–7 with 8–9, $r = 0.18$, ns; 6–7 with 10–12, $r = 0.29$, ns).

Audio-visual reactivity

The final personality factor was audio-visual reactivity, an indication of the general sensory sensitivity of the infant (Figure 11.6). This factor included assessments of body tension, excitability, and vocalizations. The optimal score in this factor was 5, as 9 was scored for rigidity and hyper-reactivity. There were no differences between the groups in the reactivity factor. Consideration of individual items revealed group differences (see Table 11.3). Responsive care infants were less tense compared both with standard care infants ($F(1,22) = 4.76, p < 0.05$) and with late arrival infants ($F(1,22) = 5.41, p < 0.05$). The groups did not differ in the amount of vocalization or in general excitability. These items do not evaluate reactions in a positive or negative manner. In other words, these items are neutral in emotional tone. Thus, we did not find it surprising that they do not distinguish the groups.

Chimpanzee infants exhibited high and significant stability in reactivity in consecutive tests (month 3–5 with 6–7, $r = 0.65, p < 0.001$; 6–7 with 8–9, $r = 0.75, p < 0.001$;

Influences on development in infant chimpanzees 249

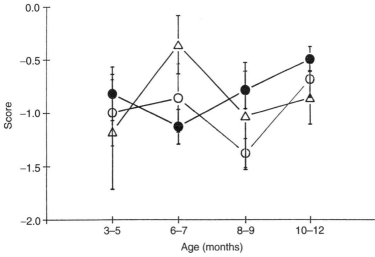

Figure 11.5. Coordination factor scores for responsive care, standard care, and late arrival chimpanzee subjects. Symbols as for Figure 11.1.

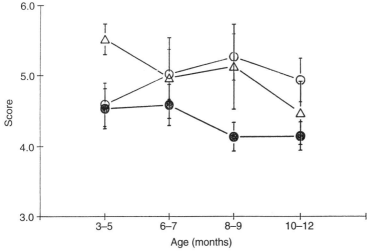

Figure 11.6. Audio-visual reactivity factor scores for responsive care, standard care, and late arrival chimpanzee subjects. Symbols as for Figure 11.1.

8–9 with 10–12, $r = 0.60$, $p < 0.001$) and across 2 monthly sessions (month 3–5 with 8–9, $r = 0.37$, $p < 0.05$; 6–7 with 10–12, $r = 0.60$, $p < 0.001$). Audio-visual reactivity at 10–12 months, however, could not be predicted by reactivity at 3–5 months ($r = 0.28$, ns).

DISCUSSION

Young chimpanzees raised under different "cultural" conditions exhibited few differences in cognition, but many differences in personality. There was a tendency for the group that received extra nurturing of species-typical chimpanzee behaviors to exhibit cognitive performances at marginally higher levels than the group that experienced early stress in the biological mother–infant relationship (stress sufficient to necessitate placement of the infant in the great ape nursery to ensure survival) and than the group that received standard nursery care.

A cross-species consideration of the effects of parenting practices lends support to the view that chimpanzees may become enculturated, either in human settings with human beliefs and values or in chimpanzee settings with chimpanzee intentions and views on the world (Bard, 1994a,d; Menzel et al., 1970). Clear evidence exists of cross-cultural differences in many behavioral patterns of human infants and toddlers (e.g. in aspects of prelinguistic communication, and mother–infant attachment patterns). These differences include differential emphasis on the value of mutual eye gaze with adults, differential ratios of verbal to physical stimulating contact, and differential responsiveness to crying, which may be apparent even in interactions with newborns. For example, American mothers may express great admiration and enthusiastic vocalizations for their 30 day old infants who merely look in response (Brazelton et al., 1974). In contrast, Kaluli mothers (of Papua New Guinea) hold their baby facing away as it is impolite to gaze while communicatively engaged, and, rather than speak to them, mothers may speak for their babies to social interactants (Ochs & Schieffelin, 1984). Clearly, by 1 year of age, infants' behavior patterns are influenced by tacit cultural expectations: for example, German ideals toward infants include early independence and large interpersonal space whereas traditional Japanese ideals include a unity between infant and mother, expressed in physical and emotional closeness. Therefore, when the quality of attachment is compared across cultures, we find that German infants act less securely attached to their mothers whereas Japanese infants act with a clinging-type of attachment compared with American "norms" (Ainsworth & Wittig, 1969; Grossman et al., 1985; Miyake et al., 1985). One must use caution in attributing these cross-cultural differences solely to enculturation. In order to be labeled a cultural difference, it may be required that the differences are transmitted across generations. Extra caution must be applied when comparisons are conducted across species, especially considering that culture, in the currently accepted view, concerns variation within a single species.

We believe that the full extent of cognitive abilities in the chimpanzee has not been tapped due, in part, to a lack of attention to early socializing influences on developing behavior. For instance, human newborns engage in facial and vocal imitation from birth, which may predispose them to engage more productively in imitation of actions on objects at 1 year of age (Kuguimutzakis, 1993). Typically, chimpanzees do not begin to work in "language" projects before they are 6–18 months of age (Savage-Rumbaugh, 1986).

The original goal of using the standardized situation of the Bayley test, however, was to measure individual differences in emotional responsiveness or temperamental reactions (Rothbart & Derryberry, 1981). It was, in fact, in the personality measures of the Bayley that the largest effects were found. The responsive care infants were most persistent in attaining goals, exhibited the longest attention span, and had the highest levels of manual contact with objects (measures of the demands of the task), whereas the late arrival (early stress) infants exhibited the lowest levels. Emotional responses, such as happiness or fear of the test situation were also different between groups after the first few tests.

Infants raised in a more emotionally responsive environment exhibited levels of cooperation, happiness, and acceptance of the test that increased throughout the year of testing. In contrast, late arrival infants exhibited increased fear at 6–7 months. The standard care infants, overall, exhibited unchanging emotional responses throughout the first year. With regard to activity, coordination and reactivity, similar patterns were found. Infants raised with species-typical nurturing exhibited the optimal emotional and motor responses whereas those infants who experienced the greatest amount of stress early in life exhibited the least positive emotional and coping responses.

Responsive care infants exhibited significantly lower coping ability than did standard care infants during the first month of life. That the standard care infants exhibited better coping was surprising. However, data from mother-reared chimpanzees (Hallock *et al.*, 1989) suggest that chimpanzee infants raised with their biological mothers show significantly less ability to comfort themselves and less independent coping capacity than either group of nursery-reared chimpanzee infants. These findings from the neonatal period suggest that expectations based on interactions with the environment are formed as early as the first 30 days of life. Standard care infants come to expect to cope by themselves. Responsive care infants expect others, when they are present, to facilitate.

Responsive care infants compared to standard care infants coped less well in the first month of life. More responsive care infants exhibited anger compared with standard care infants who exhibited distress. Responsive care infants were happier, better able to seek out comfort when needed and were rated as better copers compared with standard care infants later in the first year (J. Ashley & K.A. Bard, unpublished results). Infants who experienced severe distress in the first weeks of life appeared to exhibit long-term effects that were evident in increased emotional reactivity and deficits in cognitive performance that lasted throughout the first year of life.

When we compare the infants who were stressed early in life with the responsive care infants who came to the nursery at birth, we find significant and long-lasting decrements in the performance of the late arrivals. Remember that the Bayley tests were conducted from 3–12 months of age and that these late arrival infants stayed with their mother *less than* 25 days. These results suggest a long-term effect of very early stress. In fact, the sum of these results point to long-lasting differences, primarily in emotional responses and suggestively in cognitive performance, based on early rearing.

Along with these differences in temperament that we found on the standardized Bayley tests, we expect that the biggest differences between differentially reared infants will be found in general social responsiveness. We expect that chimpanzees who are raised with individuals knowledgeable about chimpanzee communicative signals will learn to display species-typical behavior patterns including appeasement, reconciliation, and consolation behavior (e.g. Veira & Bard, 1994).

There are many alternative views in contrast to our "cultural" interpretation of early environmental influences on behavioral development. These include the more traditional view of the influence of early learning on changes in behavioral expressions; the currently held nature–nurture interactionalist view; and the cross-cultural perspective that searches both for universals and explicitly taught cultural variants in behavior. Experimental research both with human infants and rhesus monkey infants suggest that similar environmental variables may affect individuals differently based on inborn temperamental characteristics, for example, of inhibition and extraversion (e.g. Kagan et al., 1988; Suomi, 1987). Recent discussions of individual differences in emotional regulation suggest that socialization processes may directly influence the neuro-anatomical bases of emotion (e.g. Dawson, 1994). Early research with nonhuman primates demonstrated convincingly the dramatic impact of social deprivation, in fact, deprivation can cause psychopathologies in development (e.g. Seay et al., 1962). Thus, there may be many mechanisms by which socialization may affect biology. The extent to which biologically based differences between groups may mediate emotional reactions or may masquerade as cultural differences is unknown. The causes of differential emotional responsiveness are far from being completely understood.

We suggest that individual chimpanzees raised in a "culture" consisting of early maternal-rearing, under stressful conditions, followed by a standard environment may have dysfunctions in emotional reactiveness compared with individuals raised in a "culture" consisting of standard-rearing conditions from birth. The difficulties in emotional regulation may result, in part, from the culturally or socially determined patterns of early mother–infant interactions, which differ in humans and chimpanzees. For instance, most intuitive parenting behaviors in chimpanzees are based on tactile communications (Bard, 1994a; similar to those parenting behaviors of humans who are deaf or blind: Fraiberg, 1974; Koester, 1994) whereas parenting behaviors of many groups of humans are mostly based on audio-visual communication (Papousek & Papousek, 1987). Early stress, however, complicates the picture. The dysfunctions in emotional reactiveness caused by early stress are incompletely understood. A relation between arousal and laterality suggests emotional reactiveness could be affected by environmental stimuli at a neuro-anatomical level (Dawson, 1994; Hopkins & Bard, 1993).

The practical implications of the comparison between standard nursery care and an experimental intervention project relate to changes in coping capacity, and behavioral well-being based on early experience. The important aspects of this study are (a) documentation of minimal to moderate environmental influences on cognition, (b) evidence of flexibility in the behavioral and temperamental repertoires of chimpanzee infants in response to social/"cultural" influences, and (c) evidence of comparability in cognitive performance between chimpanzee and human infants. The implications of this research relate to long-term effects of early experiences and early learning. We expect to find that individual differences in temperament and cognition influence the expression of individual differences in adulthood. The scientific implications of this research relate to quantifying both the early learning processes and the long-term effects of early experiences.

Consideration of the continuity of characteristics across primate species can be controversial, especially when those characteristics are defined to be uniquely human. Research with primates both in the field and in the laboratory has steadily discovered evidence of continuity in many of these tightly held unique characteristics, i.e. language, tool use, tool manufacture, and teaching. Can definitions be revised to emphasize evolutionary continuities rather than dichotomies? Can we consider the changes in behavioral organization that may occur as a result of differential early experiences, part of a cultural package? Will the altered style of mutual eye gaze, for example, experienced when chimpanzees are raised by American (human) caregivers be evident in the next generation when these chimpanzees raise their own infants? Chimpanzees, like humans, are very long-lived; cross-generational effects, therefore, are difficult to study. We must plan now in order to investigate the extent to which these behaviors are transmitted. It is certainly true that if we do not study the possibility we will never have definitive answers.

ACKNOWLEDGEMENTS

Supported in part by NIH Grants RR-00165 to the Yerkes Regional Primate Research Center of Emory University, RR-03591 to R.B. Swenson, and RR-06158 to K.A. Bard. Grateful appreciation is extended to Benjamin Jones, Stacy Bales, Jennifer Fineran, Elisabeth Bell, Sheryl Williams, Carolyn Fort, Marie Mize, Debbie Custance, Michelle Russell, Pam Gewirtz, Rick Lippman, Brian Just, and, especially, Yvette Veira, Bernice Pelea, and Kelly McDonald for their participation in the behavioral intervention project. Dara Johnston, Sheryl Williams, Mike Ringer, Curtis Skinner, Lisa Jones-Engel, Vicky Blackwell, Jeanne des Islets, Callie Wilson, Beth Jackson, Jennifer Ashley, Josh Schneider, and Dr Kathleen A. Platzman assisted with the normative behavioral research. Special appreciation is extended to Barbara McElduff for cheerfully typing the manuscript in its various forms.

REFERENCES

Ainsworth, M.D.S. & Wittig, B.A. (1969). Attachment and exploratory behavior of one-year-olds in a strange situation. In *Determinants of Infant Behavior*, vol. 4, ed. B.M. Foss, pp. 113–36. London: Methuen.

Bard, K.A. (1991). Distribution of attachment classifications in nursery chimpanzees. *American Journal of Primatology*, 24, 88.

Bard, K.A. (1994a). Evolutionary roots of intuitive parenting: Maternal competence in chimpanzees. *Early Development and Parenting*, 3, 19–28.

Bard, K.A. (1994b). Similarities and differences in the neonatal behavior of chimpanzee and human infants. In *The Role of the Chimpanzee in Research*, ed. G. Eder, F. Kaiser & F.A. King, pp. 43–55. Basel: Karger.

Bard, K.A. (1994c). Very early social learning: The effect of neonatal environment on chimpanzees' social responsiveness. In *Current Primatology, vol. II: Social Development, Learning, and Behavior*, ed. J.J. Roeder, B. Thierry, J.R. Anderson & N. Herrenschmidt, pp. 339–46. Strasbourg: Université Louis Pasteur.

Bard, K.A. (1994d). Developmental issues in the evolution of the mind. *American Psychologist*, **49**, 760–761.

Bard, K. A. (1995a). Parenting in primates. In *Handbook of Parenting, vol. II: Ecology and Biology of Parenting*, ed. M. Bornstein, pp. 27–58. Hillsdale, NJ: Lawrence Erlbaum.

Bard, K.A. (1995b). Social-experiential contributions to imitation and emotional expression in chimpanzees. In *Intersubjective Communication and Emotion in Ontogeny*, ed. S. Bråten. Oslo: Norwegian Academy of Science and Letters.

Bard, K.A. & Fort, C.L. (1993). Ontogeny of emotions in neonatal chimpanzees. *Society for Research in Child Development Abstracts*, p. 247.

Bard, K.A., Gardner, K. & Fort, C.L. (1993). The effects of behavioral interventions on nursery-reared chimpanzees. *American Journal of Primatology*, **30**, 296–7.

Bard, K.A., Gardner, K. & Platzman, K.A. (1991). Young chimpanzees score well on the Bayley Scales of Infant Development. *American Journal of Primatology*, **24**, 89. (abstract).

Bard, K.A., Platzman, K.A., Lester, B.M. & Suomi, S.J. (1992). Orientation to social and nonsocial stimuli in neonatal chimpanzees and humans. *Infant Behavior and Development*, **15**, 43–56.

Bayley, N. (1969). *Bayley Scales of Infant Development*. New York: Psychological Corporation.

Brazelton, T. B. (1984). Neonatal behavioral assessment scale. In *Clinics in Developmental Medicine*, No. 88. Philadelphia: Lippincott.

Brazelton, T.B., Koslowski, B. & Main, M. (1974). The origins of reciprocity: The early mother–infant interaction. In *The Effect of the Infant on its Caregiver*, ed. M.Lewis & L.A. Rosenblum, pp. 49–76. New York: Wiley.

Dawson, G. (1994). Frontal EEG correlates of individual differences in emotion expression in infants: A brain systems perspective on emotion. *Monographs of the Society for Research in Child Development*, **59**, 135–51.

Fraiberg, S. (1974). Blind infants and their mothers: An examination of the sign system. In *The Effect of the Infant on its Caregiver*, ed. M. Lewis & L.A. Rosenblum, pp. 215–32. New York: Wiley.

Francis, P.L., Self, P.A. & Horowitz, F.D. (1987). The behavioral assessment of the neonate: An overview. In *Handbook of Infant Development*, 2nd edn, ed. J.D. Osofsky, pp. 723–79. New York: Wiley.

Fritz, J. & Fritz, J. (1985). The hand-rearing unit: Management decisions that may affect chimpanzee development. In *Clinical Management of Infant Great Apes*, ed. C.E. Graham & J.A. Bowen, pp. 1–34. New York: Alan R. Liss.

Goodall, J. (1973). Cultural elements in a chimpanzee community. In *Precultural Primate Behavior*, ed. E. Menzel, vol. 1, pp.138–59. Fourth International Primatology Congress Symposium Proceedings. Basel: Karger.

Grossman, K., Grossman, K.E., Spangler, S., Suess, G. & Unzner, L. (1985). Maternal sensitivity and newborn orientation responses as related to quality of attachment in Northern Germany. *Monographs of the Society for Research in Child Development*, **50**, 233–56.

Hallock, M.B., Worobey, J. & Self, P.A. (1989). Behavioral development in chimpanzee (*Pan troglodytes*) and human newborns across the first month of life. *International Journal of Behavioral Development*, 12, 527–40.

Hopkins, W.D. & Bard, K.A. (1993). Hemispheric specialization in infant chimpanzees (*Pan troglodytes*): Evidence for a relation with gender and arousal. *Developmental Psychobiology*, 26, 219–35.

Kagan, J., Reznick, J.S. & Snidman, N. (1988) Biological bases of childhood shyness. *Science*, 240, 167–71.

Kawai, (1965). Newly acquired pre-cultural behavior of the natural troop of Japanese monkeys on Koshira Islet. *Primates*, 6, 1–30.

Koester, L.S. (1994). Early interactions and the socioemotional development of deaf infants. *Early Development and Parenting*, 3, 51–60.

Korner, A.F. (1973). Sex differences in newborns with special reference to differences in the organization of oral behavior. *Journal of Child Psychology and Psychiatry*, 14, 19–29.

Kugiumutzakis, G. (1993). Intersubjective vocal imitation in early mother–infant interaction. In *New Perspectives in Early Communicative Development*, ed. J. Nadel & L. Camaioni, pp. 23–47. London: Routledge & Kegan Paul.

Maccoby, E. (1966). *The Development of Sex Differences*. Stanford, CA: Stanford University Press.

Matheny, A.P. (1980). Bayley's Infant Behavior Record: Behavioral components and twin analyses. *Child Development*, 51, 1157–67.

Matheny, A.P. (1983). A longitudinal twin study of stability of components from Bayley's Infant Behavior Record. *Child Development*, 54, 356–60.

McGrew, W. & Tutin, C. (1978). Evidence for a social custom in wild chimpanzees? *Man*, 13, 234–51.

McGrew, W., Baldwin P.G. & Tutin, C. (1979). Chimpanzees' tools and termites: Cross-cultural comparisons of Senegal, Tanzania, and Rio Muni. *Man*, 14, 185–214.

Menzel, E.W., Davenport, R.K. & Rogers, C.M. (1970). The development of tool use in wild-born and restriction-reared chimpanzees. *Folia Primatologica*, 12, 273–83.

Miyake, K., Chen, S. & Campos, J.J. (1985). Infant temperament, mother's mode of interaction, and attachment in Japan. *Monographs of the Society for Research in Child Development*, 50, 276–97.

Nishida, T. (1987). Local traditions and cultural transmission. In *Primate Societies*, ed. B. Smuts, D. Cheney, R.M. Seyfarth, R.W. Wrangham & T.T. Struhsaker, pp. 462–74. Chicago: University of Chicago Press.

Nugent, J.K., Lester, B.M. & Brazelton, T.B. (eds.) (1989). *The Cultural Context of Infancy*, vol. 1. Norwood, NJ: Ablex.

Ochs, E. & Schieffelin, B. (1984). Language acquisition and socialization. Three developmental stories and their implications. In *Culture Theory*, ed. R. Shweder & R. LeVine, pp. 276–320. Cambridge: Cambridge University Press.

Papousek, H. & Papousek, M. (1987). Intuitive parenting. In *Handbook of Infant Development*, 2nd edn, ed. J.D. Osofsky, pp. 669–720. New York: Wiley.

Rothbart, M. & Derryberry, D. (1981). Development of individual differences in temperament. In *Advances In Developmental Psychology*, ed. M.E. Lamb & A.L. Brown, pp. 37–86. Hillsdale, NJ: Lawrence Erlbaum.

Savage-Rumbaugh, E.S. (1986). *Ape Language: From Conditioned Response to Symbol*. New York: Columbia University Press.

Seay, B.M., Hansen, E.W. & Harlow, H.F. (1962). Mother–infant separation in monkeys. *Journal of Child Psychology and Psychiatry*, 3, 123–32.

Suomi, S.J. (1987). Genetic and maternal contributions to individual differences in rhesus monkey biobehavioral development. In *Perinatal Development: A Psychobiological Perspective*, ed. N.A. Krasnegor, E.M. Blass, M.A. Hofer & W.P. Smotherman, pp. 397–419. San Diego, CA: Academic Press.

Tabachnick, B.G. & Fidell, L.S. (1983). *Using Multivariate Statistics*. New York: Harper & Row.

Tomasello, M., Kruger, A.C. & Ratner, H.H. (1993). Cultural learning. *Behavioral and Brain Sciences*, 16, 495–511.

van der Meulen, B.F. & Smrkovsky, M. (1985). A factor analyses of Bayley's Infant Behavior Record: A Dutch replication and extension. *British Journal of Developmental Psychology*, 3, 345–52.

Veira, Y. & Bard, K.A. (1994). Developmental milestones for chimpanzees raised in a responsive care nursery. *American Journal of Primatology*, 33, 246. (abstract).

12

Heterochrony and the evolution of primate cognitive development

JONAS LANGER

Since at least the time of Piaget's (1954) research it has been established that sensorimotor interactions are sufficient to generate practical knowledge about the physical world of objects, space, time, and causality, such as object permanence and causal contingency. Thus, classical theories of cognitive development (including those proposed by Bruner, Gibson, Koffka, Piaget, Vygotsky and Werner) agree that prerepresentational sensorimotor forms of elementary physical cognition develop in human infants and in other primates (e.g. Doré & Dumas, 1987; Langer, 1988; Parker & Gibson, 1979). At the same time it has also been generally assumed, although not by Piaget (1972), that even the most elementary forms of logicomathematical cognition of classes, relations, and number, such as quantitative augmenting by adding, require grammatical symbolic representation (see e.g. Braine, 1990; Carnap, 1960; Falmagne, 1990; Macnamara, 1986).

At least two implications flow from this assumption, one phylogenetic and the other ontogenetic. The phylogenetic implication is that only primates that have evolved grammatical symbolic systems can develop logicomathematical cognition. So, the phylogenetic implication is that logicomathematical cognition is pretty much a uniquely human development unless it can be proved that other primates, perhaps great apes, acquire grammatical symbolic behavior. The ontogenetic implication for humans is that logicomathematical cognition can not develop before late infancy or early childhood. The reason is that children do not begin to develop grammatical language until their third year (e.g. Bickerton, 1990; Bowerman, 1978; Maratsos, 1983).

On this classical view, then, logicomathematical cognition is a postsensorimotor representational development by humans. This view has led to the expectation that the order of ontogenetic onset in humans is first physical cognition then logicomathematical cognition, with an expected time lag of at least 2–3 years between them. Curiously, this expected ontogenetic order, we shall see, fits the cognitive development of monkeys who remain fundamentally prerepresentational and do not develop anything approaching a grammatical language. Initial data analyses suggest that it partly fits the ontogeny of chimpanzees, who develop rudimentary representation and may acquire proto-grammatical language. The expected order of physical preceding logicomathematical cognition does not fit at all the ontogeny of human infants. They begin to develop logicomathematical cognition before grammatical language, and in parallel with their development of physical cognition (Langer, 1980, 1988, 1990, 1993, 1994a,b).

The data fit much more closely with our originalist hypothesis that logicomathematical cognition, like physical cognition, is a cognitive primitive and a primary development. The originalist hypothesis leads to a set of counter-expectations. The onset age is the same, early infancy, for both physical and logicomathematical cognition in humans. Both develop in parallel and at the same velocity throughout infancy. Both are initially prerepresentational sensorimotor constructions. Neither require grammatical language for their initiation. Therefore, both elementary physical and logicomathematical cognition may originate in other primate species as well (even if they can not acquire grammatical language) if not necessarily in the same ontogenetic order or structural organization as in humans, and if not necessarily to the same ontogenetic level of development as in humans.

Significant similarities mark the origins and early cognitive development of young human and nonhuman primates over the period when sensorimotor constructions predominate (the first 2–3 years in humans, 5 years in great apes, and 3–4 years in cebus and macaque monkeys). The increase in findings of similarities in primate cognitive origins and development, including those outlined in the next three sections, are of major importance. Considered alone, these similarities suggest a recapitulationist view of the evolution of intelligence (Langer, 1988, 1989). However, the increase in findings of significant differences (especially in ontogenetic order and structural organization), also mark the early cognitive development of primate species. An adequate model of cognitive evolution must therefore, at a minimum, account for these patterns of species similarities coupled with differences in young primates' intellectual origins and development. This led me to suggest that other evolutionary mechanisms of heterochrony were at work. In particular, I have proposed that organizational heterochronies, or phylogenetic change in the ontogenetic onset and the velocity at which cognitive structures develop in relation to each other, are a major mechanism of cognitive evolution (Langer, 1989, 1993, 1994b; for extensive history, definition and concepts of heterochrony, see Gould, 1977; McKinney & McNamara, 1991; Parker, 1995; Parker & Gibson, 1979).

My discussion of these issues starts in the next section by surveying major similarities and differences in the developing elements of primate cognition, and their development of logicomathematical and physical cognition is discussed in the following two sections (see pp. 263–8). In the fourth section (pp. 268–70) this survey provides the empirical base for testing the hypothesis that logicomathematical cognition is as much an original prerepresentational development as physical cognition, while in the fifth section (pp. 270–2) heterochrony is discussed as a mechanism of cognitive evolution. This provides the necessary background for considering mechanisms of cognitive development in the final section.

THE INITIAL ELEMENTS OF PRIMATE COGNITION

This section considers the elements that primates construct as the constants for their cognition. A set of cognitions can have no structure unless: (a) the elements of the set are themselves constant; and (b) one or more operations, functions, and/or relations on the set are defined (Langer, 1990). To illustrate partially, structuring determinate equivalence and inequivalence relations by one-to-one correspondence or by substitution requires, at a minimum, constructing invariant objects of comparison. For example, using one-to-one correspondence to determine whether two sets of objects are equal or not requires, at a minimum, that the membership of each set is stabilized at least momentarily. Human infants, we shall see, begin to do this during their second year by composing stable sets of objects in temporal overlap and spatial proximity.

Our research method with children included studying spontaneous constructive interactions with 4–12 objects. The range of objects spanned geometric shapes to realistic things such as cups (as illustrated in Figures 12.1–4). Class structures were embodied by some of the object sets presented (e.g. multiplicative classes such as a yellow and a green cylinder and a yellow and a green triangular column shown in Figure 12.1). But there was nothing in the procedures we used that required subjects to do anything about the objects' class structures. No instructions were given and no problems were presented to the subjects. They were simply allowed to play freely with the objects as they wished because our goal was to study their constructive intelligence.

Many of the findings on humans that I review in this and the next two sections have been replicated with both Aymara and Quecha Indian children in Peru (Jacobsen, 1984). The Indian children were raised in impoverished conditions as compared to the mainly Caucasian middle class San Francisco Bay Area children in my samples. Nevertheless, no differences were found in onset age, velocity, sequence, or extent of cognitive development during infancy in these different human samples.

The following discussion of primate cognitive development is based on these studies of human children (Langer, 1980, 1986, unpublished results); and on parallel studies (Antinucci, 1989; P. Potì & F. Antinucci, unpublished results; Spinozzi, 1993) on cebus monkeys (*Cebus apella*), macaques (*Macaca fascicularis*) and chimpanzees (*Pan troglodytes*) using the nonverbal and nondirective methods developed to study human children. The samples of human subjects comprised 12 children (six boys and six girls) tested at each of ages 6, 8, 10, 12, 15, 18, 21, 24, 30, 36, 48, and 60 months for a total of 144 subjects (for further details of methods, analyses, and results, see Langer, 1980, 1986). The cebus monkey sample comprised three subjects: first, one cross-sectional subject tested at age 16 months; a second, longitudinal subject at ages 16, 36, and 48 months; and a third, longitudinal subject at ages 36 and 48 months (for details, see Antinucci, 1989). The macaque sample comprised two longitudinal subjects tested at ages 22 and 34 months (for details, see Antinucci, 1989). The chimpanzee sample comprised five subjects: four cross-sectional subjects tested one at each of ages 15, 30, 41, and 54 months; and one longitudinal subject tested at ages 23, 37, and 51 months (for details, see Spinozzi, 1993).

The nonhuman primate findings, then, are based on fewer subjects than the human

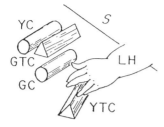

Figure 12.1. Six month old subject composing a green cylinder (GC) with a yellow triangular column (YTC) using left hand (LH). GTC, green triangular column; YC, yellow cylinder; S, side.

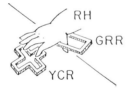

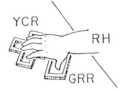

Figure 12.2. Six month old subject composing a green rectangular ring (GRR) with a yellow cross ring (YCR) using right hand (RH).

findings. As such, the nonhuman primate findings may be less secure, but they are the subject of continuing data analyses and testing of more chimpanzees, including bonobos (*Pan paniscus*), in collaboration with the research teams of F. Antinucci and E.S. Savage-Rumbaugh. Moreover, the nonhuman primate data have at least two noteworthy strengths. First, they include longitudinal data. Secondly, they are based upon very

Evolution of primate cognitive development 261

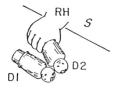

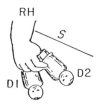

Figure 12.3. Six month old subject composing two dolls (D1 and D2) using right hand (RH). S, side.

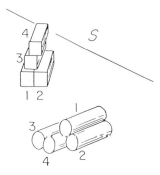

Figure 12.4. Second-order classifying by a 21 month old subject.

large numbers of both repeated and varied trials that generated very large numbers of measurable behaviors. To illustrate, the findings on set construction, to which we now turn, include those based on 4476 behaviors by chimpanzees (see Table 2 of Spinozzi, 1993).

All primates we have studied so far composed sets of objects as elements for their cognition (such as those illustrated in Figures 12.1–4). They composed sets of objects by bringing two or more objects into contact or close proximity with each other (i.e. no more than 5 cm apart). Although all primate species we have studied are quite productive, their comparative growth curves diverge. With increasing age the rate of producing sets of objects: (a) did not change in macaques (Potì & Antinucci, 1989; Spinozzi & Natale, 1989); (b) increased in cebus monkeys (Potì & Antinucci, 1989; Spinozzi & Natale, 1989); (c) increased in chimpanzees (Spinozzi, 1993); and (d) first increased (during the first year) and then, with some minor fluctuations, did not change (during the second year) in human infancy (Langer, 1980, 1986).

Sets of objects, as elements of cognition, can be measured for their constancy and power. Measures of constancy include spatio-temporal stability of the sets composed; and measures of power include the number and size of sets composed in spatio-temporal proximity. Thus, a prime measure of the constancy and power of sets as elements of cognition – as constant givens – is the spatio-temporal relations between the sets. Again the comparative growth curves differentiate. In cebus and macaque monkeys, with increasing age the proportion of temporally: (a) isolated sets decreased, but remained substantial; (b) sequential sets increased; and (c) overlapping sets was null (Potì & Antinucci, 1989; Spinozzi & Natale, 1989).

In chimpanzees, with increasing age the proportion of temporally: (a) isolated sets decreased but remained substantial; (b) sequential sets first increased (through the third year) and then did not change (up to age $4\frac{1}{2}$ years); and (c) overlapping sets increased but was small (Spinozzi, 1993). Although a decided advance, the overlapping sets that chimpanzees composed during their fifth year were nevertheless comparatively unstable or labile elements of cognition because they were almost all kinetic constructions (e.g. waving around two objects in each hand at the same time). Further, they remained minimal elements since they did not exceed two sets at a time.

In human infancy, with increasing age the proportion of temporally: (a) isolated sets decreased until it approached zero; (b) sequential sets first increased and then decreased; and (c) overlapping sets increased such that producing overlapping sets predominated by the third year (Langer, 1980, 1986, unpublished results). Not only did they predominate, but the overlapping sets that older infants composed were comparatively stable or constant elements of cognition because they were progressively static constructions (e.g. two stacks of four objects next to each other as illustrated in Figure 12.4). Further, the overlapping sets became progressively more powerful, since they increasingly exceeded two sets at a time (e.g. four separate stacks of blocks).

The elements of organisms' cognition constrain the level to which their intellectual operations can develop. Once the elements for cognition diverge in different species, then the intellectual operations that they can map onto them must inevitably also diverge. Cebus and macaque monkeys, we have seen, were limited to constructing single sets of objects or what we call first-order elements. They were therefore constrained to mapping nothing more than first-order cognitions onto these elements (as illustrated in the next two sections). Chimpanzees constructed labile second-order elements. Thereby, they constructed the minimal elements necessary to map rudimentary second-order cognitions onto these elements (such a rudimentary one-to-one correspondence between two sets is discussed in the next section). Beginning in their second year, human infants constructed progressively more robust and stable second-order elements. So, the constraints upon their second-order cognitions progressively diminished.

This, I propose, is just the beginning of a significant feature of comparative cognitive development that has fundamental consequences. At first, the elements of cognition constructed by different species overlap. But, with development, the elements increasingly diverge; and consequently, so does their cognition, as we shall see in the next two sections.

THE ORIGINS OF LOGICOMATHEMATICAL COGNITION

Logicomathematical operations are initially constructed by primates' actions. As the elements progressively approximate constant givens, they open up new and ever-growing possibilities for primates' logicomathematical operations. These operations map intensive (e.g. classifying) and extensive (e.g. substituting) part–whole transformations onto the elements (e.g. sets) of primates' cognition. These mappings produce qualitative (e.g. identity) and quantitative (e.g. equivalence) relations.

Human infants

At first human infants' logicomathematical operations are elementary and weak (e.g. constructing single category classes of identical objects). Progressively they become ever more complex and powerful mappings that increasingly approximate, but never achieve during infancy, the status of fully formed logicomathematical operations (e.g. constructing two category classifications by identities and differences)[1]. I refer to the initial operations as first-order because they comprise direct mappings onto elements. I refer to the more developed operations as second-order because they comprise mappings onto mappings. Accordingly, this section outlines the nature and development of first- and second-order logicomathematical cognition in human infants.

Composing single sets of objects originates, develops, and dominates infants' constructions during their first year. While they begin to lose their primacy during infants' second year, single sets continue to progress in power (e.g. in size). As long as infants are limited to composing single sets, they are also limited to mapping elementary operations onto them. So, logicomathematical cognition is limited to first-order operations during most of infants' first year. With progress in composing single sets comes progress in the development of first-order operations so that they become increasingly powerful even during infants' second year, when they no longer exclusively mark logicomathematical cognition. The expectation is that progress in composing single sets provides requisite elements for increasingly powerful first-order operations.

The core operations with which infants begin to map quantitative or extensive transformations onto sets of objects include substituting, replacing, and commuting (Langer, 1980, 1986). Parallel development marks all these three exchange operations in human infancy. So, I will illustrate with findings on substituting only. One third of infants at age 6 months produce quantitative equivalence within single sets by substituting objects. To illustrate, one 6 month old infant dropped a doll into a ring, quickly lifted the doll out of the ring, and dropped another doll in its place in the ring. While still limited to single collections, by age 12 months all infants substitute and invert objects within two-object sets they have constructed. In the just-mentioned illustration, inverting objects would mean lifting the substitute doll out of the ring and dropping the initial doll back into the ring again. In addition, 50% of infants at age 12 months extend substituting to three-object sets. Progress in first-order substituting continues during infants' second year. Most notably, some infants begin substituting within single four to eight-object sets they have constructed. Such exchanges comprise fundamental operations with which infants as well as adults construct quantitative relations of equivalence and nonequivalence.

Another group of fundamental operations that infants in their first year already begin to share with adults, albeit in the most rudimentary forms, are classificatory (Langer, 1980, 1986). These operations map intensive transformations onto sets to produce similarity, identity, and difference relations. The sequence of development is as follows. At age 6 months, infants consistently compose sets of objects from *different* classes with each other (as illustrated in Figure 12.2) when presented with two contrasting classes of two objects (e.g. two identical yellow cross rings and two identical green rectangular rings). So, for example, 6 month olds consistently pair crosses with rectangles, rather than crosses with crosses or rectangles with rectangles. At ages 8 and 10 months, infants no longer consistently compose sets of objects from different classes. Instead, their compositions are *random*. Thus, for example, 8 and 10 month olds are equally likely to pair crosses with rectangles as they are to pair crosses with crosses and rectangles with rectangles. By age 12 months, infants begin to compose sets of *identical* objects (e.g. red dolls with red dolls as illustrated in Figure 12.3). Somewhat varying procedures and analyses yield comparable results on classifying by identities at age 12 months (Nelson, 1973; Riccuiti, 1965; Starkey, 1981; Sugarman, 1983). By age 15 months, infants begin to compose consistently *similar* (e.g. red dolls with blue dolls), as well as identical, sets (e.g. red dolls with red dolls).

While first-order operations, such as substituting and classifying, completely dominate infants' cognitive development during their first year, they no longer do so during their second year (Langer, 1980, 1986). The rudiments of more advanced operations originate towards the end of their first year and develop steadily during their second year. These more advanced second-order operations are the developmental products of infants' integrating their first-order operations.

Fundamental to infants' constructing second-order operations is their forming elements comprising minimal compositions of compositions as manifest in their composing two sets of objects in temporal overlap. Thus, the major new feature marking the development of second-order substituting consists of exchanges between two contemporaneous sets (Langer, 1986). For instance, infants exchange blocks between two stacks they had previously constructed. Some of these exchanges include mapping substituting onto previously constructed corresponding sets (e.g. two stacks of three blocks each). Then the quantitative results are second-order relations of equivalences upon equivalences.

Infants also begin to construct second-order classification during their second year. They produce contemporaneous sets in which the membership of each set is composed of identical or similar objects, while the membership of the two sets is composed of different objects (as illustrated in Figure 12.4). Infants begin to construct two contrasting classes by age 18 months and do so consistently by age 36 months (Langer, 1986, unpublished results; Nelson, 1973; Riccuiti, 1965; Roberts & Fischer, 1979; Sinclair et al., 1982; Starkey, 1981; Sugarman, 1983; Woodward & Hunt, 1972).

Cebus and macaque monkeys

The onset age, sequence, extent, and rate of classificatory development were different in cebus and macaque monkeys from each other, from humans, and from chimpanzees. The extent did not progress beyond first-order classifying, since, as we have already seen, they did not compose two temporally overlapping sets. The limited elements that they constructed – single sets – constrained the level of their cognitive development to first-order operations. Hence, cebus and macaque monkeys did not manifest second-order classifying (Spinozzi & Natale, 1989); at least when interacting with the nonsocial objects that have been studied.

Even within their more limited extent, the rate and sequence of developing first-order classifying was different between these monkeys and humans; although all primates end up by classifying in terms of identity and similarity. For example, cebus monkeys began by mostly classifying at *random* at age 16 months, changed to mostly classifying by *differences* at age 36 months, and finally classified by *identities* and *similarities* by age 48 months.

The comparative developmental picture was similar for quantitative operations such as substituting. During their second year, cebus and macaque monkeys began to develop first-order substituting, but the extent was limited (Potì & Antinucci, 1989). For instance, it rarely exceeded substituting in single two-object sets. So even the extent of their first-order cognition was already less than that in human infants. Further, cebus and macaque monkeys did not develop any second-order quantitative cognition. As with classifying, this was not possible because they did not construct the requisite minimal elements, which are two temporally overlapping sets.

Moreover, the developing organization of their exchange operations diverged radically from that in human infants. Like human infants, cebus and macaque monkeys developed all three exchange operations, commuting, replacing, and substituting. Unlike human infants, they did not develop them in parallel. Instead, substituting lagged considerably behind the development of commuting and replacing (Potì & Antinucci, 1989).

Chimpanzees

Unlike cebus and macaque monkeys but like human infants, chimpanzees construct elements necessary for second-order cognition. Juvenile chimpanzees constructed minimal temporally overlapping sets. Although much of the data are still being analyzed, two findings are central. One is that juvenile chimpanzees developed rudimentary second-order cognition, including classifying (Spinozzi, 1993) and substituting (P. Potì & F. Antinucci, unpublished results), that is partly similar to the level attained by 18 month old human infants. Another is that the rate of juvenile chimpanzees' development was much slower than that of human infants. The onset age for second-order logicomathematical operations was not until the fifth year in chimpanzees while it is the second year in human infants.

THE ORIGINS OF PHYSICAL COGNITION

The development of physical cognition of objects, space, time, and causality complement the development of logicomathematical cognition. Nevertheless, physical constructions take different generative forms, means–ends transformations; whereas logicomathematical mappings, we have just seen, involve part–whole transformations (Langer, 1980, 1986). Means–ends transformations vary by physical category (cf. Piaget *et al.*, 1977). For instance, spatial means-ends transformations construct placement and displacement relations between objects, whereas causal means–ends transformations construct energy relations between objects, such as when one object is pushed against another. I have therefore proposed that primates' mapping of means–ends transformations constitute basic functions that construct contingent dependency relations (i.e. physical possibility and impossibility). Since causality is central to the development of physical cognition, it will be the focus of our comparative discussion.

Human infants

Causal development, like all physical cognition, parallels logicomathematical development in human infancy. A substantial proportion of the sets of objects that humans compose in early infancy involve causal properties (e.g. composing a two-object set by pushing a block into a cylinder which, as a consequence, rolls away). By age 6 months, human infants' causal compositions already include constructing two elementary types of causal relation or function, composing effects and negating effects.

Composing effects begins with infants constructing, replicating, and observing elementary effects that are directly dependent upon rudimentary causes (Langer, 1980). To illustrate, infants use one object as a means to repeatedly push or bang on another object while observing their construction (see also Piaget, 1954). Negating effects, the second type of causal function, begin when infants anticipate and observe elementary effects that directly influence their subsequent rudimentary causal constructions (Langer, 1980). To illustrate, infants use one object as a means to block and stop another object that is rolling in front of them while observing the effects of their causal constructions (for infants' catching skills, see also von Hofsten, 1983). Their causal blocking is directly dependent on the prediction of their targets' trajectories, otherwise they would miss their targets, which would then continue rolling away.

There is marked progress in constructing both direct causal functions during infants' second year (Langer, 1986). This includes generating effects that are direct linear functions of causes (e.g. objects are pushed harder and harder). The direct functional dependency may be formalized as one-way ratio-like relations, such as "moving farther is a dependent function of pushing harder." This elementary structure is what differentiates first-order from second-order causal functions that originate during infants' second year.

Second-order functions are integrative means–ends transformations. They coordinate elementary first-order means–ends mappings to each other. For example, infants may coordinate (a) composing effects using object A to launch object B with (b) negating effects by using object A to block object B, as illustrated in the next paragraph. This produces a second structural level of more powerful functions. The effects are directly dependent upon the causes in first-order functions. In contrast, the effects begin to be proportional to the causes in second-order functions. So, the basic difference is that first-order functions are featured by direct one-way ratio-like relations. Second-order functions are marked by more indirect two-way analogical or proportional-like relations that map the one-way ratios onto each other (for the formal distinction, see Langer, 1986, pp. 370–5).

Older, like younger, infants use one object as an instrument with which to push a second dependent object (i.e. to compose an effect). But beginning at age 18 months, when the effect is that the dependent object rolls away, then infants may also transform the instrument into a means with which to block the dependent object (i.e. to negate an effect). Correlatively, infants thereby transform the end or goal from rolling to stopping. As soon as the dependent object stops rolling infants transform the same instrumental object back into a means with which to make the dependent object roll away again. And so on.

Thus, older infants begin to covary their transformations of both means and ends (i.e. to covary composing and negating causes and effects). These covariations form coordinate proportional-like dependencies between causes and effects. These proto-proportions map previously constructed first-order dependencies onto each other. The products are second-order causal functions, such as "moving is a function of pushing, as stopping is a function of blocking."

Cebus and macaque monkeys

As with their logicomathematical cognitions, cebus and macaque monkeys developed first-order but not second-order physical cognition. During their first year, cebus and macaque monkeys developed rudimentary first-order causal means–ends relations, such as using a support to get an object resting on it (Spinozzi & Poti, 1989). More complex first-order causal functions, such as using a stick to get an object that is out of reach, developed by ages 18 to 20 months in cebus monkeys, but may never develop in macaques (Natale, 1989). It appears, then, that cebus and macaque monkeys' development of first-order physical cognition, including causality, preceded their development of first-order logicomathematical cognition. This is in contrast with the ontogenetic pattern in human infancy, where physical and logicomathematical cognition develop in parallel.

Chimpanzees

Preliminary analyses of the data indicate that chimpanzees' causal development is somewhere between that of monkeys and that of human infants (P. Poti & G. Spinozzi, unpublished results). Chimpanzees probably developed some second-order causal functions. Like monkeys, on the one hand, chimpanzees' physical cognition, including causality, began to develop before their logicomathematical cognition. Unlike monkeys, on the other hand, chimpanzees' logicomathematical development began well before their development of physical cognition was completed.

THE ORIGINALIST HYPOTHESIS: COGNITION AND LANGUAGE

The comparative picture of primate cognitive development emerging from the research outlined in the previous sections is partly and provisionally schematized in Figure 12.5. It is only partial because it does not reflect many important findings (e.g. differences in the cognitive development of cebus and macaque monkeys and differences in the level of second-order cognition developed by chimpanzees and humans). The schema is also necessarily provisional because it is based upon ongoing research. As such, it is as much a portrayal of hypotheses as it is facts based upon the data available so far.

The comparative primate cognitive development just reviewed and schematized in Figure 12.5 supports our originalist hypothesis that logicomathematical cognition is a primary and initial development in humans, just as physical cognition is. This comparative picture is inconsistent with what we may call the derivationist hypothesis, that grammatical language is a prerequisite to the development of logicomathematical cognition. On the contrary, second-order operations precede grammatical language in human infant ontogeny.

Logicomathematical cognition, like physical cognition, originates very early in human infancy (see also Chapter 8). Like physical cognition, logicomathematical cognition already develops through two levels in human infancy: from first-order to second-order logicomathematical operations and from first-order to second-order physical function. Further, first-order operations originate prior to any manifest language (e.g. the one-word stage); and second-order operations develop prior to anything more than the most rudimentary proto-grammatical language. Imposing a hierarchy, a keystone of grammatical language, does not begin to develop until infants' third year (cf. Bickerton, 1990; Lieberman, 1991).

Elementary logicomathematical cognition also originates in other primate species that do not develop language. Cebus and macaque monkeys develop first-order operations without developing any manifest language. Juvenile chimpanzees develop first-order and (at least) rudimentary second-order operations without developing any manifest grammatical language, though they do learn proto-grammatical language (e.g. Greenfield & Savage-Rumbaugh, 1990; Savage-Rumbaugh et al., 1993).

These findings are consistent with my proposal that second-order cognition provides axiomatic systems properties necessary for any proto-grammatical symbolic system, including language (Langer, 1982a, 1983, 1986, 1993). Second-order cognition is necessary in order to begin to produce and comprehend the arbitrary conventions used to make symbolic combinations stand for and communicate semantic referents in syntactic forms. Second-order cognition enables organisms to begin to generate syntactic constructions in which linguistic elements are progressively combinable and inter-

Evolution of primate cognitive development

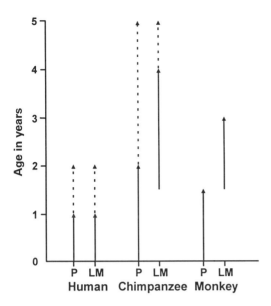

Figure 12.5. Comparative cognitive development: Vectorial trajectories of developmental onset age, velocity, sequence, and organization (but not extent or offset age). P, physical cognition; LM, logicomathematical cognition; continuous arrow, first-order; dashed arrow, second-order.

changeable, yet meaningful (for a fuller discussion, see Chapter 19 of Langer, 1986).

The basic proposal is that second-order operations and functions provide the axiomatic rewrite rules without which grammatical language is not possible. To illustrate, proto-grammatical sentences are not possible without at least rudimentary second-order substituting of elements between two sentential compositions. We have seen wide variation among primate species in the ability to substitute elements between two compositions. It develops towards the end of the second year in human infants, barely develops by the fifth year in chimpanzees, and did not develop in the cebus and macaque monkeys tested. So too, grammatical language is developed by human children and proto-grammatical language can be learned by chimpanzees, but not by monkeys.

This structural developmental proposal about the relations between cognition and language leads to a corollary hypothesis: the development of second-order cognition is a necessary prerequisite to the formation of fully grammatical language. This corollary entails three implications: (a) since human infants develop fully formed second-order cognition, they can develop grammatical language; (b) since juvenile chimpanzees develop rudimentary second-order cognition only, they are limited to acquiring proto-grammatical language; (c) since cebus and macaque monkeys do not develop second-order cognition, they cannot acquire even proto-grammatical language. Crucial testing of these hypotheses awaits, minimally, chimpanzee ontogenetic data on the correspondences, if any, between the onset ages of first-order operations, second-order operations, productive proto-grammar, and receptive proto-grammar.

Logicomathematical cognition develops earlier in humans than in other primates. It also develops simultaneously and in parallel with physical cognition (Langer, 1985, 1989, 1993). This developmental synchrony is crucial, as I will explain (see pp. 270–2). Humans' ontogenetic pattern is clearly inconsistent with the derivationist hypothesis that the development of logicomathematical cognition depends upon prior acquisition of grammatical language.

Logicomathematical cognition begins to develop after physical cognition in cebus and macaque monkeys' ontogeny. The development of their physical cognition is well underway or completed by the onset age at which they begin to develop logicomathematical cognition. These monkeys' ontogenetic sequence is the one expected by the derivationist hypothesis. Here, of course, the problem is that monkeys do not develop grammatical language.

Logicomathematical cognition begins to develop while physical cognition is in the midst of development in chimpanzee ontogeny. The development of their physical cognition is in progress by the onset age at which they begin to develop logicomathematical cognition. How this partly overlapping sequence corresponds to the ontogeny of proto-grammar has yet to be determined, as just noted.

HETEROCHRONY AS A MECHANISM OF COGNITIVE EVOLUTION

These comparative differences indicate that the developmental sequencing between cognitive domains diverges in living primates, reflecting divergent evolution. In human infants, logicomathematical cognition originates at the same age as physical cognition. Further, the development of infants' physical cognition is in progress at the same ages at which their logicomathematical cognition is in progress. Neither type of cognition begins or ends before the other during childhood. Consequently, both forms of cognition are open to similar environmental influences and to each other's influence.

In cebus and macaque monkeys, physical and logicomathematical cognition constitute consecutive developmental trajectories that are relatively independent of each other. Asynchronic developmental trajectories do not readily permit direct interaction or information flow between cognitive structures, since they are out-of-phase with each other. At most, then, the interaction between cognitive domains in the early development of cebus and macaque monkeys may be indirect. The main potential lines of information are from relatively developed physical cognition to undeveloped logicomathematical cognition.

In chimpanzees' ontogeny, physical and logicomathematical cognition also constitute consecutive developmental trajectories that are relatively independent of each other. However, chimpanzees' development of physical cognition is not completed before the manifest onset of logicomathematical cognition. Physical and logicomathematical cognition constitute partially asynchronic developmental trajectories. We can therefore expect that these two cognitive domains may eventually begin to partially interact and inform each other, but late in ontogenesis as compared with humans.

From the start of human ontogeny, physical and logicomathematical cognition constitute contemporaneous developmental trajectories that become progressively interdependent. Synchronic developmental trajectories permit direct interaction or information flow between cognitive domains. Mutual and reciprocal influence between logicomathematical and physical cognition is readily achievable, since humans develop them simultaneously and in parallel. Thus, we have found that, even in infancy, logicomathematical cognition introduces elements of necessity and certainty into physical cognition (Langer, 1985). At the same time, physical cognition introduces elements of contingency and uncertainty into logical cognition.

In primate evolution, unilinear growth (of physical followed by logicomathematical cognition) evolved into multilinear growth (of physical and logicomathematical cognition). The sequential pattern of physical followed by logicomathematical cognition in the ontogeny of cebus and macaque monkeys became "folded over" and, hence, concurrent developments: (a) first to form descendant partially multilinear development midway in chimpanzee ontogeny; and (b) eventually to form fully multilinear development from the start in human ontogeny (as illustrated in Figure 12.5). The onset age for beginning to develop physical cognition is roughly the same in all primates studied so far. In cebus and macaque monkeys the onset age for logicomathematical cognition is retarded such that its development does not overlap with the development of physical cognition. In chimpanzees the onset age for logicomathematical cognition is partially accelerated such that its development partly overlaps with the development of physical cognition. In humans the onset age for logicomathematical cognition is accelerated to the point that it becomes contemporaneous with the onset age of physical cognition.

Phylogenetic displacement in the ontogenetic onset or timing of one structural development relative to another within the same organism causes a disruption in the repetition of phylogeny in ontogeny. This is one of several kinds of evolutionary change known collectively as heterochrony (Gould, 1977; McKinney & McNamara, 1991). Since heterochronic displacement involves a dislocation of the phylogenetic order of succession, heterochrony produces a change in the rate or timing of ancestral processes. The timing may be accelerated or retarded. Thus, heterochrony is an evolutionary mechanism by which ancestral correlations between growth, differentiation, centralization, and hierarchic integration are disrupted and new descendant correlations are established. Werner (1948), it should be recalled, proposed that development is governed by the orthogenetic principle of increasing differentiation, centralization, and hierarchic integration. Then, heterochrony is an evolutionary mechanism that changes the pattern of orthogenesis governing development and all that entails for ontogenetic change.

The comparative data on the organization of and sequencing between cognitive domains are consistent with the hypothesis that heterochrony is a mechanism of the evolution of primate cognition. On this hypothesis, heterochronic displacement is a mechanism whereby consecutively developing ancestral cognitive structures were transformed in phylogenesis into simultaneously developing descendant cognitive structures in human ontogenesis. Heterochrony produced the reorganization of nonaligned ancestral cognitive structures in cebus and macaque monkeys into the partly aligned descendant structures in chimpanzees and the fully aligned descendant structural development of cognition in human infancy.

This orthogenetic reorganization opened up the possibility for full information flow between logicomathematical (e.g. classificatory) and physical (e.g. causal) cognition in human infancy (e.g. making it possible to form a "logic of experimentation"). These same domains are segregated from each other in time and, therefore, in information flow in the early development of nonhuman primates such as cebus and macaque monkeys; they are partially segregated in chimpanzees. Thus, the possibilities for cognitive development of other primates remain permanently partial, and, thus, we find the permanent possibility for further cognitive development in humans (e.g. a history of science).

RECURSIVENESS AS A MECHANISM OF COGNITIVE DEVELOPMENT

Two key features begin to structure human infants' cognitive development during their second year. First, parallel or synchronic construction of logicomathematical and physical cognition opens up the possibility of lateral integration between them (e.g. to begin constructing a logic of causal experimenting). Second, mapping first-order operations onto each other and first-order functions onto each other to form second-order operations (e.g. to begin constructing class hierarchies) and functions (e.g. to begin constructing causal covariation) opens up concurrent possibilities for hierarchic integration within their logicomathematical and physical cognitions. Both are initial features of the recursive development marking human cognition.

Two key features of recursiveness, the data are also beginning to indicate, partly structure juvenile chimpanzees' cognitive development. While still in the process of analysis, the supportive data include initial findings of rudimentary second-order classifying (Spinozzi, 1993) and exchange operations (P. Potì & F. Antinucci, unpublished results) outlined on pp. 265–6. The onset age for both is greatly retarded, not until their fifth year as compared to human infants' second year. A fundamental proximal cause of both the partial and retarded development of second-order cognition, I have hypothesized, is that the manifest onset of chimpanzees' logicomathematical cognition does not occur until the development of their physical cognition is well underway (Langer, 1994a). Partly overlapping development limits the possibilities for recursive, lateral, and hierarchic structural integration between and within juvenile chimpanzees' cognitions.

Recursiveness, we have seen, does not structure the cognitive development of cebus and macaque monkeys. A fundamental proximal cause, I have hypothesized, is that the onset ages of their logicomathematical and physical cognitions are so divergent that there is little, if any, overlap between their developments (Langer, 1989, 1993). Accordingly, neither lateral nor hierarchic structural integration is possible between and within monkeys' cognitions.

These comparative findings and hypotheses are consistent with the proposal that recursiveness changes the rules of development (Langer, 1969a, 1980, 1986, 1994a).

Recursiveness generates the onset of more advanced cognition because then the elements of human infants' cognitive mappings become other cognitive mappings as well as actual things. In this way, human infants begin to detach their logicomathematical operations and physical functions from their initial concrete objects of application. They begin to apply operations to operations as well as to concrete objects, and functions to functions as well as to concrete objects. Detaching operations and functions from their concrete object referents and, instead, mapping them onto other operations and functions is pivotal to forming abstract relations as elements of cognition. These cognitive mappings onto mappings provide the minimal conditions necessary for abstract reflecting and forming representational concepts that are hierarchical as well as integrated. On this hypothesis, partial recursiveness in juvenile chimpanzees implies that they develop only minimal representational concepts (and acquire proto-grammatical language, as discussed on pp. 268–9). Absence of recursiveness in monkeys implies that they can not develop representational concepts (and even acquire proto-grammatical language, as discussed on pp. 268–9).

Representation, on this proposal, begins with hierarchic mappings onto mappings. Its conceptual origins are subjects' mappings of operations onto operations and functions onto functions. A good illustration is the already cited development by human infants during their second year of mapping substitutions onto correspondences to produce equivalences upon equivalences. Then the referents of substituting are no longer limited to the concrete objects forming the two corresponding sets. The referents can become equivalence relations. But relations are more abstract than objects, so the referents are becoming abstract.

In general, with the formation of second-order hierarchic cognition, the referents of operations and functions are no longer limited to objects. Cognition is no longer limited to the concrete. Progressively the referents of operations and functions are becoming relations, such as equivalence and dependency, that are the product of other operations and functions mapped onto objects. By mapping operations and functions onto relations, cognition is becoming abstract and reflective or what we might call thoughtful. Then representational logicomathematical and physical concepts begin to become abstract and reflective.

This recursive model posits that, comparatively, primate cognitive development is multidirectional, multileveled, and multistructural. Progenitor structures do not disappear, are not lost, when they spawn their descendant transforms. Both progenitor and descendant cognitive structures continue to evolve and develop. Eventually, progenitor structures do so less than their descendant structures. Progenitor structures (i.e. first-order cognitions) serve as progressively powerful elements for descendant structures (second-order cognitions). In turn, they progressively integrate their progenitors.

This recursive model also specifies the course and form of cognitive development comparatively across primate species. In ontogeny, the course of first-order cognition develops in power stage-by-stage even when part of it is transformed into second-order cognition, as it is in humans and chimpanzees. Once second-order cognition arises, it also develops in power stage-by-stage, at least in humans. This is because first-order cognitions are already in the midst of their development when second-order cognitions are just beginning to develop.

Structural ontogenetic and phylogenetic continuity in this recursive model is provided by the part–whole transformations and means–ends transformations mapped by primates from their earliest infancy on through their cognitive development into adulthood. These structural developmental invariants are the basic constructive elements of all concept formation. They are material determinants or causes of logicomathematical operations and physical functions.

Structural ontogenetic and phylogenetic discontinuity in this recursive model begins to be generated by chimpanzees and humans mapping their first-order cognitions onto each other. These structural developmental transformations are the basic constructive elements of all concept formation. They are formal determinants or causes of the development of second-order cognition.

So second-order cognitions are both continuous with and discontinuous from first-order cognitions in both ontogeny and phylogeny. More advanced conceptual forms progressively integrate the very progenitor forms of which they are derivative transforms. The cognitive offspring incorporate their progenitors.

Optimally, structural continuity, discontinuity, and integration are facilitated by the greater synchrony of developing first- and second-order cognition found only in human infancy. This produces progressive equilibrium (e.g. coherence, virtuosity, and completeness) in the organization of cognition. At the same time it produces progressive manageable disequilibrium (e.g. imbalances between levels of cognition, opening of new possibilities and impossibilities, and introduction of uncertainty) that I have proposed as being a basic structural source of cognitive development (Langer, 1969b, 1980, 1986). Together, progressive structural equilibrium and disequilibrium are efficient determinants or causes of cognitive development.

The recursive model of equilibration is an open dialectical model in which (a) progressive disequilibrium is a source of progressive equilibrium, and (b) progressive equilibrium is a source of progressive disequilibrium (Langer, 1974, 1982b). So development is produced by both progressive equilibrium and disequilibrium. For instance, 18 month old human infants go on to experiment with the limits of tilting an upright cylinder once they have become experts or virtuosos at balancing it upright. Thus begins the long process of determining the parameters of what is physically possible and what is impossible.

Evolving cognition, on the present view, is a developmental construction; it is not preformed genetically or maturationally. The ontogenetic antecedents of cognitive development are elementary but already constructive mappings. Specifically, part–whole transformations are the sources of logicomathematical operations and the production of necessary equivalence, nonequivalence, and reversibility; and means–ends transformations are the sources of physical functions and the production of contingent possibility and impossibility.

Initially differentiated structures of logicomathematical and physical cognition are further differentiated with development. But they also progressively interact by informing each other. So their comparative (phylogenetic, ontogenetic, and historical) development is marked by progressive differentiation and progressive coordination into hierarchically integrated organizations. This orthogenetic principle, as Werner (1948, 1957) proposed, is at the heart of the comparative psychology of mental development.

The first major, but still transitional, transformation toward differentiation and hierarchic integration in human ontogeny is the development of second-order logicomathematical and physical cognition during late infancy. This orthogenetic process, I have proposed, initiates the construction of representational cognition by infants. These cognitions are already marked by precursory necessity and contingency. So, too, though perhaps in the more rudimentary forms we have seen, chimpanzees begin to develop representational cognition during their fifth year. It is most likely, then, that other great apes develop comparable representational cognition.

From this comparative perspective, the structural development of necessary (logicomathematical) and contingent (physical) cognition in its peculiarly human form differs from that of other primates in at least onset age, organization, velocity, extent, sequencing, and offset age. Furthermore, in line with Baldwin's (1915) hypothesis, human ontogeny provides the initial and unique impetus for the historical societal development of necessity and possibility. Their perennial individual and historical growth is an open-ended process of reciprocal amplification. The process begins by constructing natural pragmatic forms of cognition. It progresses toward constructing never-ending formal scientific knowledge that integrates, laterally and hierarchically, logicomathematical with other forms of cognition.

NOTES

1 Such an achievement requires understanding (a) that subclasses can be constructed, and (b) that when constructed they are necessarily subordinate complements of the next higher superordinate class (e.g., that apples and oranges are different and similar at the same time), such that (c) they form a genealogy of nested elements.

REFERENCES

Antinucci, F. (ed.) (1989). *Cognitive Structure and Development of Nonhuman Primates.* Hillsdale, NJ: Lawrence Erlbaum.
Baldwin, J.M. (1915). *Genetic Theory of Reality.* New York: Putnam.
Dickerton, D. (1990). *Language and Species.* Chicago: University of Chicago Press.
Bowerman, M. (1978). Structural relationships in children's utterances: Syntactic or semantic? In *Readings in Language Development*, ed. L. Bloom, pp. 217–30. New York: Wiley.
Braine, M.D.S. (1990). The "natural logic" approach to reasoning. In *Reasoning, Necessity, and Logic: Developmental Perspectives*, ed. W.F. Overton, pp. 133–57. Hillsdale, NJ: Lawrence Erlbaum.
Carnap, R. (1960). *The Logical Syntax of Language.* Paterson, NJ: Littlefield & Adams.
Doré, F.Y. & Dumas, C. (1987). Psychology of animal cognition: Piagetian studies. *Psychological Bulletin*, 102, 219–33.
Falmagne, R.J. (1990). Language and the acquisition of logical knowledge. In *Reasoning, Necessity,*

and Logic: Developmental Perspectives, ed. W.F. Overton, pp. 111–31. Hillsdale, NJ: Lawrence Erlbaum.

Gould, S. J. (1977). *Ontogeny and Phylogeny*. Cambridge, MA: Harvard University Press.

Greenfield, P.M. & Savage-Rumbaugh, E.S. (1990). Grammatical combination in *Pan paniscus*: process of learning and invention in the evolution and development of language. In *"Language" and Intelligence in Monkeys and Apes: Comparative Developmental Perspectives*, ed. S.T. Parker & K.R. Gibson, pp. 540–78. New York: Cambridge University Press.

Jacobsen, T.A. (1984). The Construction and Regulation of Early Structures of Logic. A Cross-cultural Study of Infant Cognitive Development. Ph.D. thesis, University of California at Berkeley.

Langer, J. (1969a). *Theories of Development*. New York: Holt, Rinehart & Winston.

Langer, J. (1969b). Disequilibrium as a source of development. In *Trends and Issues in Developmental Psychology*, ed. P.H. Mussen, J. Langer & J. Covington, pp. 22–37. New York: Holt, Rinehart & Winston.

Langer, J. (1974). Interactional aspects of mental structures. *Cognition*, 3, 9–28.

Langer, J. (1980). *The Origins of Logic: Six to Twelve Months*. New York: Academic Press.

Langer, J. (1982a). From prerepresentational to representational cognition. In *Action and Thought*, ed. G. Forman, pp. 37–63. New York: Academic Press.

Langer, J. (1982b). Dialectics of development. In *Regression in Development*, ed. T.G.R. Bever, pp. 233–66. Hillsdale, NJ: Lawrence Erlbaum.

Langer, J. (1983). Concept and symbol formation by infants. In *Toward a Holistic Developmental Psychology*, ed. S. Wapner & B. Kaplan, pp. 221–34. Hillsdale, NJ: Lawrence Erlbaum.

Langer, J. (1985). Necessity and possibility during infancy. *Archives de Psychologie*, 53, 61–75.

Langer, J. (1986). *The Origins of Logic: One to Two Years*. New York: Academic Press.

Langer, J. (1988). A note on the comparative psychology of mental development. In *Ontogeny, Phylogeny, and Historical Development*, ed. Strauss, S, pp. 68–85. Norwood, NJ: Ablex.

Langer, J. (1989). Comparison with the human child. In *Cognitive Structure and Development of Nonhuman Primates*, ed. F. Antinucci, pp. 163–88. Hillsdale, NJ: Lawrence Erlbaum.

Langer, J. (1990). Early cognitive development: Basic functions. In *Developmental Psychology: Cognitive, Perceptuo-Motor, and Neuropsychological Perspectives*, ed. C.A. Hauert, pp. 19–42. Amsterdam: North Holland.

Langer, J. (1993). Comparative cognitive development. In *Tools, Language and Cognition in Human Evolution*, ed. K.R. Gibson & T. Ingold, pp. 300–13. New York: Cambridge University Press.

Langer, J. (1994a). From acting to understanding: The comparative development of meaning. In *The Nature and Ontogenesis of Meaning*, ed. W.F. Overton & P. Palermo, pp. 191–213. Hillsdale, NJ: Lawrence Erlbaum.

Langer, J. (1994b). Logic. In *Encyclopedia of Human Behavior*, ed. V.S. Ramachandren, pp. 83–91. San Diego: Academic Press.

Lieberman, P. (1991). *Uniquely Human*. Cambridge, MA: Harvard University Press.

Macnamara, J. (1986). *A Border Dispute: The Place of Logic in Psychology*. Cambridge, MA: MIT Press.

Maratsos, M. (1983). Some current issues in the study of the acquisition of grammar. In *Handbook of Child Psychology*, ed. P.H. Mussen, pp. 707–86. New York: Wiley.

McKinney, M.L. & McNamara, J.K. (1991). *Heterochrony: The Evolution of Ontogeny*. New York: Plenum Press.

Natale, F. (1989). Causality II: The stick problem. In *Cognitive Structure and Development in Nonhuman Primates*, ed. F. Antinucci, pp. 121–33. Hillsdale, NJ: Lawrence Erlbaum.

Nelson, K. (1973). Some evidence for the primacy of categorization and its functional basis. *Merrill-Palmer Quarterly*, **19**, 21–39.

Parker, S.T. (1995). Using cladistic analysis of comparative data to reconstruct the evolution of cognitive development in hominids. In *Phylogenies and the Comparative Method*, ed. E. Martins. New York: Oxford University Press. (in press.)

Parker, S. T. & Gibson, K. R. (1979). A developmental model for the evolution of language and intelligence in early hominids. *Behavioral and Brain Sciences*, **2**, 367–408.

Piaget, J. (1954). *The Construction of Reality in the Child*. New York: Basic Books. (First published in French in 1937.)

Piaget, J. (1972). *Essai de Logique Opératoire*. Paris: Dunod.

Piaget, J., Grize, J. B., Szeminska, A. & Vinh Bang (1977). *Epistemology and Psychology of Functions*. Dordrecht: Reidel.

Potì, P. & Antinucci, F. (1989). Logical operations. In *Cognitive Structure and Development of Nonhuman Primates*, ed. F. Antinucci, pp, 189–228. Hillsdale, NJ: Lawrence Erlbaum.

Riccuiti, H.N. (1965). Object grouping and selective ordering behavior in infants 12 to 24 months. *Merrill-Palmer Quarterly*, **11**, 129–48.

Roberts, R.J. & Fischer, K.W. (1979). A developmental sequence of classification skills. Paper presented at the Society for Research in Child Development Annual meeting, San Francisco, USA. April.

Savage-Rumbaugh, E.S., Murphy, J., Sevcik, R.A., Brakke, K.E., Williams, S.L. & Rumbaugh, D.M. (1993). Language comprehension in ape and child. *Monographs of the Society for Research in Child Development*, **58**, 1–221.

Sinclair, M., Stambak, M., Lezine, I., Rayna, S. & Verba, M. (1982). *Les Bébés et Les Choses*. Paris: Presses Universitaires de France.

Spinozzi, G. (1993). The development of spontaneous classificatory behavior in chimpanzees (*Pan troglodytes*). *Journal of Comparative Psychology*, **107**, 193–200.

Spinozzi, G. & Natale, F. (1989). Classification. In *Cognitive Structure and Development of Nonhuman Primates*, ed. F. Antinucci, pp. 163–88. Hillsdale, NJ: Lawrence Erlbaum.

Spinozzi, G. & Potì, P. (1989). Causality I: The support problem. In *Cognitive Structure and Development of Nonhuman Primates*, ed. F. Antinucci, pp. 113–20. Hillsdale, NJ: Lawrence Erlbaum.

Starkey, D. (1981). The origins of concept formation: Object sorting and object preference in early infancy. *Child Development*, **52**, 489–97.

Sugarman, S. (1983). *Children's Early Thought: Developments in Classification*. New York: Cambridge University Press.

von Hofsten, C. (1983). Catching skills in infants. *Journal of Experimental Psychology: Human Perception and Performance*, **9**, 75–85.

Werner, H. (1948). *Comparative Psychology of Mental Development*. New York: International Universities Press. (First published in German in 1926.)

Werner, H. (1957). The concept of development from a comparative and organismic point of view. In *The Concept of Development*, ed. D.B. Harris, pp. 125–48. Minneapolis: University of Minnesota Press.

Woodward, W.M. & Hunt, M.R. (1972). Exploratory studies of early cognitive development. *British Journal of Educational Psychology*, **42**, 248–59.

13

Simon says: The development of imitation in an enculturated orangutan

H. LYN MILES, ROBERT W. MITCHELL AND
STEPHEN E. HARPER

INTRODUCTION

A party, of which he formed one, landed at Borneo, and passed into the interior on an excursion of pleasure. Himself and a companion . . . captured [an] Ourang-Outang. . . . Returning home from some sailors' frolic . . . in the morning, . . . he found the beast occupying his own bedroom, into which it had broken from a closet adjoining, where it had been, as was thought, securely confined. Razor in hand, and fully lathered, it was sitting before a looking-glass, attempting the operation of shaving, in which it had no doubt previously watched its master through the keyhole of the closet.
(Poe, 1938, pp. 165–6).

The notion of apes being human-like or imitating human activities, as in Edgar Allen Poe's account of a murderous orangutan, has persisted since our first contact with apes, yet has only recently been studied scientifically. This chapter presents scientific evidence that an orangutan can indeed imitate. Since humans develop imitative abilities in a distinctly human cultural context, to ask whether an orangutan possesses this ability is problematic, because even biologically based human abilities emerge fully only as enacted solutions to problems posed within a cultural context. And since we cannot remove a human neonate from cultural influence, we have alternatively approached this question by placing an orangutan in human culture.

If an orangutan can imitate, it is important to understand how this ability develops. In our research, we have investigated the ability of an enculturated language-trained orangutan, Chantek, to imitate both gestural signs and simple motions and sounds that were modeled for him by caregivers. These modeled actions were preceded by the request for Chantek to DO THE-SAME-THING, much as when a player says "Simon says do this" in the children's game of "Simon says." We focused on the development of Chantek's imitations and imitation-based activities, including his imitative responses to and use of the signs DO THE-SAME-THING when he was 6–8 years of age.

In his acquisition of this imitative ability, as well as his linguistic and other cognitive skills, we stressed the following as theoretically and analytically important: an understanding of the social, situational, and contextual factors in the interaction; the caregiver's as well as Chantek's behavior; and the biosocial process by which enculturation occurs. We were interested in Chantek's successful demonstrations of the capacity for imitation, but also how and why he sometimes failed in specific circumstances. We concerned ourselves with how the caregivers' culturally informed expectations of Chantek and their interaction history define the imitation that occurs; that is, we were as interested in how Chantek learned the patterns of the way we do things, as we were in documenting the cognitive abilities he possessed.

IMITATION IN ORANGUTANS

For much of the eigteenth and nineteenth centuries, few intensive observations were made of orangutans, yet instances of human-inspired imitation such as opening a lock or carrying a cane were readily described (Buffon, 1812; Cuvier, 1831; Yerkes & Yerkes, 1929), and orangutan imitativeness came to be so well publicized that Poe (1841/1938) used it in his fiction. In the wake of Darwin's (1896) evolutionary ideas, orangutan imitativeness allowed Furness (1916, p. 281) to conclude that they might be "developed to a grade of human understanding perhaps only a step below the level of the most primitive human being."

Furness (1916, pp. 281, 283) taught speech-like sounds to an orangutan who had difficulty imitating sounds, but not motions, and comprehended verbal commands such as "do this," which apparently allowed Furness (1916, p. 284) to achieve imitations on command. Other early studies and observations combined the orangutan's propensities for tool use and imitation to test for "higher order" thought, and claimed, for example, that an orangutan could imitate the use of a stick or rake to obtain food (Haggerty, 1910, 1913; Hornaday, 1922; Yerkes & Yerkes, 1929, p. 176). However, these interpretations were problematic because orangutans easily employed and invented uses for tools and some orangutans failed to imitate in problem-solving situations (Harrison, 1962, 1969; Koehler, 1993; Shepherd, 1923; Wright, 1972; Yerkes & Yerkes, 1929).

Critical evaluation of evidence for bodily and vocal imitation by animals began in the late 1980s. Imitation broadly defined was commonplace in animals' natural activities, but only apes, dolphins, and humans evinced the most complex form of imitation – communicative simulation (Mitchell, 1987). By contrast, experimental evidence for bodily imitation failed to satisfy stricter criteria such as matching another's exact motor patterns, and too closely resembled stimulus enhancement and social facilitation rather than bodily mapping (Galef, 1988). Imitation, considered as a means of spreading cultural innovations, caused researchers to be concerned with another form of imitation based on novel actions (Visalberghi & Fragaszy, 1990; Whiten & Ham, 1992; see Thorpe, 1963; Tomasello *et al.*, 1993). Monkeys seemed never to replicate novel actions, and apes did infrequently. However, field studies of orangutans had indicated that they took humans, as well as other orangutans, as their models for imitation in natural settings (Galdikas, 1982; Horr, 1977), and that rehabilitant orangutans had a keen interest in replication even when not understanding the action's significance (Russon & Galdikas, 1993; Chapter 7). Other evidence suggested that chimpanzees could be taught to imitate (Custance & Bard, 1994; Custance et al., 1994). Another wrinkle in the debate on the presence of imitation in apes occurred when Terrace et al. (1979) reported that their sign-taught chimpanzee Nim extensively imitated his caregivers' signs, and they interpreted this as a lack of intelligence. Later it was recognized that Nim's teachers' actions increased his imitations, that children's conversation also includes imitations, and that, under different conditions, the sign-taught orangutan Chantek imitated few of his caregivers' signs (Miles, 1983).

Imitation was also put forth as a cognate capacity with mirror-self-recognition in humans, great apes, and possibly dolphins, in that both of these abilities are based on kinesthetic–visual body and facial matching of self with others (Mitchell, 1993a,b; Parker, 1991). In humans, kinesthetic–visual matching is related to a variety of actions that require unidirectional or bidirectional mapping between self and other, including pretense, deferred imitation, planning, mirror-self-recognition, recognition of being imitated, and iconic gestural representation, as these require a mapping between kinesthetic feelings and visual impressions, either of others or of oneself in mental images (Piaget, 1972; Meltzoff, 1990; Mitchell, 1990, 1993b, 1994b; Parker & Milbrath, 1994). Kinesthetic–visual matching is present in the generalized imitation of the chimpanzee Viki, who was taught to imitate actions of her caregivers when they said "do this" (Hayes & Hayes, 1952). Although Viki's success was deemed to "beg replication" (Whiten & Ham, 1992, p. 255), this study has been recently replicated with chimpanzees (Custance & Bard, 1994; Custance et al., 1994), and had already been replicated with orangutans twice (by Furness over 80 years ago and with Chantek over a decade ago).

DEVELOPMENT OF IMITATION BY HUMAN INFANTS AND CHILDREN

Imitation per se is interesting to developmental psychologists because it implies the child's growing capacity for representation and symbolization. For Piaget (1962; 1972), internal representation in the form of mental images begins as interiorized imitation, which simultaneously derives from and supports external representation in the forms of deferred imitation, imitation of novel actions, and symbolic play that are prominent in children at about 18 months of age. Once imitation is interiorized, children can produce new external imitations that will further elaborate those internal images. By contrast, Meltzoff (1990) argued that representation is innately present at birth (as exemplified in, e.g. neonatal facial imitation), and that later forms of imitation are merely elaborations of an already present representational system.

In contrast to these authors, we prefer to focus on the specific perceptual modalities used in any given example of imitation, rather than on a singular representational capacity which ignores modality (cp. Parker, 1977). By far the most prominent form of imitation is based on kinesthetic–visual matching. A developmental approach to kinesthetic–visual matching illuminates its nature and its relation to other cognitive abilities (Mitchell, 1994b). Kinesthetic–visual matching presumably originates from two sources: self-imitation present soon after birth in circular reactions, and innate mapping between kinesthesis and vision as exemplified by neonatal replication of emotional and facial expressions (Baldwin, 1903; Gopnik & Meltzoff, 1994; Guillaume, 1971; Hatfield *et al.*, 1994; Meltzoff, 1990; Mitchell, 1994b; Parker, 1993; Piaget, 1962). Deferred imitation appears at 9 months of age (Meltzoff, 1988a,b); pretense, planning, and iconic gestural communications show rudimentary beginnings at 12 months which elaborate markedly by 24 months (Mitchell, 1994a,b), and recognizing that others are imitating oneself starts as early as 14 months (Meltzoff, 1990). Although facial imitation wanes soon after birth, it reappears in 50% of children by 18 months (Užgiris, 1973), just when, on average, mirror-self-recognition begins (Hart & Fegley, 1994; Lewis & Brooks-Gunn, 1979). Prior to recognizing themselves in mirrors, children tend to produce high levels of social imitation of gestures (Hart & Fegley, 1994), and caregivers' higher imitation rates increase infants' rates (Masur, 1987, 1993). Other forms of imitation are based on matching within or between perceptual modalities, as when children imitate vocal and verbal sounds based on auditory–auditory matching as well as nonverbal gestures and drawing based on visual–visual matching (Masur, 1993). Generalized imitation, such that a child can imitate a great variety of actions, likely begins around 15–18 months.

PROJECT CHANTEK

Project Chantek began in 1978 at the University of Tennessee at Chattanooga to investigate the cognitive and linguistic abilities of an orangutan, Chantek, using a comparative approach to his development within an evolutionary framework (Miles, 1986, 1990, 1991, 1994; Miles & Harper, 1994; Miles *et al.*, 1992). Beginning at 9 months of age, Chantek started his process of "enculturation," the anthropological term for immersion of an agent in a system of meaningful human relations that include language, behavior, beliefs, and material culture (Linton, 1936). The term "enculturation" is differentiated here from "cross-rearing," "cross-fostering," or "socialization" to emphasize that the goal of the project was not simply to teach Chantek a set of static rules and passive linguistic and social skills. Instead, the aim was to create the possibility of Chantek becoming an active agent in a meaningful system of relations that he would begin to embody as he acted within that system and as his understanding was affected by it (Miles, 1978). Within this context, Chantek was taught to communicate using American Sign Language (ASL) for the deaf, first by molding his hands into the shape of the sign then later by imitation, in a human-like setting in a trailer on campus (Miles, 1990).

METHODS

During an 8 year period from January 1979 through February 1986, caregivers kept a sign context log in which were recorded the sign and sign combinations communicated by Chantek, including the specific referents and contexts in which they were used. After several years of enculturation and exposure to signs, Chantek learned over 150 different signs, which he combined into multisign sequences. He developed capacities for symbol use, sign invention, and linguistic deception (Miles, 1986, 1990).

During this 8 year period, in addition to teaching him signs, caregivers also recorded longitudinal data on cognitive development, including imitation and other behaviors, to compare Chantek's development with normal and impaired children. Using a developmental life history approach, and a combination of *ad libitum* and focal animal sampling methods, caregivers recorded, throughout the day, Chantek's activities and interactions with caregivers and others. This resulted in over 35 000 hours of observation. Caregivers particularly noted the first time Chantek engaged in an activity or demonstrated (or failed to demonstrate) a skill. As is the case in *ad libitum* sampling and diary accounts in long-term studies of human children, once he performed a particular action and it became routine, caregivers did not always indicate similar instances, but rather concentrated on interesting emerging behaviors and further milestones of development. Consequently, the daily behavior log is qualitative in nature, and most helpful for providing an overall record of Chantek's development and for pinpointing when Chantek first performed a given activity.

A videotape record (two half-hour periods during each month) of his interactions and conversations with caregivers was also taken over a $6\frac{1}{2}$ year period. The purpose of videotaping was not to show all tasks and signs Chantek understood, but rather to collect a socio-linguistic discourse sample and visually capture the daily interactions in which he and his caregivers typically engaged. Videotapes also provided a quantifiable permanent record of the development and use of many of Chantek's signs and cognitive behaviors. For this study, we examined 40 half-hour videotaped sessions, from 26 January 1984, when Chantek was 6 years 1 month old, to 22 February 1986, when Chantek was 8 years 2 months old (unfortunately, two videotapes, with eight half-hour sessions, were destroyed in a fire). This was the period during which we played the sign version of the "Simon says" game with Chantek, so our focus was specifically on episodes when caregivers requested DO THE-SAME-THING with Chantek. At least two of the authors watched each videotape to detect episodes of DO THE-SAME-THING imitations. Each episode was transcribed into: a narrative depicting who signed; what was signed; what actions were modeled; what Chantek imitated, including sounds, facial gestures, bodily actions, and, when possible, the number of demonstrations and imitative repetitions; and whether Chantek mirrored the caregiver's (model's) actions, e.g. used his left hand when the caregiver used his or her right hand. Any molding of Chantek by the caregiver and other significant information were noted.

DEVELOPMENT OF IMITATION BY CHANTEK

Information on the development of imitative behaviors is taken from the daily *ad libitum* behavior log. We focused on a specific type of imitation – one in which Chantek saliently replicated some aspect of his own or another's behavior as a result of learning, pretending, or planning. Table 13.1 describes briefly the first time Chantek performed a variety of imitative behaviors. Also included are first examples of jealousy and sympathy, which suggest Chantek's identification with and attachment to his caregivers, necessary ingredients for the emergence of social imitation in humans and orangutans (Mussen, 1967; Russon & Galdikas, 1993, 1995; Valentine, 1930). Chantek showed imitation of his own and others' activities, and his self-imitations were used for both deception and communication. His kinesthetic–visual matching was used and developed in games of body-part identification, sign-learning, pretense, and mirror-self-recognition.

Chantek's development is similar to that of children, but delayed (see Table 13.1). Previous reports on Chantek's linguistic, deceptive, cognitive, and self-development (Miles, 1986, 1990, 1994) allow us to characterize Chantek up to at least his eighth year using the following three stages: *instrumental performative* (through 24 months of age), *subjective representation* (till 55 months of age), and *nascent perspective-taking* (beyond 55 months of age) (Miles, 1990, 1994).

In the *instrumental performative stage*, Chantek seems to have understood that his actions had specific consequences, and used his behavior and signs to manipulate his caregivers; his concerns seemed tied to the present. He reenacted his own behaviors to manipulate his caregivers, such as when he acted as though he needed to urinate in order to remain in the (fun-filled) bathroom. He imitated not only others' behaviors, as when he moved a door back and forth after observing his caregiver do so, but also their signs. His first sign imitation was WIPER, which referred to either a napkin or the need to wipe one's face, and was made by wiping your hand across your mouth. Chantek learned this sign without any molding. Chantek also located comparable body-parts during imitation games with caregivers, as they pointed to their own nose (or eye or ear, etc.) and signed WHERE?, after which Chantek imitated their behavior. He also showed evidence of pretense and empathy.

In the *subjective representation stage* beginning at 25 months of age, Chantek's use of signs for displaced reference indicated that he understood signs as representations. He engaged in planning, and numerous deceptions in which he replicated his own actions, emotions, and signs. For example, at 25 months he teased his caregivers by offering an object and withdrawing it when they approached it, while at 30 months he led his caregivers to one area, only to run back to obtain disallowed treats in another. Children use comparable deceptions at 19 and 24 months, respectively (Chevalier-Skolnikoff, 1986; R.W. M., personal observation). He exhibited elaboration of kinesthetic–visual matching when he imitated a photograph (a two-dimensional representation) of the sign-using gorilla Koko touching her nose. During this stage, Chantek began more frequently to exhibit delayed reproduction of signs without molding, producing new signs days after he had seen them. He also imitated caregivers' behaviors such as "cooking" his cereal (by putting it in a pot and placing that on top of the stove) as well, unfortunately, as his camera (by putting it in a pot of water). Evidence of bidirectional kinesthetic–visual matching occurred early in his second year when he groomed a mark on his face while looking at himself in the mirror, but during the rest of the second year until the middle of his fourth year he also continued to experiment with mirrors by

Table 13.1. *Developmental stages and ages for Chantek's imitative abilities, and ages for comparable activities in children*

Earliest age for Chantek (yr; month)	Children (yr; month)	Stage and examples
I. Instrumental performative		
0;10	< 0;9	Imitation (pushing bathroom door back and forth)
1;2		First self-signing (signed FOOD-EAT while stealing bread)
1;3	0;10	First imitation of sign: WIPER (during self-play)
1;5	1;1	Identification (jealousy toward caregiver)
1;5	1;1	Imitation of caregiver's body-part pointing in response to WHERE X?
1;8	1;0	Pretense (fed toy animal)
1;8	< 1;3	Lack of self-recognition (reaching behind mirror at own image)
1;10	1;3	Behavioral deception (acting as though has to urinate, to stay in bathroom)
2;0		Identification (sympathy toward caregiver: protecting from attacking toy animal)
II. Subjective representation		
2;1	1;3	Self-recognition (touched spot on forehead, but inconsistent until 3;5)
2;1	1;7	Behavioral deception (teasing – showing object, pulling it away)
2;1	1;3	Behavioral self-deception (acting afraid of nonexistent cat, to play)
2;4		Replication of situation to communicate (gave caregiver knife and nipple and stared at milk cabinet to receive milk)
2;6	2;0	Behavioral deception (led caregiver away, only to go back)
2;11	1;6	Learned signs by delayed imitation
3;2	1;6	Planning, in form of self-signing prior to acting toward displaced object (alone, signs IN MILK RAISIN before asking caregiver for MILK)
3;4	< 2;6	Imitation of 2-dimensional representations (photograph of Koko pointing to nose)
3;5	1;3	Successful double-blind sign-recognition test
3;8		Spontaneous self-grooming with mirror, opens and closes mouth in exploratory facial contingency play
3;9	3;5	Signed deception (LISTEN POINT when no sound, to avoid sleep)
4;0		Signed deception (acted afraid and signed CAT but no cat, to get attention)
4;7		Identification (sympathy for toy animal)

Table 13.1. (cont.)

III. Nascent perspective-taking		
4;10	2;6	Pretend play (clay man symbolic play)
4;10	2;0	Perspective-taking (manually directed caregiver's eyegaze before signing)
5;2	< 4;0	First invented sign (NO – TEETH)
6;4		Receptive understanding of SAME
6;8		Second invented sign (EYE – DRINK)
7;3		Use of DO THE-SAME-THING to manipulate caregiver
7;10		Used mirror to observe self as he imitates caregiver's eyelash curling

Earliest ages of human children taken from information by Bretherton (1984); Masciuch & Kienapple (1993); and Mitchell (1993a,c, 1994a,b).

reaching behind or scraping them (Miles, 1994). Similar contradictory responses toward one's mirror-image are also observed in children (Mitchell, 1993a).

In the nascent *perspective-taking stage* at 58 months of age, Chantek seemingly developed a point of view toward himself, and recognized that others had attentional processes. At this stage, he sometimes used the mirror for self-exploration, when examining a new look (putting on sunglasses and looking at his image), and when engaging in an imitation (observing himself as he curled his eyelashes, as he had seen his caregiver do). He manipulated caregivers' eyes toward himself before signing (an action they performed to get his attention), suggesting that he understood that they see through their eyes.

During this stage he invented at least five signs based on kinesthetic recreation of similarity to his own prior actions, or on visual–visual or kinesthetic–visual matching between the sign and what it represented. The first invented sign was NO-TEETH, in which he used his fingers to close his mouth, indicating that he would not bite during play. The second was EYE-DRINK, in which he made the sign for DRINK at his own eye to represent his caregiver's placing of contact lens solution into her eyes. He also invented DAVE-MISSING-FINGER, in which he held the middle digit of a bent finger that corresponded to a severed finger portion on the hand of a favorite university employee. Another invented sign was VIEWMASTER, in which he covered his eyes with his fingertips in a V-shape in the same position they would be in if he were holding a viewmaster. Finally, he invented BALLOON, in which he held his thumb and index finger together at his lips (as if blowing up a balloon) and blew.

DO THE-SAME-THING SIGNED REQUEST

Here we present Chantek's kinesthetic–visual imitations of caregivers' bodily actions, auditory–auditory imitations of sounds, and visual–visual imitations of drawings (what

Baldwin (1903, p. 86) called "tracery imitation"), when requested by his caregiver to DO THE-SAME-THING, taken from the videotaped interactions with his caregivers.

Seventeen episodes in which Chantek was asked to imitate his caregiver (the model), and one episode in which Chantek asked his caregiver to DO THE-SAME-THING, were present in the 2 year videotaped interactions. Three different caregivers initiated the imitation game, 13 episodes were with one caregiver, and the other two caregivers initiated 2 episodes each. During any given episode, Chantek was asked to imitate from 2 to 16 demonstrated actions (mean = 5.24 demonstrations/episode), of 1 to 5 different types (mean = 1.41 types/episode): overall, 88 demonstrations of 24 different types.

Chantek had been introduced to performing caregivers' actions following their signing the word DO. DO originally requested that Chantek perform a task he already understood. For example, a caregiver might ask Chantek to DO WORK, and then provide Chantek with a well-explored tool. Thus, when caregivers wanted Chantek to imitate actions, they signed DO and performed the actions. This implicit use of DO to mean "do the same thing I do" occurred in the first two videotaped episodes. The request DO THE-SAME-THING was introduced when the caregiver asked Chantek to draw what the caregiver had drawn on a piece of paper, which was enacted over the next four episodes. Following this, DO THE-SAME-THING was used to request that Chantek perform any action, even novel ones. Infrequently, the request was abbreviated to signing THE-SAME-THING alone. When Chantek failed to imitate the demonstration, either at all or in the way desired, caregivers terminated the interaction or repeated their request and/or illustration.

On the videotape, the caregiver asked Chantek to perform a series of actions, which we call an "episode." An episode was detected by the caregiver's signing DO, THE-SAME-THING, or DO THE-SAME-THING, followed by the caregiver's demonstration and, usually, Chantek's imitative attempts. (Although DO THE-SAME-THING was a common game with Chantek from 6 to 8 years of age, it was not always enacted while Chantek was recorded.) Table 13.2 shows frequencies both of requests for Chantek to imitate and of episodes containing the particular demonstration. Demonstrations are separated into 6 categories: drawing, auditory–facial, auditory–bodily, bodily, bodily–craniofacial, and facial actions.

Demonstrations used a variety of body parts. Chantek was requested to do something actively with his arm(s) (or fingers and hands) for 67% of the 24 types of demonstrations, with a part of his face (eyelids, lips, tongue) for 20.8%, with his leg(s) or feet for 12.5%, with internal organs (to produce sounds) for 12.5%, and with his whole body (i.e. headstand, jump) for 8.3%. (Four demonstrations modeled more than one body part.) He was requested to act upon his face (cheeks, teeth, tongue) for 25% of the 24 types of imitation, his arm or hand for 8.3%, his leg or foot for 8.3%, his head for 8.3%, his chest for 4.2%, and an extra-body object (i.e. paper, stairs) for 12.5%. (He was not always asked to act upon something.)

Chantek's response to each demonstration was coded as either full imitation, partial imitation, or nonimitation (see Table 13.2). For full imitation, Chantek reproduced salient aspects of the demonstration; for partial imitation, Chantek reproduced less salient aspects of the demonstrated action; for nonimitation, Chantek either failed to

Table 13.2. *Frequency of actual and implied instances of demonstrations, out of 24 types, number of episodes these occurred during, and number of full, partial or failed imitations by Chantek*

Actions performed by model	Demonstrations			Chantek's imitations			
	No.	Implied	Episodes	Full	Partial	None	% Mirror-image
Drawing							
Draw a circle on paper with pen	4		2	2	2		—
Draw a line on paper with pen	10	1	2	11			—
Auditory – facial							
Donald Duck gargle	2		2	1	1		—
Lip-popping noise	11		2	3	6	2	—
Puff of air made with mouth	1		1	1			—
Rasberry noise	1		1	1			—
Auditory – bodily							
Clap hands	4		2	4			100(4/4)
Stomp foot	7		1	1	5	1	100(6/6)
Bodily							
Grab own arm	2	1	1	3			33.3(3/3)
Put hand up (over head)	4		2	2	2		50(2/4)
Headstand	2		1	1	1		—
Jump	12	3			12	3	41.7(5/12)
Lift leg and grab foot with hand	3		2	3			100(3/3)
Slap chest	1		1	1			0(0/1)
Slap thigh	1		1	1			100(1/1)
Bodily – craniofacial							
Pat top of head	3		1	3			100(3/3)
Poke cheeks with fingers	4		1	4			100(4/4)
Put hands on head	1		1		1		0(0/1)
Put hands over ears	1		1	1			100(1/1)
Put hands over eyes	2		2	2			100(2/2)
Put hands over mouth	1		1	1			100(1/1)
Touch teeth with finger	1		1	1			100(1/1)
Touch tongue with finger	1		1	1			100(1/1)
Facial							
Blink eyes	9	3	4	6	3	3	33.3(3/9)
Totals	88	8	18[a]	54	33	9	70.2(40/57)

Implied models were present when Chantek attempted, after he had already produced or attempted an imitation of an already demonstrated action, a new sort of imitation of that previous model.

[a] Each episode included several types of actions (35 in total) but there were only 18 actual episodes.

respond or produced a saliently inaccurate imitation (e.g. the wrong sound). What is considered salient is described below for each category of imitation.

Chantek's imitations were full in 56.2% of his responses, partial in 34.4%, and nonexistent in 9.4%. In most full imitations, Chantek replicated the action closely. Cases in which there were discrepancies between the caregiver's demonstrations and Chantek's imitations are described in detail below.

Drawing
Coding

For drawing, full imitation required that Chantek draw something similar to the mark made by the caregiver – either a straight line or a circle. An approximately straight line, whether parallel to, perpendicular to, diagonal to, or traced on to the demonstrated line, was full imitation. An incomplete circle was partial imitation, but several circular markings that produced a closed circular mark was full imitation.

Results

Of the four responses to the caregivers drawing a circle on paper, Chantek twice traced over the caregiver's circle, and twice drew partial circles. Of 11 responses to the caregivers drawing a line on paper, Chantek twice traced over the caregiver's line, once made several short straight lines separate from and strikingly perpendicular to the caregiver's line, three times drew a line parallel to it, and in the rest made lines either by swiping the marker across the page or pressing the marker down hard and moving it about on the page. Of the parallel lines, Chantek drew one after the caregiver moved Chantek's marker-holding hand back and forth above the paper, another after the caregiver forced Chantek to trace over a line, and another after the caregiver requested him to do another line after he had just made one on the paper.

Auditory-facial imitations
Coding

For auditory–facial imitations, full imitation required that Chantek produce the same sound, and partial imitation required that he attempt the same sound. No sound or the wrong sound was nonimitation. The auditory rather than facial aspect of these imitations is the focus for full imitation because Chantek had to produce a facial gesture that resembled that of the caregiver simply to make the sound. Therefore, for some actions, whether he attempted the same facial gesture is unclear. The Donald Duck gargle, puff, and raspberry noises are produced respectively by pushing air from the lungs through a constricted throat and open mouth, through compressed lips abruptly, or through compressed flapping lips, respectively. (The Donald Duck gargle is so named because it sounds like the Disney character's voice.) Chantek's and the caregiver's facial gestures are similar for these probably because of how the sounds were produced, rather than from any attempt to simulate facial gestures. However, for lip-popping sounds which are produced by abruptly separating compressed lips, Chantek was probably imitating the facial gestures in order to produce the sound, or at least had to imitate the facial gesture on first reproduction, because facial gestures are the only clues to the sound's creation.

Results

Chantek easily produced raspberry and puff sounds. Although able to make the Donald Duck gargle and the lip-popping sound, he had difficulty producing these. In one sequence in which the Donald Duck gargle was demonstrated, he attempted and failed (making a similar facial gesture) but, upon a second demonstration, succeeded. He was then asked to make a lip-popping sound. Again he made a facial gesture similar to the caregiver's but failed to make a sound, and then, after the second demonstration, produced the Donald Duck gargle three times! After another demonstrated lip-pop, he produced the sound. In another sequence of demonstrated lip-pops, Chantek reproduced the caregiver's facial gesture, but produced a faint popping sound in only two of his eight responses.

Auditory–bodily imitations
Coding

For auditory–bodily imitation, full imitation required Chantek to use the same body parts in the same action to attempt the same sound as the caregiver, while partial imitation required his using similar body parts in a similar action to attempt a similar sound.

Results

When the caregiver clapped hands, Chantek produced a sound in all four cases. When the caregiver stomped her foot by slamming her shoe on a stairstep, thereby producing a sound, Chantek responded by reproducing a sound using his limbs. After three of the first four demonstrations, Chantek produced partial imitations: he clapped his hands, next slapped the bottom of his foot, then failed to respond, and then knocked his fist on the same stairstep as the caregiver. The caregiver stomped again and pointed to her right foot, and Chantek slapped his (mirrored) left foot bottom with his hand, and did it again after the next demonstrated stomp. Finally, with the next demonstrated stomp, Chantek stomped the stair with his foot, producing his first full imitation. In this sequence, Chantek appeared to view the imitation initially in terms of the sound produced, then as involving his foot to produce the sound, then as requiring that the sound be produced on the same step as the caregiver, and then again as involving his foot to produce the sound. It was only after the caregiver repeated the demonstration without being satisfied that Chantek eventually imitated the action fully.

Other bodily and facial imitations
Coding

For bodily, bodily–craniofacial, and facial actions, full imitation required that Chantek use the same body parts to produce the same action as the caregiver, and partial imitation allowed the same or different body parts to achieve a similar action. In two types of demonstration, blinking and jumping, Chantek seemed incapable of performing the caregiver's action, and so enacted additions, approximations, and/or transformations.

Results

For most of these demonstrations, Chantek produced reasonably accurate, mirror-image imitations that were fully imitative. For a few full imitations, however, he produced minor deviations from the caregiver. In slapping his chest, Chantek alternated his hands, whereas the caregiver slapped both hands simultaneously; in slapping the thigh, both Chantek and the caregiver used both hands simultaneously, but the caregiver's were open palmed, whereas Chantek slapped first with closed palms and then with more open palms. In poking both his cheeks with his fingers, Chantek used all his fingers for the first three of the four instances but only his index fingers for the last; the caregiver always used the index fingers together.

In a few instances in which he produced partial imitations, Chantek either failed to complete or changed the action. In reproducing a headstand in one instance, he moved his head to the ground, followed by falling forward onto his back. He then produced an accurate headstand after the next demonstration. When putting hands on head was demonstrated, Chantek patted his head, alternating right and left hands.

With the four types of demonstration discussed next, the sequence of Chantek's responses suggest that what he took to be salient aspects to be reproduced were sometimes at variance with what the caregiver took to be salient.

Grab arm. When the caregiver enacted grabbing her own left forearm with her right hand on top, Chantek grabbed his own left forearm with his right hand underneath. When the caregiver repeated her demonstration, Chantek produced a mirror-image of her action, but once again grabbed underneath. The caregiver grabbed Chantek's arms and held the right arm above the left arm and let go, and Chantek grabbed his left forearm with his right hand on top, which was what the caregiver required (though this was not a mirror-image). In the first two imitations, Chantek performed the demonstrated action of grabbing his arm, but failed to use an overhand grab. His second attempt mirrored the caregiver's act, and his third, through the caregiver's intervention, produced a nonmirrored overhand grab.

Hand up. When the caregiver put her right hand over her head during one episode, Chantek put his left hand up and waved, which the caregiver accepted. During a subsequent episode, Chantek imitated the same action by putting both hands up. The caregiver repeated her action, and Chantek repeated his version of the imitation. The caregiver held Chantek's left hand up, pushed his right hand down, and released his hands. The caregiver then produced her action again, and Chantek stayed in position.

Jump. In one episode, his caregiver requested 12 times that Chantek imitate her jump. After each demonstration, Chantek produced different responses, each a partial match. In sequence, he lifted and stomped his foot, then took exaggerated steps toward his caregiver, then lifted each leg alternately (right, left, right), then after each of three demonstrated jumps lifted and dropped both legs from a sitting position. He failed to respond to the next two demonstrated jumps. After a further demonstrated jump, he alternately stomped each foot, then sat and ignored the next demonstrated jump. After another demonstrated jump, he stomped his right foot repeatedly, then lifted and dropped both legs, and then, after a pause, lifted and dropped his right foot. The caregiver produced her last jump, and Chantek lifted and dropped his right foot, and then reached back with his arms, pulled himself up to a step behind him, and lifted his feet in the air as he moved backward. All of these imitations are partial because Chantek never lifted his whole body off the ground, but his actions suggest that jumping was not possible for him. His low center of gravity may preclude vertical whole body leaping. His partial imitations indicate, however, that he interpreted the demonstration as requiring feet being off the ground and replicated that action repeatedly with variations.

Blink eyes. Because apes appear to be without voluntary control over their eyelids (see Discussion), Chantek sometimes used his finger to push down his eyelid when an eye blink was demonstrated. This action fulfilled criteria for full imitation in that movement of his eyelid and those of the caregiver were the same (although clearly produced by different means). In one episode, the caregiver blinked both her eyes and pointed to her right eye. Chantek responded, with his left hand, by touching the caregiver's right eye and then manually moving his own left eyelid up and down. The caregiver blinked again, and Chantek repeated his action on his eyelid. The caregiver blinked again, and Chantek kissed her eyelids. In an episode on the next day, Chantek twice moved his eyes toward the caregiver's eyes after her blinks at the beginning and end of the episode and, after the middle demonstrated eyeblink, pushed his left eyelid down with his left hand. In a later episode, after the demonstrated blink, Chantek opened his mouth in partial imitation; the caregiver signed SAME EYES SAME, and Chantek moved his left eyelid down with his left hand. The caregiver blinked again, and Chantek again opened and closed his lips in imitation, and then moved his left eyelid up with his left thumb. In the final episode, the caregiver again blinked both eyes, and Chantek simultaneously pushed his left and right eyelids down with the corresponding hands.

Imitation and mirror-image responses

During most bodily and facial imitations, Chantek faced the caregiver and produced mirror-images of demonstrations (see Table 13.2). A response was counted as a mirror-image if Chantek used his left or right appendage when the caregiver used his or her right or left appendage (respectively), or if Chantek used both appendages when the caregiver used both as well. (The appendages here include legs, feet, arms, hands, eyelids, and cheeks.) Under these criteria, 70.2% of Chantek's responses mirrored the caregiver's actions (see Table 13.2). In some cases, for example eyeblinks, Chantek used an appendage that did not have a left–right dimension – his mouth – to imitate the motion of the appendages, and thereby could not have produced a mirror-image.

Note, however, that Chantek is left-handed. Most of the mirrored actions involved the caregiver using his or her right side and Chantek using his left, which suggests that he may not have been mirroring in his responses. One way to find out if Chantek was reproducing mirrored aspects is to compare his responses when the caregiver used both appendages. In 24 of 31 instances (77.4%) in which the caregiver used both appendages, Chantek used both appendages in his response. In five of these responses, Chantek alternated use of appendages which the caregiver used simultaneously: put hands on head (once), jump (thrice), and slap chest (once).

Chantek's use of DO THE-SAME-THING

Also captured on videotape was one episode in which Chantek, at 7 years 3 months of age, requested that his caregiver DO THE-SAME-THING. This episode followed a few minutes after his caregiver had requested that he imitate lip-pops. According to the criterion that a sign is acquired only after it has been used spontaneously on 15 days over a 1 month period (Terrace et al., 1979), Chantek had not yet acquired the sign SAME at this time. Moreover, he was never taught specifically to produce the combination DO THE-SAME-THING, nor had he used the combination before this event.

In this episode, interrupting the lip-popping and whistle-blowing imitation sequence, Chantek broke off the imitation interaction with the caregiver, and taking the whistle, induced the caregiver to chase him to reobtain the whistle and his attention, without getting into a wrestling or grappling contest with Chantek. After being chased around the yard, Chantek climbed on to a jungle gym and faced his caregiver. During this standoff, Chantek resumed the imitation interaction by signing DO THE-SAME-THING, and slapping a bar on the jungle gym. When the caregiver failed to respond, Chantek repeated his slap of the gym, and the caregiver slapped the same metal bar but at a distance from where Chantek slapped, avoiding Chantek's subsequent grab for his hand. A game ensued in which Chantek slapped the gym, and the caregiver contacted the gym but avoided Chantek's grab for his hand. The caregiver attempted to stop the game by telling Chantek to retrieve the whistle from the ground where he had dropped it while climbing onto the jungle gym. Chantek picked up the whistle, and then with a long sweeping motion of his arm and hand as if he were throwing the whistle toward the caregiver, Chantek flipped the whistle in an arc back toward himself. As the caregiver reached for the whistle, Chantek successfully grabbed the caregiver's hand.

DISCUSSION

Like other recent examples of orangutan imitation (Russon & Galdikas, 1993), Chantek's imitative activities support claims since the eighteenth century that orangutans readily imitate humans. Chantek capably imitated his caregivers' actions, including their signed gestures to request imitation, and he developed his imitative skills years prior to being directed by his caregivers to imitate them with the signs DO and DO THE-SAME-THING. His imitative behaviors fulfill most requirements for "true" or advanced imitation: his facial ("invisible") imitations, spontaneous deferred imitations, and pretend play satisfy Meltzoff's (1988a) and Piaget's (1962) requirements for representation and symbolic activities, his replication of meaningless actions satisfies Guillaume's (1971) criteria for "true" imitation, his communicative simulations fulfill Mitchell's (1987) requirements for the most developed form of imitation, his imitations of bodily actions satisfy Galef's (1988) criteria for "true" imitation, and his spontaneous production of novel signs and other imitations of novel behaviors fulfill the criteria for "true" imitation outlined by Thorpe (1963), Visalberghi & Fragaszy (1990) and Whiten & Ham (1992).

Chantek's abilities for bodily (including facial) imitation and mirror-self-recognition also support the idea that kinesthetic–visual matching is necessary for both activities (Mitchell, 1993a; Parker, 1991). Against this idea, Custance & Bard (1994) have argued that imitation and mirror-self-recognition in chimpanzees develop independently. Their support for this argument is that a chimpanzee had difficulty learning to enact imitations upon command several years after he consistently showed mirror-self-recognition. However, this evidence does not argue against kinesthetic–visual matching as the basis for both bodily imitation and mirror-self-recognition, because reproducing bodily imitation, and the recognition that bodily imitation is what is desired by another (coupled with the desire to satisfy that desire), are distinct psychologically, with the latter dependent upon existence of the former (Mitchell, 1993b). Chantek's understanding that he was to replicate demonstrations when caregivers signed DO THE-SAME-THING came into being much later than his spontaneous bodily imitations, which, like those of children (Hart & Fegley, 1994), were present earlier than his consistent recognition of himself in a mirror.

Chantek's responses to caregivers' requests to DO THE-SAME-THING indicate that his imitations were not slavish repetitions of actions, but involved selection of salient components at each repetition, much like imitations by rehabilitant orangutans (Russon & Galdikas, 1993; and see Chapter 7). Sometimes he replicated auditory phenomena, sometimes visual phenomena (via kinesthetic–visual or visual–visual matching). Sometimes he reproduced an action's result but with another body part (e.g. making a sound with his hand rather than his foot). Sometimes he achieved imitation circuitously because a direct route was impossible. For example, in replicating his caregiver's jump in the last episode, instead of vertically leaping, he eventually lifted both feet off the ground by "standing" on his hands. In another example, to imitate eyeblinking, Chantek pushed down his eyelid with his finger because he did not have voluntary control over eyelids. the home-reared chimpanzee Viki reacted similarly by "putting a finger to her eye, which caused it to close," when asked to imitate her caregivers' eyeblinks (Hayes & Hayes, 1952, p. 452). Chantek's spontaneous imitation of DO THE-SAME-THING is consistent with other spontaneous sign imitations, but the sequence of his actions in this instance suggests some development in his planning abilities and social understanding. The two instances of opening and closing his mouth in replication of the caregiver's opening and closing his eyelids calls to mind Piaget's (1953, pp. 337–8) description of his daughter Lucienne's attempts, at 16 months of age, to understand how to open a matchbox further by opening her mouth further and further. Chantek's other imperfect and abstract imitations suggest as well that his understanding developed beyond simple bodily mapping to a more analogical understanding (Mitchell, 1994b).

Although his imitative behaviors tended to develop later than those of children, the developmental sequencing of imitative abilities for both shows strong similarities, suggesting cognitive homologies (Miles, 1990, 1994; Parker, 1991; Piaget, 1962). Also like children (as well as other human-reared orangutans), Chantek's close bonds and intensely emotional relationships with caregivers probably increased his desire to imitate and "impersonate" them (Mussen, 1969; Russon & Galdikas, 1993, 1995; Valentine, 1930; Chapter 7). However, because he was enacting bodily imitations, the division between emulation and impersonation (Tomasello, 1990) is blurred: reproducing the goal (emulation) and doing exactly what the person does (impersonation) produce the same response.

Although Chantek duplicated novel actions in his imitation of new signs, in the DO THE-SAME-THING context it is unclear whether Chantek's imitations were of novel actions: no account of all actions produced by Chantek was or could be produced. Some researchers might argue that these instances may not be "true" imitation, which for them requires production of novel actions (Thorpe, 1963; Visalberghi & Fragaszy, 1990; Whiten & Ham, 1992). Against this view, we note that Chantek's responses were usually not formulaic and well-trained reproductions. Rather, in many cases Chantek appeared to be attempting either to reproduce an action that was difficult for him or to figure out what the caregiver wanted him to imitate, indications that his matching behavior was sensitive to the demonstration. Like Chantek, young children often struggle to perfect their imitations and fail to achieve perfection even after several attempts (Speidel & Nelson, 1989; Chapter 7). In more recent requests to DO THE-SAME-THING when he was in his mid-teens, Chantek easily imitated strange actions that were intended to be novel, although he apparently failed to imitate actions employing complex tool use, and devised his own solutions (M. Tomasello, personal communication to H.L.M., February 1993). In these tool use tasks, however, Chantek was not asked to DO THE-SAME-THING; he was simply shown a problem and solution by a human, and then offered a chance to solve the problem. Given no request to DO THE-SAME-THING and no close bond with the human, Chantek not surprisingly solved problems by emulation rather than impersonation.

It is important to recognize that the "capacity for imitation" should not be viewed as simply a result of maturation; that is, we should be careful not to reify imitation as something an animal does or does not "possess." Imitation typically involves a transaction or interaction between agents. Furthermore, a reified concept of imitation disregards the important aspects of the practice of imitation that allow it potentially to reference cultural or social meanings not immediately present and that give imitations their significance (see Ortner, 1984).

To be sure, cognitive capacities have biological foundations, but those cognitive capacities are developed only as the biological heritage of agents are acted upon through the agent's actions within culturally and/or socially defined contexts. Our emphasis has been placed, not on *ad hoc* responses to stimuli, but on agency within the organizational and evaluative framework of cultural and social systems. Thus, the most parsimonious explanation of our results is that these behaviors emerged from a combination of Chantek's enculturation and each specific circumstance in which he found himself.

More concretely, in some of the interactions with the caregivers we see a progressive movement of Chantek's behavior toward the target behavior that the caregivers will accept as an imitation. This progression is especially apparent in the requests that he jump and that he stomp his foot. Chantek tried several modalities in an attempt to identify which of the behaviors is the socially salient and culturally meaningful "imitative" behavior. This cooperative effort is similar to the "guided reinvention" process Lock has described for human language development (Lock, 1980). The episodes are constituted by a process of negotiation within the transaction between both Chantek and the caregivers. Not only does Chantek move closer to the target behavior but the caregivers will sometimes adjust the target behavior toward Chantek's last action. This adjustment means that the caregivers are not only indexing the target behavior in reference to the larger cultural system, but to their expectations for Chantek (taking into account his most recent attempts at imitation), and in reference to the specific context of their practice (the social dynamic between the two parties that must capture the attention of both to be momentarily sustained). Indeed, many of Chantek's "failures" at imitation involve his lack of attention or boredom with the interaction, as the caregiver must encourage Chantek to perform while maintaining his interest. Practically, Chantek will often want to "play" while the caregiver wants him to "work."

This is the case in the episode in which Chantek's request that the caregiver DO THE-SAME-THING accomplished a momentary role reversal, which he used as a deceptive ploy to engage the caregiver in play. Chantek successfully resisted by transforming the meaning of the existing social relations from DO THE-SAME-THING as work to DO THE-SAME-THING as play. The ability to change the meaning of existing relations on a microlevel through the imitation of the surface form of a cultural convention, while simultaneously adopting a covert intention and meaning in one's self-interest, has the potential to both transform and replicate cultural convention (Sahlins, 1981). Thus, this episode on a microlevel evokes some of the same cultural processes of humans in practice with an ape, in a very rudimentary way.

Like children's imitations, Chantek's imitations are those of an enculturated organism – one who has extensive experience with a particular linguistic and cultural context. As Hayes & Hayes (1952) noted about the imitations of the home-reared chimpanzee Viki, cultural contexts have within them an implicit foundation for imitation: a home-reared ape, more than a caged ape, expects to do what others do because he or she has extensive experience doing what others do. In making comparisons between imitations by children and apes, modern-day primatologists tend to take children's imitative activities for granted, forgetting that these activities are developed through a socio-cultural ambiance and cultural expectations acting on native abilities.

ACKNOWLEDGMENTS

This research was supported by the National Institute of Child Health and Human Development grant NICHD 14918, National Science Foundation grant BNS 8022260 and grants from the UC Foundation. The loan of Chantek was provided by the Yerkes

Regional Primate Research Center supported by National Institutes of Health grant RR 00165. We thank the members of Project Chantek for their assistance in the enculturation of Chantek and the collection of data.

REFERENCES

Baldwin, J.M. (1903). *Mental Development in the Child and the Race*, 2nd edn. New York: Macmillan. (First published in 1896).

Bretherton, I. (1984). *Symbolic Play*. New York: Academic Press.

Buffon, G. (1812). *Natural History, General and Particular*, vol. X. London: T. Cadell & W. Davies.

Chevalier-Skolnikoff, S. (1986). An exploration of the ontogeny of deception in human beings and nonhuman primates. In *Deception: Perspectives on Human and Nonhuman Deceit*, ed. R.W. Mitchell & N.S. Thompson, pp. 205–20. Albany, NY: State University of New York Press.

Custance, D. & Bard, K. (1994). The comparative and developmental study of self-recognition and imitation: The importance of social factors. In *Self-awareness in Animals and Humans: Developmental Perspectives*, ed. S.T. Parker, R.W. Mitchell & M. L. Boccia, pp. 207–26. New York: Cambridge University Press.

Custance, D., Whiten, A. & Bard, K. (1994). The development of gestural imitation and self-recognition in chimpanzees (*Pan troglodytes*) and children. In *Current Primatology: Selected Proceedings of the XIVth Congress of the International Primatological Society, Strasbourg*, vol. 2. *Social Development, Learning and Behaviour*, ed. J.J. Roeder, B. Thierry, J.R. Anderson & N. Herrenschmidt, pp. 381–7. Strasbourg, France: Université Louis Pasteur.

Cuvier, G. (1831). *The Animal Kingdom Arranged in Conformity with its Organization*. New York: G. & C. & H. Carvill.

Darwin, C. (1896). *Descent of Man and Selection in Relation to Sex*. New York: D. Appleton & Co. (First published in 1871).

Furness, W.H. (1916). Observations on the mentality of chimpanzees and orangutans. *Proceedings of the American Philosophical Society*, **55**, 281–90.

Galdikas, B.M.F. (1982). Orang-utan tool use at Tanjung Puting Reserve, Central Indonesian Borneo (Kalimantan Tengah). *Journal of Human Evolution*, **10**, 19–33.

Galef, B.G., Jr (1988). Imitation in animals: History, definition, and interpretation of data from the psychological laboratory. In *Social Learning*, ed. T. Zentall & B.G. Galef, Jr, pp. 3–28. Hillsdale, NJ: Lawrence Erlbaum.

Gopnik, A. & Meltzoff, A.N. (1994). Minds, bodies and persons: Young children's understanding of the self and others as reflected in imitation and theory of mind research. In *Self-awareness in Animals and Humans: Developmental Perspectives*, ed. S.T. Parker, R.W. Mitchell & M.L. Boccia, pp. 166–86. New York: Cambridge University Press.

Guillaume, P. (1971). *Imitation in Children*, 2nd edn. Chicago: University of Chicago Press. (First published in French in 1926).

Haggerty, M.E. (1910). Preliminary studies on anthropoid apes. *Psychological Bulletin*, **7**, 49.

Haggerty, M.E. (1913). Plumbing the minds of apes. *McClure's Magazine*, **41**, 151–4.

Harrison, J. (1962). *Orang-utan*. London: Collins.

Harrison, J. (1969). The nesting behaviour of semi-wild juvenile orang-utans. *Sarawak Museum Journal*, **17**, 336–84.

Hart, D. & Fegley, S. (1994). Social imitation and the emergence of a mental model of self. In *Self-awareness in Animals and Humans: Developmental Perspectives*, ed. S.T. Parker, R.W. Mitchell & M.L. Boccia, pp. 149–65. New York: Cambridge University Press.

Hatfield, E., Cacioppo, J.T. & Rapson, R.L. (1994). *Emotional Contagion*. New York: Cambridge University Press.

Hayes, K.J. & Hayes, C. (1952). Imitation in a home-raised chimpanzee. *Journal of Comparative and Physiological Psychology*, **45**, 450–9.

Hornaday, W. (1922). *The Minds and Manners of Wild Animals*. New York: Scribners.

Horr, D.A. (1977). Orang-utan maturation: Grouping up in a female world. In *Primate Biosocial Development: Biological, Social and Ecological Perspectives*, ed. S. Chevalier-Skolnikoff & F.E. Poirier, pp. 289–322. New York: Garland Publishing, Inc.

Koehler, W. (1993). The mentality of orangs. *International Journal of Comparative Psychology*, **6**, 189–229. (First published in German in 1920).

Lewis, M. & Brooks-Gunn, J. (1979). *Social Cognition and the Acquisition of Self*. New York: Plenum Press.

Linton, R. (1936). *The Study of Man*. New York: Appleton-Century-Crofts.

Lock, A. (1980). *The Guided Reinvention of Language*. New York: Academic Press.

Masciuch, S. & Kienapple, K. (1993). The emergence of jealousy in children 4 months to 7 years of age. *Journal of Social and Personal Relationships*, **10**, 421–35.

Masur, E.F. (1987). Imitative interchanges in a social context: Mother–infant matching behavior at the beginning of the second year. *Merrill-Palmer Quarterly*, **33**, 453–72.

Masur, E.F. (1993). Transitions in representational activity: Infants' verbal, vocal, and action imitation during the second year. *Merrill-Palmer Quarterly*, **39**, 437–56.

Meltzoff, A.N. (1988a). Infant imitation and memory: Nine-month-olds in immediate and deferred tests. *Child Development*, **59**, 217–25.

Meltzoff, A.N. (1988b). Infant imitation after a 1-week delay: Long-term memory for novel acts and multiple stimuli. *Developmental Psychology*, **24**, 470–6.

Meltzoff, A.N. (1990). Foundations for developing a concept of self: The role of imitation in relating self to other and the value of social mirroring, social modeling and self practice in infancy. In *The Self in Transition: Infancy to Childhood*, ed. D. Cicchetti & M. Beechly, pp. 139–64. Chicago: University of Chicago Press.

Miles, H.L. (1978). Conversations with apes: The use of sign language with two chimpanzees. *Dissertation Abstracts International*, **39**, 11A.

Miles, H.L. (1983). Apes and language: The search for communicative competence. In *Language in Primates*, ed. J. de Luce & H. Wilder, pp. 43–61. New York: Springer-Verlag.

Miles, H.L. (1986). How can I tell a lie? Apes, language, and the problem of deception. In *Deception: Perspectives on Human and Nonhuman Deceit*, ed. R.W. Mitchell & N.S. Thompson, pp. 245–66. Albany, NY: State University of New York Press.

Miles, H.L. (1990). The cognitive foundations for reference in a signing orangutan. In *"Language" and Intelligence in Monkeys and Apes: Comparative Developmental Perspectives*, ed. S.T. Parker & K.R. Gibson, pp. 451–68. New York: Cambridge University Press.

Miles, H.L. (1991). The development of symbolic communication in apes and early hominids. In *Studies in Language Origins*, ed. J.W. von Raffler-Engel, J. Wind & A. Jonker, vol. 2, pp. 9–20. Amsterdam: John Benjamins.

Miles, H.L. (1994). ME CHANTEK: The development of self-awareness in a signing orangutan. In *Self-awareness in Animals and Humans: Developmental Perspectives*, ed. S.T. Parker, R.W. Mitchell & M.L. Boccia, pp. 254–72. New York: Cambridge University Press.

Miles, H.L. & Harper, S. (1994). "Ape language" studies and the study of human language origins. In *Hominid Culture in Primate Perspective*, ed. D. Quiatt & J. Itani, pp. 253–78. Boulder, CO: University of Colorado Press.

Miles, H.L., Mitchell, R.W. & Harper, S. (1992). Imitation and self-awareness in a signing orangutan. Paper presented at the XIVth Congress of the International Primatological Society, Strasbourg, France, 16–21 August.

Mitchell, R.W. (1987). A comparative developmental approach to understanding imitation. In *Perspectives in Ethology*, ed. P.P.G. Bateson & P.H. Klopfer, vol. 7, pp. 183–215. New York: Plenum Press.

Mitchell, R.W. (1990). A theory of play. In *Interpretation and Explanation in the Study of Animal Behavior*, ed. M. Bekoff & D. Jamieson, vol. 1, pp. 197–227. Boulder, CO: Westview Press.

Mitchell, R.W. (1993a). Mental models of mirror-self-recognition: Two theories. *New Ideas in Psychology*, 11, 295–325.

Mitchell, R.W. (1993b). Recognizing one's self in a mirror? A reply to Gallup and Povinelli, Anderson, de Lannoy, and Byrne. *New Ideas in Psychology*, 11, 351–77.

Mitchell, R.W. (1993c). Animals as liars: The human face of nonhuman duplicity. In *Lying and Deception in Everyday Life*, ed. M. Lewis & C. Saarni, pp. 59–89. New York: Guildford Press.

Mitchell, R.W. (1994a). Multiplicities of self. In *Self-awareness in Animals and Humans: Developmental Perspectives*, ed. S.T. Parker, R.W. Mitchell & M. L. Boccia, pp. 81–107. New York: Cambridge University Press.

Mitchell, R.W. (1994b). The evolution of primate cognition: Simulation, self-knowledge, and knowledge of other minds. In *Hominid Culture in Primate Perspective*, ed. D. Quiatt & J. Itani, pp. 177–232. Boulder, CO: University of Colorado Press.

Mussen, P. (1967). Early socialization: Learning and identification. *New Directions in Psychology*, 3, 51–110.

Ortner, S.B. (1984). Theory in anthropology since the sixties. *Comparative Studies in Society and History*, 26, 126–66.

Parker, S.T. (1977). Piaget's sensorimotor series in an infant macaque: A model for comparing unstereotyped behavior and intelligence in human and nonhuman primates. In *Primate Biosocial Development: Biological, Social and Ecological Perspectives*, ed. S. Chevalier-Skolnikoff & F.E. Poirier, pp. 43–113. New York: Garland Publishing, Inc.

Parker, S.T. (1991). A developmental approach to the origins of self-recognition in great apes. *Human Evolution*, 6, 135–19.

Parker, S.T. (1993). Imitation and circular reactions as evolved mechanisms for cognitive construction. *Human Development*, 36, 309–23.

Parker, S.T. & Milbrath, C. (1994). Contributions of imitation and role-playing games to the construction of self in primates. In *Self-awareness in Animals and Humans: Developmental Perspectives*, ed. S.T. Parker, R.W. Mitchell & M.L. Boccia, pp. 108–28. New York: Cambridge University Press.

Piaget, J. (1972). *Psychology of Intelligence*. Totowa, NJ: Littlefield, Adams & Co. (First published in French in 1947).

Piaget, J. (1962). *Play, Dreams and Imitation in Childhood*. New York: Norton. (First published in French in 1951).

Piaget, J. (1953). *The Origin of Intelligence in the Child*. London: Routledge & Kegan Paul.

Poe, E.A. (1938). The murders in the rue Morgue. In *The Complete Tales and Poems of Edgar Allan Poe*, pp. 141–68. New York: Modern Library. (First published in 1841 in Graham's Lady's and Gentleman's Magazine).

Russon, A.E & Galdikas, B.M.F (1993). Imitation in free-ranging rehabilitant orangutans (*Pongo pygmaeus*). *Journal of Comparative Psychology*, 107, 147–61.

Russon, A.E. & Galdikas, B.M.F. (1995). Constraints on great ape imitation: Model and action selectivity in rehabilitant orangutan (*Pongo pygmaeus*) imitation. *Journal of Comparative Psychology*, **109**, 5–17.

Sahlins, M. (1981). *Historical Metaphors and Mythical Realities: Structure in the Early History of the Sandwich Islands Kingdom*. Ann Arbor, MI: University of Michigan Press.

Shepherd, W.T. (1923). Some observations and experiments on the intelligence of the chimpanzee and ourang. *American Journal of Psychology*, **34**, 590–1.

Speidel, G.E. & Nelson, K.E. (eds). (1989). *The Many Faces of Imitation in Language Learning*. New York: Springer-Verlag.

Terrace, H., Petitto, L., Sanders, R. & Bever, T. (1979). Can an ape create a sentence? *Science*, **206**, 809–902.

Thorpe, W.H. (1963). *Learning and Instinct in Animals*, 2nd edn. Cambridge, MA: Harvard University Press.

Tomasello, M. (1990). Cultural transmission in the tool use and communicatory signalling of chimpanzees? In *"Language" and Intelligence in Monkeys and Apes: Comparative Developmental Perspectives*, ed. S.T. Parker & K.R. Gibson, pp. 274–311. New York: Cambridge University Press.

Tomasello, M., Kruger, A. & Ratner, A. (1993). Cultural learning. *Behavioral and Brain Sciences*, **16**, 495–552.

Užgiris, I.C. (1973). Patterns of vocal and gestural imitation in infants. In *The Competent Infant*, ed. L.J. Stone, H.T. Smith & L.B. Murphy, pp. 599–604. New York: Basic Books.

Valentine, C.W. (1930). The psychology of imitation with special reference to early childhood. *British Journal of Psychology*, **21**, 105–32.

Visalberghi, E. & Fragaszy, D. (1990). Do monkeys ape? In *"Language" and Intelligence in Monkeys and Apes: Comparative Developmental Perspectives*, ed. S.T. Parker & K.R. Gibson, pp. 247–72. New York: Cambridge University Press.

Whiten, A. & Ham, R. (1992). On the nature and evolution of imitation in the animal kingdom: Reappraisal of a century of research. In *Advances in the Study of Behavior*, ed. P.J.B. Slater, J.S. Rosenblatt, C. Beer & M. Milinski, vol. 21, pp. 239–83. New York: Academic Press.

Wright, R.V.S. (1972). Imitative learning of a flaked stone technology – The case of an orangutan. *Mankind*, **8**, 296–306.

Yerkes, R.M. & Yerkes, A.W. (1929). *The Great Apes*. New Haven, CT: Yale University Press.

14

Imitation, pretense, and mindreading: Secondary representation in comparative primatology and developmental psychology?

ANDREW WHITEN

INTRODUCTION

Can we talk of "the great ape mind?" An increasing number of studies point to sophisticated mental abilities that may be shared by the great apes, but are unmatched by other primates including monkeys and perhaps the lesser apes. This book contains an impressive catalog of these achievements, three of which are highlighted in my title. A question that naturally arises, then, is whether the various abilities rest on some central and rather special cognitive process which drives them all. At the other extreme, the abilities might be cognitively unrelated, even if at some basic level they share a common cause: for example, they may be cognitively unrelated consequences of the brains of these species incorporating a primate cortical network that has exceeded a certain threshold size.

There are of course various possible truths intermediate between these two extremes. It is also important to acknowledge that postulating a "great ape mind" is not to deny significant mental differences between the species of great ape: rather, there is first a question about whether the great apes share a significant number of special abilities; and then a second question about whether any such array rests in turn on some more fundamental, but no less special, cognitive foundation.

This latter possibility has been explored by myself and by others before. But there are good reasons for the reexamination which this chapter offers: a rich interplay of new findings and new theories both in comparative primatology and in developmental psychology suggests that some of the big "cognitive linkages" alluded to in the paragraphs above may be part of an even wider group of psychological patterns that are extending our understanding of normal and abnormal development in children. Successfully reaching into the minds of great apes will be an achievement with broad-ranging implications.

COGNITIVE LINKS BETWEEN MINDREADING, PRETENSE AND IMITATION?

Mindreading

In the early 1980s ethologists developed theoretical arguments suggesting a radical new vision of animal communication: it was proposed that, far from being selected to transmit good honest information about motivations or intentions, animals' communication signals would be designed most fundamentally to manipulate others to the signal-sender's ultimate genetic advantage. Signals might thus not send true information about signallers' intentions: this would lead in turn to selection pressures for protagonists to become more skilled at discerning the true state of mind of signallers, in the face of behavioral subterfuge. This would set up an escalating evolutionary arms race, leading to refinement of an ability to "mindread" (Dawkins & Krebs, 1978; Krebs & Dawkins, 1984). Humphrey (1980), building on the earlier hypothesis that primate intellect has been honed by selection for social skill in complex societies (Humphrey, 1976; Jolly, 1966), suggested that mindreading was likely to be of particular benefit. In interactions that were soon to be described as "political" (de Waal, 1982), one individual could get one step ahead in the game if they were able more accurately to discriminate what was going on in the mind of others.

Humphrey talked of such mindreaders as "natural psychologists," and the growing literature on this topic has spawned yet other synonyms: perhaps the most common in current usage are "theory of mind" and "mentalizing" (for a more comprehensive guide, see Whiten, 1994). Although there is a case for retaining "mindreading" as a more inclusive category than the others in the spirit in which Krebs & Dawkins used it, I have in this chapter followed common practice and used it as a neat term synonymous with the others, all defined by *the ability to recognize states of mind (mental states) in oneself and/or others*.

Even before the ethologists developed these functional hypotheses, Premack & Woodruff (1978) had attempted ground-breaking experiments to test whether their captive chimpanzee, Sarah, utilized such a "theory of mind." For example, Sarah was shown the videotaped actions of a human, frustrated in various different ways in his attempts to obtain out-of-reach food. Sarah showed a remarkable ability to select, from amongst a set of photographs, the one depicting the correct solution to each problem. Premack & Woodruff argued that she could only do this if she saw the person as desiring or intending to obtain the goal at stake. Nevertheless, for a long period there was no empirical work forthcoming for those *natural societies* of primates in the wild that the ethological theorizing of N. Humphrey, J. Krebs and R. Dawkins suggested would be the most fertile niches for the evolution of mindreading. The philosopher Daniel Dennett (1983) was actually one of the first to delve into what an observer should be looking for. Dennett focused attention on the now famous observation by Seyfarth & Cheney of a vervet monkey giving an apparently false leopard alarm call, which, by manipulating the minds of the caller's protagonists into a false belief, got the caller off the hook in an aggressive confrontation (see also Cheney & Seyfarth, 1990b).

At this time Richard Byrne and I observed incidents of what we subsequently described as "tactical deception" in baboons (Byrne & Whiten, 1985). Tactical deception occurs where acts from the normal repertoire are deployed by an individual such that another individual is likely to misinterpret what the acts signify, to the first individuals advantage. Such deception by its nature being a relatively rare and subtle occurrence, we went on to gain more meaningful sample sizes by collating evidence for tactical deception from many different primatologists on many different species (Whiten & Byrne, 1986). Although our principal aim at that time was modest – the construction of a taxonomy to systematize the variety of functional consequences that deception achieves – we also found interspersed in the records signs of what we thought primate mindreading might look like in the context of natural social interactions (Whiten & Byrne, 1988a,b). What soon emerged – the critical point for the present discussion – was that this evidence was most impressive in chimpanzees, showing some consistencies with the conclusions of the earlier captive research by Premack & Woodruff (1978). But as Whiten & Byrne (1988a) and Mitchell (1986) made clear, not all deception implies mindreading by any means. Although records of deception were plentiful for some monkey taxa, their mindreading appeared to extend only so far as a facility to discriminate what another individual might be seeing (or not seeing). The records for chimpanzees, by contrast, suggested deeper mindreading: notably, in the case of counter-deception, an ability to discriminate the intentions of others who were engaging in deception themselves (Byrne & Whiten, 1988). Nobody was able to say much about other apes, possibly because of a lack of equivalent sampling.

In summary then, pre-1990s work on primate mindreading involved much creative theorizing, but really just two exploratory, empirical attacks on the subject: the one involving laboratory experiments, the other, naturalistic observations. More empirical research has been published in the 1990s than in the whole period beforehand. This work has been reviewed and criticized by Heyes (1993a) and Whiten (1993). Space does not permit a detailed analysis here, but Table 14.1 presents an overview of the animal research, and some key findings are as follows.

Povinelli and his colleagues have published a series of experiments that are particularly important in the present context because of their comparative nature: they used human children, chimpanzees and rhesus monkeys as subjects. One series of experiments concerned role reversal, in which each subject was teamed with a human partner (Povinelli et al., 1992a,b). One individual of the pair knew which of several food trays was baited, but only the ignorant partner had access to the lever that gave both of them access to one of the trays. Thus, the attainment of reward rested on successful communication between the knowledgeable and ignorant partners, through gestures like pointing. Two chimpanzees successfully learned to do this as informants, and two as information-receivers. Each of these subjects was then put into the reverse role with their human partners. The two chimpanzees who moved into the informant role were reported to show "immediate evidence of comprehension of their new social role" (Povinelli et al., 1992a, p. 633), now gesturing so as to inform the human partner. One of the chimpanzees who became an information-receiver also showed appropriate behavior

Table 14.1. *The study of mindreading in animals: A summary history*

Recognition of a possibility
- Thorndike (1911)
- Lloyd-Morgan (1930)

Theoretical statements on functional plausibility
- Humphrey (1980)
  Skilled social gamesmanship would be enhanced by "natural psychology"
- Krebs & Dawkins (1984)
  Communication should evolve to be manipulative and mindreading would support this

Empirical studies
- Premack & Woodruff (1978)
  Chimpanzee attributes wants, intentions
- Savage-Rumbaugh (1986)
  Chimpanzees recognize ignorance (need for information)
- Whiten & Byrne (1988a,b)
  Chimpanzees distinguish true and apparent intention: Baboons discriminate what others can see
- Premack (1988)
  Chimpanzees recognize seeing-knowing linkage, fail to attribute false belief
- Cheney & Seyfarth (1990a,b)
  Macaque monkeys do not discriminate knowledge versus ignorance
- Byrne & Whiten (1990, 1991, 1992)
  Great apes distinguish true and apparent intention: Baboons, macaques discriminate what others can see
- Boesch (1991)
  Chimpanzee recognizes need for information
- Gómez (1991)
  Gorilla understands perception-action linkage
- Povinelli *et al.* (1990, 1991)
  Chimpanzees and 4 year olds discriminate knowledge and ignorance: Macaques and 3 year olds do not
- Povinelli (1991)
  Chimpanzee discriminates intentional from accidental acts
- Povinelli *et al.* (1992a,b)
  Chimpanzees show empathy in role reversal: Macaques do not
- Gómez & Teixidor (1992)
  Orangutan recognizes ignorance

in the first set of role-reversed trials. By contrast, four rhesus monkeys tested in the same way did not show the rapid transfer that the chimpanzees did.

The authors concluded that the experiments offered evidence of "some form of cognitive empathy" in the case of the chimpanzees. With respect to the subject of this present chapter, these results can be more specifically expressed as an ability to translate between one's own, and another individual's, intentional or representational state. Thus, the chimpanzee who successfully switches to the role of informant apparently recognizes that the other individual is in a state in which it lacks information – the state it had itself experienced before the roles were reversed. Conversely, the chimpanzee who successfully switches to the role of information-receiver may, as Povinelli et al. suggest, have "demonstrated role-taking by learning the intentional significance of such pointing from the perspective of the opposite role" (Povinelli et al., 1992a, p. 638).

At this point we may anticipate later discussion of a possible linkage between mindreading and imitation, because the translation from the other individual's perspective for action to one's own which occurs in role-reversal also underlies imitation. Indeed, the role-reversal shown by the chimpanzees could be seen as incorporating a type of delayed imitation: the subject performs the same act its partner did before the reversal.

A more direct test of chimpanzees', rhesus monkeys', and childrens' ability to mindread, in discriminating ignorance versus knowledge in others, used the same apparatus (Povinelli et al., 1990, 1991; Povinelli & de Blois, 1992). Now, however, the subject had to choose between the gestured "advice" of two potential informants as to which tray to choose. One informant was knowledgeable, because he hid the food, so he gave good advice. The other was ignorant because he left the room while the food was hidden, and this individual gave bad advice: he pointed to the wrong container. The chimpanzees, but not the monkeys, learned to discriminate in favor of the knowledgeable adviser. To establish if this was more than a simple discrimination (of who hid the food, or who was present when it was hidden) a "critical transfer test" was performed in which ignorance was created in one individual not by them leaving, but by covering the eyes with a bag. Three of the chimpanzees quickly transferred their discrimination to this very different situation. Although it remains questionable whether the potential effects of the blindfold itself were well enough controlled (Heyes, 1993a; Whiten, 1993), the experiment points to some appreciation by the subjects of the state of ignorance versus knowledge in the partner, occasioned by different types of circumstance.

By contrast with the chimpanzees, the monkeys failed even to learn the initial discrimination. The 3 year old children likewise failed to make the discrimination between knowledgeable and ignorant advisers, and an attempt to train one child to do so using the primate procedures met with no success over 80 trials. Thus, the young children and monkeys performed in similar fashion. By contrast the 4 year old children did succeed, as we would expect from previous research, and the performance of the chimpanzees therefore appears to parallel a developmental advance occurring between 3 and 4 years of age in children.

A convergence of findings on monkeys' lack of discrimination of knowledge states comes from experiments conducted by Cheney & Seyfarth (1990a,b): for example, macaque mothers did not alter their rate of giving food calls or alarm calls (with respect to food or dangers, respectively) according to whether their offspring were either knowledgeable or ignorant of the relevant circumstances.

Further consistency comes from the more recent observational data on spontaneous tactical deception in primates. The first survey and catalog by Whiten & Byrne (1988a) was followed up by a second, which expanded the database to 101 journal pages (Byrne & Whiten, 1990). The earlier suggestion of a difference between chimpanzees and the several species of monkey well represented in the database continues to be supported: evidence for mindreading that goes beyond visual perspective-taking has accumulated for the chimpanzees but not the monkeys. In addition, a small number of records for orangutans and gorillas can be classed with those from chimpanzees as indicative of mindreading (Byrne & Whiten, 1992: for analysis of these records, see also Byrne & Whiten, 1991).

Finally, evidence that we may be dealing with a feature of great ape mind comes from an extension of experimentation to the orangutan. Gómez & Teixidor (1992) trained an orangutan to indicate by gesture which of two baskets should be unlocked to obtain the bait that the ape, but not the human helper, had seen hidden. The key was always taken from the same hook on the wall. In the critical test, another person entered before the helper did and moved the key to a quite different location. When the helper entered, the orangutan gestured towards the new key location, evidencing an appreciation that the helper was ignorant of the transfer.

The overall picture is thus currently one of convergence of evidence for some degree of mindreading in great apes, which goes beyond the visual perspective-taking suggested by data on monkeys. This pattern remains tentative because of the small number of studies and the complexity of interpretation in this area (e.g. Heyes, 1993a; Tomasello & Call, 1994; Whiten, 1993, 1994, 1996); but it is strengthened by virtue of the very different quarters from which the evidence comes and I adopted it as a working hypothesis for the remainder of this chapter.

The relationship between mindreading and pretense

The Premack & Woodruff (1978) paper, together with its accompanying commentaries, played a major part in spawning parallel experimental investigations of the development of a theory of mind in young children (Astington et al., 1988; Whiten, 1991a; Wimmer & Perner, 1983). One of the most powerful theoretical analyses of the nature and origins of childrens' mindreading came in a paper by the developmental psychologist Alan Leslie (1987). Its reverberations continue today within the now overwhelming literature on the child's theory of mind (see particularly Hobson, 1990; Leslie & Frith, 1990; Leslie & Roth, 1993; Perner, 1991, 1993). The part of Leslie's analysis relevant to us here concerns the proposed linkage between theory of mind and pretense. Leslie suggested that pretend play points to the origins of the child's developing theory of mind, because these two abilities share an important psychological operation, for which he used the term *metarepresentation*.

The essential argument runs as follows. To operate a theory of mind is to mentally represent the mental representations of others. For example, if individual A thinks there is fruit in a tree, then A has a mental representation of the world which is effectively "fruit in tree." If B thinks that A is thinking "fruit in tree," B has a second-order representation, or metarepresentation (i.e. a representation of a representation).

Now consider pretend play. At first glance, this is a poor candidate to share the metarepresentational character of theory of mind: cases of pretend play may be nonsocial, where there is no other mind to read. Leslie argued instead that what is relevant in the case of pretense is that the two representations necessary to diagnose metarepresentation coexist in the child's head. Consider Leslie's example of a child who uses a banana as a telephone. We may be tempted to say simply that she represents it mentally as a telephone. But if she does this, her primary mental representation of the world, which functions to know the world as it really is, will be subject to confusing distortion. In short, she must not believe the banana to *be* a telephone. Instead, the child must effectively maintain the first-order representation of the banana as banana, deriving from it a second-order representation, mentally marked as a pretense, of the banana-as-telephone.

So much for the theory (summarized all too baldly given the sophistication of the original – for details, see Leslie, 1987; Whiten & Byrne, 1991). What about the data? One database derives from normal child development, where Leslie argued that the emergence of pretend play in the second year is closely accompanied by the first signs of theory of mind. This is consistent with the theory but really represents only a rough developmental correlation – a similar case could be made for the significance of many other milestones achieved in cognitive and communicative development at this time. More powerful is the double deficit established for children with the severe social difficulties that typically characterize autism (Leslie, 1987). These children have repeatedly been shown to perform extremely poorly in tests for theory of mind (Baron-Cohen *et al.*, 1985; for a recent overview, see Baron-Cohen *et al.*, 1993). But in support of Leslie's theory, they also showed a noticeable lack of spontaneous pretend play (Baron-Cohen, 1987; Lewis & Boucher, 1988; Sigman & Ungerer, 1984; Wing *et al.*, 1977).

Is it possible that a third database consistent with the theory is provided by patterns of mindreading and pretense in nonhuman primates? Before tackling this question, it is important to recognize the quite severe criteria that episodes of pretense must meet according to Leslies analysis. These are best illustrated by an example. If a child is pretending an empty glass is full, and his arm is nudged so he "spills" some, his reaching down to "mop up" the spillage is a clear demonstration that he is operating in a pretend mode in which novel logical implications are worked through. It is such accommodations that convince us the child is acting within a pretend play world, distinct from primary reality.

Whiten (1988) and Whiten & Byrne (1991) pointed out that correlated with the evidence for theory of mind in chimpanzees is observational data on different captive chimpanzees, offered by different observers, which meet Leslie's criterion for pretend play in a way not matched in observations of either wild or captive monkeys. Thus, the chimpanzee Viki went through a series of appropriate movements to untangle an apparently nonexistent cord she was pulling on, and when it was "freed" she went off with "a little jerk" (Hayes, 1951); the bonobo Kanzi not only acted as if eating a nonexistent fruit, but appeared to spit out bad parts and signalled "bad" on his keyboard (Savage-Rumbaugh & McDonald, 1988).

Pretense of this character is difficult to elicit at will in experimental tests. Thus, while efforts devoted to experiments on imitation and theory of mind have produced the steady stream of results discussed elsewhere in this chapter, the same is not true for pretense. My previously expressed interest in pretense has, however, prompted others to draw my attention to further records, often buried in nonmainstream, in-house publications and newsletters. Such records are often difficult to interpret, at least partly because they are not presented so as to convince a skeptic about the depth of pretense involved. For example, Tanner (1985) described the sign-using gorilla Koko playing with a lobster claw. The transcript includes the following: "she plays with it, signing to herself: 'Smoke' – she smokes it like a cigarette. 'Hear' – she puts it to her ear, as if listening to a seashell . . . 'Earring that' (to the claw as she puts it to her ear)" (Tanner, 1988, p. 3). Given that pretend acts are frequently performed by humans in front of the gorilla, the possibility that those listed above did not merely represent imitations needs to be ruled out before we can be sure the acts represent spontaneous pretense of the kind at issue. (It is ironic that imitation, itself now disputed as a primate ability – see below – has historically been invoked so readily to demote potential evidence of other abilities, from sign-language acquisition to the case of pretense discussed here!)

The evidence for pretense in primates, although it appears quite convincing in some cases for great apes, is thus still fragile. Perhaps now that its theoretical significance has been highlighted, researchers will take care to establish and publish the evidential basis for excluding other explanations for candidate actions they observe. In the meantime, the most convincing records are available for apes rather than monkeys (for further records from an orangutan, see Chapter 13). Thus, the exciting prospect is that capacities for pretense and mindreading fall into systematic patterns observable in such diverse comparisons as younger versus older children, autistic versus nonautistic children, and great apes versus other, simian primates.

Imitation

Why might imitation be a third piece of the jigsaw?
Imitation has been defined concisely as a process in which one individual "learns some aspect(s) of the form of an act" from another individual (Whiten & Ham, 1992). Whiten (1988) and Whiten & Byrne (1991) noted that, as for pretense and mindreading, recent reevaluations of the evidence point to such imitation being within the abilities of chimpanzees, yet not (surprisingly, given conventional wisdoms) proven in monkeys (for more recent reviews, see Galef, 1990; Visalberghi & Fragaszy, 1990; Whiten & Ham, 1992). This raised the question of whether imitation might be a third ability, somehow cognitively linked to the proposed mindreading–pretense complex in great apes. Observing human infants, Piaget (1962) had earlier discussed such links between pretense and imitation, as does Mitchell (1994) in his own attempt to chart the "larger patterns" of ability scrutinized here.

The existence of such links would suggest that imitation in apes shares the metarepresentational basis suggested to underlie those other two abilities, pretense and mindreading. This may sound unrealistically convoluted, insofar as an imitator can be said to copy the *behavior* it sees the model perform, rather than the model's state of mind (Heyes, 1994): so why should any process of second-order representation be involved? The answer may be that imitation seems deceptively simple if described in this way, and yet we are now discovering that it is a difficult thing for a monkey to do: in fact to produce a good imitation, the imitator will have to operate a program of action that replicates the program of action which controlled the model's behavior. In terms akin to those in which we described mindreading, imitation can thus be seen as involving two representations: a first-order one (a program, or action-plan) which drives the actions of the model, and then a second one in the imitator, which, to the extent that it replicates the model's first-order representation, is effectively a second-order representation. Different processes through which the second representation may get built are discussed on pp. 316–7.

The empirical reasons for exploring such ideas go beyond the putative ape–monkey contrast. Whiten & Byrne (1991) suggested that if there is something in these theoretical linkages, imitation might also be particularly impaired in the case of autism, and they noted two studies that appeared to support this. In parallel, autism researchers Rogers & Pennington (1991) thoroughly reviewed the literature on imitation and autism, finding an early study by DeMeyer *et al.* (1972) that showed autistic subjects to be significantly poorer than controls on imitation of both body movements and actions involving objects. These subjects were unimpaired when instructed to carry out similar actions, showing that a motor deficit cannot account for the negative imitation results. Rogers & Pennington reviewed a further five studies completed since 1985 that generated comparable results. Four of the studies used nonsymbolic, sensorimotor tasks (i.e. acts not requiring any appreciation of the pretense involved in, for example, acting out combing one's hair), so the imitative deficits are not explicable merely as secondary to those established in pretend play or theory of mind. In short, they concluded that "a wider sampling of the imitation literature suggests specific deficits (i.e. worse than mentally retarded MA and CA matched comparison groups) in imitation of simple motor movements and affect expressions, as well as in higher level symbolic imitations in autistic persons" (Rogers & Pennington, 1991, p. 141).

Seeking to make sense of the patterns appearing to emerge here, Rogers and Pennington went on to argue that "early capacities involving imitation, emotion-sharing, and theory of mind are primarily and specifically deficient in autism. Further, these three capacities involve forming and coordinating social representations of self and other at increasingly complex levels via representational processes that extract patterns of similarity between self and other" (Rogers & Pennington, 1991, p. 137). The authors proposed a model of the development of these processes in which imitation plays a primary constructive role in the generation of theory of mind, pretense and other capacities (Figure 14.1). Thus, while Whiten & Byrne suggested the linkage of imitation with these capacities as a relatively tentative afterthought, Rogers & Pennington argued

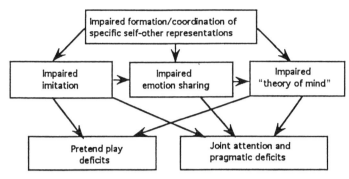

Figure 14.1. The cascade theory of autism proposed by Rogers & Pennington (1991). As in Whiten & Byrne's (1991) analysis of primate cognition, imitation, theory of mind and pretense are linked. In Rogers & Pennington's model, an impairment in imitation lies close to the root of the cascade of effects that include later deficits in mentalizing and pretense.

that it had fundamental significance from an ontogenetic perspective (a proposal further developed by Meltzoff & Gopnik 1993, discussed on pp. 318). What, then, have we discovered about the occurrence and correlates of imitation in animals since these ideas were developed?

Are great apes imitators?
Considerable skepticism arose about the claims made by earlier field and laboratory studies for imitation in monkeys, because other mechanisms of social learning can generally account for the results (Boinski & Fragaszy, 1989; Fragaszy & Visalberghi, 1989; Galef, 1988, 1990; Visalberghi & Fragaszy, 1990; Whiten, 1989; Whiten & Ham, 1992). Chief amongst the alternatives is some process of *stimulus enhancement*, in which the observer's attention is drawn to relevant environmental features by the actions of the model, thus facilitating independent acquisition by the observer of the successful technique, and creating an illusion of imitative acquisition.

Whiten & Ham's (1992) review concluded that because of such methodological difficulties, imitation was unproven in monkeys and most apes. By contrast they argued that chimpanzees appear able to imitate: but this conclusion was based largely on the scale, accuracy, and complexity of copying described in the spontaneous actions of chimpanzees reared in captive conditions where novelty in behavioral repertoires can be more easily recognized than in the wild. The only systematic experimentation, however, was at odds with this (Tomasello *et al.*, 1987): chimpanzees learned something from observing others using a tool to get food, but they did not imitate the technique used.

More recent studies have used a variety of more refined methods to produce convincing evidence of imitation in apes. Custance *et al.* (1994, 1995) returned to an approach originally attempted by Hayes & Hayes (1952), in which the subject is literally trained to imitate to order, given a request like "DO THIS!". The technique involves a potentially lengthy period of training in which a finite set of actions is modeled for the subject, and successful replication is rewarded. This phase can be thought of as a case of learning set, in which a successful outcome is the subject recognising that it is required to attempt to copy any act modeled in front of it. The second, critical phase then follows, in which the subject's ability to imitate a variety of different novel actions can be systematically examined. This is potentially a powerful technique, because as the sample of actions rises in number, the probability of the subject achieving matches by chance falls very rapidly. Hayes & Hayes (1952) reported a positive conclusion in their application of this approach to the home-reared chimpanzee Viki.

Unfortunately, the 1952 paper did not carry a systematic account of the actions demonstrated and reputedly copied, and no statistical tests were applied, so that 40 years later it is, sadly, near impossible to evaluate. Custance et al. therefore set up a systematic "DO AS I DO" experiment in which the training and testing actions were clearly defined and independent observers were required to select which test action they judged each chimpanzee action matched. The resulting overall performance was evaluated statistically. The number of correct imitative matches was significantly greater than chance for two $4\frac{1}{2}$ year old chimpanzees. Stimulus enhancement can be ruled out, because the actions were gestures and expressions that did not involve objects or environmental locations.

Two other findings in these studies should be highlighted. First (putting chimpanzee imitation into perspective), performance was unimpressive when compared with that of young children. Of 48 novel actions modeled, the two chimpanzees imitated only 13 and 20, respectively, whereas most 4 year old children performed all but a handful (Custance et al., 1995). Secondly however, on a more positive note, the chimpanzees did show success in the copying of actions involving parts of the body out of the subject's sight. This included facial expressions (e.g. lip-protrusion, teeth-chattering, cheek-puffing), and acting on hidden body parts (e.g. lip-wobbling, mouth-popping with finger, touching back or top of head, touching nose; Custance et al., 1994). This is particularly important in demonstrating the relatively abstract level of those "representational processes that extract patterns of similarity between self and others," which Rogers & Pennington's (1991, p. 137) analysis highlighted.

A quite similar approach was taken by Tomasello et al. (1993), who requested chimpanzees, bonobos and children to "DO WHAT I DO" with respect to a battery of actions on largely novel objects, ranging from simple (e.g. placing a sifter on the head) to complex (e.g. reeling in an object with a reel). Three of the ape subjects were highly socialized through human-like rearing ("enculturated"), the other three much less so. However, both groups were said to require little explicit training to cooperate with the request. The important result with respect to the present discussion is that the clearest experimental evidence for imitation by chimpanzees was obtained: the enculturated chimpanzees imitated approximately two thirds of the simple acts, and one third of the complex ones – a performance indistinguishable from the child subjects, who were either 18 or 30 months old. By contrast the mother-reared apes showed relatively little imitation. Tomasello et al. concentrated their statistical analysis on establishing these group differences, rather than on whether imitative matching differed from chance, so it is difficult to judge whether the mother-reared chimpanzees should be credited with imitating per se. The tables record 10 imitations for them, out of a possible 38 simple acts, and 7 imitations out of a possible 29 complex acts. Putting all the studies together, an ability to imitate can now be said to have been demonstrated clearly and experimentally for chimpanzees (Whiten & Custance, 1995; see also Whiten et al., 1996).

A very different approach has been taken by Russon & Galdikas (1993), who documented spontaneous imitation of both human and novel conspecific actions by rehabilitant orangutans. No experimental manipulations were performed. The evidence is compelling, however, because of the authors' systematic recording of actions that by their nature make explanations other than imitation exceedingly implausible: such acts were not previously in the subject's repertoire, but were acquired rapidly and/or exhibited features not explicable by individuals being shaped by intrinsic task demands or social reinforcers (for detailed criteria and records, see Chapter 7).

In summary, chimpanzees and orangutans have demonstrated imitative capabilities that, despite the opportunities offered them in earlier experiments (Visalberghi & Fragaszy, 1990; Chapters 2 and 3), monkeys have failed to exhibit. As things stand, the difference appears quite striking. There are, however a number of notes of caution that should be sounded. First, it may be that imitative proficiency in apes is associated with particular types of human contact, rather than a generalized ability, as suggested by Tomasello et al. (1993; Chapter 17; for further discussion of this issue, see Russon & Galdikas, 1995). Secondly, if the conclusions drawn about orangutan imitation on the basis of observational evidence are correct, it should be possible to more convincingly confirm such capacities with the right experiments. In addition to these concerns, it must be noted that the conclusion that "great apes can imitate" is premature until specific evidence for the species phylogenetically intermediate between orangutan and chimpanzee – gorillas – is forthcoming. Differences in the skills of resident and immigrant mountain gorillas are consistent with a role for imitation (Byrne & Byrne 1993; and see Chapter 5) and some anecdotes can be gleaned from the literature like those available for chimpanzees, but stronger evidence does not yet exist (Whiten & Custance, 1996). Virtually nothing is known about the lesser apes, gibbons and siamangs.

Other jigsaw pieces?

Whiten & Byrne (1991) also speculated that other apparent ape–monkey differences might be part of a metarepresentational pattern. One of these is apes' capacity to use remote representations of themselves – mirror reflections and video pictures – to guide examination of parts of their face they cannot see (Gallup, 1970, 1982; Menzel et al., 1985; Patterson & Cohn, 1994). This contrasts with monkeys' tendency to treat such images not as self but as another monkey (Gallup, 1979). When we ourselves look in a mirror, our literal or primary perception is also that of another individual: but unlike the monkeys, we and other great apes are able to see the image as a representation of ourselves, and in this sense we demonstrate a second-order representation. Note that it would not be true to say apes see the image as themselves – in fact they show that they distinguish it from themselves, because they do not attempt to groom the face in the mirror, but instead use it to guide manipulations on their own face.

At the time we speculated on the sense in which mirror-self-recognition might fit the explanatory principle of metarepresentation, we had not realized that Gallup had earlier (1982) suggested that mirror-self-recognition predicted a cluster of other capacities including mental attribution and deception. A detailed comparison with Gallup's analysis, summarized in Table 14.2, is, however, beyond the scope of this chapter, as are other analyses which focus on the linkage of imitation and theory of mind with mirror-self-recognition (see, e.g. Gergely, 1994; Gopnik & Meltzoff, 1994; Mitchell, 1993; Parker, 1991; Parker et al., 1994; Chapter 16). Key points of contrast for present purposes are that Gallup suggested the heart of the complex lies in *self-awareness* rather than metarepresentation, and he did not include imitation as part of the pattern. The main difficulty I see with Gallup's proposal is the assumption of a necessary link between two meanings of "self-awareness." The first is that demonstrated in ape mirror-use, where the subject shows it is "aware" that the visual image it sees can be mapped to its kinesthetic experience of "self" (Heyes, 1994; Mitchell, 1993). The second meaning of "self-awareness" is that embodied in such self-reflection as thinking about one's own states of mind, which is the level at which Gallup makes the link to the other cognitive abilities like mindreading. However, it is not obvious why the first sense of "self-awareness" should necessarily generate the second (Heyes, 1994). Thus, we still need to examine alternative explanations for the cluster of cognitive abilities that exist in the great ape mind.

INTERPRETING THE COGNITIVE CORRELATIONS

So far, I have outlined evidence for cognitive correlations, and explained why, a few years ago, a capacity for metarepresentation looked promising as an explanation. The basic idea is that, while monkeys may be limited to primary or literal representations of the world, the great apes can go beyond these to generate second-order representations: metarepresentations. Since 1991, this idea has appeared to gain strength by the accumulation of data reviewed above, which support the cognitive correlations and the ape–monkey contrasts. However, other recent developments urge reevaluation of the ideas. Some of these are empirical, particularly concerning imitation. Others are theoretical and I have next examined one of these in particular.

Metarepresentation and secondary representation

In referring to metarepresentation, Whiten & Byrne (1991) followed Leslie (1987), who used the term interchangeably with second-order representation to cover all cases of mental attribution, such as my believing you are ignorant of a certain fact.

However, Perner (1991; see also Whiten & Perner, 1991) took issue with Leslie's usage, as a result of which we need to distinguish two meanings of metarepresentation. This, on the face of it, is an unfortunate and confusing state of affairs, but to be more positive, we might celebrate the more refined distinctions that have thus been generated.

Perner's thinking can be introduced through summarizing three levels of mental representation that he proposed children achieve sequentially in their development.

Table 14.2. *Gallup's scheme of mental and behavioral correlates of self-awareness (after Gallup, 1982)*

Traits	Hard-wired analogs	Self-awareness instances
Attribution	Unlearned reactions to conspecific threat postures and predators	Attribution of intent[a] and responsibility[b]; anthropomorphism[c]
Deception	Mimicry	Intentional distortion and/or withholding of information[d]
Reciprocal altruism	Alarm calls	Reciprocal aid-giving[e]; selectively withholding aid from cheaters and stealers[d]
Empathy	Responses to appeasement gestures and reactions to infant distress calls	Providing solace to injured conspecifics[b,f]
Reconciliation		Preferential contact between opponents folowing an aggressive encounter[f]
Pretending	Injury feigning; death feigning	Certain forms of deception[g]

[a] Hebb (cited in Gallup, 1982) (1979)
[b] Goodall (1968)
[c] Premack & Woodruff (1978)
[d] Woodruff & Premack (1979)
[e] Savage-Rumbaugh et al. (1978)
[f] de Waal & van Roosmalen (1979)
[g] Menzel (cited in Gallup, 1982) (1975)

First, there is primary representation, which as Leslie stressed, performs the fundamental job of representing the world as faithfully or literally as possible, such that we can operate successfully in it: Perner suggested that, in their first year, children are constrained to this level.

If we jump to the 5 year old, however, we find they have reached the third level, which is the one Perner is prepared to call metarepresentational. Only at this level do children easily succeed in tests requiring some recognition of the implications of false beliefs in others. For Perner, this signifies the achievement by the child of a crucial *understanding* of the *nature of mental representation*: that is, children at this age appreciate that others may have different thoughts about reality than they do themselves. In short, they "understand that the mind represents" (Perner, 1991, p. 10). There are similar achievements in their understanding of external referents such as pictures.

In between the primary and metarepresentation stages, when children start to show pretend play, Perner suggests they are capable of generating secondary representations.

Going beyond representing things only as they are (primary representation), the child can represent how they *might* be – he/she can imagine hypothetical situations.

Thus, in the case of pretending (that this empty cup is full, for example), the child needs a primary representation of the real situation (cup empty), and a secondary representation of an alternative situation (cup full), which is in some way mentally marked as "pretend." Turning to mindreading, children at this level "conceive of thinking not as a representational activity but as a preoccupation with the thought-about situation" (Perner, 1991, p. 10), and so Perner calls them "situation theorists." These children have empathic reactions to others' distress and start to talk about others' mental states, demonstrating what Perner calls a "mentalistic theory of behavior." But Perner agued that not until the later, third phase, when they understand that beliefs may be held about the world that can be false, do they grasp that the mind represents. Then they have a "representational theory of mind" and, since they can now represent the *representing relationship* existing between mind and world, they can be said to show true metarepresentation.

Another way of drawing out the distinction at stake is to say that Perner is wishing to emphasize one particular meaning of metarepresentation, where a subject can be said to "represent representations *as* representations." We might call this the strong sense of metarepresentation, and contrast it with the weaker sense where we say just that a subject "represents certain representations." This latter sense is the one used by Leslie (1991) and by Whiten & Byrne (1988a, 1991), to refer to all cases of mindreading, where by definition the mindreader is mentally representing anothers' mental states, and thus effectively their mental representations. This usage would encompass the reading of all mental states, including, for example, desire and knowledge. To earn Perner's strong version of the term, a mindreader has to have the level of understanding of the nature of representation that underlies the possibility of another individual holding a false belief. The 3 year old child does not have this understanding, and thus does not metarepresent in the strong sense, but typically does attribute other mental states and thus metarepresents in the weak sense.

So, where do such theoretical niceties lead us with respect to the idea that metarepresentation may underlie the great ape mind?

Apes and secondary representation
With Perner's distinction before us, I suggest that it is secondary representation, rather than metarepresentation in the strong sense, which is the appropriate level to which we should assign the patterns of ape mindreading and pretense to which Whiten & Byrne (1991) drew attention. Thus, all the evidence for ape mindreading concerns the recognition of mental states known to be achieved in childhood prior to the 4 year old watershed of false belief attribution that marks the achievement of metarepresentation in Perner's terms (Whiten, 1991b). Admittedly, this may reflect absence of evidence as much as evidence of absence: there has been only one experiment published that explicitly attempted to test for the recognition of false belief, with a negative, if slightly ambiguous, result (Premack, 1988). The result of a recent alternative approach developed by

S. O'Connell (unpublished results) can be summed up in the same way. Gómez & Teixidor's experiment with an orangutan, noted earlier, might be interpreted as evidencing recognition of a false belief (the helper falsely believing the key was in its usual location), but recognition merely of the *ignorance* of the helper would suffice for the ape's change in behavior. As things stand, the picture conforms to Premack's (1988) rule of thumb, that: "capacities that do not appear in the 3 1/2 year old child will not be found in the ape" (Premack, 1988, p. 173). The occurrence of the types of pretense described for apes is also consistent with this: by their nature they go beyond primary, to secondary representation, but as for 2 year old children's pretend play, they do not require metarepresentation in the strong sense in which Perner distinguishes it, as understanding the nature of representation.

Secondary representation and imitation

Perner has directly related mindreading and pretense to secondary representation, but he has little to say about imitation. How does it fit with the theoretical distinctions outlined in the last paragraphs? Recall that the basis of secondary representation is the ability to coordinate multiple models, which represent different situations. One could argue that this is intrinsic to imitation. Thus, in *pretending* to be you performing action X, I act AS-IF I am you doing X (Bruner, 1972; Fein, 1981) and the similarity of *imitation* to this description is obvious; when I imitate you doing X, I act very much AS-IF I am you doing X. In both cases, the secondary situation of, say, *my brushing my teeth* is translated from the primary situation of *your brushing your teeth*. Likewise, turning from pretense to mindreading, we can note that mindreading resembles this translation function of imitation insofar as, for example, I am able to *think what you are thinking*. Empathy is a particularly imitative form of mindreading, when it is defined as one individual acquiring a copy of another's emotions.

As outlined earlier, apparent patterns found in ape–monkey and normal–autistic child contrasts have appeared to confirm such proposed linkages of mindreading, pretense and imitation. However, there are important empirical data that do not fit easily with any straightforward correlation with imitation. First, the pretense and mindreading abilities at stake in the human case emerge by about the middle of the child's second year, whereas an ability to imitate quite complex behavior can be demonstrated at half this age (Meltzoff, 1988). In addition there is evidence for a form of imitation in human neonates (Meltzoff & Moore, 1977; for a review, see Meltzoff & Gopnik, 1993), although this continues to be disputed (Anisfeld, 1991; Moore, 1992).

Secondly, since Whiten & Byrne (1991) sketched an ape–monkey cognitive pattern with imitation as a rather special ape capacity, evidence of imitation in rodents and birds has been offered that, although not yet replicated, appears on the face of it to be more convincing than that available for monkeys. Moore (1992) has described a parrot as imitating quite elaborate human actions associated with particular utterances, such as waving while saying "ciao" (cheerio), dropping an object and saying "whoops," and gesturing with the head while saying "get back!". Heyes *et al.* (1992) showed that rats were more likely to push a lever to the side they had seen a demonstrator rat prefer.

These latter authors went on claim that there is in fact no primate data (in either monkeys or apes) that shows imitation with comparable rigor.

Clearly if these results are accepted, they do not lie easily beside the ideas that (a) imitation is found in apes, but not monkeys; and (b) this is because imitation is a high-level cognitive achievement underwritten by powers of secondary representation associated also with mindreading and pretense. This contradiction might be resolvable in a number of ways.

1. *The nonprimate results do not actually demonstrate imitation.* Although the non-primate cases appear impressive, there are reasons for caution. In the case of the parrot, we have a single individual studied by a single individual. Cases of purported imitation occurred at quite separate times from the modeled actions they were thought to copy, and the claim that each case *was* an imitative copy was not subject to the rigor of being coded by observers blind to the predicted matches, as done by Custance et al. (1995), for example. In the case of the rats, there was no evidence of imitation in the animals' first actions on the lever, and their actual behaviour at the lever site has not been described. The fact that when the lever apparatus was turned through 90°, no decrement in matching of push-direction was observed, is so impressive as to create a worry that some artifactual process might be at work. This is not to impune an important contribution to the field, but rather to emphasize that interpretation must needs be provisional until replications are made of these studies that analyze and present data in comparable ways for nonprimate and primate research.

2. *The triadic pattern extends to nonprimates.* Contrary to nonprimate data being misleading, it may be both correct and consistent with the triadic pattern described earlier: namely that imitation will be associated with mindreading and pretense capacities. Does the parrot have a theory of mind? Whatever one's skepticism about such prospects, it is nevertheless true that nobody has checked.

3. *Some forms of imitation fit the triad hypothesis and others do not.* Whatever the outcome with respect to (1) and (2) above, they do not deal with the occurrence of clear imitation in human infancy, ahead of the emergence of the other two capacities. This suggests that only some levels of imitation and not others are candidates for the triadic pattern, and this is plausibly the case for the data from comparative psychology also.

What might link certain special aspects or levels of imitation to mindreading or pretense, consistent with an underlying process of secondary representation? Two principal types of link have been discussed (Whiten, 1992). The first is the perspective change involved in both mindreading and imitation. A mindreader might recognize that another individual will see the same social situation differently from the way they themselves do. An imitator must translate from its view of the action when performed by the other, to what it will be for it to perform the act itself. This involves more than mere visual perspective change in that the imitator has to generate a motor plan corresponding to the act, as originally seen from a different viewpoint. Others have suggested that the "problem" (i.e. what makes imitation difficult and perhaps rare in animals) lies in translating between the act as perceived in the model's performance, and kinesthetic/proprioceptive counterparts in the imitating self (Heyes, 1994; Mitchell, 1993, 1994). These alternatives are not necessarily exclusive, because such feedback could be used to refine the implementation of an action-plan (which might be visually guided), rather than operating in a trial-and-error fashion. And each of these analyses of what makes imitation difficult has been suggested to apply particularly to the case of copying seen actions, as opposed to the vocal imitation that is so common in song birds (Heyes, 1994; Moore, 1992; Whiten & Ham, 1992). However, it is this sort of translation that both the rats (Heyes *et al.*, 1992) and the parrot (Moore, 1992) referred to earlier have been reported to manage. The rats were more likely to push a lever to their left as had the model, even when the lever was rotated through 90° before they approached it. The ability of the parrot appeared the more remarkable, because the translation was from human exemplars: one case, for example, involved the parrot accompanying the vocalization "ciao" with a wave of the wing on some occasions, and with a wave of the foot on others (the human model had waved goodbye whilst vocalizing "ciao"). Thus, we are currently presented with the challenging question of whether reputedly "intelligent" monkeys really cannot compute such translations, while these nonprimate species (as well as apes) can. Whatever the answer to that question, the nonprimate results suggest that the imitative translation process can proceed without a necessary linkage to mindreading or secondary representation.

A second type of linkage between imitative processes and mindreading has been suggested to be an ability to recognize others' goals – what others are *trying to do*, in the acts which elicit imitation (Cheney & Seyfarth, 1990b; Tomasello *et al.*, 1993a). Cheney & Seyfarth suggest this may explain the evidence of imitation in apes as opposed to monkeys, and Tomasello *et al.* make a similar argument with respect to children who reach a certain stage of development. Insofar as recognition of goals *is* a form of mindreading, it falls into the concept of secondary representation and offers a very direct "explanation" for any clustered mindreading-imitation-pretense contrasts between apes and monkeys. Byrne (1994; and see Chapter 5) elaborated this theme to describe "program-level" imitation of the hierarchical, subgoal structure of a model, but not necessarily the precise means used to attain the goals.

The rat and parrot imitations can be interpreted as outside these categories, if classed as merely *mimicking*, in the terminology of Tomasello *et al.* – copying the form of the acts without any representation of their goal. This is possible, even plausible, although Heyes (1993b) noted that as yet no good empirical demonstrations exist that discriminate mimicking from imitation in the strong "mindreading" sense advocated by Tomasello *et al.* (1993b). A mimic of the original act, applied to the same part of the environment, is likely to achieve the original goal also, as a matter of course. Conversely, individuals may replicate the subgoal structure of acts performed by their companions because this is simply the optimal structure in the environment concerned, and not because they are imitating the subgoals. Experimental manipulation of the sequencing of components within a complex act performed by a model would be one way to test whether "goal emulation" (Whiten & Ham, 1992) is in operation, but such work has yet to be published (see Whiten & Custance, 1995).

Thus, the comparative study of imitation and the relation of imitation to processes of secondary representation is in considerable and interesting turmoil. But this is because of the vigor and variety of both empirical and theoretical work in recent years. To summarize: data suggesting ape superiority over monkeys in imitation led in the 1980s to speculation that this might be linked cognitively to correlated abilities in mindreading and pretense. It is possible that the imitation more recently described in rats and parrots may be of a different character, consistent with the idea that some imitation may rest on processes of secondary representation and other imitation not. We are not able to resolve such questions empirically as yet, but research on the great ape mind has played a major part in generating the continuing ferment of theory and comparative analysis (Mitchell, 1994).

The idea that some forms of imitation may be related to secondary representation, and others not, takes on special significance within developmental and evolutionary perspectives. Thus, it is consistent with the view put forward by Rogers & Pennington (recall Figure 14.1) that infant imitation may act as a precursor for the more advanced capacities of pretense and mindreading. In this model, imitation is cognitively linked to mindreading and pretense on the principle that each incorporates a facility in self–other representations, but imitation is a simpler and developmentally prior achievement. Meltzoff & Gopnik (1993) suggested more specifically that what may be developmentally important is that in infant imitation, externally perceived behavior of others is mapped on to a set of self's proprioceptive sensations, motor intentions, and plans, thus providing a device for beginning to appreciate the connection between behavior observed in others and the appropriately matching mental states perceived in the imitating self.

Such developmental hypotheses can be translated into evolutionary ones, in which certain sophisticated forms of imitation could be linked with mindreading and pretense, and other more primitive forms of imitation might be precursors of these. Accordingly, we would not expect that imitation would always be linked to mindreading or pretense, but we would expect the reverse, namely, that where there is evidence for these latter capacities, an imitative ability would be well developed. This is what we appear to see in the great apes.

CONCLUSION

Much recent primate research tends to reinforce, rather than refute, the general lines of the arguments above for the basis of a "great ape mind" in which mindreading, pretense and at least some forms of imitation rest upon a capacity for secondary representation. However, the present tentative nature of much of the empirical base must be acknowledged. The studies that claim to demonstrate ape imitation, mindreading and pretense are few; nearly all are derived from subjects not living under natural conditions; many of the conclusions based on observations have not yet been backed up by experiments (although it would appear that in principle they could be); and interpretations of many of the experiments that have been done are disputed (Heyes, 1993a, 1994; Tomasello & Call, 1994). A principal value of proposing the existence of these cognitive patterns is thus at this stage not so much to convince, as to guide further research in a less piecemeal fashion than heretofore. At all stages of research, it is surely wise to try to see the wood and not just the trees – to shift back and forth between the narrow and the wider view, even if both are more ill-defined than we would wish. This is one reason for finishing with a sideways look at parrots and rats. Real insight into what is special about ape minds must surely hinge on a broad comparative approach, which looks both towards humans, and to other quite different species, as well as to the other primates.

REFERENCES

Anisfeld, M. (1991). Neonatal imitation. *Developmental Research*, 11, 60–97.

Astington, J.W, Harris, P.L. & Olson, D.R. (eds.) (1988). *Developing Theories of Mind*. Cambridge: Cambridge University Press.

Baron-Cohen, S. (1987). Autism and symbolic play. *British Journal of Developmental Psychology*, 5, 139–48.

Baron-Cohen, S., Leslie, A. M. & Frith, U. (1985). Does the autistic child have a "theory of mind'? *Cognition*, 21, 37–46.

Baron-Cohen, S., Tager-Flusberg, H. & Cohen, J.D. (eds.) (1993). *Understanding Other Minds: Perspectives from Autism*. Oxford: Oxford University Press.

Boesch, C. (1991). Teaching among wild chimpanzees. *Animal Behaviour*, 41, 530–2.

Boinski, S. & Fragaszy, D.M. (1989). The ontogeny of foraging in squirrel monkeys (*Saimiri oerstedii*). *Animal Behaviour*, 37, 415–28.

Bruner, J.S. (1972). Nature and use of immaturity. *American Psychologist*, 27, 687–708.

Byrne, R.W. (1994). The evolution of intelligence. In *Behaviour and Evolution*, ed. P.J.B. Slater & T.R. Halliday, pp. 223–65. Cambridge: Cambridge University Press.

Byrne, R.W. & Byrne, J.M.E. (1993). Complex leaf-gathering skills of mountain gorillas (*Gorilla g. beringei*): Variability and standardization. *American Journal of Primatology*, 31, 241–61.

Byrne, R.W. & Whiten, A. (1985). Tactical deception of familiar individuals in baboons. *Animal Behaviour*, 33, 669–73.

Byrne, R.W. & Whiten, A. (1988). Towards the next generation in data quality: A new survey of primate tactical deception. *Behavioral and Brain Sciences*, 11, 267–73.

Byrne, R.W. & Whiten, A. (1990). Tactical deception in primates: The 1990 database. *Primate Report*, 27, 1–101.

Byrne, R.W. & Whiten, A. (1991). Computation and mindreading in primate tactical deception. In *Natural Theories of Mind: Evolution, Development and Simulation of Everyday Mindreading*, ed. A. Whiten, pp. 127–41. Oxford: Basil Blackwell Ltd.

Byrne, R.W. & Whiten, A. (1992). Cognitive evolution in primates: Evidence from tactical deception. *Man*, 27, 609–27.

Cheney, D.L. & Seyfarth, R.M. (1990a). Attending to behaviour versus attending to knowledge: Examining monkey's attribution of mental states. *Animal behaviour*, 40, 742–53.

Cheney, D.L. & Seyfarth, R.M. (1990b). *How Monkeys See the World*. Chicago: Chicago University Press.

Custance, D.M., Whiten, A. & Bard, K.A. (1994). The development of gestural imitation and self-recognition in chimpanzees (*Pan troglodytes*) and children. In *Current Primatology: Selected Proceedings of the XIVth Congress of the International Primatological Society, Strasbourg*, vol. 2: *Social Development, Learning and Behaviour*, ed. J.J. Roeder, B. Thierry, J.R. Anderson & N. Herrenschmidt, pp. 381–7. Strasbourg, France: Université Louis Pasteur.

Custance, D.M., Whiten, A. & Bard, K.A. (1995). Can young chimpanzees imitate arbitrary actions? Hayes & Hayes (1952) revisited. *Behaviour*, **132**, 839–58.
de Waal, F.B.M. (1982). *Chimpanzee Politics*. London: Jonathan Cape.
de Waal, F.B.M. & van Roosmalen, A. (1979). Reconciliation and consolation among chimpanzees. *Behavioral Ecology and Sociobiology*, **5**, 55–66.
Dawkins, R. & Krebs, J.R. (1978). Animal signals: Information or manipulation? In *Behavioural Ecology: an Evolutionary Approach*, ed. J.R. Krebs & N.B. Davies, pp. 282–309. Oxford: Blackwell Scientific Ltd.
DeMeyer, M.K., Alpern, G.D., Barton, S., DeMeyer, W.E., Churchill, D.W., Hingtgen, J.N., Bryson, C.Q., Pontius, W. & Kimbirlin, C. (1972). Imitation in autistic, early schizophrenic, and non-psychotic subnormal children. *Journal of Autism and Childhood Schizophrenia*, **2**, 264–87.
Dennett, D.C. (1983). Intentional systems in cognitive ethology: The "Panglossian paradigm" defended. *Behavioral and Brain Sciences*, **6**, 343–90.
Fein, G.G. (1981). Pretend play: An integrative review. *Cognitive Development*, **52**, 1095–118.
Fragaszy, D.M. & Visalberghi, E. (1989). Social influences on the acquisition and use of tools in tufted capuchin monkeys (*Cebus apella*). *Journal of Comparative Psychology*, **103**, 159–70.
Galef, B.G., Jr (1988). Imitation in animals: History, definitions, and interpretation of data from the psychological laboratory. In *Social Learning: Psychological and Biological Perspectives*, ed. T. Zentall & B.G. Galef Jr, pp. 3–28. Hillsdale, NJ: Lawrence Erlbaum.
Galef, B.G., Jr (1990). Tradition in animals: Field observations and laboratory analyses. In *Interpretations and Explanations in the Study of Behaviour: Comparative Perspectives*, ed. M. Bekoff & D. Jamieson, pp. 74–95. Boulder, CO: Westview Press.
Gallup, G.G., Jr (1970). Chimpanzees: Self-recognition. *Science*, **167**, 417–21.
Gallup, G.G., Jr (1979). Self-awareness in primates. *American Scientist*, **67**, 417–21.
Gallup, G.G., Jr (1982). Self-awareness and the emergence of mind in primates. *American Journal of Primatology*, **2**, 237–48.
Gergely, G. (1994). From self-recognition to theory of mind. In *Self-awareness in Animals and Humans: Developmental Perspectives*, ed. S.T. Parker, R.W. Mitchell & M.L. Boccia, pp. 51–60. New York: Cambridge University Press.
Gómez, J.C. (1991). Visual behaviour as a window for reading the mind of others in primates. In *Natural Theories of Mind: Evolution, Development and Simulation of Everyday Mindreading*, ed. A. Whiten, pp. 195–207. Oxford: Basil Blackwell Ltd.
Gómez, J.C. & Teixidor, P. (1992). Theory of mind in an orangutan: a non-verbal test of false belief appreciation. Paper presented at the XIVth Congress of the International Primatological Society, Strasbourg, France, 16–21 August.
Goodall, J. (1968). The behaviour of free-living chimpanzees in the Gombe Stream Reserve. *Animal Behaviour Monographs*, **1**, 165–311.
Gopnik, A. & Meltzoff, A.N. (1994). Minds, bodies and persons: Young children's understanding of the self and others as reflected in imitation and theory of mind research. In *Self-awareness in Animals and Humans: Developmental Perspectives*, ed. S.T. Parker, R.W. Mitchell & M.L. Boccia, pp. 166–86. New York: Cambridge University Press.
Hayes, C. (1951). *The Ape in Our House*. New York: Harper & Row.
Hayes, K.J. & Hayes, C. (1952). Imitation in a home-reared chimpanzee. *Journal of Comparative Psychology*, **45**, 450–9.
Heyes, C.M. (1993a). Anecdotes, training, trapping and triangulating: Do animals attribute mental states? *Animal Behaviour*, **46**, 177–88.
Heyes, C.M. (1993b). Imitation, culture and cognition. *Animal Behaviour*, **46**, 999–1010.

Heyes, C.M. (1993c). Imitation without perspective-taking. *Behavioral and Brain Sciences*, **16**, 524–5.
Heyes, C.M. (1994). Reflections on self-recognition in primates. *Animal Behaviour*, **47**, 909–19.
Heyes, C.M., Dawson, G.R. & Nokes, T. (1992). Imitation in rats: Initial responding and transfer evidence. *Quarterly Journal of Experimental Psychology*, **45B**, 229–40.
Hobson, R.P. (1990). On acquiring knowledge about people and the capacity to pretend: Response to Leslie (1987). *Psychological Review*, **97**, 114–21.
Humphrey, N.K. (1976). The social function of intellect. In *Growing Points in Ethology*, ed. P.P.G. Bateson & R.A. Hinde, pp. 303–17. Cambridge: Cambridge University Press.
Humphrey, N.K. (1980). Nature's psychologists. In *Consciousness and the Physical World*, ed. B. Josephson & V. Ramachandran, pp. 57–80. London: Pergamon Press.
Jolly, A. (1966). Lemur social behaviour and primate intelligence. *Science*, **153**, 501–6.
Krebs, J.R. & Dawkins, R. (1984). Animal signals: Mind reading and manipulation. In *Behavioral Ecology: An Evolutionary Approach*, ed. J.R. Krebs & N.B. Davies, 2nd edn, pp. 340–401. Oxford: Blackwell Scientific Ltd.
Leslie, A.M. (1987). Pretense and representation in infancy: The origins of "theory of mind". *Psychological Review*, **94**, 84–106.
Leslie, A.M. (1991). The theory of mind impairment in autism: Evidence for a modular mechanism of development? In *Natural Theories of Mind: Evolution, Development and Simulation of Everyday Mindreading*, ed. A. Whiten, pp. 63–78. Oxford: Basil Blackwell Ltd.
Leslie, A.M. & Frith, U. (1990). Prospects for a cognitive psychology of autism: Hobson's choice. *Psychological Review*, **97**, 122–31.
Leslie, A.M. & Roth, D. (1993). What autism teaches us about representation. In *Understanding Other Minds: Perspectives from Autism*, ed. S. Baron-Cohen, H. Tager-Flusberg, & J.D. Cohen, pp. 83–111. Oxford: Oxford University Press.
Lewis, V. & Boucher, J. (1988). Spontaneous, instructed and elicited play in relatively able autistic children. *British Journal of Developmental Psychology*, **6**, 325–39.
Lloyd-Morgan, C. (1930). *The Animal Mind*. London: Edward Arnold.
Meltzoff, A.N. (1988). Infant imitation and memory: Nine-month-olds in immediate and deferred tests. *Child Development*, **59**, 217–25.
Meltzoff, A.N. & Gopnik, A. (1993). The role of imitation in understanding persons and developing a theory of mind. In *Understanding Other Minds: Perspectives from Autism*, ed. S. Baron-Cohen, H. Tager-Flusberg, & J.D. Cohen, pp. 335–66. Oxford: Oxford University Press.
Meltzoff, A.N. & Moore, M.K. (1977). Imitation of facial and manual gestures by human neonates. *Science*, **198**, 75–8.
Menzel, E.W., Savage-Rumbaugh, E.S. & Lawson, J. (1985). Chimpanzee (*Pan troglodytes*) spatial problem-solving with the use of mirrors and televised equivalents of mirrors. *Journal of Comparative Psychology*, **99**, 211–17.
Mitchell, R.W. (1986). A framework for discussing deception. In *Deception: Perspectives on Human and Nonhuman Deceit*, ed. R.W. Mitchell & N.S. Thompson. Albany, NY: State University of New York Press.
Mitchell, R.W. (1993). Mental models of mirror-self-recognition: Two theories. *New Ideas in Psychology*, **11**, 295–325.
Mitchell, R.W. (1994). Multiplicities of self. In *Self-awareness in Animals and Humans: Developmental Perspectives*, ed. S.T. Parker, R.W. Mitchell & M.L. Boccia, pp. 81–107. New York: Cambridge University Press.
Moore, B.R. (1992). Avian movement imitation and a new form of mimicry: Tracing the evolution of a complex form of learning. *Behaviour*, **122**, 231–63.

Parker, S.T. (1991). A developmental approach to the origins of self-recognition in great apes and human infants. *Human Evolution*, 6, 435–49.

Parker, S.T., Mitchell, R.W. & Boccia, M.L. (eds) (1994). *Self-awareness in Animals and Humans: Developmental Perspectives*. New York: Cambridge University Press.

Patterson, F.G.P. & Cohn, R.H. (1994). Self-recognition and self-awareness in lowland gorillas. In *Self-awareness in Animals and Humans: Developmental Perspectives*, ed. S.T. Parker, R.W. Mitchell & M.L. Boccia, pp. 273–90. New York: Cambridge University Press.

Perner, J. (1991). *Understanding the Representational Mind*. Cambridge, MA: Bradford/MIT Press.

Perner, J. (1993). The theory of mind deficit in autism: Rethinking the metarepresentation theory. In *Understanding Other Minds: Perspectives from Autism*, ed. S. Baron-Cohen, H. Tager-Flusberg & J.D. Cohen, pp. 112–37. Oxford: Oxford University Press.

Piaget, J. (1962). *Play, Dreams and Imitation in Childhood*. London: Routledge & Kegan Paul Ltd. (First published in French in 1947).

Povinelli, D.J. & de Blois, S. (1992). Young children's (*Homo sapiens*) understanding of knowledge formation in themselves and others. *Journal of Comparative Psychology*, 106, 228–38.

Povinelli, D.J., Nelson, K.E. & Boysen, S.T. (1990). Inferences about guessing and knowing by chimpanzees (*Pan troglodytes*). *Journal of Comparative Psychology*, 104, 203–10.

Povinelli, D.J., Nelson, K.E. & Boysen, S.T. (1992a). Comprehension of role reversal in chimpanzees: Evidence of empathy? *Animal Behaviour*, 43, 633–40.

Povinelli, D.J., Parks, K.A. & Novak, M.A. (1991). Do rhesus monkeys (*Macaca mulatta*) attribute knowledge and ignorance to others? *Journal of Comparative Psychology*, 105, 318–25.

Povinelli, D.J., Parks, K.A. & Novak, M.A. (1992b). Role reversal by rhesus monkeys, but no evidence of empathy. *Animal Behaviour*, 43, 269–81.

Premack, D. (1988). "Does the chimpanzee have a theory of mind?" revisited. In *Machiavellian Intelligence: Social Expertise and the Evolution of Intellect in Monkeys, Apes and Humans*, ed. R.W. Byrne & A. Whiten, pp. 160–79. Oxford: Clarendon Press.

Premack, D. & Woodruff, G. (1978). Does the chimpanzee have a theory of mind? *Behavioral and Brain Sciences*, 1, 515–26.

Rogers, S.J. & Pennington, B.F. (1991). A theoretical approach to the deficits in infantile autism. *Development and Psychopathology*, 3, 137–62.

Russon, A.E. & Galdikas, B.M.F. (1993). Imitation in free-ranging rehabilitant orangutans (*Pongo pygmaeus*). *Journal of Comparative Psychology*, 107, 147–61.

Russon, A.E. & Galdikas, B.M.F. (1995). Constraints on great ape imitation: Model and action selectivity in rehabilitant orangutan (*Pongo pygmaeus*) imitation. *Journal of Comparative Psychology* 109, 5–17.

Savage-Rumbaugh, E.S. (1986). *Ape Language: From Conditioned Response to Symbol*. New York: Columbia University Press.

Savage-Rumbaugh, E.S. & McDonald, K. (1988). Deception and social manipulation in symbol-using apes. In *Machiavellian Intelligence: Social Expertise and the Evolution of Intellect in Monkeys, Apes and Humans*, ed. R.W. Byrne & A. Whiten, pp. 224–37. Oxford: Clarendon Press.

Savage-Rumbaugh, E.S., Rumbaugh, D.M. & Boysen, S. (1978). Symbolic communication between two chimpanzees (*Pan troglodytes*). *Science*, 201, 641–4.

Sigman, M. & Ungerer, J. (1984). Cognitive and language skills in autistic, mentally retarded and normal children. *Developmental Psychology*, 20, 293–302.

Tanner, J. (1985). Koko and Michael, Gorilla gourmets. *Gorilla*, **Dec. 1985**, 3.

Thorndike, E.L. (1911) *Animal Intelligence*. New York: Macmillan.

Tomasello, M. & Call, J. (1994). Social cognition of monkeys and apes. *Yearbook of Physical Anthropology*, **37**, 273–305.
Tomasello, M., Davis-Dasilva, M., Camak, L. & Bard, K.A. (1987). Observational learning of tool use by young chimpanzees. *Journal of Human Evolution*, **2**, 175–83.
Tomasello, M., Kruger, A.C. & Ratner, H.H. (1993a). Cultural Learning. *Behavioral and Brain Sciences*, **16**, 495–552.
Tomasello, M., Savage-Rumbaugh, E.S. & Kruger, A. (1993b). Imitative learning of actions on objects by children, chimpanzees, and enculturated chimpanzees. *Child Development*, **64**, 1688–705.
Visalberghi, E. & Fragaszy, D.M. (1990). Do monkeys ape? In *"Language" and Intelligence in Monkeys and Apes: Comparative Developmental Perspectives*, ed. S.T. Parker & K.R. Gibson, pp. 247–73. New York: Cambridge University Press.
Whiten, A. (1988). From literal to non-literal social knowledge in human ontogeny and primate phylogeny. Paper presented at the Primate Society of Great Britain meeting on Social Knowledge in Primates, London, Britain, December.
Whiten, A. (1989). Transmission mechanisms in primate cultural evolution. *Trends in Ecology and Evolution*, **4**, 61–2.
Whiten, A. (ed.), (1991a). *Natural Theories of Mind: Evolution, Development and Simulation of Everyday Mindreading*. Oxford: Basil Blackwell Ltd.
Whiten, A. (1991b). The emergence of mindreading: Steps towards an interdisciplinary enterprise. In *Natural Theories of Mind: Evolution, Development and Imulation of Everyday Mindreading*, ed. A. Whiten, pp. 319–31. Oxford: Basil Blackwell Ltd.
Whiten, A. (1992). Mindreading, pretense and imitation in monkeys and apes. *Behavioral and Brain Sciences*, **15**, 170–1.
Whiten, A. (1993). Evolving a theory of mind: The nature of non-verbal mentalism in other primates. In *Understanding Other Minds: Perspectives from Autism*, ed. S. Baron-Cohen, H. Tager-Flusberg & D.J. Cohen, pp. 367–96. Oxford: Oxford University Press.
Whiten, A. (1994). Grades of mindreading. In *Origins of an Understanding of Mind*, ed. C. Lewis & P. Mitchell, pp. 47–70. Hillsdale, NJ: Lawrence Erlbaum.
Whiten, A. (1996). When does smart behaviour-reading become mindreading? In *Theories of Theories of Mind*, ed. P.Carruthers & P.K. Smith. Cambridge: Cambridge University Press. (in press.)
Whiten, A. & Byrne, R.W. (1986). The St Andrews catalogue of tactical deception in primates. *St Andrews Psychological Reports*, **10**, 1–47.
Whiten, A. & Byrne, R.W. (1988a). Tactical deception in primates. *Behavioral and Brain Sciences*, **11**, 233–73.
Whiten, A. & Byrne, R. W. (1988b). Taking (Machiavellian) intelligence apart. In *Machiavellian Intelligence: Social Expertise and the Evolution of Intellect in Monkeys, Apes and Humans*, ed. R.W. Byrne & A. Whiten, pp. 50–65. Oxford: Clarendon Press.
Whiten, A. & Byrne, R.W. (1991). The emergence of metarepresentation in human ontogeny and primate phylogeny. In *Natural Theories of Mind: Evolution, Development and Simulation of Everyday Mindreading*, ed. A. Whiten, pp. 267–81. Oxford: Basil Blackwell Ltd.
Whiten, A. & Custance, D.M. (1996). Studies of imitation in chimpanzees and children. In *Social Learning in Animals: The Roots of Culture*, ed. B.G. Galef Jr & C.M. Heyes. London: Academic Press. (in press.)
Whiten, A., Custance, D.M., Gómez, J.C., Teixidor, P. & Bard, K.A. (1996). Imitative learning of artificial fruit processing in children (*Homo sapiens*) and chimpanzees (*Pan troglodytes*). *Journal of Comparative Psychology*. (in press.)

Whiten, A. & Ham, R. (1992). On the nature and evolution of imitation in the animal kingdom: Reappraisal of a century of research. In *Advances in the Study of Behaviour*, vol. 21, ed. P.J.B. Slater, J. S. Rosenblatt, C. Beer & M. Milinski, pp. 239–83. New York: Academic Press.

Whiten, A. & Perner, J. (1991). Fundamental issues in the multidisciplinary study of mindreading. In *Natural Theories of Mind: Evolution, Development and Simulation of Everyday Mindreading*, ed. A. Whiten, pp. 1–17. Oxford: Basil Blackwell Ltd.

Wimmer, H. & Perner, J. (1983). Beliefs about beliefs: Representation and constraining function of wrong beliefs in young children's understanding of deception. *Cognition*, 13, 103–28.

Wing, L., Gould, J., Yeats, S.R. & Brierly, L.M. (1977). Symbolic play in severely mentally retarded and autistic children. *Journal of Child Psychology and Psychiatry*, 18, 167–78.

Woodruff, G. & Premack, D. (1979). Intentional communication in the chimpanzee; the development of deception. *Cognition*, 7, 333–62.

15

Self-awareness and self-knowledge in humans, apes, and monkeys

DANIEL HART AND MARY PAT KARMEL

INTRODUCTION

Whether nonhuman species have a "self" has fascinated generations of psychologists and philosophers. There are similar questions about the presence of other human qualities in nonhuman species – language, tool use, and so on – but the issue of selfhood is especially compelling. One reason for this is that having a "self" is often viewed as an essential quality for being a genuine person. A human without language capabilities, for instance, can easily be imagined to be a person, still a real human. Such is the case with severe aphasia; one might say that the individual's language capabilities have been greatly diminished without concluding that the individual is less of a real human. Stripped of a self, however, an individual is no longer considered to be a person. In the USA for example, courts rely on testimony concerning the presence or absence of "self-consciousness" or "self-awareness" in deciding whether life-support can be withdrawn from individuals who have suffered severe brain damage (e.g. those who are characterized as having Persistent Vegetative State syndrome. See, for instance, In Re Jobes, 529 A. 2d 434 [NJ 1987]).

Once having a "self" becomes widely accepted as an important criterion for determining which objects in the world receive the special treatment accorded to persons, it follows that determining which objects have this special quality becomes a project of social significance. Traditionally, most writers from industrialized Western cultures have argued that only *Homo sapiens* is capable of self-awareness and developing a self-concept, and therefore only humans satisfy the criterion for special treatment as persons. Recently, however, such claims have been disputed by many psychologists, who have argued that one can indeed find evidence of selfhood in nonhuman species, specifically in chimpanzees, gorillas, and orangutans (e.g. Gallup, 1982). Arguments for a sense of self in nonhuman species have been dismissed by at least some philosophers (e.g. Taylor, 1989) who claim that the evidence that has been presented to date simply fails to prove that nonhuman species have anything similar to the sense of self (at least in its important connotations) found in humans.

Our goal in this chapter is to review the evidence for the sense of self in great apes and monkeys, with attention paid both to the similarities and differences that might be detected across species. We begin by exploring briefly what it means to have a sense of self; this review draws heavily upon our previous research on the ways in which humans think about themselves (e.g. Damon & Hart, 1988, 1992; Hart & Fegley, 1994a, b, 1995). To outline this discussion, our argument will be that there are multiple components to the sense of self, each of which is partially independent of the others (for discussions of multiple component approaches to the sense of self, see Hart & Fegley, 1994a; Mitchell, 1994; Neisser, 1988). Consequently, it becomes difficult to determine the extent to which a species may be characterized as having a "self," because there is no tightly bound psychological entity or quality that contains all the characteristics of selfhood (Parker et al., 1994). The distinctions among components of the sense of self frame the discussion of indices of the sense of self (how can one tell that another organism has a sense of self?) as well as the consideration of research on different species. We conclude with some brief comments on the functions of the sense of self and the processes that lead to its emergence.

THE NATURE OF THE SENSE OF SELF

What experiences or phenomena constitute the sense of self or reveal its existence? An answer to this question is a necessary prelude to describing the commonalities and differences in the sense of self across humans, great apes, and monkeys. The delineation of the facets of the sense of self that we present here draws from a wide body of psychological, anthropological, and philosophical literature concerned with self-awareness and the self-concept in humans (see Hart & Fegley, 1994a, b, 1995). For the purposes of examining the sense of self across species, a set of criteria developed from the study of only humans poses the obvious interpretative problems: does the failure of a species to meet the criteria for a facet of the sense of self mean that the facet does not exist in the species, or merely that the criteria for its presence are not appropriate for nonhuman species? Is it also possible that important components of the sense of self are absent in humans and therefore omitted from the model described here, but present in nonhuman species? These and other related issues that challenge any anthropocentric approach to comparative psychology can never be fully resolved.

There is much however, to recommend the anthropocentric approach (but, for an opposing view, see Kennedy, 1992). The fascination with the sense of self in humans has persisted for centuries (Baumeister, 1987), which has resulted in a well-developed corpus of knowledge that can be drawn upon to characterize a sense of self with multiple facets. This richness and diversity of experiences constituting the sense of self could not be abstracted from research on nonhuman primates for the simple reason that investigations into the minds of great apes and monkeys have begun only relatively recently. We believe that the delineation of the sense of self presented in this chapter, based as it is upon consideration of humans, can nevertheless serve both as a framework within which currently available comparative-psychological findings can be organized, as well as a guide to future research on the sense of self in nonhuman species.

Our view is that the sense of self is composed of two basic types of experience, *self-awareness* and the *understanding of self* each of which has constituents (for more extensive descriptions of these components in humans, see Hart & Fegley, 1994a, 1995). These types of experience, and their corresponding constituents, are described below.

Self-awareness

Self-awareness is a fundamental and necessary quality of the experience of self. There are two qualities of self-awareness that have received extensive attention from psychologists, cultural anthropologists, and philosophers: these two qualities might be called *objective self-awareness* and *subjective self-awareness*. Objective self-awareness refers to an organism's ability to distinguish itself from the rest of the world and to focus its attention on the class of stimuli corresponding to itself. Objective self-awareness is likely to be found widely across the animal kingdom. As Dennett (1991) has pointed out, virtually all organisms act as if they are aware of their own boundaries; for instance, a carnivore does not eat itself. From an evolutionary perspective, it seems likely that an inability to distinguish between the class of stimuli corresponding to oneself and other types of stimuli would be an enormous handicap, as the self-consumption example suggests.

The value of distinguishing between self and other may underlie a perceptual attunement to the powerful perceptual clues or affordances in the environment that allow for the ready distinction between oneself and the world. For instance, the parts of one's body are always approximately the same distance away from one's point of view: one's hands are never further away from one's eyes than 1 m, one's nose always appears in the center of one's visual field, and so on (for valuable discussions, see Butterworth, 1990, and Neisser, 1989). Although objectively distinguishing between the self and the world is a capability likely to be distributed widely in the animal kingdom there may be wide variations in (a) the ability to focus sustained attention on the class of stimuli corresponding to oneself, and (b) the tendency or disposition to focus attention on this class of stimuli.

Subjective self-awareness refers to a facet of the sense of self most usually considered by philosophers (e.g. Nozick, 1980). In the first place, to be self-aware usually means more than simply focusing attention on a particular class of stimuli (objective self-awareness); it also involves *identification* with that class of stimuli. One usually cares about the object of self-awareness (oneself) in an emotionally involving way that is different from the way one cares about other objects of awareness. In fact, some have argued that genuine emotions occur only when subjective self-awareness is present (Rosaldo, 1984). For instance, they argue that anger may require the recognition that the achievement of one's goals, which one cares about and identifies with (subjective self-awareness), is impeded by the actions of others. This same argument can be extended to explain empathy by suggesting that the emotions of another become one's own only if one includes that other within the boundaries of oneself.

Understanding of self

The experience of self is not constituted solely of objective and subjective self-awareness; humans also infuse the self with cognitive components. These cognitive contents are of three basic types, ranging from very specific to global. At the specific level, humans possess *personal memories*, memories which for the individual are essential constituents of the self (for a thoughtful consideration of autobiographical and personal memories, see Ross, 1991 and for a discussion of memories in the construction of the sense of self, see Neisser, 1989). The birth of one's child, graduation from high school, the death of a friend, and many other events that occur over the course of a life are embedded in memory and firmly linked to the sense of self.

At an intermediate level of abstraction, humans form *representations* of themselves. These include ascriptions about the self's typical physical appearance, intelligence, social relationships, and so on.

Finally, at the most abstract level humans hold *theories* about themselves: they assume stances about which parts of themselves are central, and how the different facets of themselves are interrelated. For instance, many human adolescents explain the importance of the various characteristics they attribute to themselves in terms of a theory of self, according to which the central values are social acceptance and integration into relationships; moreover, from these theories, adolescents are able to generate lengthy stories of how they came to acquire their characteristics through social interaction (Damon & Hart, 1988, 1992). These theories reflect the constructive efforts of the individual to form coherent notions of the self, but also draw upon the models of selfhood available in the particular culture (for a description of the ways in which theories of self are influenced by culture, see Hart & Fegley, 1995).

INDICES OF THE PRESENCE OF A COMPONENT OF THE SENSE OF SELF AND EVIDENCE FOR THEIR PRESENCE IN HUMANS, GREAT APES, AND MONKEYS

Drawing distinctions among facets of the sense of self is a much easier task than determining when these phenomena are present in different species. Partly this is because what constitutes the self is presumed to vary greatly from individual to individual, as each center of conscious activity determines for itself what is, and what is not, included as part of the self. As Taylor (1989) pointed out, the authority persons are granted in determining the boundaries and content of the self makes the self unique among psychological constructs that might be described: observers must either rely on the person to communicate the content and boundaries of his or her own self or, as in the case with most psychological phenomena, must make tentative inferences based upon behavioral cues with the foreknowledge that these inferences might be wrong.

The reliance upon communication for learning about another organism's sense of self is perhaps problematic. There are three primary approaches regarding the relationship between communication and the self. It might be argued that relying upon communication is unfortunate because the restricted communicative abilities of some species may prevent researchers from learning of that species' sense of self. From this perspective, communication may reveal the sense of self, but it is not necessary for its genesis or development.

Alternatively, it is possible that the elaboration of at least some of the components to the sense of self would not be possible without communication (this idea is well-developed by Mead (1934), and is fundamental to symbolic interactionism). From a third perspective, the sense of self is a motivational prerequisite for the ontogenesis of language: Terrace & Bever (1980, p. 180) argued that "a necessary condition of human language may prove to be the ability to symbolize *oneself*." An inability to use human language according to this last perspective strongly suggests the lack of an elaborated sense of self.

Whether the absence of language suggests the lack of a sense of self (as suggested by Terrace & Bever) or means only that researchers should expect that nonverbal measures may very well reveal sophisticated self-concepts in species with limited communicative abilities cannot as yet be determined. For this chapter, we have merely pointed to the assumed centrality of language for determining the presence and nature of what is usually meant by the "sense of self." The review of evidence that follows begins with these indices of the components of the sense of self that are evident in communication through language, and then proceeds to nonlinguistic behavioral and emotional markers of the sense of self.

Linguistic markers

Table 15.1 presents some linguistic markers that seem especially promising for indicating the presence of the sense of self, and how these markers might be mapped onto specific facets of the sense of self. Three types of language use in particular seem useful. The first, *narrative, theory-construction language use* refers to the accounts and theories persons form about themselves. In humans, language use that spawns theories of self arises by the third year of life, and is particularly evident in the conversations between child and mother (Nelson, 1993). This type of language use both reveals and forms the sense of self. In particular, narrative language use would seem intimately tied to personal memories and theories of self. It is quite common in the interchanges between mother and child to make reference to events of significance that happened in the past. In recalling these episodes, the child reveals his or her personal memories. There is almost no evidence to indicate that this sort of language use occurs in language-trained great apes. However, Patterson (cited in Tuttle, 1986) reported one incident that suggests lack of evidence may not reflect the complete absence of ability. Patterson reported that Michael, one of the language-trained gorillas in her care, has described being captured by hunters in Africa. Another gorilla in Patterson's care, Koko, has related personal memories as well. When shown a picture of herself at her birthday party, Koko signed "me love happy Koko there" (Patterson & Linden, 1981, p. 86). The fact that Michael and Koko would recall these incidents suggests that these memories are of personal significance.

There is no evidence that monkeys are capable of recalling personal memories. It might be that the lack of evidence for personal memories reflects the difficulty in training

Table 15.1. *Linguistic markers of the sense of self*

Marker	Objective self-awareness	Subjective self-awareness	Personal memories	Representations	Theories	Distribution across species
Narrative or theory-construction language use	Some self-focus is required when discussing the self		Specific memories constituting the sense of self can be communicated		Cultural models of selfhood are learned through theory-construction language, and articulation of explicit theories of self may be most likely to occur in communicative contexts	Humans: probably evident in early narratives constructed between parent and children (Nelson, 1993), although the significance of these memories for the sense of self is unclear. Children are able to offer rudimentary theories of self by the age of 4 or 5 (Damon & Hart, 1988) and begin to assimilate cultural models through narrative exchanges with parents (Nelson, 1993) Chimpanzees: no evidence Orangutans: no evidence Gorillas: anecdotal evidence that particularly traumatic events may be recalled (Patterson, cited in Tuttle, 1986) Monkeys: no evidence
Linguistic self-referencing	Ability to use personal pronouns and tendency to do so are indicators of presence and extent of objective self-awareness	Use of reflexive self-referring pronouns suggests that subjective self-awareness is present, although boundaries of the self that is referred to are unclear				Humans: use of personal pronouns is high (Budwig, 1989) by 3 years of age Chimpanzees: capable of using personal pronouns (Gardner & Gardner, 1974; Temerlin, 1975) but less likely to do so than humans Orangutans: ability to use personal pronouns, but use of pronouns in spontaneous speech is rare (Miles, 1990)

Declarative, labeling speech	Ascribing a representation to the self requires a focus on the self	Subjective self-awareness usually extends to the representations that are ascribed to the self	Gorillas: capable of using personal pronouns "me," "mine," "myself" (Patterson & Linden, 1981) but may be less likely to do so than humans Monkeys: no evidence of ability
		Self-descriptions can be offered which are clear indications of representations of self (Terrace & Bever, 1980)	Humans: children as young as 2 or 3 can provide verbal descriptions of self (Damon & Hart, 1988, 1992) Chimpanzees: anecdotal; evident in descriptions of other chimpanzees as "black bugs" by chimpanzee raised by humans and attributions about others (Tuttle, 1986), but likely to be rare (Terrace, 1985) Orangutans: Anecdotal; but rare (Miles, 1990) Gorillas: anecdotal; one gorilla has described herself as "stubborn devil," "happy," and as "good" in the future (Patterson & Linden, 1981) Monkeys: no evidence

monkeys to use language. An alternative explanation is that monkeys are unable altogether to store and to recall memories of specific events (Ridley, 1992); in the terms of memory researchers (e.g. Squires, 1987), monkeys may lack episodic memories (memories of specific events in one's past), which can be consciously recalled with some knowledge of when they occurred. Because personal memories are a subclass of episodic memories, the absence of the latter necessarily means an inability to have the former.

Narrative language use also allows the child to learn of the models of selfhood that exist in the surrounding culture, and then to reformulate these theories so as to reflect their views of themselves. Certainly by the time children are 6 years old, they are able to articulate theories of themselves (see Damon & Hart, 1988); however, it is likely that an analysis of children's speech would reveal the rudiments of a theory of self at much younger ages. Although there is no evidence (to our knowledge) that the exchanges between humans and language-trained great apes contain fragments of a theory of self, it might be fruitful to inspect transactions for this evidence.

A second facet of language that reveals facets of the sense of self is *linguistic self-referencing*, reflected in using words such as "I" and "me" appropriately. Linguistic self-referencing has been studied extensively by philosophers (e.g. Zemach, 1972) and psychologists (e.g. Budwig, 1989). To use the word "I" appropriately, one must use it in such a way so as to refer to the self from "the inside" (Nozick, 1980): one must *know* that it is the self to which one is referring. The sentence "this sentence refers to itself," is self-referring, but presumably the sentence lacks the awareness that it is doing so. Moreover, because one cannot be wrong in the referent of the word "I" when it is used appropriately (Zemach, 1972), it must also be the case that the self to which one is referring is constructed in the process (Nozick, 1980): the boundaries of the self that is referred to vary from occasion to occasion.

Appropriate use of the word "I," then, reveals the presence of objective self-awareness and subjective self-awareness. Children begin using the word "I" in ways suggesting an appropriate understanding of the word in the second year of life. At about the same time that they begin using the word appropriately, other self-related phenomena begin to emerge: for instance, children become more possessive of objects (see Levine, 1983; Russon & Waite, 1991).

The evidence suggests that there are reasons to suspect that language-trained great apes are capable of reflexive self-referring. Chimpanzees, orangutans, and gorillas have all been reported to use "I" and "me" spontaneously and appropriately in transactions with humans (e.g. Gardner & Gardner, 1974; Miles, 1990, 1994; Patterson, 1978). It is, of course, difficult to *prove* unequivocally that chimpanzees, orangutans, and gorillas are referring to themselves in the same way that humans do when they use personal pronouns. It could be argued that they are using the words in a superficially appropriate fashion without appreciation of their genuine meaning. However, in the absence of experimental evidence demonstrating that the apes do not understand the meaning of personal pronouns despite using the words appropriately, it seems most reasonable to judge that their apparently correct use of "I" and "me" reveals the presence of self-awareness (this seems most reasonable, because one would make the same inference if one observed apparently correct self-referencing in a culture of which one had little knowledge).

Reflexive self-referring may not only reveal the capacity for self-awareness, but also the tendency or disposition to take the self as the object of awareness. In humans, the frequency with which a person makes use of the word "I" in conversation has been found to be related to conditions intended to heighten the focus of attention on the self (Davis & Brock, 1975). This suggests that one may use the frequency of self-referencing as one index of the salience of the self in an organism's awareness.

In human children, reflexive self-referring is quite common (Budwig, 1989). Even young children make frequent use of words such as "I," "me," and "mine" in spontaneous speech. From the published literature, it is difficult to determine the frequency with which language-trained apes self-refer, but the available evidence suggests that the rate is probably much lower than in humans. For instance, in the 2 weeks of spontaneous language production of the orangutan Chantek (he was 8 years of age at the time) recorded by Miles (1990), reflexive self-referring was very rare, occurring on only a few days. Although it appears likely that self-referring is more frequent in language-trained gorillas (Patterson, 1978) and chimpanzees (Gardner & Gardner, 1974) than in orangutans (only a few orangutans have been language-trained to date), the published accounts do not permit direct comparisons.

Finally, language permits organisms to explicitly characterize themselves, revealing their *representations*. Human children make use of speech in this way by 4 years of age, with some evidence that this occurs in much younger children as well (for reviews, see Damon & Hart, 1988, 1992). These representations include references to physical, active, social, and psychological characteristics of the self.

There is evidence to indicate that these sorts of attributions are found in the language of great apes. For instance, Washoe (a chimpanzee) described other chimpanzees as "black bugs," suggesting an ability to make attributions. Patterson & Linden (1981) reported that Koko described other gorillas she did not like as "dirty," "trouble," and herself as a "stubborn devil." Miles (1990) has reported similar ascriptions in the language of orangutans.

Cognitive-behavioral markers

Mirror-self-recognition

The most compelling nonverbal marker of sense of self is visual self-recognition in mirrors. Researchers for more than a century have presented human infants and nonhuman species with mirrors in order to observe their reactions. Preyer (1893) reported that conscious self-recognition in his human infant subject appeared at 14 months of age. Few thought the same ability could be found in the great apes. For instance, Schmidt (1878, cited in Yerkes & Yerkes, 1929) claimed that orangutans' behavior in front of mirrors suggested that they were not aware that the images they viewed in the mirror were their own; the same had been judged true of chimpanzees (Yerkes & Yerkes, 1929, p. 253) and gorillas (Yerkes & Yerkes, 1929, p. 437).

Gallup's (1970) ingenious modification to the mirror procedure demonstrated that earlier researchers had underestimated the abilities of the great apes. As in earlier studies, Gallup allowed chimpanzees to view themselves in mirrors for several days. Then he anesthetized the chimpanzees and marked their faces with dye. Upon recovery, the chimpanzees were again presented with mirrors. Gallup found that, upon seeing their images in the mirrors, the chimpanzees frequently touched the marks that were on their faces.

What does this mark-directed behavior mean? In a general sense, it is apparent that an organism capable of mark-directed behavior recognizes a correspondence between the mirror image and itself: it recognizes itself. Gallup, therefore, had demonstrated that chimpanzees indeed could recognize themselves, and subsequent research has demonstrated that mark-directed behavior occurs in orangutans (Suarez & Gallup, 1981) and gorillas (Parker, 1994; Patterson & Cohn, 1994), but not in monkeys (e.g. Anderson, 1994; Gallup et al., 1980). Table 15.2 presents the ways in which this phenomenon of self-recognition maps onto the distinctions in the sense of self developed earlier (for fuller descriptions, see Hart & Fegley, 1994a,b). Mark-directed behavior requires objective self-awareness; the organism must be able to focus attention on its mirror image, a set of stimuli corresponding to itself. It also seems likely that the organism must include physical appearance as a component of the sense of self (subjective self-awareness). The organism is drawn to inspect the mark because a part of what it considers to be itself has been changed unexpectedly. Finally, mark-directed behavior suggests that the organism has a representation of its typical appearance. When it sees its image with the unexpected mark on it, the organism is aware that the current image is inconsistent with its representation of its typical appearance and explores the specific sites where the inconsistencies are located. Neither personal memories nor theories of self are required for mark-directed behavior (Lewis & Brooks-Gunn, 1979). However, it is interesting to note that in the ontogenesis of humans, the emergence of mark-directed behavior at about 19 months of age (elicited in a slightly different paradigm – without anesthesia) is associated with the appearance of other abilities (Gopnik & Meltzoff, 1994) that may be related to the sense of self (e.g. self-referencing and possessive behavior). This suggests that there may be a mental model of self forming at about this age (see Cicchetti & Beegley, 1990; Hart & Fegley, 1994a). Similarly, those species capable of visual self-recognition may have rudimentary theories of mind, while these theories may be absent in species without this capability (Gallup, 1985; Povinelli, 1993).

Mark-directed behavior in the mirror paradigm developed by Gallup (1982) compellingly suggests self-recognition. However, it should be noted that there remain questions about the task. First, as Mitchell (1993) demonstrated in a careful review, the process through which an organism learns that its body corresponds to a mirror image is quite complex and partly unknown. Secondly, some critics point to the fact that mark-directed behavior is just that; concluding that a complex sense of self exists in an organism based upon the emission of a single behavior goes far beyond the boundaries of reasonable induction (Kennedy, 1992). These questions and cautions presumably will yield to further research (for discussions of the issues related to mark-directed behavior, see Lin et al., 1992; Parker, 1991; Parker et al., 1994).

Imitation

There is a long theoretical tradition in psychology that links imitation and the sense of self (e.g. Baldwin, 1902; Guillaume, 1971; Meltzoff, 1990; Meltzoff & Gopnik, 1993). Of particular interest are the sophisticated forms of imitation where it is clear that the organism focuses attention on itself in order to acquire new behavior that matches that of the model observed, and then makes use of the newly acquired behavior in appropriate contexts. This type of imitation is sometimes described as "true" or role-taking imitation (Russon & Galdikas, 1993), to differentiate it from less sophisticated behaviors in which the organism's behavior resembles that of a model as a result of less complex processes such as social facilitation or contagion (for excellent discussions, see Mitchell, 1987; Tomasello et al., 1993; Whiten & Ham, 1992).

Imitation reveals several facets of the sense of self. First, imitation as described above requires objective self-awareness: the organism must focus attention on its own actions, and sustain this attention for minutes at a time in order to shape its behavior to match that of the model. Consequently, frequent imitation suggests a well-developed capacity for objective self-awareness. There is abundant evidence that humans are capable of this sophisticated form of imitation (e.g. Mitchell, 1987), and that it is quite frequent from late infancy through adulthood (indeed, Baldwin (1902), believed imitation was one of the fundamental processes of learning and socialization in humans). Researchers have long been divided as to whether the nonhuman great apes are capable of imitation (e.g. from Yerkes & Yerkes, 1929, to Tomasello et al., 1993). In our judgement, the large number of investigators reporting instances of imitation for chimpanzees, gorillas, and orangutans suggests that these species probably are capable of genuine imitation (for reviews, see Whiten & Ham, 1992; Yerkes & Yerkes, 1929). At the same time, however, the various studies and reports of imitation in the great apes strongly suggest that the frequency of imitation is much lower in these species than in humans, and virtually absent in monkeys (Visalberghi & Fragaszy, 1990; see also Chapters 2 and 3).

Imitation also suggests that in at least some cases the organism has some knowledge that the useful actions displayed by the other are absent from its own behavioral repertoire. In turn, such knowledge suggests that the organism has representations of itself. Kagan (1981) has described one way in which imitation might reveal that representations of self are present. Kagan and his colleagues modeled simple and complex sequences of behaviors to human infants of different ages, and then indicated to the infants that they were to replicate the actions. Infants younger than 18 months of age attempted to imitate both the simple and complex sequences, but, as designed, failed to succeed with the complex ones. Older infants would imitate the simple sequences, but a large percentage of them failed to attempt to replicate the complex sequences. Kagan interprets the reluctance to imitate as an indication that the infants recognized that the complex modeled actions far exceeded their abilities to replicate them; therefore the infants saw little reason in attempting to do so. Indeed, many of these older infants became distressed at the researchers' expectations that they do what they could not. Failure to attempt to imitate in contexts in which many other behaviors are imitated, then, may be a sign that an organism is aware of its inability to replicate a modeled action.

Table 15.2. *Cognitive-behavioral markers of the sense of self*

Marker	Objective self-awareness	Subjective self-awareness	Personal memories	Representations	Theories	Distribution across species
Mirror-self-recognition	Demonstrates the ability to distinguish between visual stimuli corresponding to the self's body and other visual stimuli	Mirror-self-recognition may require subjective self-awareness extending to body image in order for there to be motivation to touch the mark (Hart & Fegley, 1995)	Not an index of personal memories	A stable representation of one's face is required for mirror-self-recognition	May reflect the emergence of a mental model of self (Hart & Fegley, 1995), although evidence is equivocal. Possession of a mental model may be a necessary precursor to a theory of self	Humans: present Chimpanzees: present (Gallup 1970; for exceptions, however, see Swartz & Evans, 1991) Orangutans: present (Gallup, 1980) Gorillas: present (Patterson & Cohn (1994); but for a discussion of the possible lack of generalizability of this finding, see Povinelli, 1993) Monkeys: no evidence of mirror-self-recognition (Gallup et al., 1980)
Role-taking imitation	Monitoring of self's actions in order to achieve similarity to the model demands ability to focus attention on the self (Mitchell, 1987)	Not demanded by imitation	Not demanded by imitation	Not required by imitation	Not required by imitation	Humans: present by 18–24 months of age Chimpanzees: may be present (Yerkes & Yerkes, 1929), but likely to be extremely rare Orangutans: present, but likely to be rare (Russon & Galdikas, 1993) Gorillas: present, but likely to be rare (Yerkes & Yerkes, 1927) Monkeys: not present, or exceptionally rare (Visalberghi & Fragaszy, 1990)

There are some reports that similar reactions are observed in other great apes. For instance, Yerkes described a gorilla in his care as demonstrating the same sorts of reaction as older infants:

> Congo [the gorilla] exhibits in far less degree than do the other great apes random investigative activity, copying or seemingly purposeless and almost automatic repetition of the acts of others, and intelligent imitation. At once less of an investigator and less imitative than the chimpanzee, she is also, it would appear from my varied observations, less willing or eager to work persistently in the face of discouraging conditions . . . Otherwise expressed, the gorilla, like not a few men, roughly measures or estimates the probability of success, and if the chances seem to be against her, gives up the task. What at the start appears to her impossible she will not even attempt.
> *(Yerkes & Yerkes, 1929, p. 498).*

Custance *et al.* (1994) have reported that chimpanzees may also refrain from attempting to imitate very difficult actions. While we are not aware of published reports of this phenomenon in observations of orangutans, there is little reason to imagine that it would not be found in this species. In addition, Russon & Galdikas (1995) found evidence that suggests that older orangutans choose to imitate routines that are outside the realm of their current repertoire but not too far beyond their abilities.

Emotional markers

Looking for evidence of the sense of self in the emotional responses of organisms should be a promising avenue of investigation. It is because of its presumed role in regulating emotions that clinicians find the sense of self to be important to address in psychotherapy (e.g. Beck *et al.*, 1979). Moreover, many theoretical accounts suggest that the discrete emotion experienced by individuals depends in part upon the sorts of attributions that are made about the self (e.g. Roseman *et al.*, 1990). Without these attributions, some argue that an emotion is not experienced at all; in the absence of interpretation about the self and the world, the organism is only characterized by global arousal and patterns of sensation (Lazarus, 1991).

Although for these reasons emotions should be intimately involved with the sense of self, problems arise when an emotion must (a) be differentiated from global arousal, and (b) its underlying attributions must be discerned through observations alone (i.e. without asking the individual what emotion is being experienced, and the sources for it). We believe that looking for evidence for a sense of self in most emotional displays will be fruitless; facial expressions associated with global arousal and many emotions may be innate and exhibited by many species without mediation of a sense of self. However, there are two types of emotional response, self-conscious emotions and empathy in particular, which do seem to reveal the sense of self relatively clearly.

Self-conscious emotions

Self-conscious emotions have been examined most thoroughly by Michael Lewis (see Lewis, 1992). The self-conscious emotions – pride, hubris, guilt, shame, and embarrassment

– require that the experiencing organism be able to take the self as an object of reflection. For instance, pride, in Lewis' scheme, arises from a comparison of the self's actual accomplishments with the self's goals; if one judges that the match is good, then pride results. Hubris is similar, except that the match results in a judgement that all of the self is good, not merely those facets of the self related to the accomplishments. Guilt and shame are products of the same processes. Guilt results when the individual judges that one facet of the self has failed to meet the standards the individual has set for the self; shame is experienced when one judges that every facet of the self has failed to live up to expectations. Finally, embarrassment results from the sense that the self is in some sense revealed or exposed to others; no judgement of the self's worth is made. Self-conscious emotions therefore draw upon many facets of the sense of self, as outlined in Table 15.3. Clearly, one must be able to focus attention on the self (objective self-awareness) and subjective self-awareness must extend to incorporate both the individual's actions and ideals. Standards or an ideal self are also required, and these are representational elements of the sense of self. Finally, attributions that the entire self is good (hubris) or bad (shame) would seemingly require some theory through which the various elements of self are seen to be integrated into a whole.

Based upon his research, Lewis is quite confident that self-conscious emotions are evident in human children by the end of the second year of life. For instance, in one series of studies, Lewis (Lewis *et al.*, 1989) tracked the development of embarrassment. Embarrassment in this research was suggested whenever the infant smiled and then showed gaze aversion accompanied by nervous touching of clothing and parts of the body. Their work suggested that embarrassment was rare in infants before 18 months of age.

Russon & Galdikas (1993) reported an incident that suggested that orangutans may experience pride, another of the self-conscious emotions. One orangutan, Supinah, was observed attempting to hang a hammock from two trees, apparently in imitation of humans. The task is fairly difficult, involving finding trees at the appropriate distance from each other, wrapping the ropes supporting the hammock at the right height around the trees, and securing the rope so that the hammock remains in place. After accomplishing this task quite possibly for the first time, Supinah "sat up and then threw herself back in the hammock" with apparent joy, and "hug[ged] herself with both arms," which might be interpreted as an expression of self-satisfaction and pride. A similar incident has been described by Hayes (1951, p. 188), who reported evidence of self-satisfaction in the chimpanzee Viki upon solving a difficult puzzle. Yerkes (1916/1979) reported that a chimpanzee who failed in its efforts to imitate successfully the actions of a model banged its head, and showed other signs of irritation and frustration; it is possible that these behaviors reflect shame in the self's performance (clearly this interpretation is not demanded by the evidence however).

There are anecdotal reports of other self-conscious emotions as well. For instance, Blyth (1847; cited in Yerkes & Yerkes, 1929) claimed that an orangutan he observed regularly covered its genitals to avoid embarrassment (or shame), a behavior thought to occur in chimpanzees as well (Yerkes & Yerkes, 1929). For these interpretations to be true, it would seem necessary for the observed orangutan and chimpanzees to have incorporated human standards of dress and display of sexual organs into their own views of themselves, which in turn suggests that such behavior would be limited to great apes with extensive, emotionally-involving contact with humans. We report these interpretations because they are suggestive of embarrassment and shame, but note that the evidence for them is sketchy and that the inferences underlying them require faith in the potential for identification with humans in the great ape species.

To the best of our knowledge, no evidence has been reported suggesting that self-conscious emotions are experienced by monkeys.

Empathy

Empathic emotional responding might also be an indication of the presence of some components of the sense of self. Genuine empathy requires that the individual: (a) observe an emotional response in another, (b) recognize that the observed response belongs to that other; and (c) vicariously experience the same emotion (Eisenberg et al., 1991). By its definition[1], then, empathy requires objective self-awareness (self and other must be differentiated). Most usually, differentiation is demonstrated by behavior. For instance, an empathizing organism may act to relieve distress in the other, which indicates that it is aware that the distress belongs to the other.

Empathy may also require subjective self-awareness: empathic responding might occur only when the other's welfare and goals are in some sense included within the boundaries of the self. In other words, the organism must identify with the other while at the same time recognizing that the emotion is not occurring in the self. Research with humans (for a review, see Batson, 1987) clearly demonstrates that empathy is more likely when the other experiencing the emotion is similar to, and consequently more easily assimilated in, the self of the observer (this may indicate the presence of subjective self-awareness). Moreover, this phenomenon demonstrates that empathic responding (at least in some instances) is mediated by judgements of similarity between self and other, which apparently would require comparing representations of self and other. Although neither personal memories nor theories of self are required for empathic responding, both might enhance it. For instance, one might be more empathic upon witnessing another bump his or her head on a kitchen cabinet door, if one can recall the sharp pain the self experienced in a similar situation. If one infers similarity between one's own theory of self and that of the other, then empathic responding may be more likely as well.

Empathy in the sense used here is evident in human infants by the end of the second year of life (Zahn-Waxler et al., 1992). Human infants have been observed to react to the distress of parents with emotional displays and helping behaviors that strongly suggest empathy (Zahn-Waxler et al., 1992). Chimpanzees have also been observed to act altruistically in ways that suggest empathy (see Chapter 4). For instance, Boesch described the care given to a chimpanzee suffering injuries from a leopard attack:

Table 15.3. *Emotional markers of the sense of self*

Marker	Objective self-awareness	Subjective self-awareness	Personal memories	Representations	Theories	Distribution across species
Self-conscious emotions (pride, hubris, guilt, shame)	Attention must be focused on the self's actions and self's standards (Lewis, 1992)	Subjective self-awareness must include both self's action and the self's standards (Lewis, 1992)	Not required for self-conscious emotions	Must have representation of some type of ideal self (Lewis, 1992)	Some type of hierarchy and connection among representations is necessary in order for hubris and shame, because these emotions require attributions that a unified self has succeeded or failed	Humans: present by 18–24 months of age (Lewis, 1990) Chimpanzees: anecdotal evidence Orangutans: anecdotal evidence Gorillas: anecdotal evidence Monkeys: no evidence

Empathy	Requires recognition that the emotion originates in another (Eisenberg, et al., 1991)	May require extension of self so as to include others (Rosaldo, 1984)	Empathy probably does not require retrieval of specific personal memories, although empathic arousal may be intensified if personal memories are retrieved	Empathy probably requires understanding of representations of generic emotional states, which then may be attributed to self	Theories of self are not required for empathy, but may amplify vicarious emotion	Humans: present by 18–24 months of age (Eisenberg et al., 1991) Chimpanzees: evidence for empathy in at least some social groups (Boesch, 1992), but it is probably very rare (Cheney & Seyfarth, 1990) Orangutans: little evidence Gorillas: little evidence Monkeys: little evidence of empathy despite an ability to read the emotional states of others (Cheney & Seyfarth, 1990)

After having received fresh wounds from an attack of a leopard, the injured individual is constantly looked after by group members, all trying to help by grooming and tending the wounds. Dominant adult males prevented other group members from disturbing the wounded chimp by chasing playing infants or noisy group members away from his vicinity. In addition, as wounds handicapped the movements of the injured animal, group members remained with him as long as he needed before he was able to begin to walk again; some just waited, whereas others would return to him until he started to move.
(Boesch, 1992, p. 149).

Although these incidents do suggest the presence of empathy in the great apes, it would appear that empathy is observed only rarely in these species[2], and to date has not been documented as occurring in monkeys (Cheney & Seyfarth, 1990).

CONCLUSIONS

Our goals for the chapter were to outline the components of the sense of self and to sketch the evidence for their presence in humans, chimpanzees, orangutans, gorillas, and monkeys. These are complex topics, and cannot be fully explored within the confines of a chapter. For instance, the list of markers of the sense of self proposed in the chapter may be incomplete simply because we know too little about the nonhuman primates to detect subtle signs of self-awareness and self-knowledge. Moreover, there is little research on the sense of self that focuses on great apes and monkeys, and consequently we have been forced to rely heavily on anecdotal evidence; this is clearly undesirable in an effort to contrast capabilities of different species. Many of the ambiguities in this review will be resolved by future research: conceptual distinctions will be refined as constructs become operational and systematic research will clarify the ways in which components of the sense of self are revealed in thought and behavior.

Despite these limitations, it is possible to draw several conclusions from the review. In the first place, there is little reason to suspect that monkeys have a sense of self. Monkeys have been untrainable with regard to language abilities, and consequently there are no instances of monkeys making use of linguistic markers for the sense of self. Nor have monkeys demonstrated mirror-self-recognition, role-taking or "true" imitation, self-conscious emotions, or empathy. We believe that the lack of evidence reflects a genuine deficit: monkeys probably do not have much of a sense of self. An explanation for their lack of a sense of self cannot be attempted here, but certainly a full explanation will draw upon neurological and social differences between monkeys and great apes.

While there is a qualitative distinction between monkeys and great apes, the difference between humans and other great apes appears to be largely quantitative. There is convincing evidence that chimpanzees, orangutans, and gorillas have objective self-awareness (e.g. mirror-self-recognition), subjective self-awareness (linguistic self-referencing), and representations of self (mirror-self-recognition, declarative speech). Anecdotal accounts from research with gorillas suggest that great apes might have personal memories as well. Only evidence for the existence of explicit theories of self could not be found in the literature concerned with the great apes.

Although there seem to be a number of similarities in the variety of experiences of self found among the great apes, these experiences appear to be much more salient and elaborate in humans than in chimpanzees, orangutans, or gorillas. In comparison to these other species, markers of the various components of the sense of self are more common in humans. Humans are more likely than other great apes to refer to themselves in spontaneous speech, attribute characteristics to themselves, imitate more frequently, experience self-conscious emotions, and respond empathically to the distress of others. It is, of course, difficult to know the direction of the relation of behavior to the sense of self: would the sense of self be more salient in chimpanzees if they imitated more frequently (e.g. see Hart & Fegley, 1994a,b)? Or would increasing the salience of the sense of self in chimpanzees stimulate their interest in imitation? Would empathic responding in gorillas become more likely if their subjective sense of self extended more frequently to include others? These and other questions can be answered through further research, which will cast new light on the relation of the sense of self to behavior.

We would like to return briefly to the issue that introduced the chapter. There we suggested that for at least some social issues our judgements are guided by our beliefs as to whether an organism has a sense of self or not (e.g. in instances of Persistent Vegetative State). In this chapter we have argued that the great apes show evidence of experiencing all of the facets of the sense of self except for the theory or narrative components. One might suggest that judgements that great apes are not deserving of the respectful treatment accorded to humans, because humans have selves, reflect either (a) an ignorance of the great apes' capability for self-experience, or (b) the importance of the theory/narrative facet of the sense of self in judgements of who has, and who does not have, the kind of self requisite for treatment as a person. Comparative-psychological research offers a unique vantage point for a thoughtful consideration of these sorts of judgements; discovering what great apes think about themselves sheds light not only on these species" cognitive abilities and experiences, but also on our understanding of ourselves and the role of selfhood in social judgement.

ACKNOWLEDGEMENTS

The authors thank Kim Bard, Annerieke Oosterwegel, Sue Taylor Parker, Bruce Ross, Anne Russon, Paul Vasey, and Bill Whitlow for their helpful comments on an earlier version of this chapter.

NOTES

1 There are other definitions of empathy according to which self–other differentiation is not required, and consequently is a primitive emotion that needs little or no sense of self to be experienced. However, theorists in this tradition acknowledge that empathy is transformed by the awareness that the emotion belongs to another (e.g. Hoffman, 1982, 1987). It is this latter form of empathy that draws upon the sense of self and which we consider to be genuine empathy.

2 Because empathic responding in humans is so strongly controlled by social context, it might well be the same for nonhuman primates. Consequently, it is at least possible that empathy is frequently experienced by nonhuman primates, but not translated into behavior because such behavior is inconsistent with the social context.

REFERENCES

Anderson, J.R. (1994). The monkey in the mirror: A strange conspecific. In *Self-Awareness in Animals and Humans: Developmental Perspectives*, ed. S.T. Parker, R.W. Mitchell & M.L. Boccia, pp. 315–29. New York: Cambridge University Press.

Baldwin, J.M. (1902). *Social and Ethical Interpretations in Mental Life*. New York: Macmillan.

Batson, C.D. (1987). Prosocial motivation: Is it ever truly altruistic? In *Advances in Experimental Social Psychology*, ed. L. Berkowitz, pp. 65–122. New York: Academic Press.

Baumeister, R.F. (1987). How the self became a problem: A psychological review of historical research. *Journal of Personality and Social Psychology*, 52, 163–76.

Beck, A.T., Rush, A.J., Shaw, B.F. & Emery, G. (1979). *Cognitive Therapy of Depression*. New York: Guilford Press.

Boesch, C. (1992). New elements of a theory of mind in wild chimpanzees. *Behavioral and Brain Sciences*, 15, 149–50.

Budwig, N. (1989). The linguistic marking of agentivity and control in child language. *Journal of Child Language*, 16, 263–84.

Butterworth, G. (1990). Self-perception in infancy. In *The Self in Transition: Infancy to Childhood*, ed. D. Cicchetti & M. Beeghly, pp. 119–37. Chicago: University of Chicago Press.

Cheney, D.L., & Seyfarth, R.M. (1990). *How Monkeys See the World*. Chicago: University of Chicago Press.

Cicchetti, D. & Beeghly, M. (eds.) (1990). *The Self in Transition: Infancy to Childhood*. Chicago: University of Chicago Press.

Custance, D., Whiten, A. & Bard, K.A. (1994). The development of gestural imitation and self-recognition in chimpanzees (*Pan troglodytes*) and children. In *Current Primatology*, vol. III, *Social Development, Learning and Behaviour*, ed. J.J. Roeder, B. Thierry, J.R. Anderson & N. Herrenschmidt, pp. 381–87. Strasbourg: Université Louis Pasteur.

Damon, W. & Hart, D. (1988). *Self-understanding in Childhood and Adolescence*. New York: Cambridge University Press.

Damon, W. & Hart, D. (1992). Social understanding, self-understanding, and morality. In *Developmental Psychology: An Advanced Textbook*, 3rd edn, ed. M. Bornstein & M. E. Lamb, pp. 421–64. Hillsdale, NJ: Lawrence Erlbaum.

Davis, D. & Brock, T.C. (1975). Use of first person pronouns as a function of increased objective self-awareness and performance feedback. *Journal of Experimental Social Psychology*, 11, 381–8.

Dennett, D.C. (1991). *Consciousness Explained*. Boston, MA: Little, Brown, and Company.

Eisenberg, N., Shea, C.L., Carlo, G. & Knight, G.P. (1991). Empathy- related responding and cognition: A "chicken and the egg" dilemma. In *Handbook of Moral Behavior and Development*, ed. W.M. Kurtines & J.L. Gewirtz, pp. 63–88. Hillsdale, NJ: Lawrence Erlbaum.

Gallup, G.G. Jr (1970). Chimpanzees: Self-recognition. *Science*, 167, 86–7.

Gallup, G.G. Jr (1982) Self-awareness and the emergence of mind in primates. *American Journal of Primatology*, 2, 237–48.

Gallup, G.G. Jr (1985). Do minds exist in species other than our own? *Neurosciences and Behavioral Reviews*, 9, 631–41.

Gallup, G.G. Jr, Wallnau, L.B. & Suarez, S.D. (1980). Failure to find self-recognition in mother–infant and infant–infant rhesus monkey pairs. *Folia Primatologica*, **33**, 210–9.

Gardner, B.T. & Gardner, R.A. (1974). Comparing the early utterances of child and chimpanzee. In *Minnesota Symposia on Child Psychology*, ed. A. Pick, vol. 8, pp. 3–23. Minneapolis: University of Minnesota Press.

Guillaume, P. (1971). *Imitation in Children*. Chicago: University of Chicago Press. (First published in French in 1926).

Gopnik, A. & Meltzoff, A.N. (1994). Minds, bodies, and persons: Young children's understanding of the self and others as reflected in imitation and theory of mind research. In *Self-awareness in Animals and Humans: Developmental Perspectives*, ed. S.T. Parker, R.W. Mitchell & M.L. Boccia, pp. 166–86. New York: Cambridge University Press.

Hart, D. & Fegley, S. (1994a). Social imitation and the emergence of a mental model of self. In *Self-awareness in Animals and Humans: Developmental Perspectives*, ed. S.T. Parker, R.W. Mitchell & M.L. Boccia, pp. 149–65. New York: Cambridge University Press.

Hart, D. & Fegley, S. (1994b). Social imitation and the sense of self. In *Current Primatology*, vol. III, *Social Development, Learning and Behaviour*, ed. J.J. Roeder, B. Theirry, J.R. Anderson & N. Herrenschmidt, pp. 389–96. Strasbourg: Université Louis Pasteur.

Hart, D. & Fegley, S. (1995). The development of self-awareness and self-understanding in cultural context. In *Culture, Experience, and the Conceptual self*, ed. U. Neisser & D. Jopling, New York: Cambridge University Press. (in press.)

Hayes, C. (1951). *The Ape in Our House*. New York: Harper & Brothers.

Hoffman, M.L. (1982). Development of prosocial motivation: Empathy and guilt. In *The Development of Prosocial Behavior*, ed. N. Eisenberg, pp. 281–311. New York: Academic Press.

Hoffman, M.L. (1987). The contribution of empathy to justice and moral judgment. In *Empathy and its Development*, ed. N. Eisenberg & J. Strayer, pp. 47–80. New York: Cambridge University Press.

Kagan, J. (1981). *The Second Year: The Emergence of Self-awareness*. Cambridge, MA: Harvard University Press.

Kennedy, J.S. (1992). *The New Anthropomorphism*. Cambridge: Cambridge University Press.

Lazarus, R.S. (1991). Progress on a cognitive-motivational-relational theory of emotion. *American Psychologist*, **46**, 819–34.

Levine, L. (1983). Mine: self-definition in 2-year-old boys. *Developmental Psychology*, **19**, 544–9.

Lewis, M. (1992). *Shame: The Exposed Self*. New York: Free Press.

Lewis, M. & Brooks-Gunn, J. (1979). *Social Cognition and the Acquisition of Self*. New York: Plenum Press.

Lewis, M., Sullivan, M., Stanger, C. & Weiss, M. (1989). Self development and self-conscious emotions. *Child Development*, **60**, 146–56.

Lin, A.C., Bard, K.A. & Anderson, J.R. (1992). Development of self-recognition in chimpanzees. *Journal of Comparative Psychology*, **106**, 120–7.

Mead, G.H. (1934). *Mind, Self and Society*. Chicago: University of Chicago Press.

Meltzoff, A.N. (1990). Foundations for developing a concept of self: the role of imitation in relating self to other and the value of social mirroring, social modeling, and self practice in infancy. In *The Self in Transition: Infancy to Childhood*, ed. D. Cicchetti & M. Beeghly, pp. 139–64. Chicago: University of Chicago Press.

Meltzoff, A.N. & Gopnik, A. (1993). The role of imitation in understanding persons and developing a theory of mind. In *Understanding Other Minds: Perspectives from Autism*, ed. S. Baron-Cohen, H. Tager-Flusberg & D.J. Cohen, pp. 335–66. Oxford: Oxford University Press.

Miles, H.L. (1990). The cognitive foundations for reference in a signing orangutan. In *"Language" and Intelligence in Monkeys and Apes: Comparative Developmental Perspectives*, ed. S.T. Parker & K.R. Gibson, pp. 511–39. New York: Cambridge University Press.

Miles, H.L. (1994). ME CHANTEK: The development of self-awareness in a signing orangutan. In *Self-awareness in Animals and Humans: Developmental Perspectives*, ed. S.T. Parker, R.W. Mitchell & M.L. Boccia, pp. 254–72. New York: Cambridge University Press.

Mitchell, R.W. (1987). A comparative-developmental approach to understanding imitation. In *Perspectives in Ethology*, vol. 7: *Alternatives*, ed. P.P.G. Bateson & P.H. Klopfer, pp. 183–215. New York: Plenum Press.

Mitchell, R.W. (1993). Mental models of mirror-self-recognition: Two theories. *New Ideas in Psychology*, 11, 295–325.

Mitchell, R.W. (1994). Multiplicities of self. In *Self-awareness in Animals and Humans: Developmental Perspectives*, ed. S.T Parker, R.W. Mitchell & M.L. Boccia, pp. 81–107. New York: Cambridge University Press.

Neisser, U. (1988). Five kinds of self-knowledge. *Philosophical Psychology*, 1, 35–59.

Nelson, K. (1993). The psychological and social origins of autobiographical memory. *Psychological Science*, 4, 7–13.

Nozick, R. (1980). *Philosophical Explanations*. Cambridge, MA: Harvard University Press.

Parker, S.T. (1991). A developmental approach to the origins of self-recognition in great apes. *Human Evolution*, 6, 435–49.

Parker, S.T. (1994). Incipient mirror-self-recognition in zoo gorillas and chimpanzees. In *Self-awareness in Animals and Humans: Developmental Perspectives*, ed. S.T. Parker, R.W. Mitchell & M.L. Boccia, pp. 301–7. New York: Cambridge University Press.

Parker, S.T., Mitchell, R.W. & Boccia, M.L. (eds.) (1994) *Self-awareness in Animals and Humans: Developmental Perspectives*. New York: Cambridge University Press.

Patterson, F.G. (1978). Linguistic capabilities of a lowland gorilla. In *Sign Language and Language Acquisition in Man and Ape*, ed. F.C. Peng, pp. 161–201. Boulder, CO: Westview Press.

Patterson, F.G.P. & Cohn, R.H. (1994). Self-recognition and self-awareness in lowland gorillas. In *Self-awareness in Animals and Humans: Developmental Perspectives*, ed. S.T. Parker, R.W. Mitchell & M.L. Boccia, pp. 273–90. New York: Cambridge University Press.

Patterson, F.G. & Linden, H. (1981). *The Education of Koko*. New York: Holt, Rinehart & Winston.

Povinelli, D.J. (1993). Reconstructing the evolution of mind. *American Psychologist*, 48, 493–509.

Preyer, W. (1893). *The Mind of the Child*. New York: Appleton.

Ridley, R.M. (1992). How do monkeys remember the world? *Behavioral and Brain Sciences*, 15, 166.

Rosaldo, M. (1984). Toward an anthropology of self and feeling. In *Essays on Mind, Self, and Emotion*, ed. R. Shweder & R. Levine, pp. 137–57. New York: Cambridge University Press.

Roseman, I.J., Spindel, M.S. & Jose, P.M. (1990). Appraisals of emotion-eliciting events: Testing a theory of discrete emotions. *Journal of Personality and Social Psychology*, 39, 899–915.

Ross, B. (1991). *Remembering the Personal Past*. New York: Oxford University Press.

Russon, A.E. & Galdikas, B.M.F. (1993). Imitation in free-ranging rehabilitant orangutans (*Pongo pygmaeus*). *Journal of Comparative Psychology*, 107, 147–61.

Russon, A.E. & Galdikas, B.M.F. (1994). Constraints on great ape imitation: Model and action selectivity in rehabilitant orangutan (*Pongo pygmaeus*) imitation. *Journal of Comparative Psychology*, 109, 5–17.

Russon, A.E. & Waite, B.E. (1991). Patterns of dominance and imitation in an infant peer group. *Ethology and Sociobiology*, 13, 55–73.

Squires, L. (1987). *Memory and Brain*. New York: Oxford University Press.

Suarez, S.D. & Gallup, G.G. (1981). Self-recognition in chimpanzees and orangutans but not gorillas. *Journal of Human Evolution*, **10**, 175–88.
Swartz, K. & Evans, S. (1991). Not all chimpanzees (*Pan troglodytes*) show self-recognition. *Behaviour Processes*, **17**, 396–410.
Taylor, C. (1989). *Sources of the Self: The Making of the Modern Identity*. Cambridge, MA: Harvard University Press.
Temerlin, M.K. (1975). *Lucy: Growing up Human*. Palo Alto, CA: Science and Behavior Books.
Terrace, H.S. (1985). In the beginning was the "name." *American Psychologist*, **4**, 1011–28.
Terrace, H.S. & Bever, T.G. (1980). What might be learned from studying language in the chimpanzee? The importance of symbolizing oneself. In *Speaking of Apes: A Critical Anthology of Two-way Communication with Man*, ed. T. Sebeok & J. Umiker-Sebeok, pp. 179–89. New York: Plenum Press.
Tomasello, M., Kruger, A.C. & Ratner, H.H. (1993). Cultural learning. *Behavioral and Brain Sciences*, **16**, 495–552.
Tuttle, R. (1986). *Apes of the World*. Park Ridge, NJ: Noyes Publications.
Visalberghi, E. & Fragaszy, D.M. (1990). Do monkeys ape? In *"Language" and Intelligence in Monkeys and Apes: Comparative Developmental Perspectives*, ed. S.T. Parker & K.R. Gibson, pp. 247–73. New York: Cambridge University Press.
Whiten, A. & Ham, R. (1992). On the nature and evolution of imitation in the animal kingdom: Reappraisal of a century of research. In *Advances in the Study of Behavior*, ed. P.J.B. Slater, J.S. Rosenblatt, C. Beer & M. Milinski, vol. 211, pp. 239–83. New York: Academic Press.
Yerkes, R.M. (1916/1979). *The Mental Life of Monkeys and Apes: A Study of Ideational Behavior*. Delmar, NY: Scholars, Facsimilies, and Reprints, Inc. (First published in 1916.)
Yerkes, R.M. & Yerkes, A.W. (1929). *The Great Apes: A Study of Anthropoid Life*. New Haven, CT: Yale University Press.
Zahn-Waxler, C., Radke-Yarrow, M., Wagner, E. & Chapman, M. (1992). Development of concern for others. *Developmental Psychology*, **28**, 126–36.
Zemach, E. (1972). The reference of "I". *Philosophical Studies*, **23**, 68–75.

16

Apprenticeship in tool-mediated extractive foraging: The origins of imitation, teaching, and self-awareness in great apes

SUE TAYLOR PARKER

INTRODUCTION

Rogoff (1992) used apprenticeship as a metaphor for guided participation in shared activities of a routine nature. Her contextualist model of human cognitive development focuses on the integrated social and technological structure of events and activities, particularly problem-solving activities in everyday life. It thereby situates the changing levels and kinds of guided participation that occur during development in particular socio-cultural settings.

The contextualist approach offers a new dimension for comparative developmental studies of primate cognition by focusing specifically on the integration or lack of integration of ecological and social intelligence. In contrast to narrowly social hypotheses that have been proposed to explain increased intelligence and enlarged brains of primates as compared to most other mammals (e.g. Byrne & Whiten, 1988), the apprenticeship model elaborated here is a hypothesis that explains an integrated complex of social and instrumental cognitive abilities that are unique to the great apes and hominids: intelligent tool use, true imitation, teaching by demonstration, and self-awareness. (Intelligent tool use is defined as the use of a detached object to change the position and/or state of an object with an understanding of the causal dynamics involved in the contact and movement of the tool relative to the object (e.g. Parker & Poti, 1990).)

This apprenticeship model is an elaboration of the extractive foraging model that was proposed to explain the origins of intelligent tool use and imitation in the common ancestor of the great apes (Gibson, 1986; Parker & Gibson, 1977, 1979). According to this model, extractive foraging during the dry season on a variety of embedded food sources such as hard-shelled fruits and nuts and nesting social insects favored intelligent tool use in the small, omnivorous common ancestor of living great apes. Given the cognitive demands of tool-mediated extractive foraging and the consequent delay in the development of efficient feeding in young animals, selection also favored both maternal food-sharing and the capacity for learning foraging skills, including tool use, through imitation.

According to this model, early hominids expanded the seasonal dependency on tool-mediated extractive foraging into a year-round dependency as well as expanding the range of food items exploited by this technique to include roots and tubers, fossorial animals, and, most notably, bone marrow and brains, and ultimately meat. In this chapter, the apprenticeship model is elaborated and expanded to explain the origins of teaching and self-awareness in the common ancestor of the great apes.

Considerable controversy surrounds the purported capacities for imitation, teaching, and self-awareness in great apes. Much of this controversy pivots on the definitions of imitation, teaching, self-awareness, etc. In this chapter I use broad functional definitions of these concepts in order to focus attention on adaptive functions of behaviors common to a variety of species, considering the various proximate means by which these functions are achieved separately. In other words, in these definitions I focus on the results of the actions rather than the means by which the results are achieved. The justification for these broad definitions is that they stimulate the construction of adaptive hypotheses. After proposing these definitions I go on to discuss some of the variety of proximate means by which they are achieved in various primate species.

IMITATION

In common parlance, imitation refers to any sort of similar behavior manifested by one individual who has had an opportunity to witness such a behavior by another individual. In other words, the common concept of imitation focuses on behavioral outcomes rather than on processes of behavioral acquisition. The focus on adaptive outcomes or functions is also true of biological parlance. Taking this functional approach as developed by Caro & Hauser (1992), we can define imitation as follows: an individual A, can be said to imitate another individual, B, if A modifies its behavior in such a manner that it approximates or matches some aspect of B's behavior that it had witnessed in the past. As a result A may acquire knowledge or learn a skill earlier or more rapidly or efficiently than it might otherwise have done, or that it might not have learned at all. Imitation should evolve when the benefits to the imitator exceed the cost to the imitator where cost and benefit are ultimately measured in terms of reproductive success, but can also be proximally measured in terms of energy expenditure (see e.g. Trivers, 1972). It is important to note that imitation can be subserved by a variety of proximate mechanisms. (This definition is analogous to Caro & Hauser's (1992) definition of teaching, which is quoted below. For another functional definition of imitation, see Mitchell, 1987.)

In psychological parlance, in contrast to biological parlance, the concept of imitation focuses on such proximate mechanisms as the psychological/cognitive processes underlying imitative behavior. In recent literature on animal cognition, for example, imitation is distinguished from matching behavior arising from such processes as stimulus enhancement and social facilitation (e.g. Galef, 1988; Mitchell, 1987; Nagel *et al.*, 1993; Tomasello, 1990; Visalberghi & Fragaszy, 1990; Whiten & Ham, 1992). Such selective use of the concept of imitation to mean the acquisition of novel behavior on the basis of observational learning (Galef, 1988) has concomitantly reduced the number of animal species who can be said to imitate, not only suggesting that monkeys do not "ape" but also that apes rarely ape (Visalberghi & Fragaszy, 1990).

The limited data on imitation in great apes may also be an artifact of the idealized standard of imitation employed by some comparative psychologists (e.g. Tomasello, 1990). Studies of human children reveal that imitation is more complex and variable than is generally realized: far from perfect replication of entire sequences of actions, children's imitation involves a wide variety of degrees of completeness and fidelity to the model's actions and varying kinds of insertions into, and recombinations with, other actions (Speidel & Nelson, 1989; Chapter 7).

In contrast to the biological and comparative psychological literature, the literature on child development treats imitation as a developmental phenomenon that undergoes progressive elaboration in infancy and early childhood (e.g. Bretherton, 1984; Meltzoff, 1988, 1990; Piaget, 1962). Because of their sequential epigenetic nature, developmental stages provide a meaningful scale for identifying and comparing cognitive levels as proximate mechanisms in imitation-related phenomena among primates.

Piagetian stages, which provide the foundation for most models of intellectual development, describe a progression of imitative achievements from birth through early childhood, beginning with "reflexive" contagious crying in early infancy and culminating in the capacity for delayed "trial-and-error" matching of novel action sequences at the sixth and final stage of the sensorimotor period at about 2 years[1]. The sixth stage leads into symbolic play, which culminates in increasingly elaborated sequences of pretend play by the end of the intuitive subperiod of the preoperations period at about 5 years (Piaget, 1962). In my view, the five most useful distinctions for comparative studies are the following:

1. Proto imitation: imitation of actions in own repertoire (sensorimotor (SM) stage 4).
2. Facial imitation: imitation of model's facial expression through familiar actions invisible to the imitator (SM stage 4).
3. True imitation: directed trial-and-error imitation of novel actions (SM stages 5 and 6).
4. Pretend play: using objects to represent other objects and pretending to be another individual (symbolic substage of the preoperations period).
5. Pretend role play: engaging in complementary role-playing scenarios such as playing mother and babies, teacher and pupils (intuitive subperiod of preoperations period).

True imitation (level 3) of another individual's behavior involves directed "trial-and-error" kinesthetic-visual matching (Mitchell, 1993) of some elements of the model's goals and/or actions. This matching seems to involve the following operations: (a) differentiation and recombination of actions; (b) simultaneous or alternating attentional focus on the model's actions and one's own actions; (c) comparison between the model's actions and the imitator's actions; and (d) systematic adjustment of the imitator's actions to match those of the model. Matching also seems to involve (a) simultaneous or alternating attention to the outcome of the model's actions and the imitator's actions; (b) comparison between these outcomes; (c) adjustment of actions to reduce the discrepancy (Mitchell, 1987, 1993); and often (d) repetition to further reduce the discrepancy. Contrary to Piaget's hypothesis, it seems likely that some variety of representation occurs at all levels of imitative development (Meltzoff & Moore, 1977; Mitchell, 1983; Mounoud & Vinter, 1981). In immediate imitation of some novel behaviors that the infant cannot see itself perform, e.g. facial movements invisible to the imitator, the imitator is matching a representation of its own actions to the perceptual spectacle of some aspect of the model's actions; in deferred imitation, it is matching a representation of its own actions to a representation of some aspect of the model's actions.

Piagetian stages of imitation are useful because they allow us to peg the levels of imitative abilities in nonhuman primates relative to those of human infants at various stages of development. They are doubly useful because, as one series in a larger set of related stages focusing on various aspects of physical and logicomathematical knowledge, they allow us to to compare levels of achievement in various domains (e.g. space, causality, etc.) within and across related species (Piaget, 1970, and see e.g. Antinucci, 1989; Chevalier-Skolnikoff, 1977; 1983; Mathieu & Bergeron, 1981; Mignault, 1985; Parker, 1977; Potì & Spinozzi, 1995).

In applying Piagetian stages to other species, it is important to emphasize the differences as well as the similarities between humans and other primate species. Monkeys seem to be limited to fourth stage or proto-imitation, i.e. to imitating actions that are already in their own repertoire in familiar or unfamiliar contexts (Parker, 1977; Visalberghi & Fragaszy, 1990). Matching behaviors of models at this level may be mediated by a variety of mechanisms including contagion and local or stimulus enhancement wherein the animal's attention is drawn to an object or substrate which evokes a familiar action (Galef, 1988).

Among nonhuman primates only the great apes display facial imitation, true imitation and pretense. Imitation in great apes has been studied primarily in three contexts: cross-fostered apes (e.g. Hayes & Hayes, 1952), language-trained apes (e.g. see Chapter 13), and rehabilitant apes (Russon & Galdikas, 1993b). Facial imitation has been reported in chimpanzees (e.g. Custance & Bard, 1994; Hayes & Hayes, 1952) and gorillas (Patterson & Linden, 1981). True imitation is displayed by sign language-trained apes who learn signs through a variety of mechanisms that include gestural imitation (e.g. Gardner *et al.*, 1989; Miles, 1990; Patterson, 1980; Terrace, 1979). Likewise, true imitation of a variety of novel actions with tools and other objects has been reported in cross-fostered chimpanzees (e.g. Hayes & Hayes, 1952; Temerlin, 1977), bonobos (Tomasello *et al.*, 1993), gorillas (Patterson & Linden, 1981), and orangutans (Miles, 1990; Chapter 13) as well as in rehabilitant orangutans (Russon & Galdikas, 1993a, b, 1995; Chapter 7). Imitation in great apes even develops into the rudiments of pretend play in cross-fostered and symbol-trained apes (e.g. Hayes & Hayes, 1952; Jensvold &

Fouts, 1994; Patterson & Cohn, 1994; Savage-Rumbaugh & MacDonald, 1988; Chapter 13; *contra* Mignault, 1985).

Although true imitation (level 3) occurs in great apes, it occurs far less frequently and comprehensively than in human infants. It also develops much later and more slowly than in humans. Moreover, as pointed out by Chevalier-Skolnikoff (1977), it is limited primarily to the facial and manual/gestural modalities, rarely appearing in the vocal modality; but see Laidler (1980) for reports of vocal imitation in orangutans and Savage-Rumbaugh *et al.* (1989) for reports of vocal imitation in bonobos. Finally, imitation in great apes extends to the boundaries of simple pretend play (level 4), but stops short of the more elaborated forms of pretend role play (level 5) and theory of mind characteristic of 4 year old human children (e.g. Bretherton, 1984; Gopnik & Meltzoff, 1994). The highest levels of imitative abilities reported in great apes are comparable to those of 3 year old human children. They are also compatible with the levels of intelligence achieved by great apes in other domains (e.g. Parker & Gibson, 1979). The more limited scope of imitation in great apes is consistent with their more restricted repertoire of actions as compared to humans. Great apes, like children, are primarily interested in imitating actions that they are just capable of mastering, i.e. behaviors in their "zone of proximal development" (Russon & Galdikas, 1995; Speidel & Nelson, 1989; Vygotsky, 1962; Yando, Seitz & Zigler, 1978; Chapter 7). The limited occurrence of imitation in experimental studies of great apes (e.g. Tomasello, 1990) may also reflect the fact that they, like children, tend to imitate those with whom they have emotional ties of one sort or another (Russon & Galdikas, 1995).

TEACHING

In common parlance, teaching refers to any sort of behavioral acquisition by a younger, less experienced individual that has been encouraged in any way by an older, more experienced individual. In biological parlance, the term teaching refers more narrowly to directed processes of behavioral transmission by more knowledgeable (usually older) individuals to less knowledgeable (usually younger) individuals (Barnett, 1968; Ewer, 1969).

Caro & Hauser (1992) have provided a broad functional definition of teaching:

> An individual A can be said to teach if it modifies its behavior only in the presence of a naive observer, B, at some cost or at least without an immediate benefit for itself. A's behavior thereby encourages or punishes B's behavior, or provides B with experience, or sets an example for B. As a result, B acquires knowledge or learns a skill earlier in life or more rapidly or efficiently than it might otherwise do, or that it might not learn at all *(Caro & Hauser, 1992, p. 152)*.

Within this framework they distinguish opportunity teaching from coaching: in opportunity teaching, the teacher puts the pupil in an environment conducive to learning, and in coaching the teacher directly alters the pupil's behavior by punishing or encouraging it (Caro & Hauser, 1992, p. 166). In their survey of teaching behavior, these authors found that while opportunity teaching was widespread among mammals and birds, coaching was relatively rare, occurring primarily in primates.

As the authors emphasized, this is a functional definition that does not specify the proximate mechanisms by which such effects might be achieved in various species. Indeed the utility of the definition is increased by the fact that these conditions could be met through a variety of mechanisms in different species (Caro & Hauser, 1992). Teaching in this broad sense can occur through a variety of mechanisms, instinctive and/or cognitive. It is useful to turn to the literature on human behavior to understand the role of cognition in teaching.

Teaching behaviors among humans vary strikingly with the capacities of the pupil. They typically involve demonstration with simplification appropriate to the developing level of the pupil, the teacher adjusting his or her behavior to provide ongoing challenges to the student's capacities, or pulling at the "zone of proximal development" (Vygotsky, 1962). Human teaching involves the following "scaffolding functions:" (a) recruitment, (b) reduction in degrees of freedom, (c) direction maintenance, (d) marking critical features, (e) frustration control, and (f) demonstration . . . or modeling solutions to a task (Wood et al., 1976, p. 98). Most of these features fit under the first part of Caro & Hauser's definition, i.e. modification of the more experienced individual's behavior in the presence of the less experienced individual at some cost to the more experienced individual in terms of getting food or foregoing other profitable activities. They are unique, however, in their cognitive demands on both teacher and pupil.

The concept of scaffolding or supporting structures for aiding inexperienced or immature individuals (Wood et al., 1976) provides a useful dimension for comparing cognitive aspects of teaching. The simplest forms of teaching in nonhuman primates involve virtually no scaffolding beyond that which arises from the customary proximity of mothers to their young. This proximity leads to repeated observation of, and in some cases participation in, the mother's activities. Depending on the capacities of the offspring, the teachers' effect on learning in such situations may be mediated by stimulus enhancement, social facilitation, or true imitation (Galef, 1988). This lowest level corresponds to Caro & Hauser's opportunity learning.

A somewhat more elaborate form of teaching in nonhuman primates, which uses slightly greater scaffolding involves coaching, i.e. "discouragement" through punishment or "encouragement." Both monkeys and great apes, for example, place their offspring in situations requiring climbing or locomotion and encourage them to move forward or to climb on their backs (e.g. van de Rijt-Plooij & Plooij, 1987; A. Whiten, unpublished results). This second level corresponds to Caro & Hauser's coaching.

Overt or active teaching by manual demonstration (Boesch, 1991) involves more scaffolding, specifically of the kinds outlined by Wood et al. (1976). Overt teaching with language involves still more intervention, scaffolding, and planning. Overt teaching with pedagogical materials provides yet another level of intervention and support. Teaching in Caro & Hauser's sense encompasses a variety of mechanisms for transmission of information in social contexts (for an alternative formulation, see King, 1994). Social transmission of information among monkeys seems to occur through opportunity teaching (Caro & Hauser, 1992) and/or through unplanned effects of social punishments and rewards (Premack, 1984), or coaching (Caro & Hauser, 1992). Socially-mediated rewards and punishments serve primarily to change the frequencies and contexts of behaviors.

Some social transmission among great apes seems to occur through encouragement, true imitation, and manual demonstration. Chimpanzees and other great apes have long been credited by some investigators with the ability to invent and imitate novel behaviors, but until recently no one had reported that they display the ability to teach or demonstrate novel behaviors. Recent evidence suggests, however, that chimpanzee mothers in captivity (Fouts et al., 1989) and in the wild, do indeed teach novel manual behaviors (Boesch, 1991) as well as coaching appropriate use of existing schemes (van de Rijt-Plooij & Plooij, 1987; A. Whiten, unpublished results).

A few striking examples of teaching by demonstration have been reported in symbol-trained great apes: Miles (1994) reported that the orangutan Chantek instructed one of his teachers to make certain signs; while Patterson & Cohn (1994) described the gorilla Koko molding her doll's hands in the form of a sign just as Fouts et al. (1989) had reported the chimpanzee Washoe molding her adopted son Loulis's hands in the form of a sign.

Studies of chimpanzees from the Taï Forest in Côte d'Ivoire, reveal that at least some chimpanzee mothers purposefully teach certain elements of tool-mediated nut-cracking. They bring unopened nuts to the anvil, place them in the anvil, and position the implements for opening nuts. They slow down and exaggerate their movements to demonstrate the correct positioning of the hammer for their offspring while watching their offspring's face to check their attention to the mother's actions with the tool on the nut. Juvenile chimpanzees, in turn, carefully watch their mother's actions, look at the tool set up, and imitate the grip and orientation of the tool in the hand and the direction of blows (Boesch, 1991; Chapter 18).

In other words, demonstration teaching and true imitation in the manual modality co-occur in apprenticeship in wild chimpanzees as in humans. It is notable that all of the scaffolding involved in manual demonstration (Wood et al., 1976) seems to entail the same kinds of matching operation involved in true imitation (see above). This makes sense given that the effectiveness of teaching by manual demonstration depends upon the pupil's capacity for directed "trial-and-error" imitation of novel manual schemes. Manual demonstration also entails a gauging of the pupil's attention and capacity to perform at a given level, and evaluation of his or her performance relative to a standard (Premack, 1984). Beyond this, teaching by demonstration seems to involve the teacher's expectation that the pupil will imitate his or her actions. Given that humans find the perception of being imitated pleasurable and reinforcing, the apprenticeship relationship may play an important role in social bonding (Yando et al., 1978).

Teaching by demonstration seems to be a manifestation of a more general capacity to adopt and exchange complementary roles that begins to develop during the third year in human children (e.g. Bretherton, 1984; Piaget, 1962). Recent research suggests that some chimpanzees are capable of understanding and reversing instrumental roles (e.g. Povinelli et al., 1992). Role reversal seems to be implicit in some of the deceptive behaviors reported in captive chimpanzees and in a language-trained orangutan (Miles, 1994). Capacities for symbolic role playing and role reversal and the attendant capacities for awareness of the roles of self and others seem to develop in human children during pretend play from about 3 years onward (Bretherton, 1984; Fein, 1984; Mead, 1970; Piaget, 1962). A model of stages of development of teaching behaviour in children would provide a useful framework for comparative studies of teaching. At this point we can only guess that the capacity for teaching in humans begins to develop with pretend play from about age 3 onward when children begin to use motherese in role-playing games (e.g. Bretherton, 1984; Cook-Gumperz, 1992). Given the greater cognitive and motivational complexity of demonstration teaching relative to true imitation, it seems likely that this form of teaching evolved on the back of true imitation.

SELF-AWARENESS

Using the functional approach employed throughout this chapter, we can define self-awareness as follows: an individual, A, can be said to show self-awareness when it modifies its behavior, position, or appearance in response to informational feedback from its own actions, reflected image, or echoed sound, or from the attention and/or behavior of another individual, B, in a manner that depends upon processing information about some aspect of its own behavior or appearance to which it had no direct access before receiving information from the reflection or the other individual B. As a result, individual A modifies its behavior toward itself or others in some manner that it would not have done otherwise.

As with imitation and teaching, self-awareness could, in theory, occur at many levels and be mediated by a variety of proximate mechanisms, instinctual and/or cognitive. All animals must have means for distinguishing information about their own physiological states and movements from information from the environment in order to organize their own perceptions and actions: in some cases, as with some bird species, this may involve single neurons that can detect the animal's own song (Margoliash & Fortune, 1992) or proprioceptive loops that register the animal's own movements (e.g. Eibl-Eibesfeldt, 1975). The kinds of self-awareness engendered by such single modality feedback loops probably constitute the simplest level of self-awareness at the organismic (as opposed to the cellular) level.

In contrast, a more complex level of self-awareness, which I will call true self-awareness, can be indexed by mirror-self-recognition (MSR). Rather than registering their own actions in a single modality, animals who display MSR are integrating information across visual and kinesthetic modalities (Mitchell, 1993). It is this integration that allows them to recognize parallels between their actions and those of other individuals (social mirroring (Meltzoff, 1990)) and/or their actions and external representations of these actions in a mirror or videotape. The primary means for doing this seems to be matching between the actor's kinesis and visual information about the actions of the model or mirror image (Mitchell, 1993). Since true fifth stage imitation seems to be based on kinesthetic-visual matching, it is hardly surprising that it seems to be a prerequisite for MSR (Parker, 1991; *contra* Custance & Bard, 1994).

A model for the stages of development of self-awareness in human children would be useful for comparative studies of self-awareness. It is clear that self-awareness in humans is a complex multifaceted phenomenon that develops over many years from infancy through adulthood (e.g. Cicchetti & Begley, 1990; Damon & Hart, 1988; Lewis & Brooks-Gunn, 1979; Mitchell, 1994; Stern, 1992). Some of the earliest manifestations of self-awareness occurring by about 5 months are tied to recognition of contingencies between the infant's volitional actions and his or her proprioceptive experiences (e.g. Lewis & Brooks-Gunn, 1979: Stern, 1992; Watson, 1994). Other slightly later manifestations of self-awareness apparently occur during imitation beginning at about 9 months (e.g. Meltzoff, 1990). MSR beginning as early as 15 months may be based on the kinesthetic-visual matching underlying imitation (Mitchell, 1993). Self-conscious emotions develop at approximately the same time (Lewis & Brooks-Gunn, 1979). Both are followed by recognition of photographs of the self by about 2 years (Lewis & Brooks-Gunn, 1979). Verbal self-labeling and recognition of pictorial representations of the self follow soon afterward by about $2\frac{1}{2}$ years (Lewis & Brooks-Gunn, 1979). Following that, self-evaluative emotions appear by 3 years (Lewis *et al.*, 1989). Conceptual aspects of the self including autobiography appear by 4 years of age and continue to develop through adolescence (Snow, 1990). Each successive level of self-awareness is tied to developing intellectual abilities (e.g. Damon & Hart, 1988; for summaries, see Parker *et al.*, 1994; Chapter 15).

Self-awareness in nonhuman primates has been studied primarily through Gallup's (1970, 1977) controlled mark test in which an individual's face is marked while it is unconscious and its reactions to the reflection of its marked image are compared to its earlier reactions to its unmarked mirror image. MSR studies of this kind have revealed that great apes can pass the mark test in addition to showing a variety of self-directed behaviors that depend upon feedback from the mirror. Monkeys however, rarely, if ever, pass the mark test (e.g. Anderson, 1986, 1994; Boccia, 1994; Gallup, 1977; Thompson & Boatright-Horowitz, 1994; Chapter 2) even though they can use mirrors to localize objects and other animals. Although several tests of MSR in gorillas have yielded negative results, at least two gorillas have passed the test (Patterson & Cohn, 1994; Swartz & Evans, 1994). Significantly, not only gorillas but several chimpanzees have failed to pass the mark test, suggesting that MSR is probably variable in its manifestations in all the great apes (Swartz & Evans, 1994).

In addition to MSR, symbol-trained great apes also manifest other more developmentally advanced aspects of self-awareness including verbal self-labeling and recognition of pictorial representations of self (e.g. Miles, 1994; Patterson & Cohn, 1994; Temerlin, 1977; Chapter 13). These aspects of self-awareness are characteristic of human children at 2–3 years of age (see above) and seem to depend upon symbolic capacities. Although great ape infants display developmental patterns similar to those of human children in this regard, their development is significantly slower and more impoverished than that of human infants and children (e.g. Miles, 1994; Chapter 13).

THE RELATIONSHIPS AMONG IMITATION, TEACHING, AND SELF-AWARENESS

The fact that true imitation, MSR, and teaching by demonstration all co-occur in the great apes and only in the great apes and humans strongly suggests a relationship between these abilities. The fact that true imitation and MSR seem to depend upon fifth stage imitation, and that teaching by demonstration is associated with symbolic intelligence provides a clue as to the possible nature of this relationship.

Clearly MSR is only an index of a certain level of self-awareness rather than an evolved function. Questions immediately arise regarding the proximate and ultimate causes of this level of self-awareness. Not only the co-occurrence of true imitation and MSR, but the probable dependence of MSR on imitation suggest a close connection between the two phenomena (e.g. Gopnik & Meltzoff, 1994; Hart & Fegley, 1994; Mitchell, 1993; Parker, 1991). As indicated in the section on imitation (pp. 349–52), true imitation involves self-awareness through matching actions of the self to actions of the model. Although various authors disagree about whether imitation generates self-awareness or self-awareness generates imitation or whether they are inseparable phenomena, imitation seems to imply the co-existence of self-awareness.

Teaching by demonstration is undoubtedly a higher-level ability than true imitation because it depends upon the pupil's ability to learn by true imitation. It is rare among the great apes and probably represents the very apex of their cognitive abilities. (In human children playful teaching apparently first occurs at about 3 or 4 years of age in such pretend games as mothers and babies and teacher and pupils.) In other words, even more than directed true imitation, teaching by manual demonstration seems to require both other-awareness and self-awareness on the part of the teacher and attribution of intentionality. Having outlined these cognitive mechanistic relationships, we are ready to examine their evolutionary history.

CLADISTIC ANALYSIS OF THE ORIGINS OF TRUE SELF-AWARENESS, TRUE IMITATION, AND DEMONSTRATION TEACHING

Given the phylogeny of primates (see Figure 16.1), it is possible to reconstruct the origins of true imitation, demonstration teaching, and true self-awareness, just as it is possible to reconstruct the origins of any character given sufficient comparative data among living species. This can be done by using a "cladistic method" to identify the common ancestor who first displayed the focal character state (Ridley, 1984; Wiley, 1981): (a) identifying contrasting character states, e.g. differing levels of mental abilities in this case, (b) charting the distribution of these various mental abilities by species, (c) mapping these abilities onto a previously established phylogeny to see in which common ancestor which character state originated. Mapping behaviors determines which character states are "shared derived features;" that is, features present in all the members of the ingroup (great apes and humans in this case) and few if any of the

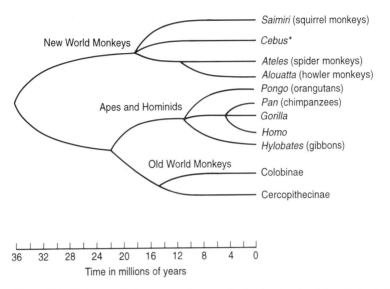

Figure 16.1. Phylogeny of primates according to molecular data (reprinted from Parker & Gibson, 1977). *There are no data on the time of the adaptive radiation of *Cebus* species.

members of the outgroup (lesser apes and monkeys in this case). Nest building, for example is present in all the living great apes and none of the lesser apes or monkeys, therefore, we can conclude that it is a shared derived character among great apes and hence must have arisen in their common ancestor.

In this analysis, the relevant character states are intelligent tool use, true imitation, teaching by demonstration, and true self-awareness (as measured by MSR). (Imitation and teaching of intelligent tool use involve self-awareness.) It is important to note that it is the ability itself (rather than the ability plus the context in which it is displayed) that is the character state in comparative studies. Like stereotyped behaviors studied by ethologists, cognitive abilities are more conservative evolutionarily than the contexts in which they are manifested. During evolution, behavior patterns often change their function and motivation as well as their frequency (e.g. Eibl-Eibesfeldt, 1975).

When these cognitive abilities are mapped onto the primate family tree, it is clear that true self-awareness, true imitation, and intelligent tool use are shared derived character states that must have arisen in the common ancestor of the great apes because they are not present in the majority of the outgroup, Old World monkeys. Given that only a few species of Old World monkeys have been reported to use tools and this only rarely and with scaffolding (see e.g. Beck, 1980; McGrew, 1992), it is also clear that intelligent tool use was not present in the common ancestor of Old World or New World monkeys and must have arisen independently in the common ancestor of cebus monkeys (Parker & Gibson, 1977). Finally, it appears from the limited information on teaching behaviors that teaching by manual demonstration also arose in the common ancestor of the great apes. See Figure 16.2 for cladistic mapping of these cognitive abilities. Given

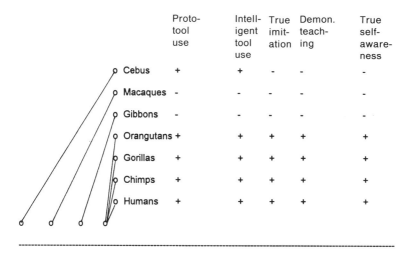

Figure 16.2. Reconstructing common ancestry of various abilities. CA, common ancestor.

the co-occurrence of this complex of cognitive abilities, it is impossible to infer the order of their evolution in earlier ancestors who left no living descendants other than those in the ingroup.

RECONSTRUCTING THE ADAPTIVE SIGNIFICANCE OF TRUE IMITATION, DEMONSTRATION TEACHING, AND TRUE SELF-AWARENESS IN TOOL-MEDIATED EXTRACTIVE FORAGING

In the previous sections of this chapter, I presented evidence that great apes, like humans, are capable of intelligent tool use, true imitation, true self-awareness, and teaching by demonstration. I have also argued that these abilities must have arisen in the common ancestor of great apes and humans. In this section I address the adaptive significance of these abilities. Unfortunately, however, reconstructing the adaptive significance of abilities is less straightforward than reconstructing their evolutionary origins. Since I have already noted that abilities are more conservative than the contexts in which they may occur, they do not automatically reveal their original provenance. The difficulty arises from the need to reconstruct past environments in which abilities may have been favored. Given sufficient comparative data on the ecology of primates and other taxa, however, it is possible to do so provisionally by (a) identifying niche correlates of these abilities across taxonomic groups, (b) mapping niche correlates onto the previously constructed behavioral phylogenies (c) identifying unique associations between a particular ability and a particular ecological niche in distantly related species that show behavioral convergences. If successful, this procedure yields a hypothesis regarding the context in which the character might have arisen in a common ancestor (see e.g. Hamilton, 1973). Of the complex of cognitive abilities described above, only intelligent tool use is found in distantly related primate species: fortunately, the small New World cebus monkeys (see Chapters 2 and 3) offer an instructive ecological equivalent for reconstructing the adaptive significance of intelligent tool use. The virtual absence of intelligent tool use among other New World and Old World monkeys indicates that this ability was not present in the common ancestor of cebus monkeys and great apes and hence must have arisen through convergent evolution (Parker & Gibson, 1977).

Like chimpanzees, New World cebus monkeys engage in extractive foraging on a variety of embedded foods. Unlike chimpanzees, however, wild cebus monkeys rarely engage in tool-mediated extraction, but rather bang hard-shelled nuts together and/or against trees to open them (a pattern that has been dubbed proto-tool use (Parker & Gibson, 1977)) and extract the meat (Izawa & Mizuno, 1977). This pattern of destructive foraging on palm nuts (Terborg, 1983) may explain the highly stereotyped banging behavior of this species. It seems likely that the capacity for intelligent proto-tool use and true tool use in cebus monkeys may have evolved as an adaptation for extractive foraging on embedded foods. If so, this parallels the context in which intelligent tool use apparently evolved in chimpanzees[2].

As indicated, the ecological comparison of cebus monkeys and chimpanzees suggests the hypothesis that intelligent tool use arose independently as an adaptation for omnivorous extractive foraging on embedded foods in the ancestors of these two taxa. But what of true self-awareness, true imitation, and demonstration teaching? What kinds of ecological comparison will help to explain the evolution of these abilities? Cebus monkeys, in contrast with chimpanzees, seem to lack true imitation and teaching by demonstration in the development of extractive foraging (Visalberghi & Fragaszy, 1990). They also fail to actively share food with their juvenile offspring as chimpanzees do. They therefore fail as an ecological model for these abilities.

Unfortunately, the search for an ecological model for true imitation, self-awareness, and demonstration teaching is stymied. We are reduced to one or at best two cases, both in sister species: comparative behavioral ecological data on primates suggest that true imitation and demonstration teaching are uniquely associated with social transmission of intelligent tool use in the context of extractive foraging in chimpanzees (e.g. Boesch, 1991; Goodall, 1986; Chapter 18) and perhaps humans (e.g. McGrew, 1992) and their ancestors. (This proposition obviously rests on the contested assertion that chimpanzees indeed display these abilities.) Tool-mediated extractive foraging involves locating, excavating, and/or breaking open shells or other containers of high quality embedded foods (Gibson, 1986; Parker & Gibson, 1977, 1979) such as fruits, ants, termites, and nuts, and locating and transporting appropriate tools for extracting these foods (e.g. Boesch & Boesch, 1984). Whereas chimpanzees rely on extractive foraging for termites and ants during the dry season and for nuts whenever they are available, archeological data on the diet of early hominids suggest that these creatures were specialists in extractive foraging of animal and vegetable foods including bone marrow (Blumenschine,

1989; Parker & Gibson, 1979). Because such foods are difficult to locate in time and space, as well as difficult to process, they require a long apprenticeship of young individuals by their mothers. This is notable among Eastern chimpanzee infants and juveniles who require 4–5 years to become proficient termite fishers (Teleki, 1974), and 6–7 years to become proficient ant fishers (McGrew, 1977). It is even more striking among Western chimpanzee infants who require 7–8 years to become proficient nut crackers in the wild (Boesch, 1993). Whereas the development of termite and ant-fishing may be partially mediated by stimulus enhancement (Galef, 1988) combined with a propensity to poke objects in holes, the development of nut-cracking apparently involves more direct tutelage.

Part of the cost of apprenticeship for chimpanzee mothers comes in sharing difficult-to-process foods with their juvenile offspring; the overall cost is reflected in delayed investment in another offspring. Exceeding chimpanzee mothers in this regard, orangutan mothers invest 7 or 8 years in each offspring, displaying interbirth intervals of at least 7 years (Galdikas & Wood, 1990). Although no teaching has been reported in wild orangutans, the following phenomena suggest that apprenticeship in tool-mediated extractive foraging may have existed in orangutan ancestors: their long interbirth interval, premastication and active food sharing by mothers with their infants (Horr, 1977), food begging by infants from their mothers (Bard, 1990), and a recent report of tool use in foraging (van Schaik & Fox, 1994).

SELECTION FOR TEACHABILITY

Chimpanzee mothers expend considerable parental effort apprenticing offspring to use tools to extract high energy food sources. This parental investment in modeling and demonstrating tool use and sharing food with juveniles could be responsible for their 5 year birth interval (Galdikas & Wood, 1990) and hence considerable reduction in reproductive output.

Because of its high costs in terms of deferred reproduction, the increased brain growth required for the various abilities involved in apprenticeship should evolve only if it is offset by increased fitness and reproductive success (Parker, 1990). This formulation is consistent with Caro & Hauser's stricture that teaching evolved only when the benefit to the pupil outweighs the cost to the teacher (and hence the benefits outweigh the costs to the teacher as well when the offspring is the pupil).

Selection would favor mothers whose offspring were capable of imitating demonstrations of how to use tools to extract difficult-to-process foods because mothers could reduce the length and intensity of their investment in that offspring, and hence afford to bear another offspring sooner. Just as active sharing of difficult-to-process foods with juveniles becomes less expensive for chimpanzee mothers than does lactation (Silk, 1978), so teaching juveniles how to process such foods may become less expensive than sharing these foods under some conditions. Reliance on nuts by western chimpanzees in the Taï Forest suggests the sort of social selection that might have favored teaching abilities in mothers that hastened the independence of their offspring.

Interestingly, and consistent with their convergent evolution of extractive foraging behavior, cebus monkeys apparently display neither true imitation nor demonstration teaching behaviors (Visalberghi & Fragaszy, 1990; Chapter 3). Although little is known about the development of extractive foraging in cebus monkeys, these indirect lines of evidence suggest that it is considerably less social than that of the chimpanzee (Robinson, 1988). More direct experimental evidence on the development of tool use in cebus monkeys suggests that it has developed by 2 years of age (Antinucci, 1989; Parker & Potì, 1990) as compared to 3 or 4 or even 7 or 8 years in chimpanzees. This makes sense given the more rapid maturation and development of cebus monkeys and the much shorter interbirth interval among cebus monkeys: a mother's parental investment in an offspring after the first year apparently does not preclude investment in a subsequent offspring.

Because hominids probably relied primarily, rather than secondarily as chimpanzees do, on tool-extractible foods, hominid mothers must have exerted even stronger selection on imitation and self-awareness in apprenticeship. During the course of human evolution, apprenticeship must have become increasingly significant in hominid infant and juvenile development. By embedding instrumental cognition in a social context, apprenticeship in great apes initiated the transition from tool use to technology, from the individual tool user to interpersonal task specialization, role complementarity, and symbolic coordination that are characteristic of human technology (Reynolds, 1993).

AN ATAVISTIC SOLUTION TO AN EVOLUTIONARY PUZZLE

Notice that the foregoing model of true imitation and teaching by demonstration as adaptations for apprenticeship in tool-mediated extractive foraging is based on chimpanzee behavior in the wild. This leads to a conundrum: how can this chimpanzee-based model explain the reports of demonstration teaching, true imitation, and true self-awareness in captive gorillas when gorillas do not use tools for extractive foraging in the wild, or in orangutans who have only recently been observed using tools in the wild (van Schaik & Fox, 1994)? Why do other great ape species display these abilities when they do little or no tool-mediated extractive foraging? This question has two related answers: the first answer is phylogenetic inertia, and the second is adaptive radiation with movement into different feeding strategies (Pilbeam & Gould, 1974; Wilson, 1975).

Regarding phylogenetic inertia, it is important to remember that abilities remain after contexts change; it is the abilities themselves rather than the abilities in context that are the character states, for example, the capacity for intelligent tool use or true imitation rather than tool-mediated extractive foraging or imitation of tool use. The occurrence of intelligent tool use, true imitation, teaching by demonstration, and symbol use in captive gorillas and orangutans (e.g. Chevalier-Skolnikoff, 1977; Miles, 1990; 1983; Patterson, 1980; Russon & Galdikas, 1993a, b) in any context is evidence of persistence of ancestral character states through phylogenetic inertia.

Contexts change rapidly in evolution; feeding contexts or niches are particularly likely to change with adaptive radiations. Descendants of a common ancestral species typically evolve a range of body sizes and associated dietary niches that differentiate them from one another and from their common ancestor. Whereas the common ancestral species typically is a small-bodied omnivore, at least some of the descendant species evolve larger body sizes and associated preference for lower-quality bulkier foods. This pattern has been repeated in many adaptive radiations in birds and mammals; it can be seen, for example, in ungulates, carnivores (Stanley, 1973), and primates, for example, guenons (Shea, 1986), lesser apes (Raemaekers, 1984), great apes, and Australopithecines (Pilbeam & Gould, 1974).

Evidence for remnants of omnivorous tool-mediated extractive foraging in gorillas can be found in recent reports of greater dietary breadth of western lowland gorillas as compared to mountain gorillas (Kuroda & Tutin, 1992). Newly discovered feeding patterns include consumption of ants and termites (Tutin & Fernandez, 1983; Watts, 1989). Possible evidence for earlier apprenticeship can be seen in the complex food processing of mountain gorillas (Bryne & Byrne, 1991). Echos of apprenticeship can be seen in the complex food processing gorilla infants apparently learn through program-level imitation, i.e. through imitation of the higher-level process but not necessarily the exact hand movements (Bryne & Byrne, 1991).

Likewise, evidence for persistence of omnivorous tool-mediated extractive foraging in orangutans can be seen in recent reports of tool-mediated extraction of social insects (van Schaik & Fox, 1994) in wild orangutans. Intimation of apprenticeship for foraging can be seen in a recent study of imitation by ex-captive orangutans that found extensive imitation of tool use (Russon & Galdikas, 1995). Indeed, if orangutans, chimpanzees, and hominids all display tool-mediated extractive foraging, adaptive radiation into a new feeding niche would be the most parsimonious model to explain the absence in this pattern in gorillas.

All of these considerations are consistent with the hypothesis that tool-mediated extractive foraging arose in a small-bodied common ancestor and persisted predominantly in chimpanzees and hominids. The associated capacities for intelligent tool use, true imitation, true self-awareness, and teaching by demonstration, which were present in the common ancestor of the great apes, were maintained in the other great apes by phylogenetic inertia, and/or secondarily used for new purposes unrelated to their original function. Despite their high cost, the K strategy life history features associated with apprenticeship (long immaturity, long interbirth interval) have persisted among all the great apes and hominids. This hypothesis is the most parsimonious one in that it explains the co-occurrence of true tool use, true imitation, true self-awareness, and demonstration teaching as an integrated complex arising in close succession within a single adaptive context. According to this interpretation, the presence of these abilities in captive gorillas and orangutans is atavistic or vestigial (for a somewhat similar interpretation of self-awareness in the gorilla, Koko, see Povinelli, 1994).

SUMMARY AND CONCLUSIONS

In this chapter I have presented broad functional definitions of imitation, teaching, and self-awareness that are useful for comparative studies. Following these broad definitions,

I have presented models for the proximate cognitive mechanisms underpinning these phenomena in great apes. These models, which are based on the stages of development of imitation and self-awareness in human infants and children, help illuminate the structural relationships among true imitation, teaching by demonstration, and true self-awareness. After outlining these relationships, using a cladistic method of character mapping, I have reconstructed the origins of these cognitive abilities in the common ancestor of the great apes. Finally, I have proposed a model for their adaptive significance in tool-mediated extractive foraging.

The extractive foraging apprenticeship model identifies a complex of abilities including true self-awareness, true imitation, and teaching by demonstration. The model postulates that this complex had emerged in the common ancestor of great apes as an adaptation for apprenticeship in foraging and feeding, most notably in tool-mediated extractive foraging on embedded foods. According to this model, the capacity to engage in true imitation was favored by selection of offspring who were capable of effective learning of extractive foraging techniques. Such offspring were favored because they became independent earlier thereby increasing the reproductive success of their mothers. Subsequent to the evolution of true imitation, the capacity to teach by manual demonstration was favored in mothers by social selection because, in conjunction with true imitation, it facilitated their offspring's learning and hence independence. Because true imitation is associated with true self-awareness and awareness of the other relative to the self, and teaching by demonstration depends on higher levels of self-awareness and awareness of the other, apprenticeship favored the evolution of self-awareness. The apprenticeship model explains the existence of the cognitive complex of intelligent tool use, true self-awareness, true imitation, and demonstration teaching in captive gorillas and orangutans as a vestige of the adaptation for apprenticeship for tool-aided extractive foraging of the common ancestor they shared with chimpanzees and humans.

ACKNOWLEDGEMENTS

I thank my coeditors, Anne Russon and Kim Bard, for their inquiries and helpful suggestions regarding earlier drafts of this chapter. Their efforts contributed significantly to the clarity of the argument presented here.

NOTES

1 As with imitation in animals, imitation in human infants is a subject of active controversy. Piaget's sensorimotor stages of imitation have been challenged by several investigators including Meltzoff (1988, 1990), who has demonstrated both deferred imitation and imitation of novel behaviors in infants at 9–14 months of age, several months earlier than the ages described by Piaget. Because the behaviors imitated by younger infants are simpler than those described by Piaget, and because their imitation does not involve directed "trial-and-error" matching with the model's behaviors, these accomplishments can be viewed as elaborations on Piaget's stages.

2 It is important to note that extractive foraging with anatomical manipulators on a variety of embedded foods has evolved in many species of primates including baboons and macaques and may have been

present in the common ancestor of Old World monkeys, while extractive foraging with specialized anatomical manipulators on single food sources has evolved in several species including parrots and other strong-beaked birds (Gibson, 1986). In other words, extractive foraging can occur with or without tool use and tool-mediated extractive foraging can be either instinctive or intelligent. Some Old World monkeys do show intelligent extractive foraging with anatomical manipulators.

REFERENCES

Anderson, J. (1986). Monkeys with mirrors: Some questions for primate psychology. *International Journal of Primatology*, 5, 81–97.

Anderson, J. (1994) The monkey in the mirror: A strange conspecific. In *Self-awareness in Animals and Humans: Developmental Perspectives*, ed. S.T. Parker, R.W. Mitchell & M.L. Boccia, pp. 315–329. New York: Cambridge University Press.

Antinucci, F. (ed.) (1989) *Cognitive Structure and Development in Nonhuman Primates*. Hillsdale, NJ: Lawrence Erlbaum.

Bard, K. (1990). "Social tool use" by free ranging orangutans: A Piagetian and developmental perspective on the manipulation of an animate object. In *Self-Awareness in Animals and Humans: Developmental Perspectives*, ed. S.T Parker, R.W. Mitchell & M.L. Boccia, pp. 356–378. New York: Cambridge University Press.

Barnett, S.A. (1968). The instinct to teach. *Nature*, 220, 747.

Beck, B. (1980). *Animal Tool Behavior: The Use and Manufacture of Tools by Animals*. New York: Garland STPM Press.

Blumenschine, R. (1989). A landscape taphonomic model of the scale of prehistorical scavenging opportunities, *Journal of Human Evolution*, 18, 345–71.

Boccia, M. (1994). Mirror behavior in macaques. In *Self Awareness in Animals and Humans: Developmental Perspectives*, ed. S.T. Parker, R.W. Mitchell & M.L. Boccia, pp. 350–60. New York: Cambridge University Press.

Boesch, C. (1993). Transmission aspects of tool use in wild chimpanzees. In *Tools, Language and Cognition in Human Evolution*, ed. K.R. Gibson & T. Ingold, pp. 171–84. Cambridge: Cambridge University Press.

Boesch, C. & Boesch, H. (1984). Mental maps in wild chimpanzees: An analysis of hammer transports for nut cracking. *Primates*, 25, 160–70.

Boesch, C. (1991). Teaching among wild chimpanzees. *Animal Behaviour*, 41, 530–32.

Bretherton, I. (1984). Introduction: Piaget and event representation in symbolic play. In *Symbolic Play*, ed. I. Bretherton, pp. 3–41. New York: Academic Press.

Byrne, R.W. & Byrne, J.M.E. (1991). Hand preferences in the skilled gathering tasks of mountain gorillas (*Gorilla g. berengei*). *Cortex*, 27, 521–46.

Byrne, R.W. & Whiten, A. (eds.) (1988). *Machiavellian Intelligence: Social Expertise and the Evolution of Intellect in Monkeys, Apes, and Humans*. Oxford: Clarendon Press.

Caro, T. M. & Hauser, M.D. (1992). Is there teaching in nonhuman animals? *Quarterly Review of Biology*, 67, 151–74.

Cicchetti, D. & Beeghly, M. (eds.) (1990). *The Self in Transition: Infancy to Childhood*. Chicago: University of Chicago Press.

Chevalier-Skolnikoff, S. (1977). A Piagetian model for describing and comparing socialization in monkey, ape and human infants. In *Primate Biosocial Development: Biological, Social and Ecological Determinants*, ed. S. Chevalier-Skolnikoff & F. Poirier, pp. 159–87. New York: Garland STPM Press.

Chevalier-Skolnikoff, S. (1983). Sensorimotor development in orangutans and other primates. *Journal of Human Evolution*, 12, 545–63.
Cook-Gumpertz, J. (1992). Gendered contexts. In *The Contextualization of Language*, ed. P. Auer & A. di Luzio, pp. 177–98. New York: Benjamin Publishing Co.
Custance, D. & Bard, K. (1994). The comparative and developmental study of self-recognition and imitation: The importance of social factors. In *Self-awareness in Animals and Humans: Developmental Perspectives*, ed. S.T. Parker, R.W. Mitchell & M.L. Boccia, pp. 207–25. New York: Cambridge University Press.
Damon, W. & Hart, D. (1988). *Self-understanding in Childhood and Adolescence*. New York: Cambridge University Press.
de Waal, F. (1983). *Chimpanzee Politics*. New York: Harper and Row.
Eibl-Eibesfeldt, I. (1975). *Ethology: The Biology of Behavior*, 2nd edn. New York: Holt, Rinehart & Winston.
Ewer, R.F. (1969). The "instinct to teach." *Nature*, 222, 698.
Fein, G. G. (1984). The self-building potential of pretend play, or "I got a fish all by myself." In *Child's play: Developmental and Applied*, ed. T.D. Yawkey & A.D. Pelligrini, pp. 125–41. Hillsdale, NJ: Lawrence Erlbaum.
Fouts, R., Fouts, D. & van Cantfort, T.E. (1989). The infant Loulis learns signs from cross-fostered chimpanzees. In *Teaching Sign Language to Chimpanzees*, ed. R.A. Gardner B.T. Gardner & T.E. van Cantfort, pp. 280–90. New York: State University of New York Press.
Galdikas, B. & Wood, J.W. (1990). Birth spacing in humans and apes. *Primates*, 83, 185–91.
Galef, B.G., Jr (1988). Imitation in animals: History, definition and interpretation of data from the psychological laboratory.In *Social Learning: Psychological and Biological Perspectives*, ed. T. R. Zentall & B.G. Galef Jr, pp. 1–28. Hillsdale, NJ: Lawrence Erlbaum.
Gallup, G.G., Jr (1970). Chimpanzees: Self-recognition. *Science*, 167, 86–7.
Gallup, G.G., Jr (1977). Self-recognition in primates. *American Psychologist*, 32, 329–38.
Gardner, R.A., Gardner B.T. & van Cantfort, T.E. (eds.) (1989). *Teaching Sign Language to Chimpanzees*, New York: State University of New York Press.
Gibson, K.R. (1986). Cognition, brain size and the extraction of embedded food resources. In *Primate Ontogeny, Cognition and Social Behaviour*, ed. J.C. Else & P.C. Lee, pp. 93–105. Cambridge: Cambridge University Press.
Goodall, J. (1986). *The Chimpanzees of Gombe: Patterns of Behaviour*. Cambridge, MA: The Belknap Press.
Gopnik, A. & Meltzoff, A. (1994). Minds, bodies and persons: Young children's understanding of the self and others as reflected in imitation and theory of mind research. In *Self-awareness in Animals and Humans: Developmental Perspectives*, ed. S.T. Parker, R.W. Mitchell & M.L. Boccia, pp. 166–85. New York: Cambridge University Press.
Hamilton, W.J., III (1973). *Life's Color Code*, New York: McGraw Hill.
Hart, D. & Fegley, S. (1994). Social imitation and the emergence of a mental model of self. In *"Language" and Intelligence in Monkeys and Apes: Comparative Developmental Perspectives*, ed. S.T. Parker & K.R. Gibson, pp. 149–65. New York: Cambridge University Press.
Hayes, K. & Hayes, C. (1952). Imitation in a home-raised chimpanzee. *Journal of Comparative and Physiological Psychology*, 45, 450–59.
Horr, D. (1977). Orangutan maturation: Growing up in a female world. In *Primate Biosocial Development: Biological, Social and Ecological Determinants*, ed. S. Chevalier-Skolnikoff & F. Poirier, pp. 289–322. New York: Garland STPM Press.
Izawa, K. & Mizuno, A. (1977). Palm nut cracking behavior of wild black-capped capuchins. *Primates*, 18, 773–92.

Jensvold, M.L.A. & Fouts, R. S. (1994). Imaginary play in chimpanzees (*Pan troglodytes*). *Human Evolution*, 8, 217–27.
King, B.J. (1994). *The Information Continuum*, Seattle: University of Washington Press.
Kuroda, S. & Tutin, C. (1992). Field studies of African apes in tropical rain forest. Paper presented at the XIVth Congress of the International Primatological Society. Strasbourg, France, 16–21 August.
Laidler, K. (1980). *The Talking Ape*. New York: Stein & Day.
Lewis, M. & Brooks-Gunn, J. (1979). *Social Cognition and the Acquisition of Self*. New York: Plenum Press.
Lewis, M., Sullivan, M.W., Stanger, C. & Weiss, M. (1989). Self development and self-conscious emotions. *Child Development*, 60, 146–56.
Margoliash, D. & Fortune, E.S. (1992). Temporal and harmonic combination sensitive neurons in the Zebra finch's Hvc. *Journal of Neuroscience*, 12, 4309–26.
Mathieu, M. & Bergeron, G. (1981). Piagetian assessment on cognitive development in chimpanzee (*Pan troglodytes*). In *Primate Behavior and Sociobiology*, ed. A.B. Chiarelli & R.S. Corruccini, pp. 143–7. Berlin: Springer-Verlag.
McGrew, W. (1977). Socialization and object manipulation of wild chimpanzees. In *Primate Biosocial Development: Biological, Social, and Ecological Determinants*, ed. S. Chevalier-Skolnikoff & F. Poirier, pp. 261–88. New York: Garland STPM Press.
McGrew, W. (1992). *Chimpanzee Material Culture*. New York: Cambridge University Press.
Mead, G. H. (1970) *Mind, Self and Society*. Chicago: University of Chicago Press.
Meltzoff, A. N. (1988). The human infant as *Homo imitans*. In *Social Learning: Psychological and Biological Perspectives*, ed. T.R. Zentall & B.G. Galef Jr, pp. 319–41. New York: Lawrence Erlbaum.
Meltzoff, A. N. (1990). Foundations for developing a concept of self: The role of imitation in relating self to other and the value of social mirroring, social modeling, and self practice in infancy. In *The Self in Transition: Infancy to Childhood*, ed. D. Cicchetti & M. Beeghly, pp. 139–64. Chicago: University of Chicago Press.
Meltzoff, A.N. & Moore, M.K. (1977). Imitation of facial and manual gestures by human neonates. *Science*, 198, 75–8.
Mignault, C. (1985). Transition between sensorimotor and symbolic activities in nursery-reared chimpanzees (*Pan troglodytes*). *Journal of Human Evolution* 14, 747–58.
Miles, H.L. (1990). The cognitive foundations for reference in a signing orangutan. In *"Language" and Intelligence in Monkeys and Apes: Comparative Developmental Perspectives*, ed. S.T. Parker & K.R. Gibson, pp. 511–39. New York: Cambridge University Press.
Miles, H.L. (1994). ME CHANTEK: The development of self-awareness in a signing orangutan, In *Self-awareness in Animals and Humans: Developmental Perspectives*, ed. S.T Parker, R.W. Mitchell & M.L. Boccia, pp. 254–72. New York: Cambridge University Press.
Mitchell, R. W. (1987). A comparative-developmental approach to understanding imitation. In *Perspectives in Ethology*, ed. P.P.G. Bateson & P.H. Klopfer, vol. 7, pp. 183–215. New York: Plenum Press.
Mitchell, R. W. (1993). Mental models of mirror-self-recognition: Two theories. *New Ideas in Psychology*, 11, 295–325.
Mitchell, R. W. (1994). Multiplicities of self. In *Self-awareness in Animals and Humans: Developmental Perspectives*, ed. S.T. Parker, R.W. Mitchell & M.L. Boccia, pp. 81–107. New York: Cambridge University Press.
Mounoud, P. & Vinter, A. (1981). Representation and sensorimotor development. In *Infancy and Epistemology*, ed. G. Butterworth, pp. 200–35. Brighton: Harvester Press.

Nagel, K., Olguin, R. & Tomasello, M. (1993). Processes of social learning in the tool use of chimpanzees (*Pan troglodytes*) and human children (*Homo sapiens*). *Journal of Comparative Psychology*, 107, 174–86.

Parker, S.T. (1977). Piaget's sensorimotor series in an infant macaque: A model for comparing unstereotyped behavior and intelligence in human and nonhuman primates. In *Primate Biosocial Behavior: Biological, Social, and Ecological Determinants*, ed. S. Chevalier-Skolnikoff & F.E. Poirier, pp. 43–113. New York: Garland STPM Press.

Parker, S.T. (1990). Why big brains are so rare. In *"Language" and Intelligence in Monkeys and Apes: Comparative Developmental Perspectives*, ed. S.T. Parker & K.R. Gibson, pp. 129–55. New York: Cambridge University Press.

Parker, S. T. (1991). A developmental approach to the origins of self-recognition in great apes. *Human Evolution*, 6, 435–49.

Parker, S.T. & Gibson, K.R. (1977). Object manipulation, tool use, and sensorimotor intelligence as feeding adaptation in cebus monkeys and great apes. *Journal of Human Evolution*, 6, 623–41.

Parker, S.T. & Gibson, K.R. (1979) A developmental model for the evolution of language and intelligence in early hominids. *Behavioral and Brain Sciences*, 2, 367–408.

Parker, S.T. & Poti, P. (1990). The role of innate motor patterns in ontogenetic and experiential development of intelligent use of sticks in cebus monkeys. In *"Language" and Intelligence in Monkeys and Apes: Comparative Developmental Perspectives*, ed. S.T. Parker & K.R. Gibson, pp. 219–46. New York: Cambridge University Press.

Patterson, F. (1980). Innovative uses of language by a gorilla: A case study. In *Children's Language*, ed. K.E. Nelson, vol. 2, pp. 497–561. New York: Gardner Press.

Patterson, F. & Cohn, R. (1994). Self-recognition and self-awareness in lowland gorillas. In *Self-awareness in Animals and Humans: Developmental Perspectives*, ed. S.T. Parker, R.W. Mitchell & M.L. Boccia, pp. 373–90. New York:Cambridge University Press.

Patterson, F. & Linden, E. (1981). *The Education of Koko*. New York: Holt, Rinehart & Winston.

Piaget, J. (1962). *Play, Dreams and Imitation in Children*. New York: Norton.

Piaget, J. (1970). *Genetic Epistemology*. New York: Norton.

Pilbeam, D. & Gould, S.J. (1974). Size and scaling in human evolution. *Science*, 186, 892–901.

Poti, P. & Spinozzi, G. (1995). Early sensorimotor development in chimpanzees (*Pan troglodytes*). *Journal of Comparative Psychology*. (in press.)

Povinelli, D. (1994). How to create a self-recognizing gorilla (but don't try it on a macaque). In *Self-awareness in Animals and Humans: Developmental Perspectives*, ed. S.T. Parker, R.W. Mitchell & M.L. Boccia, pp. 291–300. New York: Cambridge University Press.

Povinelli, D., Nelson, K.E. & Boysen, S. T. (1992). Comprehension of social role reversal by chimpanzees: Evidence of empathy? *Animal Behaviour*, 44, 269–81.

Premack, D. (1984). Pedagogy and aesthetics as sources of culture. In *Handbook of Cognitive Neuroscience*, ed. M. Gazzaniga, pp. 15–35. New York: Plenum Press.

Raemaekers, J. (1984). Large vs small gibbons: Relative roles of bioenergetics and competition in their ecological segregation. In *The Lesser Apes: Evolutionary and Behavioral Biology*, ed. H. Preuschoft, D. Chivers, W. Brockelman & N. Creel, pp. 209–219. Edinburgh: Edinburgh University Press.

Reynolds, P. (1993). The complementation theory of language and tool use. In *Tools, Language and Cognition in Human Evolution*, ed. K.R. Gibson & T. Ingold, pp. 407–21. Cambridge: Cambridge University Press.

Ridley, M. (1984). *Classification and Evolution*. New York: Longman.

Robinson, J. (1988). Demography and group structure in wedge caped capuchin monkeys (*Cebus ninigrivaittohus*). *Animal Behaviour*, 29, 1036–56.
Rogoff, B. (1992). *Apprenticeship in Thinking*. New York: Oxford University Press.
Russon, A. & Galdikas, B. (1993a). Imitation in free-ranging rehabilitant orangutans (*Pongo pygmaeus*). Paper presented at the annual meeting of the American Anthropological Association. Washington, DC.
Russon, A. & Galdikas, B. (1993b). Imitation in free-ranging rehabilitant orangutans (*Pongo pygmaeus*). *Journal of Comparative Psychology*, 107, 147–61.
Russon, A. & Galdikas, B. (1995). Constraints on great apes' imitation: Model and action selectivity in rehabilitant orangutan (*Pongo pygmaeus*) imitation. *Journal of Comparative Psychology*, 109, 5–17.
Savage-Rumbaugh, S. & MacDonald, K. (1988). Deception and social manipulation in symbol-using apes. In *Machiavellian Intelligence: Social Expertise and the Evolution of Intellect in Monkeys, Apes and Humans*, ed. R.W. Byrne & A. Whiten, pp. 224–37. Oxford: Clarendon Press.
Savage-Rumbaugh, S., Romski, M.A., Hopkins, W.D. & Sevik, R. (1989). Symbol acquisition and use by *Pan troglodytes, Pan paniscus, Homo sapiens*. In *Understanding Chimpanzees*, ed. P. Heltne & L. Marquardt. pp. 266–95. Cambridge, MA: Harvard University Press.
van Schaik, C. P. & Fox, E.A. (1992). Tool use in wild Sumatran orangutans (*Pongo pygmaeus*). Paper presented at the XVth Congress of the International Primatological Society, Kuta-Bali, Indonesia, 3–8 August.
Shea, B. (1986). Ontogenetic approaches to sexual dimorphism in anthropoids. *Journal of Human Evolution*, 1, 97–110.
Silk, J. (1978). Patterns of food sharing among mother and infant chimpanzees at the Gombe Stream National Park in Tanzania. *Folia Primatologica*, 29, 129–41.
Snow, C. (1990). Building memories: The ontogeny of autobiography. In *The Self in Transition: Infancy to Childhood*, ed. D. Cicchetti & M. Beeghly, pp. 213–41. Chicago: University of Chicago Press.
Speidel, G. E. & Nelson, K.E. (1989). A fresh look at imitation in language learning. In *The Many Faces of Imitation in Language Learning*, ed. G.E. Speidel & K.E. Nelson, pp. 1–22. New York: Springer-Verlag.
Stanley, S.M. (1973). An explanation for Cope's rule, *Evolution*, 27, 1–26.
Stern, D. (1992). *The Interpersonal World of the Infant*. New York: Basic Books.
Swartz, K. & Evans, S. (1994). Social and cognitive factors in chimpanzee and gorilla mirror behavior and self-recognition. In *Self-awareness in Animals and Humans: Developmental Perspectives*, ed. S.T Parker, R.W. Mitchell & M.L. Boccia, pp. 189–206. New York: Cambridge University Press.
Teleki, G. (1974). Chimpanzee subsistence technology: Materials and skills. *Journal of Human Evolution*, 4, 125–84.
Temerlin, M.K. (1977), *Lucy: Growing up Human*. New York: Bantam. (First published in 1975.)
Terborgh, J. (1983). *Five New World Primates*. Princeton: Princeton University Press.
Terrace, H.S. (1979). *Nim*. New York: Knopf.
Thompson, R. L. & Boatright-Horowitz, S. (1994). The question of mirror-mediated self-recognition in apes and monkeys: Some new results and reservations. In *Self-awareness in Animals and Humans: Developmental Perspectives*, ed. S.T Parker, R.W. Mitchell & M.L. Boccia, pp. 330–49. New York: Cambridge University Press.
Tomasello, M. (1990). Cultural transmission in the tool use and communicatory signaling of chimpanzees? In *"Language" and Intelligence in Monkeys and Apes: Comparative Developmental*

Perspectives, ed. S.T. Parker & K.R. Gibson, pp. 274–311. New York: Cambridge University Press.

Tomasello, M., Savage-Rumbaugh, E.S. & Kruger, A. (1993). Imitative learning of actions on objects by children, chimpanzees, and enculturated chimpanzees. *Child Development*, **64**, 1688–705.

Trivers, R. (1972). Parental investment and sexual selection. In *Sexual Selection and the Descent of Man*, ed. B. Campbell, pp. 136–79. New York: Aldine.

Tutin, C. & Fernandez, M. (1983). Gorillas feeding on termites in Gabon, West Africa. *Journal of Mammology*, **6**, 530–1.

Visalberghi, E. & Fragaszy, D. (1990) Do monkeys ape? In *"Language" and Intelligence in Monkeys and Apes: Comparative Developmental Perspectives*, ed. S.T. Parker & K.R. Gibson, pp. 247–73. New York: Cambridge University Press.

Vygotsky, L. (1962). *Thought and Language*. Cambridge, MA: MIT Press.

Watson, J.S. (1994). Detection of self: The perfect algorithm. In *"Language" and Intelligence in Monkeys and Apes: Comparative Developmental Perspectives*, ed. S.T. Parker & K.R. Gibson, pp. 131–48. New York: Cambridge University Press.

Watts, D. P. (1989). Ant eating behaviour of mountain gorillas. *Primates*, **30**, 121–5.

Whiten, A. & Ham, R. (1992). On the nature and evolution of imitation in the animal kingdom: Reappraisal of a century of research. In *Advances in the Study of Behavior*, ed. P.J.B. Slater, J.S. Rosenblatt, C. Beer & M. Milinski, vol. 21, pp. 239–83. New York: Academic Press.

Wiley, E.O. (1981). *Phylogenetics: Theory and Practice of Phylogenetic Systematics*. New York: Wiley-Liss.

Wilson, E.O. (1975). *Sociobiology*. Cambridge, MA: Harvard University Press.

Wood, D., Bruner, J. & Ross, G. (1976). The role of tutoring in problem solving. *Journal of Child Psychology and Psychiatry*, **17**, 89–100.

Yando, R., Seitz, V. & Zigler, E. (1978). *Imitation: Developmental Perspectives*. New York: Lawrence Erlbaum.

17

The effect of humans on the cognitive development of apes

JOSEP CALL AND MICHAEL TOMASELLO

Homo sapiens is the only animal species whose ontogeny normally takes place in an environment containing both artifacts designed to assist and enhance cognition and adult conspecifics eager to demonstrate and instruct the individual in all kinds of cognitive activities including the use of artifacts. *Homo sapiens* is also the only animal species to acquire during its ontogeny certain cognitive skills, for example skills associated with complex object manipulations and language. Perhaps this is not an accident. Perhaps the presence of a cultural environment, complete with material and symbolic artifacts and adult instruction, is an essential ingredient in the cognitive development of *Homo sapiens* (Vygotsky, 1978).

In this chapter we explore the possibility that the human cultural environment is also an important ingredient in the cognitive development of some nonhuman individuals, specifically some individual great apes (hereafter, simply "apes"). It is quite common in studies of ape cognitive development for observers to note marked individual differences in the cognitive skills of particular apes – sometimes within studies but also across different studies of the same phenomenon. Whereas there are undoubtedly many factors responsible for these differences, one factor may be the degree to which, or the manner in which, particular individuals have been in contact with the human cultural environment. Indeed, most primatologists would probably not be surprised to find different cognitive skills among adult apes who have spent their ontogenies in species typical environments, in human zoos, in human laboratories where they have been taught specific skills, in human-staffed nurseries, and in human homes where they have been treated as if they were human children. But precisely what are the effects of these different environments is very much an open question. Opinions range from those of Boesch (1993), who has argued that apes display their most sophisticated cognitive skills in the natural conditions in which their evolution has taken place and that captive conditions, however rich, only diminish or distort those abilities, to those of Premack (1983), who has argued that when apes are raised with all of the education and training received by human children, especially in the use of symbols, they display a set of abstract cognitive skills that they show under no other conditions.

The major problem in attempting to review evidence related to this issue is that humans may intervene in the lives of apes in many and varied ways that do not always fall neatly into categories. In addition, there are many ape individuals who have spent different parts of their lives in different settings. To make matters even more complex, in many published reports scientists do not report, or report only briefly, on the histories of the individuals being studied. Although there is no perfect solution to these problems, for the purposes of this review, we will use the following very general descriptive terms, but will qualify them with more explicit detail when necessary:

WILD. Apes who have spent their entire lives in their natural habitats.

CAPTIVE. Apes in human captivity who have interacted directly with humans and their artifacts only minimally; this includes many zoo and some laboratory settings.

NURSERY-RAISED. Apes raised from a young age with peer conspecifics and a good deal of contact with humans and their artifacts, but without human training aimed at specific behavioral outcomes.

LABORATORY-TRAINED. Apes raised mostly in human captivity who have been trained in particular tasks, sometimes multiple tasks over many years (some of which might be symbolic).

HOME-RAISED. Apes raised by humans in something like a human cultural environment (sometimes including exposure to or training in symbolic skills); the environment need not literally be a home but must include something close to daily contact with humans and their artifacts in meaningful interactions.

Our goal in this review is to determine as specifically as possible the nature of the effects these different human environments may have on the cognitive development of the four great ape species: chimpanzees (*Pan troglodytes*), bonobos (*Pan paniscus*), orangutans (*Pongo pygmaeus*), and gorillas (*Gorilla gorilla*). We are interested in apes' knowledge both of objects and their causal and logicomathematical relations and of other animate beings and how they operate in the world. After reviewing the available data in each of these broad domains of cognition, we conclude by identifying four very specific mechanisms of human influence that may affect the cognitive development of our nearest primate relatives.

KNOWLEDGE OF OBJECTS

In this section we investigate whether and in what ways apes come to interact with objects in more complex or human-like ways as a function of their social experience with humans. We approach this issue in subsections dealing with each of the major domains in which ape physical cognition has been studied: object permanence, object manipulation, tool use, and categorization. Table 17.1 lists the studies we rely on for this analysis, classifying each of the studies used in terms of the cognitive domain involved and the category of human contact its subjects experienced.

Object permanence

Existing data on the object permanence skills of apes provide little evidence for the significance of rearing environment. Apes seem to have a basic knowledge of the existence and location of objects when the latter are out of sight, whether they are foraging for food in their natural environments (Boesch & Boesch, 1984) or whether they are tracking the invisible displacements of objects in the laboratory. Although Piagetian tasks have not been given to apes in the wild, of course, in the laboratory any number of chimpanzees and gorillas who have had many different types of experience with humans, some very limited, have solved all of the object permanence problems presented to them, including stage 6 invisible displacements. In the one study in which individuals with different backgrounds were explicitly compared on object permanence tasks (Hallock & Worobey, 1984; Table 17.1), a small developmental advantage was found for nursery-reared over captive chimpanzees in the age of emergence of basic search skills. These subjects, however, were not followed longitudinally to assess their final attainments.

Object manipulation

In their natural habitats the four great ape species engage in object manipulation mostly in the context of instrumental situations involving food and nesting materials. These can sometimes be quite complex (see Chapter 5), but there does not seem to be a human-like motivation to manipulate objects as an end in itself. When they have been raised with humans and their artifacts, however, apes seem to become more interested in objects solely as objects. For example, Gómez (1989) found that a nursery-raised gorilla exposed to extensive human interaction on a daily basis, often in concert with human objects, displayed a number of complex object manipulations, including object construction, not typically displayed by gorillas with less human contact – although it should be noted that Gómez's subject was tested at an age slightly older than that of the other gorillas (4 versus 3 years of age). Similarly, a home-raised orangutan exposed to extensive human interaction (Miles, 1990), as well as some rehabilitant orangutans raised by humans (Russon & Galdikas, 1993), have been observed to display object construction skills and human-conventional manipulations not typically observed in orangutans with less human contact (Chevalier-Skolnikoff, 1983; Laidler, 1980). And finally, complex object constructions and conventional manipulations have also been observed in chimpanzees and bonobos that have had extensive contact with humans (e.g. K.E. Brakke & E.S. Savage-Rumbaugh, unpublished results; Hayes & Hayes, 1951), whereas they have been reported for nursery-raised chimpanzees only infrequently and not until the juvenile period.

Symbolic play

A special type of object manipulation is symbolic or pretend play. Symbolic play is very difficult to identify precisely, especially when human artifacts are not involved. For example, Goodall (1986) reported that a young chimpanzee "fished" for ants with a twig at a place where no ants were present – raising the possibility that the chimpanzee was imagining the presence of ants – but many other interpretations of this behavior are also

Table 17.1. *Studies relevant to ape knowledge of objects as a function of domain of knowledge and rearing history of subjects*

	Wild	Captive	Nursery-raised	Laboratory-trained	Home-raised	Other
Object permanence	Boesch & Boesch, 1984; Goodall, 1986 Ages: various	Hallock & Worobey, 1984 Ages: i	Hallock & Worobey, 1984; Mathieu & Bergeron, 1981; Natale et al., 1986; Redshaw, 1978; Spinozzi & Poti, 1993 Ages: i, I, J		Wood et al., 1980 Ages: I	
Object manipulation	Byrne & Byrne, 1993; McGrew, 1977 Ages: various	Chevalier-Skolnikoff, 1983; Vauclair & Bard, 1983 Ages: i, I, J, S	Chevalier-Skolnikoff, 1983; Gómez, 1989; Jacobsen et al., 1932; Knobloch & Pasamanick, 1959; Laidler, 1980; Mignault, 1985; Perinat & Dalmau, 1988; Redshaw, 1978; Spinozzi & Natale, 1989 Ages: i, I, J		K.E. Brakke & E.S. Savage-Rumbaugh, unpublished results; Hayes & Hayes, 1951; Kellogg & Kellogg, 1933; Miles, 1990; Vauclair & Bard, 1983 Ages: i, I, J	Galdikas, 1982; Russon & Galdikas, 1993 Ages: I, J, S, A
Symbolic play	Goodall, 1986 Ages: I	Wimmer & Ettlinger, 1977 (cited in Ettlinger, 1983) Ages: I	Mignault, 1985 Ages: I, J	Savage-Rumbaugh, 1986 Ages: J	Gardner & Gardner, 1978, 1989; Hayes, 1951; Jensvold & Fouts, 1993; Miles, 1990; Patterson & Linden, 1981; Savage-Rumbaugh & McDonald, 1988 Ages: I, J, S, A	
Tool use	Boesch & Boesch, 1990; Goodall, 1986; McGrew, 1989, 1992 Ages: I, J, S, A	Chevalier-Skolnikoff, 1983; Jordan, 1982; Köhler, 1927; Lethmate, 1982; Menzel et al., 1972; Menzel, 1969; Parker, 1969; Savage & Snowdon, 1982; Schiller, 1952; Visalberghi et al., 1995; Yerkes, 1927	K.A. Bard, D.M. Fragaszy & E. Visalberghi, 1993; Chevalier-Skolnikoff, 1983; Natale, 1989a,b; Spinozzi & Natale, 1989 Ages: i, I, J. S, A	Limongelli et al., 1995; Visalberghi et al., 1995 Ages: J, S, A	Hayes & Nissen, 1971; Miles, 1990; Patterson, 1978; Savage-Rumbaugh, 1986; Visalberghi et al., 1995 Ages: I, J, S, A	Galdikas, 1982; Menzel et al., 1970; Rijksen, 1978; Russon & Galdikas, 1993 Ages: I, J, S, A

Use of mirrors		Gallup, 1970; Gallup et al., 1971; Hyatt & Hopkins, 1994; Lethmate & Dücker, 1973; Povinelli et al., 1993; Riopelle et al., 1973; Suarez & Gallup, 1981; Swartz & Evans, 1991, 1994 Ages: I, J, S, A	Lin et al., 1991; Povinelli et al., 1993; Suarez & Gallup, 1981 Ages: I, J, S, A	Savage-Rumbaugh, 1986; Swartz & Evans, 1994 Ages: J, A	Hayes, 1951; W.D. Hopkins, personal communication; Miles, 1994; Patterson & Cohn, 1994 Ages: I, J, A	Gallup et al., 1971; Hill et al., 1970 Ages: J
Categorization	Boesch & Boesch, 1990 Ages: various	Garcha & Ettlinger, 1979; Oden et al., 1990; Rensch, 1973 Ages: i, I, J	Braggio et al., 1979; Mathieu, 1982; Savage-Rumbaugh et al., 1992; Spinozzi, 1993 Ages: i, I, J, A	Matsuzawa, 1990; Premack, 1983; Yerkes & Petrunkevitch, 1925 Ages: J, S, A	Hayes & Nissen, 1971; Miles, 1990; Patterson & Linden, 1981; Savage-Rumbaugh, 1986; Savage-Rumbaugh et al., 1992 Ages: I, J, S	
Numerical skills				Boysen & Bernston, 1990; Matsuzawa, 1990; Rumbaugh et al., 1989 Ages: J, S, A		

i, young infant (0–1 year old); I, old infant (1–3 year old); J, juvenile (3–7 year old); S, subadult (7–12 year old); A, adult (12 year old or older); *italic* references, additional human contact.

possible. Observations of captive apes have not revealed many candidates for pretense either, even when human artifacts are present. For example, Mignault (1985) reported that, with the exception of some limited manipulation of dolls, symbolic play was not observed in four nursery-reared juvenile chimpanzees who had received regular exposure to humans and a variety of toys (see also Wimmer & Ettlinger, 1977, cited in Ettlinger, 1983). Observations of home-raised apes have provided by far the widest variety of possible instances of symbolic play. All four ape species have been observed to play with dolls, engaging in a number of different activities including such things as bathing, feeding, and tickling. Patterson & Linden (1981) reported that the home-raised gorilla Koko pretended to be an elephant by using a fat rubber tube as a substitute for the trunk; Hayes (1951) reported that the home-raised chimpanzee Viki played with a nonexistent pull toy; and the home-raised bonobo Kanzi was observed to engage others in an eating game involving imaginary food (Savage-Rumbaugh & McDonald, 1988). On the other hand, Miles (1990) reported that symbolic play was not observed in her home-raised orangutan Chantek. It is unclear in all of these cases what these behaviors meant for these individuals, and in many cases they may have been simply playing with objects that humans see as symbolic (e.g. dolls), or else mimicking a past act of human pretense. Nevertheless, it still seems important that the only serious candidates for symbolic play are all provided by home-raised apes.

Tool use

Of the four great ape species, only chimpanzees use tools on a regular basis in their natural habitats (McGrew, 1989). Chimpanzees use, and in some cases modify, more than a dozen different kinds of object as tools to solve problems, most of them having to do with the acquisition of food (McGrew, 1992). Given chimpanzees' extensive use of tools in the wild, it is not surprising that in captivity they use tools extensively as well. Chimpanzees deprived of experience with either conspecifics or objects do not use tools efficiently, although this deficiency can be reversed with experience (Menzel et al., 1970, Schiller, 1952). Orangutans, bonobos, and gorillas show a different and very interesting pattern of tool use. Gorillas and orangutans with very little human contact use various human-like tools in a variety of problem-solving situations (see Table 17.1). Rehabilitant orangutans – raised with various types of human contact – also display a rich variety of tool-using skills (Russon & Galdikas, 1993), and there is some evidence that captive bonobos have some of these skills as well (Jordan, 1982). Home-raised orangutans, gorillas, and bonobos have also learned to use a wide variety of human tools. There are no studies that directly compare the tool use performance of apes with different rearing backgrounds, but Visalberghi and colleagues have recently completed a series of studies (see Table 17.1) in which the performance of chimpanzees and bonobos with different rearing histories can be compared, and they found that different amounts of experience with humans was not associated with different levels of proficiency in their tool use task.

Use of mirrors

Apes have also been observed and tested in their ability to use a somewhat special human tool, the mirror (Gallup, 1970). Of interest are self-directed behaviors, in which individuals use a mirror to explore body parts that would not otherwise be visually accessible, including especially self-directed behaviors in the so-called mark test in which subjects seek to remove a painted spot that has been surreptitiously placed on a visually inaccessible part of their face. For obvious reasons, apes in the wild have not been systematically observed in interaction with mirrors, and so we do not know how they might use them. However, numerous studies have found that individuals of all four ape species with very limited human contact display self-directed behaviors and pass the mark test (see Captive column in Table 17.1). Subjects raised with extensive human contact – several chimpanzees, one bonobo, two gorillas, and one orangutan – have displayed these behaviors as well. Interestingly, two studies that investigated socially-deprived chimpanzees (deprived of both conspecific and human contact) found that they uniformly failed the mark test (Gallup et al., 1971), although this effect can be reversed with later social contact with conspecifics (Hill et al., 1970). It is thus conceivable that socialization plays an important role in the development of self-recognition in chimpanzees, but not necessarily human socialization.

The conclusion would thus seem to be that rearing environment has a definite effect on the tool-using skills of apes, as three of four species use tools extensively only in captivity. The effect does not seem to be a profound one, however, since simple exposure to situations in which tools might be useful is enough to elicit tool use. Ape use of mirrors to perform self-directed behaviors does not seem to be affected to any significant degree by the amount of human contact. Because socially deprived chimpanzees have failed the mark test, it is possible that social interaction in general, though not necessarily of the human kind, is important to the development of this skill.

Categorization

If categorization is defined as the sorting of objects into distinct spatial groupings on the basis of their perceptual or functional features, then there are only a few instances of ape categorization in the wild. However, apes do a number of things in their natural environments that indicate a kind of categorization, for example the gathering together of particular types of food item (e.g. nuts for cracking, plants ripe for eating, etc.), or materials for making tools (Boesch & Boesch, 1990). Wild apes thus clearly show some natural categorization abilities in a number of different functional contexts.

When placed in front of human objects, captive chimpanzees do not spontaneously sort them into groups on the basis of their perceptual or functional features (Garcha & Ettlinger, 1979; Mathieu, 1982), nor do bonobos (Savage-Rumbaugh et al., 1992). They may, however, interact with them in an order that implies a recognition of common features, for example by touching *seriatim* all of the objects with a certain shape (Rensch, 1973; Spinozzi, 1993). Home-raised apes and laboratory-trained apes, however, often do sort objects into distinct groups; for example, Matsuzawa (1990) found several cases of spontaneous sorting in a laboratory-trained, language-trained chimpanzee. Spontaneous classification of different items according to certain morphological features such as material, color, shape, or size has also been reported in one home-raised chimpanzee (Hayes & Nissen, 1971) and one home-raised bonobo (Savage-Rumbaugh, 1986).

With regard to the other two ape species, a home-raised gorilla (Patterson & Linden, 1981) and a home-raised orangutan (Miles, 1990) have also been reported to display categorization skills. It is relevant here that chimpanzee subjects who do not spontaneously classify objects can in many cases be taught to do so (Garcha & Ettlinger, 1979), although perhaps not with the skill of the human-raised ape of Hayes & Nissen (1971).

Premack (1983) found that only laboratory-trained, language-trained chimpanzees were capable of solving analogies (i.e. categorizing the relations of pairs of stimuli) in a match-to-sample procedure; captive chimpanzees failed these tasks despite extensive training. Premack (1983) originally interpreted these results as evidence for the necessity of an abstract code (i.e. language) to develop complex relational categories. More recently, however, Oden et al. (1990) found that very young captive infant chimpanzees spontaneously perceived relations between relations of stimuli (i.e. analogies between pairs of stimuli) when this was measured more generously, i.e. when experimenters simply noted the time individuals played with objects related to one another in different ways (a finding consistent with the studies on serial touching of Rensch (1973) and Spinozzi (1993)). The revised conclusion is thus that the presence of an abstract code is not necessary for the perception of relations of same/different between objects and between pairs of objects (i.e. natural categorization), but it might still be the case that using this perceptual ability to physically group objects (i.e. classification) may require some additional attention-management skills in which humans play a role.

Numerical skills
An important set of behaviors related to categorization skills comprises the numerical skills of chimpanzees. In this case, the only way of testing individuals is by giving them laboratory training in how to respond in the presence of particular types of stimuli. Research reported by Rumbaugh et al. (1989), Matsusawa (1990), Boysen & Bernston (1990), and Boysen (Chapter 8) show that laboratory-trained chimpanzees display a variety of numerical skills involving the comparing of quantities, the use of Arabic numerals, and so forth. It is interesting to note, however, that monkeys with little human contact show very rudimentary numerical skills in the form of subitizing (perceptual discrimination of different numerosities; Thomas, 1992).

Overall, then, rearing environment does seem to have some effect on apes' categorization and classification skills. Apes with extensive human contact show more human-like classificatory skills in physically sorting objects into groups than nursery-raised or other captive apes. However, in other tasks related to natural categorization skills – e.g. touching similar objects serially or perceiving categories as measured by manipulative time – captive apes perform at the same level as human-raised apes. It is thus likely that all apes have some skills of natural categorization at the perceptual level, but that explicit classification on a more conceptual level (coordinating similarities and differences simultaneously and systematically) requires some additional attention-management skills and that these might develop only in the context of human instruction.

SOCIAL KNOWLEDGE

A number of scientists have argued that the cognitive capacities of primates are most naturally deployed when they are attempting to solve social, not physical problems (e.g. Humphrey, 1976; Jolly, 1966). In their natural environments apes must cooperate and compete with conspecifics over access to resources, communicating with and learning from them in the process (de Waal, 1982). In this section we review existing data for the effect of humans on the development of ape social cognition in subsections dealing with each of the major domains in which it has been studied: social attention, intentional communication, social learning, cooperation, and "theory of mind." Table 17.2 lists the studies we rely on for this analysis, classifying each of the studies used in terms of the socio-cognitive domain involved and the category of human contact its subjects experienced.

Social attention
Using the Neonatal Behavioral Assessment Scale, Hallock *et al.* (1989) found that two human-reared chimpanzees were more responsive in their orientation to social stimuli (e.g. vocalizations) than were seven mother-reared conspecifics. Bard *et al.* (1992) compared 13 nursery-raised chimpanzee neonates to Hallock *et al.*'s seven mother-reared chimpanzees and also found significant differences in attention to human stimuli.

Social referencing
Social referencing is a form of social attention in which an individual actively monitors the reaction of another individual to some outside entity or event and then uses information about this reaction in formulating its own response to that entity or event. There is no question that all four great ape species in their natural environments are affected in some generic way by the behaviors and emotional reactions of groupmates; when a fearful or exciting event is occurring fear and excitement seem to "spread" among individuals (Sugiyama, 1972). With regard to captive apes, Evans & Tomasello (1986) observed that the infant and juvenile chimpanzees of a captive colony interacted more frequently with their mothers' preferred social partners than with other adults (irrespective of mothers' physical distance), raising the possibility of social referencing. Menzel *et al.* (1972) provoked fearful or cautious responses to two novel objects in some young captive and isolate chimpanzees and then watched this spread to other chimpanzees as they added them to the group. Miller *et al.* (1990) introduced nursery-reared infant chimpanzees to a novel, potentially frightening human in the presence of a familiar human care, and 2 year old males (as opposed to other subjects) increased their looking time to the caregiver when compared to baseline levels. With regard to home-raised chimpanzees, Kellogg & Kellogg (1933) and Hayes (1951) observed that their home-raised chimpanzees looked to them in many situations where they were frightened or uncertain (for the same phenomenon in laboratory-trained chimpanzees, see Savage-Rumbaugh *et al.*, 1983).

Joint attention
Joint attention is similar to social referencing – they are both triadic interactions involving two individuals and some outside entity – but it mainly concerns the visual

Table 17.2. *Studies relevant to ape social knowledge as a function of domain of knowledge and rearing history of subjects*

	Wild	Captive	Nursery-raised	Laboratory-trained	Home-raised	Other
Social attention		Hallock et al., 1989 Ages: i	Bard, 1992a; Bard et al., 1992; Hallock et al., 1989 Ages: i			
Social referencing	Sugiyama, 1972 Ages: I	Evans & Tomasello, 1986; Menzel et al., 1972 Ages: I, J	Miller et al., 1990 Ages: I, J	Savage-Rumbaugh et al., 1983 Ages: J	Hayes, 1951; Kellogg & Kellogg, 1933 Ages: i, I, J	Menzel et al., 1972 Ages: I, J
Joint attention	Plooij, 1978 Ages: I, J	Bard & Vauclair, 1984; Carpenter et al., 1995; de Waal, 1986; Menzel, 1971, 1973a Ages: i, I, J, S, A	Bard & Vauclair, 1984 Ages: i		Bard & Vauclair, 1984; Carpenter et al., 1995; Hayes, 1951 Ages: i, I, J, S, A	
Intentional communication	Boesch, 1991a; Galdikas & Vasey, 1992; Goodall, 1986; Hauser, 1990; McGinnis, 1979; Plooij, 1978 Ages: I, J, S, A	Carpenter et al., 1995; de Waal & van Hooff, 1981; Savage-Rumbaugh, 1984; Tomasello et al., 1985, 1989, 1994; Yerkes & Nissen, 1939 Ages: I, J, S, A	Berdecio & Nash, 1981; Call & Tomasello, 1994a; Gómez, 1989, 1990 Ages: I, J, S	Povinelli et al., 1992; Premack, 1988; Savage-Rumbaugh et al., 1983; Woodruff & Premack, 1979 Ages: J, S, A	Call & Tomasello, 1994a; Carpenter et al., 1995; Gardner & Gardner, 1989; Miles, 1986, 1990; Patterson, 1978; Patterson & Linden, 1981; Savage-Rumbaugh, 1986, 1988; Savage-Rumbaugh & McDonald, 1988; Savage-Rumbaugh et al., 1985, 1986 Ages: I, J, S, A	Bard, 1992b Ages: I, J

Social learning	Boesch, 1992; Goodall, 1986; McGrew & Tutin, 1978; Nishida, 1980 Ages: various	Call & Tomasello, 1994b; de Waal, 1982; Ettlinger, 1983; Haggerty, 1913; Nagell et al., 1993; Tomasello et al., 1987; Wright, 1972; Yerkes, 1927 Ages: J, S, A	K.A. Bard, D.M. Fragaszy & E. Visalberghi, unpublished results; Call & Tomasello, 1994b; Custance & Bard, 1994; Gómez, 1989; *Mathieu, 1982*; Nagell et al., 1993; Perinat & Dalmau, 1988; *Tomasello et al., 1987, 1993b* Ages: I, J, S, A	Fouts, 1972; Fouts et al., 1989; Gardner & Gardner, 1978; Hayes, 1951; Hayes & Hayes, 1952; Miles, 1990; Miles et al. (Chapter 13); Patterson, 1978; Sanders, 1985; Savage-Rumbaugh et al., 1986, 1992; Temerlin, 1975; Terrace, 1979; Tomasello et al., 1993b Ages: I, J, S, A	Galdikas, 1982; Russon & Galdikas, 1993 Ages: I, J, S, A
Teaching	Boesch, 1991b; Goodall, 1986; van Lawick-Goocall, 1968; van de Rijt-Plooij & Plooij, 1987 Ages: A		K.A. Bard, unpublished results; Crawford, 1937; *Savage-Rumbaugh, 1984*; Yerkes & Tomilin, 1935 Ages: A	Fouts et al., 1982; Patterson & Linden, 1981 Ages: J, A	
Cooperation	Boesch & Boesch, 1989; Nishida, 1979; Teleki, 1973 Ages: A	Chalmeau, 1994; Crawford, 1937, 1941; de Waal, 1982; Köhler, 1927; Menzel, 1972, 1973b Ages: I, J, S, A		Povinelli et al., 1992; Savage-Rumbaugh et al., 1978a Ages: J, S, A	Hayes, 1951; Savage-Rumbaugh et al., 1986; Terrace, 1979 Ages: I, J
Theory of mind		de Waal, 1982; Savage-Rumbaugh, 1984; Tanner & Byrne, 1993; Tomasello et al., 1994 Ages: I, J, S, A	Call & Tomasello, 1994a; Fernández & Gómez, 1993; D.J. Povinelli & H.K. Perilloux, unpublished results; Povinelli et al., 1994 Ages: I, J, S	Povinelli et al., 1990; Premack, 1986,1988; Premack & Woodruff, 1978 Ages: S, A	Call & Tomasello, 1994a; Miles, 1990; Tomasello et al., 1993b Ages: J, S, A

i, young infant (0–1 year old); I, old infant (1–3 year old); J, juvenile (3–7 year old); S, subadult (7–12 year old); A, adult (12 year old or older); *italic* references, additional human contact.

attention of others rather than their emotional reactions. Chimpanzees in the wild have been observed on many occasions to follow the gaze of others (Plooij, 1978). De Waal (1986) reported that a captive male vocalized loudly and looked to a particular location on several occasions, seemingly to attract the attention of other group members to that location (see also Menzel, 1971, 1973a). With regard to home-raised chimpanzees, Hayes (1951) observed the effectiveness of focusing her attention on a particular place to make the home-raised chimpanzee Viki approach and inspect the same place. The two studies that have explicitly compared ape joint attention as a function of human experience have found some differences as a function of rearing history. Bard & Vauclair (1984) found that a nursery-reared chimpanzee attended and responded appropriately to human attempts to engage him in object manipulation in a way that a captive chimpanzee and home-raised bonobo did not – although all subjects were too young to engage in extensive object manipulation. Along these same lines, Carpenter et al. (1995) found that home-raised bonobos and chimpanzees engaged more frequently in relatively long episodes of joint attention with humans and attended to the human's actions on objects to a greater extent than did captive conspecifics (who were slightly, though not significantly, younger in age).

In general, then, apes are socially attentive to others and their feelings and visual attention, with or without human contact. There do seem to be some minor quantitative increases in these behaviors with increasing human contact, but it is not likely that these are of great significance.

Intentional communication

In contrast to social attention processes, in which one individual follows into another's attention or behavior, intentional communication concerns an individual's attempts to get others to follow into its attention or behavior. According to Bates (1976) there are two basic functions of intentional communication. On the one hand, an individual may use a communicative signal to get another individual to help in attaining a goal; these are called imperatives. On the other hand, an individual may draw another's attention to something in the environment merely for the sake of sharing attention to it (e.g. by holding it up and showing it); these are called declaratives.

The use of imperatives is by far the most common form of intentional communication that has been observed in apes. For example, in a captive chimpanzee colony Tomasello and colleagues have observed requests for nursing, grooming, food, play, and travel (e.g. Tomasello et al., 1985, 1989, 1994). Gestures in most of these categories are seen in the wild as well. Bard (1992b) observed a number of imperative gestures in both juvenile rehabilitant and wild orangutans. Other more complex kinds of imperative are seen only in apes with more extensive human contact, e.g. dragging a human by the arm to a location where he or she might provide some assistance (e.g. the nursery gorilla with extra human contact described by Gómez (1989, 1990)), or giving a nut to a person who was supposed to crack it open, even slapping the nut and placing a stone on top of it to indicate what should be done (e.g. the home-raised bonobo described by Savage-Rumbaugh *et al.* (1986)). Apes with human contact are also the only apes to produce imperative pointing. Pointing is learned rather easily by apes in human interaction, either in the context of specific experiments (Povinelli *et al.*, 1992; Woodruff & Premack, 1979) or in a more "natural" way in home-raised apes who often use pointing to request objects or indicate locations to which they wish to travel (e.g. Miles, 1990; Savage-Rumbaugh *et al.*, 1986). In terms of the comprehension of imperative pointing, apes with extensive human contact are also the only apes to show human-like skills. Yerkes & Nissen (1939) reported that their captive chimpanzees did not comprehend human pointing, and Savage-Rumbaugh (1984) observed that a captive adult bonobo ignored the pointing of her home-raised offspring who had learned to point for humans. In contrast, Woodruff & Premack (1979) and Povinelli *et al.* (1992) reported that laboratory-trained chimpanzees comprehended human pointing with very little training. Call & Tomasello (1994) compared the performance of two orangutans with different histories of human contact and found that only the home-raised orangutan comprehended human pointing.

The use of declarative gestures is much less frequent in apes, and may be confined to apes with extensive human contact. Plooij (1978) observed that wild chimpanzee infants used objects to get the attention of social partners, but this behavior was not a declarative as it was used as a way of getting the other to begin travel. Tomasello *et al.* (1994) observed a number of these types of gestures in captive chimpanzees as well (e.g. holding out a stick to another, and then clutching it to the chest, as a way of initiating play). However, the function of the gesture in these cases is to attract attention to the self, or to initiate play with the self, not to share attention to the object as in prototypical declaratives such as "showing." However, there are some reports of declarative gestures in home-raised apes, although in all cases interpretation is an issue. Possible instances of "showing" in home-raised gorillas, chimpanzees, and bonobos have been reported by Patterson (1978), Savage-Rumbaugh (1988) and Savage-Rumbaugh *et al.*, (1985). Carpenter *et al.* (1995) also observed two instances of what seemed to be the showing of objects to humans by a bonobo. It should be noted, however, that some of the subjects that use declaratives have been trained to perform them (Savage-Rumbaugh *et al.*, 1983), and thus there is some question about their underlying motivation. It is also noteworthy that home-raised apes also sometimes use their symbols for commenting on current activities, anticipating a future action, or commenting on past events (Savage-Rumbaugh *et al.*, 1985), although it is possible in these cases that the subjects are simply associating their signs with specific stimuli.

Another phenomenon of intentional communication is deception. Many anecdotal instances of deception have been reported for the four great apes in a variety of environments (Byrne & Whiten, 1990). These involve both the suppression of communicative behavior that would be produced under normal circumstances and the conveying of false information. Whether these are instances of deception requiring some sort of social cognition, however, or whether they are procedures for effecting desired outcomes, is still an open question. The best evidence is the experimental study of Woodruff & Premack (1979) in which chimpanzees, with varied backgrounds of association with humans, learned to suppress information such as body orientation or glances to a baited container when a competitive trainer who would not share food was present; they used all of these when a cooperative trainer was present. Some subjects also reliably misdirected the competitive trainer. It is important to point out that in this study it took subjects more than 20 trials to begin to differentiate between the two types of trainer, however, and thus they might have learned during the course of these early trials what the most effective procedures might be.

In summary, imperatives are by far the dominant form of communication for apes across all rearing conditions. The number and sophistication of imperatives increases with human contact. Activities such as placing the other's hand on the object to be manipulated, demonstrating actions, and pointing only appear in apes with human contact, and only home-raised or laboratory-trained apes seem to comprehend human pointing. The only clear instances of prototypical declarative gestures, in which the only motivation seemed to be to share attention to an object with another, have been produced by home-raised apes – both in their natural behavior and in their use of human-like symbols. Communicative signals and symbols used for deceptive purposes have been reported for apes of all types, although their interpretation is in all cases problematic.

Social learning

In their natural habitats all four great ape species acquire information in social contexts. The precise social learning mechanisms employed by the different species, however, are not clear. Researchers have identified at least four different types of social learning, requiring different levels and types of social cognition: mimicking (reproduction of sensorimotor acts), local enhancement (fortuitous reproduction due to common attraction to stimuli), emulation (reproduction of changes of state in the environment that others have produced), or imitative learning (reproduction of intentional strategies of others). The social learning of apes has been studied in four domains: arbitrary actions, object manipulations, problem-solving, and communication.

Some wild chimpanzees have been observed anecdotally to reproduce unusual actions of conspecifics such as nut-cracking postures (Boesch, 1992) or scratching postures (Goodall, 1986, 1990) – without any obvious reason for doing so (for a similar observation of a captive chimpanzee, see de Waal, 1982). Some of the human-like behaviors of the rehabilitant orangutans observed by Russon & Galdikas (1993) may also fall into this category. There is some evidence in these cases that the baseline probability of an individual producing the behavior of its own accord is near zero. These types of behavior are also evident in home-raised chimpanzees, for example the human-raised chimpanzee of Temerlin (1975) made vomit attempts after seeing a human vomit. In addition, some studies have trained subjects with various backgrounds to mimic arbitrary actions on command. Hayes & Hayes (1952) found that after extensive training (17 months) a home-raised chimpanzee learned to reproduce novel actions on command. Similarly, Miles (1990) found that the home-raised orangutan also reproduced actions on command after some training. Recently, Custance & Bard (1994) reported mimicking on command after several months of training in two nursery-raised juvenile chimpanzees.

Object manipulation is another domain where imitative learning has been reported.

Spontaneous imitation of numerous household activities such as cleaning, cooking, or dish washing have been reported in many studies of home-raised apes (see e.g. Hayes, 1951). Russon & Galdikas (1993) and Russon (Chapter 7) reported that rehabilitant orangutans – having various background experiences with humans – also displayed imitation of some object manipulations (although there are no baseline measurements of what the subjects would do with the objects naturally). In addition, Mathieu (1982) found that four nursery-raised chimpanzees with some human contact reproduced only a limited number of human activities, and Perinat & Dalmau (1988) and Gómez (1989) reported only a few possible imitative actions on objects by three nursery-raised gorillas one with more extensive human contact. In a study directly comparing the social learning of apes with different experiential backgrounds in this domain, Tomasello et al. (1993b) found that home-raised chimpanzees and bonobos displayed imitative learning of actions on objects comparable to that of 2 year old children, whereas chimpanzees and bonobos that had been raised mostly with conspecifics with less human contact reproduced very few of the actions that were modeled by a human demonstrator (almost none of the complex actions).

Numerous studies have investigated the social learning in apes in problem-solving situations. When appropriate experimental controls have been used, no evidence of imitative learning in problem-solving contexts has been found either in chimpanzees or orangutans with limited human contact. For example, Nagell et al. (1993) presented captive chimpanzees with a rake-like tool that could be used in either of two ways leading to the same end-result (to control for stimulus enhancement). One group of subjects observed one method of tool use and another group of subjects observed the other method of tool use. It was found that chimpanzees used the same method or methods no matter which demonstration they observed – indicating attention to the change of state in the world observed (emulation learning), not the behavioral means used to effect that change of state (for a similar result with nursery-raised orangutans, see Call & Tomasello, 1994b). In terms of home-raised apes, Hayes & Hayes (1952) found that a home-raised chimpanzee solved a number of problems after observing a human demonstrating the appropriate solution, whereas a chimpanzee with limited human contact failed to solve the problem after observing a human model.

In the domain of communication, there are in the wild some population differences in chimpanzee gestural signaling (Goodall, 1986; McGrew & Tutin, 1978). The best-known of these are "leaf-clipping" (Mahale K (Tanzania) and Bossou (Guinea) communities), and "grooming hand-clasp" (Mahale K, Kibale (Uganda), and Yerkes (captive colony, USA) communities). The basis for these group differences, however, is unknown. In a series of studies of a captive colony, Tomasello and co-workers have studied the acquisition of gestures by infant and juvenile chimpanzees and have found that learning through ontogenetic ritualization (mutual social shaping, as in many prelinguistic gestures of human infants; Tomasello, 1995a) rather than imitative learning was more likely to be responsible for gesture acquisition. Among the pieces of evidence used to make this inference were: there was much individual variability both within and between generations in gestures used, there were a number of gestures used by single individuals, and shared gestures were the same as those that may be observed in chimpanzee peer groups who have had no exposure to adults. With regard to home-raised apes, on the other hand, Fouts et al. (1989) reported that two chimpanzees with extensive human contact acquired some signs by imitative learning (see also Chapter 13), although some studies show that shaping is a more effective procedure for teaching apes sign language (e.g. Fouts, 1972; Sanders, 1985). Also, modeling (in combination with other socialization techniques) has proven effective for teaching home-raised bonobos and chimpanzees to use a communication keyboard (Savage-Rumbaugh et al., 1986, 1992).

In summary, it seems clear that apes raised by humans display much more human-like social learning abilities than apes raised by conspecifics. This is true in all of the domains investigated: arbitrary actions, object manipulations, problem-solving, and communication. It is interesting to note that in the review of Whiten & Ham (1992) in which ape imitative abilities were noted, all of the well-documented examples (with the exception of the several mimicking examples reported here) were from home-raised apes.

Teaching

Another behavior relevant to social learning skills is teaching. A number of researchers report instances of apes in the wild doing things that lead to the learning of others, especially offspring, but it is unclear if this is what the "teacher" intended to accomplish. The classic example is van Lawick-Goodall's (1968) report that chimpanzee mothers "teach" their infants to walk by moving some distance away, turning, and vocalizing to them. Another possibility, however, is that the mother wishes to go somewhere and is simply encouraging the infant to come along. Goodall (1986) also classifies as teaching instances in which a mother chimpanzee takes a poisonous plant away from her infant. But this is just as likely a case of a mother simply preventing her child from doing something dangerous. Recently, Boesch (1991b) reported a number of observations of chimpanzee mothers "teaching" their youngsters a tool use behavior, two of which are possible instances of intentional instruction. In one, the mother slowed down her technique as she cracked nuts (as her youngster watched), and in the other she repositioned the nut into the correct position for cracking after the youngster had been struggling. Although these are very suggestive observations, because there are only two instances and no baseline data for comparison, there are alternative interpretations based on the possibility that the adults are behaving for their own ends (see Tomasello et al., 1993a). Home-raised chimpanzees have not shown strong evidence of teaching skills either. There are only two reported examples of a home-raised ape instructing another. In one a chimpanzee mother demonstrated for her infant a sign-language sign a total of five times (Fouts et al., 1982), and again it is possible her intention was something other than instruction. In the other a gorilla signed to a naive human and when he got no response took her hands, molded the sign, and pushed her to make her move (Patterson & Linden, 1981). Again, whether or not this is intentional instruction is open to debate.

Cooperation

Cooperation in coalitions and alliances for purposes of defense against aggressors and dominance rank acquisition have been extensively documented in wild and captive apes of all four species (Harcourt & de Waal, 1992). Another domain in which cooperation has been documented in the wild is the hunting behavior of chimpanzees in the Taï forest (Boesch & Boesch, 1989), although this same behavior has not been observed at Mahale (Nishida, 1979) or Gombe (Teleki, 1973). To what degree individuals are actually monitoring the behavior of others and adjusting their own behavior accordingly however, is not entirely clear from the published descriptions. Intentional communication about the cooperative effort was not reported. In some captive colonies chimpanzee individuals have been observed to spontaneously support a pole being used by another chimpanzee to climb into a tree (de Waal, 1982; Menzel, 1972, 1973b), but again it is unclear to what degree there was active coordination and adjustment of the individuals to one another or if there was any intentional communication.

In problem-solving situations, Köhler (1927) reported that chimpanzees with limited human contact engaged more in competition than in cooperation. In the most extensive study to date, Crawford (1937, 1941) studied the development of cooperation in a task requiring a pair of captive juvenile chimpanzees to pull simultaneously on a rope. The pair's initial attempts to jointly operate the apparatus were always uncoordinated, and they succeeded only after human instruction and shaping. Once cooperating, however, one subject actively recruited the other to operate the apparatus. Surprisingly, they did not generalize their skills to a very similar task some time later. More recently, Chalmeau (1994) exposed captive chimpanzees to a similar task and observed some spontaneous coordination of behavior, and there was again some active recruitment of a partner. Also very recently, Povinelli *et al.* (1992) reported cooperation between human–ape pairs (using laboratory-trained chimpanzees) in a task that required one subject in a pair (the informant) to indicate to his or her partner which of four locations contained a piece of food. In turn, the other subject (the operator) had to operate one of four handles so that both subjects could obtain a piece of food. Subjects achieved a level of proficiency in their cooperative efforts without extensive training, and furthermore, when roles were reversed three of the four chimpanzees adjusted to their new roles with little difficulty. Povinelli *et al.* argued that this finding demonstrated that the subjects spontaneously understood the role of the human partner they were cooperating with during the interaction, although it could also be interpreted to mean that they were able to learn their new roles quickly based on what they had learned previously (Tomasello & Call, 1994). Savage-Rumbaugh *et al.*, (1978a) reported that their laboratory-raised chimpanzees were able to collaborate in solving a complex task requiring different roles for each performer; however, this cooperation was not spontaneous because before engaging in this task both subjects were first trained by humans in each of the two collaborative roles.

In summary, ape cooperation has been observed in one form or another in all rearing environments, but the nature of this cooperation is not always clear. Although the evidence of cooperation in the wild is suggestive, the clearest cases of cooperation in which individuals coordinate with one another's roles and actively communicate about this coordination are all from apes with extensive human contact. Even in these cases, however, there are few data on precisely how apes are interacting with one another, and thus it is not clear whether they are engaging in a human-like form of cooperation with mutual role-taking and intentional communication.

Theory of mind
Many of the just reviewed social skills may be seen as relying on the perception or understanding of the mental states of others. A number of different observations are relevant to this general topic, including the understanding of gaze/attention, intention, and knowledge. Some of these phenomena are very difficult to observe outside experimental situations.

With regard to the understanding of visual gaze or attention, Tomasello *et al.* (1994) found that in a captive colony of chimpanzees visually based gestures (e.g. arm-raise) were performed only when potential recipients were oriented toward the performer; when they were not so oriented, tactilely based (e.g. poke) or auditorily based (e.g. ground-slap) gestures were used. Tanner & Byrne (1993) reported several instances of a captive gorilla hiding her play-face so that others could not see it (see also de Waal, 1982). However, Premack (1988) reported that when captive juvenile chimpanzees needed a human to help open a container, but the human's eyes were covered by a blindfold, only one of the four subjects removed the blindfold from the human. Some similar observations have been made of apes with more extensive human contact. For example, Fernández & Gómez (1983) found that nursery-raised gorillas with a moderate amount of human contact made sure that a human was attending before they performed gestures. In a comparison of two orangutans with different rearing histories, Call & Tomasello (1994) found that a nursery-raised orangutan pointed toward an out-of-reach drink it wanted even when the human's eyes were closed, whereas a home-raised orangutan significantly decreased his frequency of pointing in this situation (as compared with the eyes-open condition).

With regard to the perception of intention, D.J. Povinelli and H.K. Perilloux (unpublished results) presented captive chimpanzees with two scenes: one human spilling juice accidentally and another pouring it out intentionally. When chimpanzees were later asked to choose which of these humans should bring them juice, they showed no preference – presumably indicating an inability to distinguish intentional from accidental behavior when the outcomes were the same. In studying apes with more extensive human contact, Savage-Rumbaugh (1984) reported two anecdotal observations in which a female bonobo refrained from attacking a human that had harmed her infant, presumably because the harm was produced unintentionally. Also in this category, Premack & Woodruff (1978) presented the laboratory-trained, language-trained chimpanzee Sarah with pictures of humans in problem-solving situations (e.g. an out-of-reach banana). Among several alternative pictures presented as solutions, Sarah quite often correctly chose the one that represented an appropriate solution. It should be noted, however, that Sarah may have been choosing solutions on the basis of some criterion other than the attribution of intentions (for this argument, see Savage-Rumbaugh *et al.*, 1978b), and that Premack (1986) briefly reported on the failure to train Sarah to discriminate between other picture sequences that depicted intentional and nonintentional actions. Tomasello *et al.* (1993b) argued that the imitative learning of the behavioral strategies of others by home-raised apes demonstrated their ability, in contrast to captive apes, to perceive the intentions of others.

Finally, with regard to the attribution of knowledge to others, Premack (1988) reported negative results of a false belief task for laboratory-raised Sarah. In this experiment, Sarah did not modify her behavior to take account of a difference in knowledge states between herself and a human experimenter. Premack (1988) also reported a study in which one human baited one of two boxes in the presence of another human that had visual access to the box that was being baited. Four juvenile chimpanzees (captive, with unreported experiential histories) observed this process but were kept ignorant about which container was being baited. Another human remained outside the room and consequently did not know which container was being baited. After this naive human entered the room, subjects could choose between the naive and the knowledgeable human. After they had chosen, the human pointed to one of the containers, and subjects made their choices. Results indicated that two of the four chimpanzees came to correctly choose the experimenter who knew the food's location. Povinelli *et al.* (1990) reported similar results, and in addition found that chimpanzees learned to generalize to a situation in which a human covered his head while the apparatus was being baited. They did not generalize to this situation immediately (Povinelli, 1994a), however, and it is thus possible that, in both the Premack and Povinelli *et al.* experiments, what subjects were doing was learning to respond to specific discriminative cues. Support for this view is the recent failure to replicate these findings with nursery-raised juvenile chimpanzees (Povinelli *et al.*, 1994).

In summary, all apes seem to have some knowledge that visual gaze or attention is important for some interactions. Only apes with more extensive human contact, however, have demonstrated that they understand others in terms of their intentions – although the evidence in this domain is far from perfect. The evidence for an effect of extent of human contact on apes ability to infer the knowledge states of others is mixed, and indeed there is some question at this point of whether apes infer knowledge states at all (Povinelli, 1994b).

MECHANISMS OF HUMAN INFLUENCE

Looking across all of these domains of ape cognitive development, several patterns may be discerned (see Table 17.3). With regard to physical cognition, it may be said that after a few years of age apes of all four species, no matter how they are raised, have a sense of the permanence of objects that would seem to be very similar to that of human infants from the middle of their second year of life. When it comes to the specific properties of particular objects, however, it seems that different individuals know more or less depending on their exposure to them: apes who have been exposed to a wider range of objects and tools during early development know more things to do with them in terms of object manipulation, tool use, mirror use, and categorization. In addition, if individual apes are trained to do so they can learn to sort objects into groups on the basis of

Table 17.3. *Summary of the effects of humans on the development of apes in different cognitive domains*

Domain	Effect of interaction with humans
Object permanence	Interaction with humans not necessary
Object manipulation, tool use, mirror use, symbolic play	*Exposure* to human artifacts and *emulation* of their use leads to quantitative increases in knowledge of objects and their properties and dynamic affordances. May occur in many types of captive environments
Categorization/classification, quantitative skills	Interaction with humans not necessary for natural categorization, but some *training* may be necessary for explicit classification and quantification based on abstract properties (e.g. training in attention management skills). Typically occurs in laboratory environments
Social attention, social referencing, understanding visual gaze	Interaction with humans not necessary
Intentional communication, imitative learning, understanding intentions	Being *enculturated* by humans may lead to an understanding of others as intentional and thus to qualitatively more human-like skills of social learning and intentional communication. Typically occurs only in home-raised environments
Cooperation, teaching, understanding beliefs	Human interaction has no significant effect because human-like skills in these domains may not be attainable by apes of any kind

relatively abstract perceptual dimensions and to make judgements based on numerical quantity.

We see three possible mechanisms of these effects. First, it is possible that simple exposure to many and varied objects during ontogeny is sufficient for apes to learn many of their interesting properties, and perhaps to develop some generalized skills for exploring new objects. But secondly, it is also likely that much of ape learning about objects in human environments occurs when they observe humans manipulate and use these objects, and they learn some particular affordances of particular objects that they would not have discovered on their own. This is what we have previously called emulation learning: by observing the object manipulations of others an individual learns about changes of states in the environment that may potentially be brought about (Tomasello, 1990). This same effect is most likely what accounts for the few anecdotal observations of symbolic play in home-raised apes as well. Thirdly and finally, specific training with particular kinds of material may lead apes to attend to certain abstract properties of objects, including dimensions on which they may be classified or quantified. We thus believe that together these three mechanisms of human influence – exposure, emulation, and training – lead apes to learn much more about objects and their properties than they could possibly learn if left totally to their own devices. We do not believe, however, that such exposure has any fundamental, qualitative effect on apes' understanding of the physical world. The effect is merely quantitative in the sense that the effective variable is the amount of exposure that different individuals have had to objects and their affordances, and the outcome is learning more about the types of thing that all apes are capable of learning about.

In the social domain, the picture is more complex. Apes of all four species, however they are raised, show a basic level of social attention in the sense that they are attached to and attend to their mothers, siblings, peers, and other conspecifics. Interaction with humans is not needed for the development of these behaviors. It is also very likely that all types of ape follow the gaze of others to outside entities and use their reactions as a guide as to how they should interact with the entity, mostly in situations of emotional arousal. Again, interaction with humans is not needed for the development of these skills. At the other end of the spectrum are sociocognitive skills that humans also do not influence, but in this case it is because apes are not capable of those skills under any circumstances. Thus, we do not believe that the evidence supports the view that apes of any type understand the mental states of others in terms of their beliefs, and we do not believe that apes of any type engage in the most human-like forms of cooperation and teaching (which may rely on an understanding of the beliefs of others). Thus interaction with humans does not seem to play a role in these domains either, but for different reasons.

The sociocognitive domains in which humans seem to have the greatest effect on apes are intentional communication and social learning. Thus, referential pointing and more elaborate forms of imperative communication in general (and possibly the use of declarative forms of communication), not to mention the comprehension and use of linguistic symbols, have been observed only in home-raised apes. Similarly, the most clearly demonstrated cases of imitative learning, in a number of different behavioral domains, are all from apes who have interacted extensively with humans. Why this should be the case is not known at this time. It is possible, however, that these effects result from home-raised apes acquiring a deeper understanding of others in terms of their intentions, i.e. in terms of the means–ends structure of their behavior (Tomasello, 1995b).

Intentional behavior may be understood at different levels, and so to be more precise about ape understanding of intentions, and how humans might influence this understanding, we must look more closely at intentionality. First of all, it is clear that all apes understand something about the directedness of behavior. A mother ape watching her youngster come toward her and pull at her arm in some sense knows that it is coming to nurse, and an adult ape can tell when another adult is getting ready to mate or attack. These things are understood because they are behavioral sequences with which the mother or adult is familiar from past experience; they understand that specific acts lead to specific results with a high degree of predictability. This is the simplest level of the understanding of intentionality. At the other end of the spectrum, there is the understanding of intentional behavior that relies on the distinction between intentional and accidental actions. There is no good evidence that apes of any type understand this distinction, and in fact some evidence that they do not (D.J. Povinelli & H.K. Perilloux, unpublished results). The reason for this failure may be that in many accidents and intentional actions the outcome is the same, and apes may be predisposed to understand as similar all actions that result in the same outcome.

But there is a middle level in which the understanding of intentions goes beyond the understanding of simple directedness, or even a more generalized understanding of what an individual is likely to do next in a behavioral sequence, but still stopping short of distinguishing intentional and accidental actions that have the same result. This middle level involves the understanding of behavior as intentional in the sense of differentiating means from ends. This distinction relies on an understanding of such things as that there are multiple means to the same end, that obstacles may be overcome by intermediate means, that the same behavior may be either a means or an end in appropriate circumstances, and that others have ends (intentions) that may not match one's own (Piaget, 1952; Tomasello, 1995b).

In this case we think there is evidence for different skills in apes as a function of human contact, and this evidence consists precisely of behavior in the domains of intentional communication and social learning. Thus, in the wild, there is no evidence that apes point to outside entities for one another, nor do they imitatively learn new instrumental behaviors from one another. Our hypothesis is that they do not do these things because they do not understand that others have independent intentions and attention that differ from their own that they can follow into or direct or reproduce. Apes raised by humans, however, often do develop such skills. They point to things for others (and may use linguistic symbols), create relatively elaborate new means of imperative communication with others, and they imitatively learn new skills from others, reproducing both means and ends in the process. The hypothesis is that individuals can learn to do these things when they come to understand that others have intentions and attention that differ from their own and that may be achieved by different alternative means.

The question thus arises as to how human interaction may lead apes to an understanding of others as intentional. Our hypothesis is that, to understand others as intentional, individuals must be raised in cultural environments in which they themselves are treated as intentional by others (Kaye, 1982), perhaps from a very early age. Thus, adult human beings attempt to make their children, and sometimes apes, look at things and do things. When the children and apes do so, the adults reward them for doing so. In their interactions with children and apes, human enculturators thus structure and encourage triadic interactions (involving adult, child/ape, and outside object) in a way that adult chimpanzees in their natural environments do not do for their young. It is possible that during the enculturation of apes this "scaffolding" and intentional instruction serves to "socialize the attention" of the ape in much the same way that human children have their attention socialized by adults (Vygotsky, 1978). This process makes possible all kinds of triadic interactions involving the "referential triangle," including various forms of cultural learning and related sociocognitive skills (Tomasello *et al.*, 1993a). In terms of motivation, it is also important to note that human environments make many desirable objects inaccessible to apes, and because humans control access apes learn to use their triadic skills to help them in gaining control over the material and the social environment they inhabit (Savage-Rumbaugh, 1990).

It is important to remain aware of the fact that being treated intentionally does not turn apes into humans. There are two main differences in our view. The first concerns social and communicative motivation (Gómez *et al.*, 1993; Savage-Rumbaugh *et al.*, 1983; Terrace *et al.*, 1979). Home-raised apes learn to use imperatives to request behaviors of others, and they show very human-like skills of imperative pointing and other forms of requesting. The use of declaratives is rare in apes, however, and moreover, the purported declaratives of apes differ from human declaratives in some important ways. What have been identified as ape declaratives all occur either in noncommunicative contexts (e.g. self-signing) or else in communicative contexts in which the ape is calling attention to a specific object or person for purposes of spurring a desired action, e.g. by pointing out an object with which an action is desired. In contrast, when human infants use declaratives the goal is simply to share attention with someone to something. No apes, not even home-raised apes, seem to enjoy showing objects to others simply in order to point out to them some interesting feature or features.

The implications of this motivational difference between apes and humans are enormous. Motivational differences may not only explain the paucity of declaratives in the intentional communication of home-raised apes, but it also may be responsible for the differences found between humans and apes in other domains such as joint attention, imitation, teaching, and cooperation. Although in human ontogeny each of these abilities follows its own developmental course and timing, it is still possible that they each have (at least in some contexts) the same "purely social" motivational component, i.e. to share attention, knowledge, or skills, simply for the sake of sharing them. The possibility is thus that home-raised apes learn some triadic skills through the process of being treated intentionally, but they apply these skills in imperative contexts only – because this is the nature of their species-specific motivational structure. This motivational structure may not be subject to serious modification.

The second important difference between humans and home-raised apes is that home-raised apes do not create cultural environments for one another. Even though they respond in interesting ways to a human cultural environment, as we have shown in this chapter, it is quite another thing to create such an environment. Despite one anecdotal observation of one home-raised ape teaching another (Fouts *et al.*, 1982) it is doubtful in our view whether if all home-raised apes were suddenly placed in a single group and left to their own devices, they would create cultural environments of the human kind. The problem may be cognitive since certain forms of intentional instruction depend on being able to understand others in terms of their beliefs (Cheney & Seyfarth, 1990; Tomasello *et al.*, 1993a), and we have argued previously in this chapter that apes, whatever their upbringing, do not conceive of others in terms of their beliefs. Or it may be motivational, as we have just argued: adult apes are not motivated to instruct or share knowledge with their young because it has no immediate consequences for themselves.

Our hypothesis is thus that being treated intentionally by others, i.e. being enculturated into a cognitive community, is an integral part of the ontogeny of certain sociocognitive abilities, especially the ability to understand behavior intentionally. It may be especially important to have such experiences early in development. This then leads to a number of specific ways in which apes behave with humans and conspecifics after they have come to understand behavior in this way, mostly involving access to the referential triangle in which the intentions of the two participants are mutually understood. There are still important differences between humans and apes, however they are raised, and we have speculated that these involve the use of the referential triangle for purely social motives and the ability to create cultural environments for others.

CONCLUSION

The cognitive development of apes is affected by contact with humans. We have outlined four mechanisms that may be responsible. First, apes learn new things about objects and their properties and relationships by simple *exposure* to them, and, in the case of tools, the need to use them. When apes are raised by humans, even in simple captivity, they are typically exposed to more different types of object and artifacts crafted for specific uses than when they grow up in their species-typical environments. Secondly, apes engage in *emulation learning* about objects: when they see an object being manipulated or used they learn something about that object's affordances or relations to other objects that they might not have discovered on the basis of their own explorations. Again, when apes are raised with humans – in some captive and many nursery environments, for example – they very likely observe more of these kinds of object manipulations. Thirdly, apes may learn some specific skills through explicit *training* in which humans, typically in laboratory settings, assist them in identifying what to pay attention to and what to do in solving particular cognitive tasks such as the categorization of objects on the basis of abstract features. Fourthly and finally, we have hypothesized that the experience of being treated intentionally by others in home-raised environments, what we have called in other places *enculturation* (Tomasello *et al.*, 1993a), may lead to a fundamental change in the social cognition of apes such that they begin to differentiate between means and ends in the behavior of others and thus view these others as intentional agents. This leads to a whole host of changes in other (though not all) areas of socio-cognitive development, especially in the domains of intentional communication and social learning.

It is important to emphasize that the conclusions we have reached in this review are limited in a number of ways. First, much of the information reported in several domains is based almost exclusively on anecdotal accounts, and this is especially true of home-raised apes. Although we do not reject anecdotes as potentially valuable information, anecdotes alone are insufficient for drawing firm conclusions: "The plural of anecdote is not data" (Bernstein, 1988). Secondly, the seeming sophistication found in home-raised subjects in several domains such as teaching or symbolic play may be a result of the fact that they are observed for a greater amount of time and in more human-like environments than are other apes; consequently, interesting behaviors are much more likely to be observed, perhaps especially for behaviors of low frequency such as teaching or symbolic play. Thirdly, there are some variables such as age or age of exposure to certain rearing conditions that make it difficult to reach definitive conclusions about the role of humans in certain domains, not to mention the fact that the rearing histories of many individuals are not reported in enough detail to reconstruct what precisely were their experiences with humans.

Nevertheless, we still believe that we have uncovered some patterns of difference between apes raised in different environments, and that these deserve future research attention. We also believe that there are two rather immediate consequences of our review that are vitally important to current thinking about ape cognitive development. The first is that the supposedly fundamental differences in various aspects of the cognitive development of monkeys and apes may not be as great as many scientists have claimed (e.g. Povinelli *et al.*, 1992; Whiten & Ham, 1992). Many monkey–ape differences in cognition may result from the fact that the apes producing unique behaviors or skills have been raised in human-like cultural or laboratory environments of a type that monkeys have not been systematically raised or studied in (Tomasello & Call, 1994). We thus do not know what enculturated monkeys would do in some of the cognitive tasks at which home-raised apes are so proficient. This question awaits future research.

The second consequence of our review is a highlighting of the need for a truly developmental perspective. Whereas some researchers might argue that attributing a cognitive capacity to a species is appropriate if at least one subject, whatever its ontogenetic history, is able to display the requisite skills, we believe that this argument basically trivializes the whole notion of ontogeny and the ontogenetic environment. Organisms inherit their species-typical environments as much as they inherit their genomes, and for many species the species-typical ontogenetic sequence depends in fundamental ways on that environment (Schneirla, 1966). If individuals of the same species end up with one set of skills in one environment and another set of skills in another environment, we may say that the species has two sets of capacities, but it is more informative to say that in one ontogenetic environment one set of skills emerges and in another ontogenetic environment another set of skills emerges. In the particular case of interest here, we would simply argue that when great apes spend their early ontogenies in something resembling a human cultural environment, rather than in their species-typical environments, some of their cognitive skills develop in more human-like ways. We cannot ignore the active role of the environment, especially the social environment, in the cognitive development of primates.

ACKNOWLEDGEMENTS

The authors thank the editors Anne Russon, Kim Bard, and Sue Taylor Parker for their helpful editorial suggestions.

REFERENCES

Bard, K.A. (1992a). Very early social learning: The effect of neonatal environment on chimpanzees' social responsiveness. Paper presented at the XIVth Congress of the International Primatological Society, Strasbourg, France, 16–21 August.

Bard, K.A. (1992b). Intentional behavior and intentional communication in young free-ranging orangutans. *Child Development*, **63**, 1186–97.

Bard, K.A., Platzman, K.A., Lester, B.M. & Suomi, S.J. (1992). Orientation to social and non-social stimuli in neonatal chimpanzees and humans. *Infant Behavior and Development*, **15**, 43–6.

Bard, K.A. & Vauclair, J. (1984). The communicative context of object manipulation in ape and human adult-infant pairs. *Journal of Human Evolution*, **13**, 181–90.

Bates, E. (1976). *Language and Context*. New York: Academic Press.

Berdecio, S. & Nash, V.J. (1981). Chimpanzee visual communication. In *Anthropological Research Papers*, vol. 26, pp. 1–159. Arizona State University.

Bernstein, I.S. (1988). Metaphor, cognitive belief, and science. *Behavioral and Brain Sciences*, **11**, 247–8.

Boesch, C. (1991a). Symbolic communication in wild chimpanzees. *Human Evolution*, **6**, 81–90.

Boesch, C. (1991b). Teaching among wild chimpanzees. *Animal Behaviour*, **41**, 530–32.

Boesch, C. (1992). Aspects of transmission of tool-use in wild chimpanzees. In *Tools, Language, and Cognition in Human Evolution*, ed. K.R. Gibson & T. Ingold, pp. 171–83. Cambridge: Cambridge University Press.

Boesch, C. (1993). Towards a new image of culture in wild chimpanzees? *Behavioral and Brain Sciences*, **16**, 514–15.

Boesch, C. & Boesch, H. (1984). Mental map in wild chimpanzees: an analysis of hammer transports for nut cracking. *Primates*, **25**, 160–70.

Boesch, C. & Boesch, H. (1989). Hunting behavior of wild chimpanzees in the Taï National Park. *American Journal of Physical Anthropology*, **78**, 547–37.

Boesch, C. & Boesch, H. (1990). Tool use and tool making in wild chimpanzees. *Folia Primatologica*, **54**, 86–99.

Boysen, S.T. & Bernston, G.G. (1990). The development of numerical skills in the chimpanzee (*Pan troglodytes*). In *"Language" and Intelligence in Monkeys and Apes: Comparative Developmental Perspectives*, ed. S.T. Parker & K.R. Gibson, pp. 435–50. New York: Cambridge University Press.

Braggio, J.T., Hall, A.D., Buchanan, J.P. & Nadler, R.D. (1979). Cognitive capacities of juvenile chimpanzees on a Piagetian-type multiple-classification task. *Psychological Reports*, **44**, 1087–97.

Byrne, R.W. & Byrne, J.M.E. (1993). Complex leaf-gathering skills of mountain gorillas (*Gorilla g. beringei*): Variability and standardization. *American Journal of Primatology*, **31**, 241–61.

Byrne, R.W. & Whiten, A. (1990). Tactical deception in primates: the 1990 database. *Primate Report*, **27**, 1–101.

Call, J. & Tomasello, M. (1994a). The production and comprehension of pointing in orangutans. *Journal of Comparative Psychology*, **108**, 307–17.

Call, J. & Tomasello, M. (1994b). The social learning of tool use by orangutans (*Pongo pygmaeus*). *Human Evolution*, **9**, 297–313.

Carpenter, M., Tomasello, M. & Savage-Rumbaugh, E.S. (1995). Joint attention and imitative learning in children, chimpanzees and enculturated chimpanzees. *Social Development*. (in press.)

Chalmeau, R. (1994). Do chimpanzees cooperate in a learning task? *Primates*, **35**, 385–92.

Cheney, D.L. & Seyfarth, R.M. (1990). *How Monkeys See the World*. Chicago: University of Chicago Press.

Chevalier-Skolnikoff, S. (1983). Sensorimotor development in orang-utans and other primates. *Journal of Human Evolution*, **12**, 545–61.

Crawford, M.P. (1937). The cooperative solving of problems by young chimpanzees. *Comparative Psychologycal Monogaphs*, **14**, 1–88.

Crawford, M.P. (1941). The cooperative solving by chimpanzees of problems requiring serial responses to color cues. *Journal of Social Psychology*, **13**, 259–80.

Custance, D. & Bard, K.A. (1994). The comparative and developmental study of self-recognition and imitation: The importance of social factors. In *Self-awareness in Animals and Humans: Developmental Perspectives*, ed. S.T. Parker, R.W. Mitchell & M.L. Boccia, pp. 207–26. New York: Cambridge University Press.

de Waal, F.B.M. (1982). *Chimpanzee Politics*. London: Jonathan Cape.

de Waal, F.B.M. (1986). Deception in the natural communication of chimpanzees. In *Deception: Perspectives on Human and Nonhuman Deceit*, ed. R.W. Mitchell & N.S. Thompson, pp. 221–44. Albany, NJ: State University of New York Press.

de Waal, F. & van Hooff, J. (1981). Side-directed communication and agonistic interactions in chimpanzees. *Behaviour*, **77**, 164–98.

Ettlinger, G. (1983). A comparative evaluation of the cognitive skills of the chimpanzee and the monkey. *International Journal of Neuroscience*, **22**, 7–20.

Evans, A. & Tomasello, M. (1986). Evidence for social referencing in young chimpanzees (*Pan troglodytes*). *Folia Primatologica*, **47**, 49–54.

Fernández, P. & Gómez, J.C. (1983). Desarrollo de pautas prelingüísticas de comunicación en un grupo de gorilas jóvenes. Paper presented at the 1st Symposium on Perspectives Actuales en Psicología Evolutiva y de la Educación, Madrid, 19–20 April.

Fouts, R.S. (1972). Use of guidance in teaching sign language to a chimpanzee (*Pan troglodytes*). *Journal of Comparative Psychology*, **80**, 515–22.

Fouts, R.S., Fouts, D.H. & Van Cantfort, T.E. (1989). The infant Loulis learns signs from cross-fostered chimpanzees. In *Teaching Sign Language to Chimpanzees*, ed. R.A. Gardner, B.T. Gardner & T.E. Van Cantfort, pp. 280–92. Albany, NY: State University of New York Press.

Fouts, R.S., Hirsch, A.D. & Fouts, D.H. (1982). Cultural transmission of a human language in a chimpanzee mother-infant relationship. In *Child Nurturance. Studies of Development in Nonhuman Primates*, ed. H.E. Fitzgerald, J.A. Mullins & P. Gage, pp. 159–93. New York: Plenum Press.

Galdikas, B.M.F. (1982). Orang-utan tool-use at Tanjung Puting Reserve, Central Indonesian Borneo (Kalimantan Tengah). *Journal of Human Evolution*, **10**, 19–33.

Galdikas, B.F. & Vasey, P. (1992). Why are orangutans so smart? Ecological and social hypotheses. In *Social Processes and Mental Abilities in Nonhuman Primates*, ed. F. Burton, pp. 183–224. Lewiston, NY: The Edwin Mellen Press.

Gallup, G.G., Jr (1970). Chimpanzees: self-recognition. *Science*, **167**, 86–7.

Gallup, G.G., Jr, McClure, M.K., Hill, S.D. & Bundy, R.A. (1971). Capacity for self-recognition in differentially reared chimpanzees. *Psychological Record*, **21**, 69–74.

Garcha, H.S. & Ettlinger, G. (1979). Object sorting by chimpanzees and monkeys. *Cortex*, **15**, 213–24.

Gardner, R.A. & Gardner, B.T. (1978). Comparative psychology and language acquisition. *Annals of the New York Academy of Sciences*, **309**, 37–76.

Gardner, B.T. & Gardner, R.A. (1989). A test of communication. In *Teaching Sign Language to Chimpanzees*, ed. R.A. Gardner, B.T. Gardner & T.E. Van Cantfort, pp. 181–97. Albany, NY: State University of New York Press.

Gómez, J.C. (1989). La comunicación y la manipulación de objetos en crías de gorilas. *Estudios de Psicología*, **38**, 111–28.

Gómez, J.C. (1990). The emergence of intentional communication as a problem-solving strategy in the gorilla. In *"Language" and Intelligence in Monkeys and Apes: Comparative Developmental*

Perspectives, ed. S.T. Parker & K.R. Gibson, pp. 333–55. New York: Cambridge University Press.

Gómez, J.C., Sarriá, E. & Tamarit, J. (1993). The comparative study of early communication and theories of mind: Ontogeny, phylogeny, and pathology. In *Understanding Other Minds: Perspectives From Autism*, ed. S. Baron-Cohen, H. Tager-Flusberg & D.J. Cohen, pp. 397–426. Oxford: Oxford University Press.

Goodall, J. (1986). *The Chimpanzees of Gombe: Patterns of Behavior*. Cambridge, MA: The Belknap Press.

Goodall, J. (1990). *Through a Window*. Boston, MA: Houghton Mifflin.

Haggerty, M. (1913). Plumbing the minds of apes. *McClure's Magazine*, **41**, 151–4.

Hallock, M.B. & Worobey, J. (1984). Cognitive development in chimpanzee infants (*Pan troglodytes*). *Journal of Human Evolution*, **13**, 441–7.

Hallock, M., Worobey, J. & Self, P. (1989). Behavioral development in chimpanzee (*Pan troglodytes*) and human newborns across the first month of life. *International Journal of Behavioral Development*, **12**, 527–40.

Harcourt, A.H. & de Waal, F.B.M. (eds) (1992). *Coalitions and Alliances in Humans and Other Animals*. Oxford: Oxford University Press.

Hauser, M.D. (1990). Do chimpanzee copulatory calls incite male–male competition? *Animal Behaviour*, **39**, 596–7.

Hayes, C. (1951). *The Ape in Our House*. New York: Harper & Row.

Hayes, K.J. & Hayes, C. (1951). The intellectual development of a home-raised chimpanzee. *Proceedings of the American Philosophical Society*, **95**, 105–9.

Hayes, K.J. & Hayes, C. (1952). Imitation in a home-raised chimpanzee. *Journal of Comparative and Physiological Psychology*, **45**, 450–9.

Hayes, K.J. & Nissen, C.H. (1971). Higher mental functions of a home-raised chimpanzee. In *Behavior of Nonhuman Primates*, ed. A.M. Schrier & F. Stollnitz, pp. 59–115. New York: Academic Press.

Hill, S.D., Bundy, R.A., Gallup, G.G., Jr & McClure, M.K. (1970). Responsiveness of young nursery-reared chimpanzees to mirrors. *Proceedings of the Louisiana Academy of Sciences*, **33**, 77–82.

Humphrey, N.K. (1976). The social function of intellect. In *Growing Points in Ethology*, ed. P.J.B. Bateson & R.A. Hinde, pp. 303–21. Cambridge: Cambridge University Press.

Hyatt, C.W. & Hopkins, W.D. (1994). Self-awareness in bonobos and chimpanzees: A comparative perspective. In *Self-awareness in Animals and Humans: Developmental Perspectives*, ed. S.T. Parker, R.W. Mitchell & M.L. Boccia, pp. 248–53. New York: Cambridge University Press.

Jacobsen, C.F., Jacobsen, M.J. & Yoshioka, J.G. (1932). Development of an infant chimpanzee during her first year. *Comparative Psychology Monographs*, **9**, 1–94.

Jensvold, M.L.A. & Fouts, R.S. (1993). Imaginary play in chimpanzees (*Pan troglodytes*). *Human Evolution*, **8**, 217–27.

Jolly, A. (1966). Lemur social behavior and primate intelligence. *Science*, **153**, 501–6.

Jordan, C. (1982). Object manipulation and tool-use in captive pygmy chimpanzee (*Pan paniscus*). *Journal of Human Evolution*, **11**, 35–9.

Kaye, K. (1982). *The Mental and Social Life of Babies*. Chicago: University of Chicago Press.

Kellogg, W.N. & Kellogg, L.A. (1933). *The Ape and the Child*. New York: McGraw-Hill.

Knobloch, H. & Pasamanick, B. (1959). The development of adaptive behavior in an infant gorilla. *Journal of Comparative Psychology*, **52**, 699–704.

Köhler, W. (1927). *The Mentality of Apes*. New York: Harcourt Brace & Company.

Laidler, K. (1980). *The Talking Ape*. London: Collins.

Lethmate, J. (1982). Tool-using skills of orang-utans. *Journal of Human Evolution*, 11, 49–64.
Lethmate, J. & Dücker, G. (1973). Untersuchungen zum selbsterkennen im spiegel bei orang-utans und einigen anderen affenarten. *Zeitscrift für Tierspsychologie*, 33, 248–69.
Limongelli, L., Boysen, S.T. & Visalberghi, E. (1995). Comprehension of cause-effect relations in a tool-using task by chimpanzees (*Pan troglodytes*). *Journal of Comparative Psychology*, 109, 18–26.
Lin, K.C., Bard, K.A. & Anderson, J.R. (1992). Development of self-recognition in chimpanzees (*Pan troglodytes*). *Journal of Comparative Psychology*, 106, 120–7.
Mathieu, M. (1982). Intelligence without language: Piagetian assessment of cognitive development in chimpanzee. Paper presented at the joint meeting of the IXth Congress of the International Primatological Society and the meeting of the American Society of Primatologists, Atlanta, GA, 8–13 August.
Mathieu, M. & Bergeron, G. (1981). Piagetian assessment on cognitive development in chimpanzee (*Pan troglodytes*). In *Primate Behavior and Sociobiology*, ed. A.B. Chiarelli & R.S. Corruccini, pp. 142–7. Berlin: Springer-Verlag.
Matsuzawa, T. (1990). Spontaneous sorting in human and chimpanzee. In *"Language" and Intelligence in Monkeys and Apes: Comparative Developmental Perspectives*, ed. S.T. Parker & K.R. Gibson, pp. 356–78. New York: Cambridge University Press.
McGinnis, P.R. (1979). Sexual behavior in free-living chimpanzees: Consort relationships. In *The Great Apes*, ed. D.A. Hamburg & E.R. McCown, pp. 429–39. Menlo Park, CA: Benjamin/Cummings.
McGrew, W.C. (1977). Socialization and object manipulation of wild chimpanzees. In *Primate Biosocial Development: Biological, Social, and Ecological Determinants*, ed. S. Chevalier-Skolnikoff & F.E. Poirier, pp. 261–88. New York: Garland STPM Press.
McGrew, W.C. (1989). Why is ape tool use so confusing? In *Comparative Socioecology: The Behavioural Ecology of Humans and Other Mammals*, ed. V. Standen & R.A. Foley, pp. 457–72. Oxford: Blackwell Scientific Publications Ltd.
McGrew, W.C. (1992). *Chimpanzee Material Culture: Implications for Human Evolution*. Cambridge: Cambridge University Press.
McGrew, W.C. & Tutin, C.E.G. (1978). Evidence for a social custom in wild chimpanzees? *Man*, 13, 234–51.
Menzel, E.W. (1971). Communication about the environment in a group of young chimpanzees. *Folia Primatologica*, 15, 220–32.
Menzel, E.W. (1972). Spontaneous invention of ladders in a group of young chimpanzees. *Folia Primatologica*, 17, 87–106.
Menzel, E.W. (1973a). Leadership and communication in young chimpanzees. In *Precultural Primate Behavior*, ed. E.W. Menzel, pp. 192–225. Fourth International Primatological Congress Symposium Proceedings. Basel: Karger.
Menzel, E.W. (1973b). Further observations on the use of ladders in a group of young chimpanzees. *Folia Primatologica*, 19, 450–7.
Menzel, E.W., Davenport, R.K. & Rogers, C.M. (1970). The development of tool using in wild-born and restriction-reared chimpanzees. *Folia Primatologica*, 12, 273–83.
Menzel, E.W., Davenport, R.K. & Rogers, C.M. (1972). Protocultural aspects of chimpanzees' responsiveness to novel objects. *Folia Primatologica*, 17, 161–70.
Mignault, C. (1985). Transition between sensorimotor and symbolic activities in nursery-reared chimpanzees (*Pan troglodytes*). *Journal of Human Evolution*, 14, 747–58.
Miller, L.C., Bard, K.A., Juno, C.J. & Nadler, R.D. (1990). Behavioral responsiveness to strangers in young chimpanzees (*Pan troglodytes*). *Folia Primatologica*, 55, 142–55.

Miles, H.L. (1986). How can I tell a lie?: Apes, language and the problem of deception. In *Deception: Perspectives on Human and Nonhuman Deceit*, ed. R.W. Mitchell & N.S. Thompson, pp. 245–66. Albany, NY: State University of New York Press.

Miles, H.L. (1990). The cognitive foundations for reference in a signing orangutan. In *"Language" and Intelligence in Monkeys and Apes: : Comparative Developmental Perspectives*, ed. S.T. Parker & K.R. Gibson, pp. 511–39. Cambridge University Press.

Miles, H.L.W. (1994). ME CHANTEK: The development of self-awareness in a signing orangutan. In *Self-awareness in Animals and Humans. Developmental Perspectives*, ed. S.T. Parker, R.W. Mitchell & M.L. Boccia, pp. 254–72. New York: Cambridge University Press.

Nagell, K., Olguin, K. & Tomasello, M. (1993). Imitative learning of tool use by children and chimpanzees. *Journal of Comparative Psychology*, 107, 174–86.

Natale, F. (1989a). Stage 5 object-concept. In *Cognitive Structure and Development in Nonhuman Primates*, ed. F. Antinucci, pp. 89–95. Hillsdale, NJ: Lawrence Erlbaum.

Natale, F. (1989b). Patterns of object manipulation. In *Cognitive Structure and Development in Nonhuman Primates*, ed. F. Antinucci, pp. 145–61. Hillsdale, NJ: Lawrence Erlbaum.

Natale, F., Antinucci, F., Spinozzi, G. & Poti, P. (1986). Stage 6 object concept in nonhuman primate cognition: A comparison between gorilla (*Gorilla gorilla gorilla*) and Japanese macaque (*Macaca fuscata*). *Journal of Comparative Psychology*, 100, 335–9.

Nishida, T. (1979). The social structure of chimpanzees of the Mahale Mountains. In *The Great Apes*, ed. D.A. Hamburg & E.R. McCown, pp. 73–121. Menlo Park, CA: Benjamin/Cummings.

Nishida, T. (1980). The leaf-clipping display: A newly-discovered expressive gesture in wild chimpanzees. *Journal of Human Evolution*, 9, 117–28.

Oden, L.O., Thompson, R.K.R. & Premack, D. (1990). Infant chimpanzees spontaneously perceive both concrete and abstract same/different relations. *Child Development*, 61, 621–31.

Parker, C.E. (1969). Responsiveness, manipulation, and implementation behavior in chimpanzees, gorillas, and orang-utans. In *Proccedings of the Second International Congress of Primatology: vol. 1 Behavior*, ed. C.R. Carpenter, pp. 160–6. Basel: Karger.

Patterson, F. (1978). Linguistic capabilities of a lowland gorilla. In *Sign Language and Language Acquisition in Man and Ape*, ed. F.C.C. Peng, pp. 161–201. Boulder, CO: Westview Press.

Patterson, F.G.P. & Cohn, R.H. (1994). Self-recognition and self-awareness in lowland gorillas. In *Self-awareness in Animals and Humans: Developmental Perspectives*, ed. S.T. Parker, R.W. Mitchell & M.L. Boccia, pp. 273–90. New York: Cambridge University Press.

Patterson, F. & Linden, E. (1981). *The Education of Koko*. New York: Owl Books.

Perinat, A. & Dalmau, A. (1988). La comunicación entre pequeños gorilas criados en cautividad y sus cuidadoras. *Estudios de Psicología*, 33–34, 11–29.

Piaget, J. (1952). *The Origins of Intelligence in the Child*. New York: Basic Books.

Plooij, F.X. (1978). Some basics traits of language in wild chimpanzees?. In *Action, Gesture and Symbol: The Emergence of Language*, ed. A. Lock, pp. 111–31. London: Academic Press.

Povinelli, D.J. (1994a). Comparative studies of animal mental state attribution: a reply to Heyes. *Animal Behaviour*, 48, 239–41.

Povinelli, D.J. (1994b). What chimpanzees might know about the mind. In *Chimpanzee Cultural Diversity*, ed. R.W. Wrangham, W.C. McGrew, F.B.M. de Waal & P. Heltne, pp. 239–41. Chicago: Chicago Academy of Sciences.

Povinelli, D.J., Nelson, K.E. & Boysen, S.T. (1990). Inferences about guessing and knowing by chimpanzees (*Pan troglodytes*). *Journal of Comparative Psychology*, 104, 203–10.

Povinelli, D.J., Nelson, K.E. & Boysen, S.T. (1992). Comprehension of role reversal in chimpanzees: evidence of empathy? *Animal Behaviour*, 43, 633–40.

Povinelli, D.J., Rulf, A.B. & Bierschwale, D.T. (1994). Absence of knowledge attribution and self-recognition in young chimpanzees (*Pan troglodytes*). *Journal of Comparative Psychology*, **108**, 74–80.

Povinelli, D.J., Rulf, A.B., Landau, K.R. & Bierschwale, D.T. (1993). Self-recognition in chimpanzees (*Pan troglodytes*): Distribution, ontogeny, and patterns of emergence. *Journal of Comparative Psychology*, **107**, 347–72.

Premack, D. (1983). The codes of man and beasts. *Behavioral and Brain Sciences*, **6**, 125–67.

Premack, D. (1986). *Gavagai!* Cambridge, MA: The MIT Press.

Premack, D. (1988). "Does the chimpanzee have a theory of mind?" revisted. In *Machiavellian Intelligence. Social Expertise and the Evolution of Intellect in Monkeys, Apes, and Humans*, ed. R.W. Byrne & A. Whiten, pp.160–79. Oxford: Clarendon Press.

Premack, D. & Woodruff, G. (1978). Does the chimpanzee have a theory of mind? *Behavioral and Brain Sciences*, **4**, 515–26.

Redshaw, M. (1978). Cognitive development in human and gorilla infants. *Journal of Human Evolution*, **7**, 133–41.

Rensch, B. (1973). Play and art in apes and monkeys. In *Precultural Primate Behavior*, ed. E.W. Menzel, pp. 102–23. Basel: Karger.

Rijksen, H.D. (1978). *A Field Study on Sumatran Orangutans (*Pongo pygmaeus abellii *Lesson 1827): Ecology, Behaviour and Conservation*. Medodelirgen landbouwhogeschool Wageningen. Wageningen: H. Veenman and Zonen.

Riopelle, A.J., Jonch Cuspinera, A., Nos De Nicolau, R., Luera, R. & Sabater Pi, J. (1973). Development and behavior of the white gorilla. *National Geographic Research Reports*, **00**, 355–67.

Rumbaugh, D.M., Hopkins, W.D., Washburn, D.A. & Savage-Rumbaugh, E.S. (1989). Lana chimpanzee learns to count by "Numath": A summary of videotaped experimental report. *Psychological Record*, **39**, 459–70.

Russon, A.E. & Galdikas, B.M.F. (1993). Imitation in free-ranging rehabilitant orangutans (*Pongo pygmaeus*). *Journal of Comparative Psychology*, **107**, 147–61.

Sanders, R.J. (1985). Teaching apes to ape language: explaining the imitative and nominative signing of a chimpanzee (*Pan troglodytes*). *Journal of Comparative Psychology*, **99**, 197–210.

Savage, A. & Snowdon, C.T. (1982). Mental retardation and neurological deficits in a twin orangutan. *American Journal of Primatology*, **3**, 239–51.

Savage-Rumbaugh, E.S. (1984). *Pan paniscus* and *Pan troglodytes*. Contrasts in preverbal communicative competence. In *The Pygmy Chimpanzee*, ed. R.L. Susman, pp. 131–77. New York: Plenum Press.

Savage-Rumbaugh, E.S. (1986). *Ape Language*. New York: Columbia University Press.

Savage-Rumbaugh, E.S. (1988). A new look at ape language: Comprehension of vocal speech and syntax. In *Nebraska Symposium on Motivation*, vol. 35, *Comparative Perspectives in Modern Psychology*, ed. D.W. Leger, pp. 201–56. Lincoln, NB: University of Nebraska Press.

Savage-Rumbaugh, E.S. (1990). Language as a cause–effect communication system. *Philosophical Psychology*, **3**, 55–76.

Savage-Rumbaugh, E.S., Brakke, K.E. & Hutchins, S.S. (1992). Linguistic development: contrasts between co-reared *Pan troglodytes* and *Pan paniscus*. In *Topics in Primatology. Human Origins*, ed. T. Nishida, W.C. McGrew, P. Marler, M. Pickford & F.B.M. de Waal, pp. 51–66. Tokyo: Tokyo University Press.

Savage-Rumbaugh, E.S. & McDonald, K. (1988). Deception and social manipulation in symbol-using apes. In *Machiavellian Intelligence. Social Expertise and the Evolution of Intellect in Monkeys, Apes, and Humans*, ed. R.W. Byrne & A. Whiten, pp. 224–37. Oxford: Clarendon Press.

Savage-Rumbaugh, E.S., McDonald, K., Sevcik, R.A., Hopkins, W.D. & Rubert, E. (1986). Spontaneous symbol acquisition and communicative use by pygmy chimpanzees (*Pan paniscus*). *Journal of Experimental Psychology: General*, **115**, 211–35.

Savage-Rumbaugh, E.S., Pate, J.L., Lawson, J., Smith, S.T. & Rosenbaum, S. (1983). Can a chimpanzee make a statement? *Journal of Experimental Psychology: General*, **112**, 457–92.

Savage-Rumbaugh, E.S., Rumbaugh, D.M. & Boysen, S. (1978a). Symbolic communication between two chimpanzees. *Science*, **201**, 641–4.

Savage-Rumbaugh, E.S., Rumbaugh, D.M. & Boysen, S.T. (1978b). Sarah's problems in comprehension. *Behavioral and Brain Sciences*, **1**, 555–7.

Savage-Rumbaugh, E.S., Rumbaugh, D.M. & McDonald, K. (1985). Language learning in two species of apes. *Neurosciences and Biobehavioral Reviews*, **9**, 653–65.

Schiller, P.H. (1952). Innate constituents of complex responses in primates. *Psychological Review*, **59**, 177–91.

Schneirla, T. C. (1966). Behavioral development and comparative psychology. *Quarterly Review of Biology*, **41**, 283–303.

Spinozzi, G. (1993). Development of spontaneous classificatory behavior in chimpanzees (*Pan troglodytes*). *Journal of Comparative Psychology*, **107**, 193–200.

Spinozzi, G. & Natale, F. (1989). Early sensorimotor development in Gorilla. In *Cognitive Structure and Development in Nonhuman Primates*, ed. F. Antinucci, pp. 21–38. Hillsdale, NJ: Lawrence Erlbaum.

Spinozzi, G. & Poti, P. (1993). Piagetian stage 5 in two infant chimpanzees (*Pan troglodytes*): The development of permanence of objects and the spatialization of causality. *International Journal of Primatology*, **14**, 905–17.

Suarez, S.D. & Gallup, G.G., Jr (1981). Self-recognition in chimpanzees and orangutans, but not gorillas. *Journal of Human Evolution*, **10**, 175–88.

Sugiyama, Y. (1972). Social characteristics and socialization of wild chimpanzees. In *Primate Socialization*, ed. F.E. Poirier, pp. 145–63. New York: Random House.

Swartz, K.B. & Evans, S. (1991). Not all chimpanzees (*Pan troglodytes*) show self-recognition. *Primates*, **32**, 483–96.

Swartz, K.B. & Evans, S. (1994). Social and cognitive factors in chimpanzee and gorilla mirror behavior and self-recognition. In *Self-awareness in Animals and Humans: Developmental Perspectives*, ed. S.T. Parker, R.W. Mitchell & M.L. Boccia, pp. 189–206. New York: Cambridge University Press.

Tanner, J. & Byrne, R.W. (1993). Concealing facial evidence of mood: Perspective-taking in a captive gorilla? *Primates*, **34**, 451–7.

Teleki, G. (1973). *The Predatory Behavior of Wild Chimpanzees*. Brunswick, NJ: Bucknell University Press.

Temerlin, M.K. (1975). *Lucy: Growing up Human*. Palo Alto, CA: Science and Behavior Books.

Terrace, H.S. (1979). *Nim. A Chimpanzee Who Learned Sign Language*. New York: Washington Square Press.

Terrace, H.S., Petitto, L.A., Sanders, R.J. & Bever, T.G. (1979). Can an ape create a sentence?. *Science*, **206**, 891–902.

Thomas, R.K. (1992). Primates' conceptual use of number: Ecological perspectives and psychological processes. In *Topics in Primatology. Human Origins*, ed. T. Nishida, W.C. McGrew, P. Marler, M. Pickford & F.B.M. de Waal, pp. 305–14. Tokyo: University of Tokyo Press.

Tomasello, M. (1990). Cultural transmission in the tool use and communicatory signaling of chimpanzees? In *"Language" and Intelligence in Monkeys and Apes: Comparative Developmental*

Perspectives, ed. S.T. Parker & K.R. Gibson, pp. 274–311. New York: Cambridge University Press.
Tomasello, M. (1995a). Do apes ape? In *Social Learning and Imitation in Animals: The Evolution of Culture*, ed. B. Galef Jr & C. Heyes. New York: Academic Press. (in press.)
Tomasello, M. (1995b). Joint attention as social cognition. In *Joint Attention: Its Origins and Role in Development*, ed. C. Moore & P. Dunham, pp. 103–30. Hillsdale, NJ: Lawrence Erlbaum.
Tomasello, M. & Call, J. (1994). The social cognition of monkeys and apes. *Yearbook of Physical Anthropology*, 37, 273–305.
Tomasello, M., Call, J., Nagell, K., Olguin, R. & Carpenter, M. (1994). The learning and use of gestural signals by young chimpanzees: A trans-generational study. *Primates*, 35, 137–54.
Tomasello, M., Davis-Dasilva, M., Camak, L. & Bard, K. (1987). Observational learning of tool-use by young chimpanzees. *Human Evolution*, 2, 175–83.
Tomasello, M., George, B., Kruger, A., Farrar, M. & Evans, A. (1985). The development of gestural communication in young chimpanzees. *Journal of Human Evolution*, 14, 175–86.
Tomasello, M., Gust, D. & Frost, G.T. (1989). The development of gestural communication in young chimpanzees: A follow up. *Primates*, 30, 35–50.
Tomasello, M., Kruger, A. & Ratner, H. (1993b). Cultural learning. *Behavioral and Brain Sciences*, 16, 495–592.
Tomasello, M., Savage-Rumbaugh, S. & Kruger, A. (1993b). Imitative learning of actions on objects by children, chimpanzees, and enculturated chimpanzees. *Child Development*, 64, 1688–705.
van Lawick-Goodall, J. (1968). Behaviour of free-living chimpanzees of the Gombe Stream area. *Animal Behaviour Monographs*, 1, 163–311.
van Rijt-Plooij, H.H.C. & Plooij, F.X. (1987). Growing independence, conflict and learning in mother–infant relations in free-ranging chimpanzees. *Behaviour*, 101, 1–86.
Vauclair, J. & Bard, K.A. (1983). Development of manipulations with objects in ape and human infants. *Journal of Human Evolution*, 12, 631–45.
Visalberghi, E., Fragaszy, D.M. & Savage-Rumbaugh, E.S. (1995). Performance in a tool-using task by common chimpanzees (*Pan troglodytes*), bonobos (*Pan paniscus*), an orang utan (*Pongo pygmaeus*), and capuchin monkeys (*Cebus apella*). *Journal of Comparative Psychology*, 109, 52–60.
Vygotsky, L. (1978). *Mind in Society*. Cambridge, MA: Harvard University Press.
Whiten, A. & Ham, R. (1992). On the nature and evolution of imitation in the animal kingdom: Reappraisal of a century of research. In *Advances in the Study of Behavior*, ed. P.J.B. Slater, J.S. Rosenblatt, C. Beer & M. Milinski, vol. 21, pp. 239–83. New York: Academic Press.
Wood, S., Moriarty, K.M., Gardner, B.T. & Gardner, R.A. (1980). Object permanence in child and chimpanzee. *Animal Learning and Behavior*, 8, 3–9.
Woodruff, G. & Premack, D. (1979). Intentional communication in the chimpanzee: The development of deception. *Cognition*, 7, 333–62.
Wright, R.V.S. (1972). Imitative learning of a flacked stone technology: The case of an orangutan. *Mankind*, 8, 296–306.
Yerkes, R.M. (1927). The mind of a gorilla. *Genetic Psychology Monographs*, 2, 1–193.
Yerkes, R.M. & Nissen, H.W. (1939). Pre-linguistic sign behavior in chimpanzee. *Science*, 89, 585–7.
Yerkes, R.M. & Petrunkevitch, A. (1925). Studies of chimpanzee vision by Ladygina-Kohts. *Journal of Comparative Psychology*, 5, 99–108.
Yerkes, R.M. & Tomilin, M.I. (1935). Mother-infant relations in chimpanzee. *Journal of Comparative Psychology*, 20, 321–59.

18

Three approaches for assessing chimpanzee culture

CHRISTOPHE BOESCH

Gombe National Park, Tanzania, May, 1990:
Prof notices the entrance to a nest of driver ants and, before touching it, he breaks a sapling, turns it upwards in a move and removes the leaves by biting them away, thus producing a straight stick 80 cm long. Standing on an outcropping stone just above the nest entrance, Prof dips about 25 cm of the stick inside the nest entrance, watching the ants climbing up his tool. Once they reach about a third the way up the stick, he rapidly pulls it out and turns it sideways, dexterously pulling the end of the stick through thumb and index finger of his left hand. Having gathered all the ants in a ball, he hastily brings them to his mouth, rapidly chewing them while suddenly removing two of them by biting his lips. The ants, alerted, swarm on the ground around the nest entrance. Prof continues for 10 minutes, eating easily 760 ants per minute in this way.
(my personal observation).

Taï National Park, Côte d'Ivoire, October, 1985:
Nova inserts her index finger in the soft soil and then sniffs it carefully. She starts removing slowly a layer of the soil so as not to alert the driver ants in their nest. She breaks a small branch of a sapling and cuts it with her teeth to produce a stick about 30 cm long. She starts dipping it quietly up to 2–3 cm into the nest entrance, watching as the ants climb up the tool. Once they have reached about 10 cm up the tool, she rapidly turns the end of the stick upwards into her mouth, hastily chewing the ants, and then dipping further. It takes more than 20 minutes for the ants to really become alert and to swarm on the ground around the nest entrance. Nova dips for ants for 30 minutes, but is interrupted twice at the end by Kendo, an adult male, who plunges his arm almost to the shoulder into the nest in order to remove some handfuls of grubs from the depths. Nova eats about 180 ants per minute in this way.
(my personal observations).

These descriptions raise two puzzling questions about the chimpanzee world: how did such population-specific behaviors appear? And do they represent chimpanzee cultures? Why do individuals belonging to two different populations dip for the same species of ant (*Dorylus nigricolus*) with such different techniques and pay-offs? Why do individuals from one population remove the grubs directly from the nest with their hands, while those of the other do not? Further examples of population-specific differences have been documented, i.e. the selection of food by chimpanzees in two populations from Tanzania (Nishida *et al.*, 1983), the ant-dipping techniques in Taï and Gombe chimpanzees (Boesch & Boesch, 1990; McGrew, 1974), the oil-palm nut feeding differences throughout Africa (McGrew, 1992) or the way Taï and Gombe chimpanzees kill and eat their red colobus monkey quarry (Boesch & Boesch, 1989; Goodall, 1986).

I want to document in this chapter the evidence that has been gathered from the wild supporting the notion of a culture in chimpanzees. The concept of culture is variously and often ambiguously defined (McGrew, 1992; Segall *et al.*, 1990), partly because ethnologists, cultural and biological anthropologists as well as primatologists have different approaches to it. Some definitions of culture refer specifically to its human nature and therefore would exclude nonhuman animals including early hominids such as *Australopithecus* sp. and *Homo habilis* (McGrew, 1992). One biological definition of culture is the transfer of information by behavioral means, most particularly by the process of teaching and learning (Bonner, 1980). Such a definition is not too far from a definition given by anthropologists, which proposed that culture encompassed the learned behaviors and attitudes that are characteristic of a particular society (Ember & Ember, 1985). The mode of transmission of information and the population-specific nature of behavior patterns are important criteria in those definitions. Primatologists have proposed another definition: cultures are behavioral variants, induced by social modifications, that create individuals who will in turn modify the behaviors of others in the same way (Kummer, 1971). Under this definition, the behavior of two groups with the same gene pool and with the same type of habitat can only differ in their culture (Kummer, 1971). Adaptations to different ecological conditions are abundant in animals and the resulting behavioral diversity should not be taken as proof of cultural diversity. It is within Kummer's framework that I intend to assess the presence of culture in chimpanzees.

However, I wish to mention two more specific definitions that I shall discuss later using chimpanzee examples. First, an elaborate operational definition of culture (McGrew, 1992; McGrew & Tutin, 1978) with the following eight criteria has been proposed: innovation, dissemination, durability, standardization, diffusion, tradition, nonsubsistence, and naturalness. The two last criteria are the most rigid: nonsubsistence excludes food-oriented behaviors (difficult for a French person to understand!) and naturalness excludes all behaviors arising in artificially provisioned chimpanzee populations. McGrew (1992) concluded that no single behavioral pattern in chimpanzees satisfies all conditions of culture. Secondly, Tomasello *et al.* (1993a) proposed three essential characteristics of human culture to be (a) that cultural behavior should be performed by all group members, (b) that its form should be a faithful reproduction of that of the model and (c) that an accumulation of modifications should exist. Cultural anthropologists would, however, strongly disagree with the notion that a cultural behavior is present in all individuals within a society, as variation according to age, sex, and status of the individuals are frequent (Ingold, 1993). Innovation, therefore, is an important aspect of a cultural process and is the key process that makes cultural transmission more rapid than genetic transmission (Bonner, 1980).

The examples given at the beginning of the chapter would fulfill the criteria of Kummer's definition of culture if ecological parameters could be excluded. For example, the difference observed between two neighboring chimpanzee communities living in the Mahale Mountains in their propensity for using tools to capture termites proved to be habitat specific: a precise analysis revealed that climatic differences between the range of these two chimpanzee communities lead to differences in the soil and vegetation, which in turn lead to differences in termite availability (Nishida & Uehara, 1980). Some critics of the notion of culture in animals would argue, however, that subtle and even past ecological differences could be at the origin of the observed differences in chimpanzees (Tomasello, 1990). Such differences are difficult to exclude completely, especially for differences reported between distant populations. However, such arguments fall short of explaining differences observed in purely social gestures, for instance hand-grasping (McGrew & Tutin, 1978) and leaf-clipping (Boesch, 1995; Nishida, 1987), and those open the way for cultural differences.

The uncertainties of inferring culture from the distribution of the behavior between populations have led some authors to propose that it is the particular social transmission mechanism of the behavior concerned between individuals that is decisive in attributing culture (Galef, 1990; Tomasello, 1990; Tomasello et al., 1993a; Whiten & Ham, 1992). It has been proposed that culture requires social learning through imitation, teaching or collaborative learning (Galef, 1990; Heyes, 1993; Tomasello et al., 1993b). Other transmission mechanisms, involving aspects of individual learning or simpler social learning, would not qualify as cultural learning processes. Imitation is the process under which an observer attends to, and attempts to copy, the *behavior* of another individual, rather than to try to reach the *same goal* by whatever means it may find (called goal emulation), or to use the *same object* after its attention was attracted to it by the activity of the other individual (called stimulus enhancement: Thorpe, 1956; Tomasello, 1990; Whiten & Ham, 1992).

Imitation has been often attributed to chimpanzees in order to explain the learning of population-specific behaviors (e.g. Goodall, 1986; Sugiyama & Koman, 1979b). However, recent conceptual and experimental studies indicate that it might be more difficult for them or nonhuman primates in general than previously thought: cebus monkeys and baboons proved unable to acquire the behavior of other skilled cagemate tool users, despite a prolonged opportunity to observe them (Visalberghi, 1988; Visalberghi & Fragaszy, 1990; Whiten, 1989). In some captive studies, chimpanzees were seemingly unable to copy precisely a given behavior but only able to learn that the involved tool may be used to obtain a certain food (Tomasello et al., 1987; but for data supporting the opposite view, see Custance & Bard, 1994; Tomasello et al., 1993a). If one accepts the claim that imitation is fundamental to culture, these results challenge previous claims that population-specific differences in wild chimpanzees derive from cultural transmission processes (Boesch & Boesch, 1990; Goodall, 1986; McGrew & Tutin, 1978; Nishida, 1987; Nishida et al., 1983). Still, they may be based on the inability of the experimenters to show evidence of imitation in captive young chimpanzees (considerable work is required to demonstrate the lack of an ability Hauser, 1993). For an individual, the motivation to copy or influence the behavior of a close kin or a familiar companion is greater than that with an unrelated or unfamiliar model. An example of how a difference in the quality of the model (chimpanzee versus human) can lead to completely opposite results comes from a study where nut-cracking behavior was taught to a juvenile female chimpanzee in a zoo; a human demonstrator failed to elicit much interest in the chimpanzee pupil in 30 sessions of 1 hour each, whereas, when exposed to a successful adult female chimpanzee as demonstrator, the juvenile female tried and succeeded in opening the nut on her own after the very first session (Sumita et al., 1985). Similarly striking are the observations made with orangutans that virtually never imitated strangers but regularly imitated kin or affiliates (Russon & Galdikas, 1995).

Some have proposed that imitation should be studied in captivity (Tomasello, 1990) but if we are interested in cultural transmission we *must* study patterns of behaviors in the field that might have the potentiality to be cultural. The seeming lack of imitation by chimpanzees for sand-throwing or using reaching-sticks (Tomasello, 1990; Tomasello et al., 1987) does not tell us much about the transmission process for hand-grasping, leaf-clipping, nut-cracking or other patterns of behavior proposed to be cultural in wild chimpanzees (Boesch & Boesch, 1990; Goodall, 1986; McGrew & Tutin, 1978; Nishida & Hiraiwa, 1982). Experiments under controlled conditions have advantages but the drawbacks should not be ignored because they may make captive conditions inappropriate for studying such processes as cultural transmission and learning.

Finally, the claim that cultural transmission is only possible with imitation or teaching (Tomasello, 1990; Tomasello et al., 1993a) might well be too restrictive (Boesch, 1993; McGrew, 1992; Whiten & Ham, 1992). I do not deny that imitation plays a role in human cultures but I want to emphasize that lower forms of social learning are also part of the cultural transmission process in humans. Wynn (1993) presented examples of tasks considered as cultural that are acquired by simple social learning (stimulus enhancement or emulation). Similarly, we can all think of typical cultural behaviors in humans that do not require the acquisition of a new behavior (which is part of the definition of imitation). For example, hand-shaking, embracing or hat-lifting used in greeting in some cultures does not require the acquisition of any movement that is not already used by young children when playing. It is the context of behavior production that is new and this can be learned through local enhancement or simple social learning processes. Thus, even if captive studies reveal the inability of chimpanzees to imitate, this does not prove that they lack cultural processes (for another argument against the importance of imitation in cultural transmission, see Heyes, 1993). Other forms of transmission have been demonstrated in captive chimpanzee (Tomasello et al., 1987) that suggested that this species is thus potentially capable of cultural learning.

In conclusion, I propose that the presence of culture in an animal species needs to be investigated from three complementary approaches: first, we need the *ecological approach*, in which we ascertain that the behavior under consideration is population specific and not directly related to ecological differences. I feel that this would be easier if we can find a precise limit in the distribution of a given behavior in order to study groups that are geographically so close that we can assume genetic and habitat differences to be too small to explain the observed variations. By definition, culture is population-specific and field work is needed to exclude the possibility that the environmental conditions influence the behavior pattern. Secondly, we need to adopt a *transmission mechanism approach*, in which we analyze the transmission mechanisms of these population-specific behavior patterns, to determine whether they are adequately canalized by social learning processes to guarantee the fidelity of transmission. Regardless of the learning mechanism, it is fidelity in transmission that is important. Thirdly, we may adopt an *innovation approach*, because under the social transmission hypothesis we have to expect recent innovations in the use of behavior that look like human "fashions" and seem to be purely influenced by social group. This last approach should, however, not be considered as a condition for deciding whether a behavior is cultural, as it is extremely difficult to evaluate the rate at which we should expect innovation to occur within a social group. I shall discuss each of these approaches with new data from wild chimpanzees.

ECOLOGICAL APPROACH: THE DISTRIBUTION PATTERNS OF CERTAIN BEHAVIORS IN WILD CHIMPANZEES

Many authors were struck by differences in the distribution of some behaviors in wild chimpanzees that tend to concur with the limits of the three subspecies of chimpanzees (i.e. the nut-crackers of West Africa versus the termite-fishers of Central and East Africa (Nishida, 1973; Struhsaker & Hunkeller, 1971; Sugiyama, 1985), or the termite-fishers of East Africa and the termite diggers or termite probers of Central Africa (McGrew, 1992; Teleki, 1974)). However, new observations show a much more complicated pattern. Table 18.1 reviews the present evidence on the distribution of some behaviors that have been proposed to be culturally distributed. Of the 13 behavior patterns considered, one has a regional distribution: leaf-grooming, which is limited to the Tanzanian chimpanzees. Two behavior patterns, marrow-pick and self-tickle, are limited to one population (Taï and Gombe, respectively). Otherwise, the distribution of cultural behaviors is rather irregular and the pattern does not correspond to any obvious phylogenetic differences.

Some of these behavior patterns have a strong social component and the ecological component seems difficult to find; hand-clasping, which happens between two individuals grooming each other, and leaf-clipping or leaf-grooming, which is done with whatever leaves are available. For others, environmental factors may be responsible for the distribution of these differences and a more detailed analysis should be performed to exclude subtle but nevertheless influential differences.

The map presented in Figure 18.1 showing the detailed distribution of nut-cracking behavior and the following analysis may allow us to exclude ecological factors as a possible explanation.

Nut-cracking behavior (from Boesch *et al.*, 1994)
The distribution of nut-cracking behavior can be ascertained in detail because it has the advantage that its existence can be confirmed without observing the chimpanzees. The

Table 18.1. *Review of the present behavioral differences that have been proposed to be cultural and their distribution within populations of chimpanzees*

Pattern	West Africa			East Africa	
	Bossou	Taï	Other sites	Gombe	Mahale
Ant-dipping	+	+	Assirik	+	−
Fly-whisk	+	+		+	−
Hand-clasping	−	+	Kibale	−	+
Honey-dipping	−	+		+	−
Leaf-clipping	+	+		−	+
Leaf-grooming	−	−		+	+
Leaf-napkin	−	−	Kibale	+	−
Leaf-sponge	+	+		+	−
Marrow-pick	−	+		−	−
Nut-cracking	+	+	Figure 18.1/18.2	−	−
Play-start	−	+		+	+
Self-tickle	−	−		+	−
Termite-fishing	−	−	Assirik	+	+

To exclude the most obvious bias (length of the study period), I am presenting positive and negative results for long-term studies in which chimpanzees can be directly observed (more than 8 years), for shorter studies only the positive observations are given. (Assirik is located in Senegal (West Africa) and Kibale in Uganda (East Africa).)

Bossou, Sugiyama (1981); Sugiyama & Koman (1979);
Taï, Boesch & Boesch (1990); Boesch (1993);
Gombe, Goodall (1986); McGrew (1992);
Mahale, Nishida (1973, 1987); McGrew (1992);
Assirik (Senegal), McGrew *et al.* (1979);
Kibale (Uganda), Wrangham & Isabirye-Busata, cited in McGrew (1992).

artifacts, anvils (horizontal roots or rocks with traces of wear due to nut-pounding) with nut remains lying around them and sometimes with a hammer (a log or a stone) on top or nearby, remain visible for years to experienced observers (Anderson *et al.*, 1983; Beatty, 1951; Boesch & Boesch, 1983; Rahm, 1971; Struhsaker & Hunkeler, 1971). On the basis of the presence of such traces, nut-cracking behavior is thought to be restricted to the forest chimpanzees of West Africa, with its presence positively confirmed in Côte d'Ivoire, Guinea, Liberia, and Sierra Leone (see Figure 18.1: Boesch & Boesch, 1983; Kortlandt & Holzhaus, 1987; Savage & Wyman, 1843/4; Sugiyama, 1981; Sugiyama & Koman, 1979a; Whitesides, 1985). It is reported to be absent from the forests inhabited by chimpanzees in Cameroon (Sugiyama, 1985) and Gabon (Tutin & Fernandez, 1983)

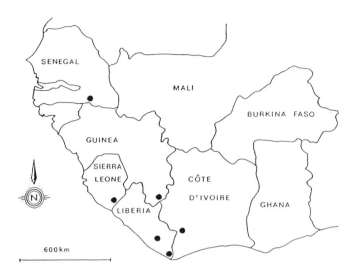

Figure 18.1. Situation of confirmed nut-cracking sites (marked by ●) made by chimpanzees in West Africa.

– forests, where, however, the same nut species occur. These patterns tend to exclude any obvious ecological differences and many authors proposed nut-cracking behavior in chimpanzees to be cultural (Boesch & Boesch, 1983; Goodall, 1986; McGrew, 1992; Sugiyama, 1990). However, we see from Table 18.2 that not all populations crack the same species of nuts or use the same kinds of tools, indicating that environmental factors are somehow involved. In addition, there are more subtle ecological differences that could affect the benefits of nut-cracking, such as a too low availability of potential tools, too low density of the nut-producing trees, and an abundance of alternative food types; these have not been studied so far (see Sugiyama, 1985).

One of the aims of the chimpanzee survey we completed in Côte d'Ivoire was to clarify the situation of nut-cracking behavior and to search for its distinct limit, because some information (C. Martin, personal communication) indicated that chimpanzees in Ghana did not crack nuts and that the limit of the behavior might be within Côte d'Ivoire. We visited 35 sites in Côte d'Ivoire, most of them situated within the Guinean belt (an area covering about 110 000 km² in the southern part of the country and comprising the evergreen and semi-deciduous forests (Figure 18.2)). The presence of chimpanzees was confirmed in 24 of these sites, whereas nut-cracking sites were found in only seven of them. Figure 18.2 shows that these seven sites are all situated in the southwestern part of the country. In Côte d'Ivoire, the *Nzo-Sassandra river constitutes the eastern limit of nut-cracking behavior* as, in spite of a careful search, no nut-cracking site has been found on the eastern side of these rivers. After intensive prospecting in the Mount Nimba and Mount Dan regions (points 1–6 of Figure 18.2), only two Coula (*Coula edulis*) nut-cracking sites have been found in Mount Nimba. However,

Table 18.2. *Nut species cracked and tools used by chimpanzees in different regions of West Africa*

		Nut species												Tools			
		Coula		Detarium		Elaeis		Panda		Parimari		Sacoglottis		Wood		Stone	
		P	C	P	C	P	C	P	C	P	C	P	C	H	A	H	A
Sierra Leone																	
Tiwai Island	1	+	−	+	+	−	−	−	−	+	−	?	−	+	+	?	?
Liberia																	
South-west	2	+	+	?	?	+	−	?	?	+	−	?	?	−	−	+	+
Unknown	3	?	?	?	?	+	+	?	?	?	?	?	?	−	−	+	+
Sapo NP	4	+	+	−	−	+	−	+	+	+	+	−	+	+	+		
Guinea																	
Bossou	5	?	−	?	−	+	+	−	−	+	−	−	−	−	−	+	+
Côte d'Ivoire																	
Taï NP	6	+	+	+	+	+	−	+	+	+	+	+	+	+	−	+	+

P, tree present; C, tree present and nut-cracking confirmed; H, hammer; A, anvil; +, yes; −, no; ?, not known.

Sources: 1, Whitesides (1985); 2, Savage & Wyman (1843/44); 3, Beatty (1951); 4, Anderson *et al.* (1983); 5, Sugiyama & Koman (1979a); 6, Boesch & Boesch (1983).

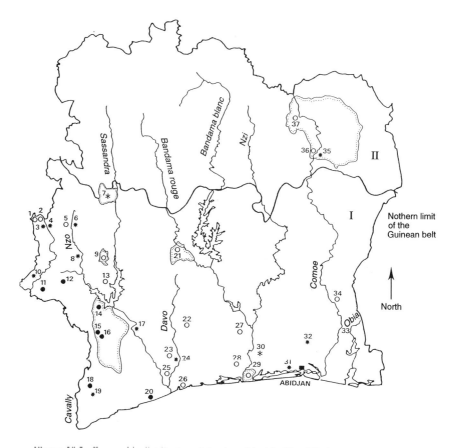

Figure 18.2. Geographic distribution of the sites visited in Côte d'Ivoire:
(●) sites with nut-cracking; (o) sites without nut-cracking; (✻) sites without indications of chimpanzees presence; (∗) 2 other sites where chimpanzees have been mentioned recently by credible observers; (-) northern limit of the Guinean belt; (I) Guinean belt; (II) Soudanian belt; NP, National Park; CF, Classified forest;
1, Mt Nimba NP; 2, Gbapleu (Tiapleu CF); 3, Tiapleu CF; 4, Mt Nieton CF; 5, Blépleu (Sangouiné CF); 6, Mt Tonkoui CF; 7, Mt Sangbé NP; 8, Tyonlé CF; 9, Mt Péko NP; 10, Goulaleu CF; 11, Mt Bétro CF; 12, Mt Zoa (Scio CF); 13, Duékoué CF; 14, Nzo reserve; 15, Taï NP -Audrenisrou; 16, Taï NP -Nipla; 17, Mt Kourabahi CF; 18, Mt Kopé; 19, Haute Dodo CF; 20, Monogaga CF; 21, Marahoué NP; 22, Nizoro CF; 23, Guiniadou (Niegré CF); 24, Davo; 25, Kouadiokro (Niegré CF); 26, Dagbégo (Dassiékro CF); 27, Mopri CF; 28, Gô CF; 29, Azagny NP; 30, Irobo CF; 31, Agnéby; 32, Yapo CF; 33, Songan CF; 34, Bossematié CF; 35, Comoé NP -Gansé; 36, Comoé NP -Amaradougou; 37, Comoé NP -Kolonkoko.

Y. Sugiyama (personal communication) recently saw some Coula nut-cracking sites in the higher part of the Guinean side of Mount Nimba, which he judges were most probably made by chimpanzees, confirming the possibility that chimpanzees may crack nuts on this mountain, but at a rather low frequency.

What ecological factors might explain such a limited distribution? We tested four factors that could directly affect the presence of the behavior (for more details, see Boesch *et al.*, 1994).

1. Chimpanzee density: as part of a survey in Côte d'Ivoire, we observed that the density of chimpanzee populations is influenced both by the type of habitat and the level of human impact but not by geographical locations (Marchesi *et al.*, 1995). Nut-cracking populations west and east of the Sassandra river have densities ranging from 0.45 to 6.39 individuals per km^2 with no difference between the two sides of the river (Marchesi *et al.*, 1995). Thus, an especially low density of chimpanzees, which could eventually lead to a very low density of nut-cracking anvils, cannot explain the absence of the nut-cracking behavior east of the Sassandra river.
2. Density of nut-producing tree species: we compared the density of the three main species of nuts cracked by Taï chimpanzees (*C. edulis*, *Parinari excelsa* and *Panda oleosa*) in 19 of the 24 sites with confirmed chimpanzee populations. For all sites, no significant difference existed in the density of trees between regions with nut-cracking sites and without (Boesch *et al.*, 1994). Hence, differences in the density of the nut tree species cannot explain the absence of nut-cracking behavior.
3. Availability of potential hammers and potential anvils: we compared the availability of potential tools for ten sites (for details on the methodology, see Boesch *et al.*, 1994). No difference in the availability of potential hammers or that of potential anvils existed between sites within or outside the known nut-cracking regions. The availability of potential hammers was even lower in the sites visited in Taï forest (points 15 and 16 in Figure 18.2), where nuts are regularly cracked. For the two areas we visited in the Taï forest and for Dagbégo (point 26) and Duékoué (point 13) most of the potential tools were of wood, whereas stones predominated in the other sites. Thus, the absence of nut-cracking behavior cannot be explained by differences in tool availability between these sites.
4. Floristic description: to evaluate the possible role of the general food availability in all sites, we compared the floristic characteristics of each of them. All the cracking sites are localized in the Sassandrian evergreen forest (the *Diospyros* spp. and *Mapania* spp. type and the *Eresmopatha macrocarpa* and *Diospyros mannii* type, as defined by Guillaumet & Adjanohoun (1971)). This is also the case for the Niégré (points 23 and 25) and Dagbégo (point 26 in Figure 18.2) regions situated east of the Sassandra river. When comparing all of them, we noticed that there was always at least one location *without* nut-cracking in the same floristic zone east of the river as sites *with* nut-cracking west of the Sassandra river.

A very precise eastern limit in the distribution of the nut-cracking behavior was found in Côte d'Ivoire. The analyses of the different sites showed that none of the ecological parameters considered could explain the presence or absence of this behavior. In addition, all the different communities belong to *Pan troglodytes verus*, most certainly also ruling out a genetic explanation as well. Ecological differences are very difficult to exclude completely and the present level of deforestation in Côte d'Ivoire prevented us from selecting sites exactly contiguous on both sides of the Sassandra river. However, the Taï (points 15 and 16) and the Niégré (point 23) forests on either side of this river are at most 50 km apart, within the same floristic division, at the same latitude (5° 30′ N) and with a similar altitude (about 150 m above sea level). The same applies for Monogaga (point 20) and Dagbégo (point 26) forests situated on either side of the Sassandra river, both being coastal forests at most 30 km apart. The two examples that we investigated especially carefully demonstrated that ecological differences cannot explain the behavioral differences observed in nut-cracking.

We concluded that the evidence is strong that nut-cracking is a cultural behavior and we proposed that its present distribution may be related to the history of the region. During the Pleistocene, a serious drought occurred (about 17 000 years ago) and the desert reached the sea, cutting West Africa into two forest refuges, one west from the Sassandra near the border between Côte d'Ivoire and Liberia, the other east from the Sassandra in Ghana (Hamilton, 1982). The nut-cracking behavior might have appeared during this period on the western side of the Sassandra river. Once the humidity increased and the forests expanded, the Sassandra river in the south increased in size and prevented the chimpanzees from crossing it. This river also represents the geographic limit for five rainforest monkey subspecies (Haltenorth & Diller, 1977). Further north, near Duékoué, it is possible to cross the river, as is obvious from the fact that there are many hybrids of these monkeys whose distribution is restricted in the southern part of the Sassandra river (Booth, 1958). However, as this northern region is within the semi-deciduous forest zone, which is poorer in nut-producing trees, the propagation of nut-cracking behavior to the forests east of the Nzo-Sassandra river has been prevented.

Thus, the nut-cracking behavior has a very restricted distribution that seems compatible with a single invention many thousands of years ago and a cultural propagation limited to the tropical rainforests west of the Sassandra river.

TRANSMISSION MECHANISM APPROACH

Two approaches are possible: the direct one in which we observe the transmission mechanism itself and the indirect one in which we predict what should be expected from a given transmission mechanism. Social transmission can come from the naive individual reproducing a model's performance or from the knowledgeable individual directly instructing a naive one. I shall provide indirect evidence for the first process and direct evidence for the second one in the context of nut-cracking in Taï chimpanzees.

Naive individuals "copying" a model

The ecological approach to nut-cracking supports the possibility that it could be a cultural behavior. However, such a proposition implies that its propagation within the population is based on a social learning process. If this proposition is true, we should expect that the individual learning the technique does not try all possible methods to solve the problem but only certain methods, which are observed in the proficient nut-cracker (the model). In other words, individual learning possibilities should be canalized by social learning to guarantee the fidelity of transmission. However, here we face the major problem of identifying *all the possibilities available to a young chimpanzee who is learning to crack nuts*. Ecological conditions are expected to constrain these possibilities (Galef, 1990; Tomasello, 1990), but *only chimpanzees can tell us what these possibilities really are* within the physical and psychological constraints of that species. Luckily, a study that tried to introduce nut-cracking behavior to captive chimpanzees listed all the different methods that the subjects used. Individuals that had never had the opportunity to crack nuts were provided with all the necessary materials (nuts, wooden hammers, and anvils) and were observed as they freely interacted with them. Despite a second phase with human demonstrators, these chimpanzees did not succeed in opening the nuts and thus any obvious form of canalization by successful models was absent (Funk, 1985).

Table 18.3 lists the possible methods that these zoo chimpanzees and the wild Taï chimpanzees were observed to use in order to try to open a nut. I did not limit myself to efficient ways of opening the nuts, because the apprenticeship for nut-cracking is so long that we should not expect youngsters of 2–4 years old to be able to judge the efficiency of any method. Four year old infants are the first ones that regularly succeed in opening a nut and therefore the first ones able to judge that hitting with a hammer is the best method to open nuts (Boesch & Boesch, 1984). Obviously, the ecological conditions in the tropical rainforest and in a zoo are far from being similar, but, as can be seen from Table 18.3, the methods used by the chimpanzees involved mainly the individual, the nut and sometimes an object, for which an equivalent can be found in the forest. Of the 14 methods that the Zürich chimpanzees used to try to open the nuts, only six were seen in Taï chimpanzees (Table 18.3). This is intriguing because some of these unused methods were actually observed in Taï chimpanzees but in other contexts; sometimes infants threw the hammer on the nut/anvil or threw sticks against conspecifics but they never threw the nut against a hard surface. Stabbing with a stick was observed against a leopard and rubbing was observed when they fed on other kinds of fruit (e.g. the hairy fruits of *Diospyros ivorensis*). This indicates that some very strong canalization is at work in Taï that limits the individual learning potentialities. Of the six methods used by Taï chimpanzees, five have elements commonly observed in adults cracking nuts, four include the hitting movement used when they crack nuts and one, biting, is used to completely open the broken nuts. Thus, the movement to crack nuts seems to be the paradigm used by youngsters and the variations observed were concerned mainly with the object used. In other words, social canalization through imitation is at work and it confines the individual learning possibilities (stimulus enhancement) to the different types of object that could be used to hit the nut.

In another experimental study of captive chimpanzees, where the nut-cracking behavior was acquired by five individuals (Sumita *et al.*, 1985), some manipulations in relation to the task are mentioned, including rubbing and pressing the nut. Contrary to unsuccessful individuals, the behavior of the successful chimpanzees tended to change

Table 18.3. *All possible methods to attempt to open nuts that have been seen to be used by two populations of chimpanzees*

Method	Zurich chimpanzee	Taï chimpanzee
Pound the nut against hard surface[a]	+	Regular
Hit the nut with hand	+	Regular
Hit with other body part[b]	+	–
Hit with an object[c]	+	Rare
Hit with a hammer	+	Often
Bite the nut	+	Often
Throw the hammer on nut	–	Regular
Throw the nut against hard surface[a]	+	–
Shake the nut	+	–
Press nut against teeth[d]	+	–
Sit on nut	+	–
Scratch the nut with fingers	+	–
Rub the nut against hard surface[a]	+	Rare
Press on the nut	+	–
Stab with a stick	+	–

[a] Chimpanzees can rub, pound or throw the nut directly with the hand against the ground, a stone, a tree trunk or a root.

[b] By other body parts: those used are the back of the hand or the elbow.

[c] By object: that is material that could not make a hammer such as a piece of cloth, small twigs or in the Taï forest another nut, a piece of termite mound or a hard-shelled fruit.

[d] Chimpanzees pressed the nut with the hand against the teeth with the mouth kept open.

The study on the captive chimpanzees of the Zurich zoo was performed by M. Funk (1985). A plus sign indicates that the method was used in the population, whereas a minus sign indicates that it was never observed.

after demonstration by a skillful nut-cracker, in the sense that striking movements were more frequently used. Thus, here again the behavior of the model tends to be copied.

Teaching

In my opinion, an individual teaches another, if, on the one hand, it influences the likelihood of naive individuals using given objects or, on the other hand, it provides information about how to solve the tasks. Taï chimpanzee mothers have been observed to attempt to accelerate the acquisition of the nut-cracking technique by their youngsters in three different ways (Boesch, 1991).

Assessing chimpanzee culture 417

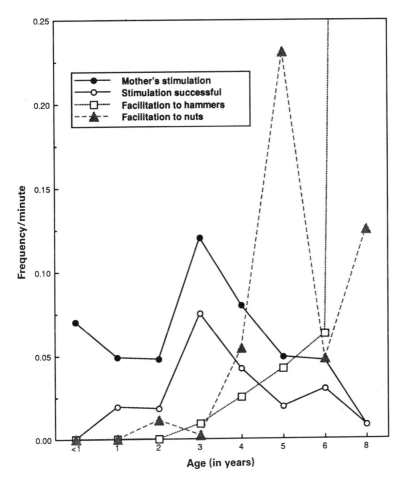

Figure 18.3. Stimulation and facilitation by the mother of the nut-cracking attempts of her offspring.

Stimulation

Mothers may stimulate the nut-cracking attempts of their infants by, first, *leaving the hammer* on the anvil, when they leave to collect more nuts. Infants, remaining near the anvil, may get interested in the hammer and begin to handle it (Figure 18.3). Secondly, mothers may *leave nuts* near the anvil, while infantless chimpanzees were never observed leaving good, intact nuts near the anvil before searching for more. In all observed cases, the infant would at least give some hits with the hammer after placing a nut on the anvil. Stimulations are more frequent when 3 year old infants start to use the hammer (Figure 18.3). However, mothers may go further than just momentary loaning of their tools.

Facilitation

Mothers may facilitate the nut-cracking of their infants by *giving their good hammers* to their youngsters or by *providing intact nuts* for them to crack. Facilitations, like the stimulations, are more frequently observed when the infants have made some progress in their nut-cracking skills. While stimulation starts for 3 year old infants, facilitation starts when the 4–5 year olds have begun to be successful. Thus, the mothers' acts are adjusted to the level of skills attained by their infants and facilitation occurs later ($r_S = -0.48$, $p < 0.05$) than stimulation.

Active teaching

Infants, despite their efforts and the use of convenient tools may face technical difficulties in nut-cracking that they are unable to overcome. In two cases, the mother, resting near her nut-cracking offspring, noticed its technical difficulties and was seen to make a clear demonstration of how to solve them (see description in Boesch, 1991). In the first example, after her 6 year old son cracking nuts positioned a piece of nut badly on the anvil, the mother possibly anticipated the consequence of his action and corrected the positioning of this piece. In the second example, the mother, seeing the difficulties of her 5 year old daughter, corrected very slowly and conspicuously an error in her daughter's way of holding an irregularly shaped hammer and then proceeded to demonstrate to her how the hammer worked with a good grip. Her daughter successfully learned the lesson, since, when cracking further, she *strictly maintained the grip demonstrated to her* despite increasing difficulties that she attempted to solve by varying her position and that of the nut but not the grip on the hammer. In these two cases, the mothers were not cracking nuts at the time but changed the course of their behavior, walking to the place of their infant to teach them. Further, they both resumed their previous behavior after teaching, while the infants went on cracking.

Thus, teaching is part of the repertoire of chimpanzees and the question may be: why is it so rare? But this does not disqualify the observed cases as teaching. The assumption that, if it exists, it should be more frequent because it is a more efficient way to learn a task (Caro & Hauser, 1992) may well be true only under some special circumstances. Under a certain age, the teaching action may not be understood conceptually by the youngster and thus teaching would be useless, and by the time they could understand it, they may at the same time be able to understand enough of the task itself to solve it on their own by individual and social learning, including imitation. In human cultures, teaching may be extremely rare or even absent (Olson & Astington, 1993; Rogoff *et al.*, 1993). For example, in a study of spontaneous interactions between mother–infant humans belonging to two different cultures (Whiten & Milner, 1984), teaching occurrences for all spontaneous behavioral interactions were observed only once for every 364 hours (2 out of 43 800 minutes). I propose that teaching will remain rare in most societies. Only when tasks have more than one solution and only one is socially acceptable might teaching become more common. In other words, the rigidity of some cultures may force teaching to become more important to assure the standardization of the technique. This might account for the fact that in some hunter–gatherer societies (e.g. !Kung bushmen) teaching is absent, whereas the educational systems of "modern" societies rely totally on teaching (Olson & Astington, 1993).

INNOVATION APPROACH (ADAPTED FROM BOESCH, 1995)

Innovation consists in the introduction of a new behavior in the repertoire of one or more individuals, which may be for example a solution to a new problem, an ecological discovery or an existing signal used for a new purpose (Kummer & Goodall, 1985). Some innovations might develop independently of any ecological influences and if they disseminate within the social group, this would by definition involve a cultural process (Kummer, 1971; Kummer & Goodall, 1985; McGrew 1992).

The best-known examples of innovation in animals come from one community of Japanese macaques that invented and purportedly copied the behavior of potato-washing and wheat-throwing (Kawai, 1965) with food provided to them by human observers. A recent and critical look at these examples has led some to doubt that imitation was at work due to the slow dissemination rate of the behavior between group members (Galef, 1990; Visalberghi & Fragasy, 1990). However, the assumption that a behavior learned through imitation disseminates more quickly than one learned through, for example, emulation has to be proved. In addition, we should expect the speed of dissemination of a new behavior to depend strongly on the social interactions observed within the group.

In chimpanzees, examples of innovation are reported from Gombe, where, for example, a juvenile female, Fifi, performed a new gesture "wrist-shaking" in an aggressive context for a period of 10 months. Another young female used the same movement within a week of the first performance by Fifi, but the gesture disappeared progressively from the repertoire of both individuals (Goodall, 1973; Kummer & Goodall, 1985). In Mahale, in one case a newly acquired behavior was transmitted to a larger number of individuals: a new style of courtship display, the "cushion-making" display, was observed to be used by ten immature males of the M-group (Nishida, 1987). These "incipient" cultures (Nishida, 1987) in the social sphere were acquired by a limited number of group members and gradually disappeared without disseminating to other individuals. This raises the question about the conditions necessary for an innovation to be taken up by other group members and to become established.

If innovation is followed by social transmission (which would constitute a cultural propagation), we should expect to find variations in behavioral occurrences within populations that cannot be explained by variations in the habitat. These variations in behavior should be found within a same group for different periods of time, in a pattern mimicking human "fashions." I shall first present the leaf-clipping behavior of Taï chimpanzees in this context and then discuss it along with leaf-grooming as cultural variants.

A possible cultural variant: Leaf-clipping

This behavior was first described in the Mahale chimpanzees in Tanzania, as follows:

> A chimpanzee picks one to five stiff leaves, grasps the petiole between the thumb and the index finger, repeatedly pulls it from side to side while removing the leaf blade with the incisors, and thus bites the leaf to pieces. In removing the leaf blades, a

ripping sound is conspicuously and distinctly produced. When only the midrib with tiny pieces of the leaf blade remains, it is dropped and another sequence of ripping a new leaf is often repeated
(Nishida, 1987, p. 466).
Note that nothing of the leaves is eaten. It has also been observed at Bossou quite regularly (Sugiyama, 1981) but only twice at Gombe (J. Goodall personal communication, cited in Nishida, 1987). This leaf-clipping behavior is regularly observed among Taï chimpanzees as well. However, the forms of the behavior are not totally identical in that, contrary to Mahale chimpanzees, Taï chimpanzees take the leaf blades from both sides between their lips at the same time and remove them in one movement, not repeatedly biting small pieces away.

Innovation in the use of leaf-clipping
In Taï, leaf-clipping is as a rule one element of the drumming sequence of the adult males and, since the beginning of our study, this was the normal use of this behavior (data were collected systematically only from 1990 onwards). All males were observed to use it at the very onset of the drumming display before they start to pant-hoot (129 of 132 observations). Due to the context of production, no individuals other than the adult males were seen to perform this behavior. Individual males, however, differ in the rate of its use. It is possible that leaf-clipping is a kind of displacement behavior to release some tensions. Only three observations were related to sexual frustration, when several males competed for an estrous female, and the males that did not succeed in mating started to leaf-clip.

In late January 1991, a new context for using the leaf-clipping behavior suddenly appeared: individuals started to leaf-clip while resting on the ground (32 of the 183 observations made since January 1990, see Figure 18.4). I immediately looked for a potential reason, possibly related to frustration. However, the individuals were just resting on the ground or interrupting a period of sleep to perform such leaf-clipping. Furthermore, other age–sex classes started to perform it, i.e. juveniles and adult females. Conditions as described above for sexual frustration were not fulfilled during these resting periods, as no estrous females were around. In addition, for females and juveniles, the cause would have had to be different. The role of leaf-clipping during these resting periods is unclear to me. The behavior, however, occasionally differed in form from the leaf-clipping used by the males in the drumming context, i.e. the ripping of the leaf was sometimes done directly with the fingers while the leaves were still attached to their saplings and not, so far, with the teeth on detached leafs. The leaf-clipping performed in the drumming context remained constant in its form (120 of the 183 observations made since December 1990) and equally frequent, but leaf-clipping for frustration in sexual conflict situations became more frequent ($\chi^2 = 8.0$, df = 1, $p < 0.01$) after December 1990.

This change appeared suddenly within a month but the inventor of the change could not be ascertained. During this month, ecological conditions remained very stable as it was the dry season and the social composition of the community remained the same except for the disappearance of one adult male.

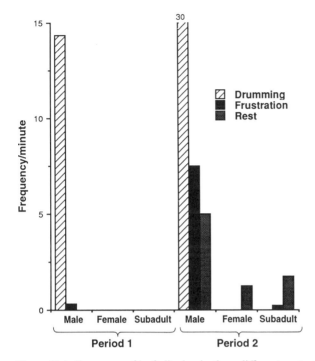

Figure 18.4. Frequency of leaf-clipping in three different contexts over two different time periods in Taï chimpanzees.

Idiosyncrasy in the drumming sequence prevails among adult chimpanzee males, but the form of the leaf-clipping behavior itself remained constant for years for all of them, satisfying Tomasello et al.'s (1993a) criteria of stability and performance across group members. For example, each adult male has his own way of preparing a drumming bout during the "warming-up" phase, e.g. while silently expelling air before starting the pant-hoots, Macho would repeatedly bend his head low toward the ground, while Kendo would shake a sapling violently forward and backward, and Brutus would horizontally and rhythmically move all his body forward and backward. After the warming-up phase, all would run toward a buttress to drum on. But despite these differences, for all these years leaf-clipping remained constant for all males. Modification in the context or function of use (frustration in sexual competition, distraction while resting, drumming sequence) is also observed. Thus, the three essential characteristics of human culture seem to be demonstrated by chimpanzees in leaf-clipping (Boesch, 1993). The new observations of leaf-clipping in Taï, also satisfy all of the criteria for culture proposed by McGrew (1992), as Taï chimpanzees were never provisioned and leaf-clipping was not food-related.

Leaf-clipping as cultural variants in chimpanzee populations
Leaf-clipping has been described in three chimpanzee populations (Bossou, Mahale,

and Taï) and is known to be absent from the Gombe chimpanzees' repertoire (Goodall, 1986). The purpose of this behavior seems arbitrary. In Mahale, chimpanzees perform it most commonly as a herding/courtship behavior in sexual contexts (23 out of 41 observations: Nishida, 1987): young adult males and adult estrous females would apparently perform it to attract the attention of group members of the other sex (M. Huffman personal communication). In Bossou, it has been observed mostly in apparent frustration or in play (41 out of 44 observations: Sugiyama, 1981, personal communication): during the habituation period individuals surprised in trees would leaf-clip while looking at the observer. Once the habituation was more complete, this form of leaf-clipping disappeared and it is now observed only in youngsters as a form of play.

An ecological difference might explain why leaf-clipping is present in only three out of four populations. But it is very difficult to find ecological reasons to explain why each population uses leaf-clipping in a different context. The changes in the context of use observed in Taï chimpanzees suggested that leaf-clipping is a cultural behavior whose context of use is locally determined by group members. A kind of a social convention that is imitated by group members seems to be the best explanation to account both for the different functions of this behavior in three chimpanzee populations and for the new function that it acquired in Taï chimpanzees.

Cultural variants in chimpanzee populations

Have other examples of modifications in cultural behaviors been observed in other chimpanzee populations? Innovations have been reported from Gombe and Mahale but did not propagate to many group members, the only exception being the spread of the use of sticks as levers to try to open banana boxes, which should rather be looked at as an adaptation to new environmental conditions (Goodall, 1986; Kummer & Goodall, 1985).

On the basis of my observations, I suggest *leaf-grooming in Gombe chimpanzees* is another example of a spontaneous cultural change. Gombe chimpanzees were observed in the grooming context to suddenly pick up one or more leaves and, peering closely at them while holding them between both hands, move both thumbs as if they were grooming the leaves often with great intensity (Goodall, 1986). After some minutes, they would discard the leaves on the ground. The function of this behavior has remained a puzzle even though it has been observed for years (Goodall, 1986). When I worked in Gombe in the spring of 1990, I observed that the chimpanzees had developed a new function for this behavior as they were placing on these leaves ectoparasites[1] that they found while grooming another individual or themselves. They would apparently chew the very small and hard parasites while detaching some leaves from a sapling, then place them with the lips directly on these leaves and try to squash them with their thumbs. Repeatedly, they would take up the parasites with the lips, bite on them and replace them on the leaves which they had flattened with their thumbs. It is important to note that earlier observers of leaf-grooming in Gombe never saw the chimpanzees actually place anything with the lips on the leaves nor take anything from the leaves to chew (J. Goodall personal communication; W.C. McGrew personal communication). Twenty-four individuals of the Kasakela community were observed to perform leaf-grooming and in only 10% of all observed leaf-grooming ($n = 42$) did the chimpanzees do it as previously, i.e. without placing and removing something on to the leaves with their lips. In three cases, youngsters placed "objects" on the leaves, once smearing from the eye and twice pus pressed out from an infected foot. In addition, this new function was stable during the 2 year interval between my visits in Gombe. Thus, generalization in the use across all group members, stability in its form, and the ability to modify it are also shown by chimpanzees in leaf-grooming.

Social conventions in cultural variants

Intriguingly, this new function of leaf-grooming in Gombe, namely squashing ectoparasites, is performed by Taï chimpanzees but in another way. Taï chimpanzees place the ectoparasites on their forearm and squash them by giving them small hits with the tip of the index finger of the other hand. After many such hits they will take it back in the mouth to chew further. Thus, the same ecological function, in this case squashing small parasites, may produce different behaviors in different chimpanzee populations. Such *analogous behaviors* can be found in other situations as well. For example, in Mahale, young male chimpanzees may attract the attention of estrous females by leaf-clipping (see above), which is less conspicuous than the common sapling-waving used for this purpose by older males. Similarly in Taï, males that may want to be inconspicuous would knock branches or tree trunks with their knuckles to attract the attention of an estrous female. Leaf-clipping in Mahale and knuckle-knocking in Taï are analogous behaviors with a similar function, attracting the attention of sexual partners, while leaf-clipping in Mahale and in Taï would be *homologous behaviors* with different functions (Table 18.4). In all cases, a *social convention* determines the function of the behavior that is reproduced by most group members. Thus, the potentialities of the chimpanzees allows them to respond to an ecological or a social problem in different ways following different social conventions, and within the same social group the solution is the same and stable for years.

DISCUSSION

The *ecological approach* emphasizes the existence of a number of behavioral variants that are independent of obvious genetic differences and remote from any simple ecological explanation. This is especially the case for nut-cracking, leaf-clipping, and leaf-grooming. The *transmission approach* supports the hypothesis that nut-cracking might be cultural by providing observations supporting the role of imitation and teaching. Finally, the *innovation approach* provides evidence that two of the behaviors noted in the ecological approach seem to be ruled by social conventions. Behavioral variations in leaf-clipping and leaf-grooming, with flexibility of use within and between communities and the arbitrariness of function, satisfy the criteria for several more stringent definitions of culture and should thus be accepted as cultural variants.

Social learning versus individual learning

The central criticism leveled against the presence of cultural behavior in chimpanzees has been that what we observe might be a similar response in animals of all populations

Table 18.4. *Some social cultural behaviors observed in wild chimpanzees*

Analogous behavior

	Behavior		
Function	Gombe	Mahale	Taï
Courtship	—	Leaf-clipping	Knuckle-knock
Squash ectoparasite	Leaf-grooming	—	Index-hit

Homologous behavior

	Function		
Behavior	Bossou	Mahale	Taï
Leaf-clip	Play	Courtship	Drumming + resting

to similar ecological stimuli; if only one way of solving the task is possible given the sensorimotor capacities of the animal, then individual learning leads to standardization (Galef, 1990; Tomasello, 1990; Tomasello *et al.*, 1993a; Visalberghi & Fragaszy, 1990). However, no one has demonstrated that this applies to the population-specific behavior patterns observed. How can we prove that only one solution is possible within the given ecological stimuli? Only chimpanzees can provide us with an answer to this question. Luckily, in some cases, we may find an answer to this central question and it supports the fact that standardization is not the result of ecological stimuli. First, Mahale and Taï chimpanzees do not leaf-clip in exactly the same way: in Mahale, they repeatedly nibble away small pieces from the leaf blade (Nishida, 1987, personal communication), whereas in Taï, after taking the leaf blades from both sides of the petiole between their lips, they remove them whole in one movement. Thus, with the same ecological stimuli and the same sensorimotor abilities, chimpanzees use two different ways of leaf-clipping. However, Taï chimpanzees were never observed to nibble small pieces away and Mahale chimpanzees were never seen to remove the whole leaf blades in one movement (T. Nishida personal communication). Thus, the standardization is population-specific, which supports the social learning hypothesis.

Secondly, the ant-dipping movement is strikingly different between Gombe and Taï chimpanzees: in Gombe, the stick is introduced deep into the ant nest and once the ants have swarmed half way up, the chimpanzee sweeps the stick through the other fist and transfers the mass of ants collected in a heap rapidly to the mouth (for more details, see Goodall, 1986), whereas in Taï, the stick just touches the ants guarding the nest entrance and once they have swarmed 10 cm up the stick they directly sweep off the ants with the teeth (Boesch & Boesch, 1990). Here again, chimpanzees have found two different responses for dipping ants. The ecological stimuli are the same because, first, Gombe and Taï chimpanzees dip for the same species of ant (*Dorylus nigricans*), and secondly, both techniques worked in both sites: I personally tested to see whether sticks could be inserted deeper than they are into the ant nests in Taï. However, Taï chimpanzees were never observed to use the Gombe technique and vice versa (Goodall, 1986). Thus, again the standardization is population-specific, which supports the social learning hypothesis. In this case, the social learning process responsible for this standardization makes Taï chimpanzees at least four times less efficient than Gombe chimpanzees in ant-dipping, which strengthens the view that whatever the contribution of individual learning is, it is strongly *socially canalized and this actually prevents the best solution from being adopted*.

Fidelity in social transmission is required for culture to exist (Heyes, 1993; Tomasello *et al.*, 1993a). The canalization of learning we reported in chimpanzees shows such fidelity. We saw this also in "squashing of ectoparasites" and "courtship," where different responses are possible but each population uses only one of them, as well as in learning to crack nuts in Taï chimpanzees (see pp. 414–6). Thus, whatever the importance of individual learning for these behavior patterns, the social canalization is so strong that it does guarantee the fidelity of transmission typical of cultural transmission.

Dichotomy between captive and wild chimpanzee observations

Most captive studies conclude that cultural transmission mechanisms are not part of the abilities of primates, including chimpanzees. Why do field data contradict results from captive studies? In my opinion, the negative effects of captivity on social animals has been largely underestimated. Not only space and opportunity to explore and play but also the absence of large and complex social interactions have a negative influence on the psychological development of the captive individuals. It is striking in this context that each time an enrichment of the captive conditions is offered to the individuals, the cognitive performance tends to increase (rhesus monkeys, Mason, 1978; chimpanzees, Savage-Rumbaugh *et al.*, 1985; Tomasello *et al.*, 1993). Imitation in captive chimpanzees was demonstrated only in those who had human language training (Tomasello *et al.*, 1993b), indicating that small artificial social groups in captivity are not enough to elicit such abilities (Tomasello *et al.*, 1987). The complexity of some social rules required during language training may be comparable to some of the challenges wild chimpanzees face and may elicit imitative abilities. The "enculturation" of chimpanzees is a long process that may be totally disrupted in captive conditions (Whiten, 1993). In my opinion, rules in the wild are more stringent, with survival and reproductive success acting as "teachers," so that the cognitive development of wild chimpanzees may be more advanced than that of their captive counterparts. Culture is first and foremost a social expression (Ingold, 1993; Rogoff *et al.*, 1993) and it is thus not a surprise that complex social mechanisms cannot be proven in captive animals that all live and/or have lived in poorer social environments than their wild counterparts (most wild chimpanzees live in ever fluctuating parties within a community of 40–80 individuals including all age–sex classes). Further, problems related to motivation and the presence of trustful models may all hamper captive studies. The potentialities of captive chimpanzees are fascinating to study (de Waal, 1992) but the potential capabilities of captive animals are not the same as those in the wild, and making inferences from one condition to the other one is not always straightforward, especially in the case of negative results. I am fully aware of the difficulties in studying such cognitive aspects in the wild but I think the challenge is worthwhile, if we want to know the propensity of primates and chimpanzees to develop cultural behaviors. The alternative is to improve the captive conditions so that the captive environment more closely approximates field conditions.

In conclusion, new field data provide further support to the idea of chimpanzee culture in the sense that the behavioral differences observed seem not only independent of environmental variables, but also their function tends to follow social conventions that are learned by all group members and that may change over time.

ACKNOWLEDGEMENTS

I thank the Ministère de la Recherche Scientifique, the Ministère de l'Agriculture et des Resources Animales of Côte d'Ivoire, the Tanzania Commission for Science and Technology, the Serengeti Wildlife Research Institute as well as the Tanzania National Parks for permitting this study, and the Swiss National Foundation, the L.S.B. Leakey Foundation and the Jane Goodall Institute for financing it. In Côte d'Ivoire, this project is integrated in the UNESCO project Taï-MAB under the supervision of Dr Henri Dosso. I want to thank especially Jane Goodall for supporting the idea of this comparative study. I am most grateful to A. Aeschlimann, F. Bourlière and H. Kummer for their constant encouragement, L. and E. Ortega of the CSRS in Abidjan and T. Tiépkan of the station IET in Taï for their logistic support, as well as to A. Collins, J. Goodall, C. Uhlenbroek and J. Wallis for their friendly support in Gombe and Dar es Salaam. I thank H. Boesch-Achermann and the editors of this volume for commenting upon and correcting the manuscript.

NOTES

1 I could confirm the presence of an ectoparasite on these leaves in only nine cases but this is not surprising, as ectoparasites may be exceptionally small. For example, I had difficulty seeing some small ticks sucking blood from my own forearms. Similarly, eggs of lice that Gombe chimpanzees so regularly remove from each others hair with their teeth were impossible to see even when I was only 1 m away.

REFERENCES

Anderson, J.R., Williamson, E.A. & Carter, J. (1983). Chimpanzees of Sapo Forest, Liberia: Density, nests, tools and meat-eating. *Primates*, 24, 594–601.
Beatty, H. (1951). A note on the behavior of the chimpanzee. *Journal of Mammalogy*, 32, 118.
Boesch, C. (1991). Teaching in wild chimpanzees. *Animal Behaviour*, 41, 530–2.
Boesch, C. (1993). Toward a new image of culture in chimpanzees? *Behavioral and Brain Sciences*, 16, 514–5.
Boesch, C. (1995). Innovation in wild chimpanzees. *International Journal of Primatology*. (in press).

Boesch, C. & Boesch, H. (1983). Optimization of nut-cracking with natural hammers by wild chimpanzees. *Behaviour*, 83, 265–86.
Boesch, C. & Boesch, H. (1984). Possible causes of sex differences in the use of natural hammers by wild chimpanzees. *Journal of Human Evolution*, 13, 415–40.
Boesch, C. & Boesch, H. (1989). Hunting behavior of wild chimpanzees in the Taï National Park. *American Journal of Physical Anthropology*, 78, 547–73.
Boesch, C. & Boesch, H. (1990). Tool use and tool making in wild chimpanzees. *Folia Primatologica*, 54, 86–99.
Boesch, C., Marchesi, P., Marchesi, N., Fruth, B. & Joulian, F. (1994). Is nut cracking in wild chimpanzees a cultural behavior? *Journal of Human Evolution* 26, 325–38.
Bonner, J. (1980). *The Evolution of Culture in Animals*. Princeton: Princeton University Press.
Booth, A.H. (1958). The zoogeography of West African primates: A review. *Bulletin de l'Institut Fondamental d'Afrique Noire*, 20, 587–622.
Caro, T.M. & Hauser, M.D. (1992). Is there teaching in nonhuman animals? *Quarterly Review of Biology*, 67, 151–74.
Custance, D. & Bard, K. (1994). The comparative and developmental study of self-recognition and imitation: The importance of social factors. In *Self-awareness in Animals and Humans: Developmental Perspectives*, ed. S.T. Parker, R.W. Mitchell & M.L. Boccia, pp. 207–26. New York: Cambridge University Press.
de Waal, F.B.M. (1992). Aggression as a well-integrated part of primate social relationships: A critique of the Seville statement on violence. In *Aggression and Peacefulness in Humans and Other Primates*, ed. J. Silverberg & P. Gray, pp. 37–56. Oxford: Oxford University Press.
Ember, C.R. & Ember, M. (1985). *Anthropology*. Englewoods Cliffs, NJ: Prentice Hall.
Funk, M. (1985). Werkzeuggebrauch beim öffnen von Nüssen: Unterschiedliche Bewältigungen des Problems bei Schimpansen und Orang-Utans. Diplomarbeit, University of Zürich.
Galef, B. (1990). Tradition in animals: Field observations and laboratory analyses. In *Interpretation and Explanation in the Study of Animal Behavior*, ed. M. Bekoff & D. Jamieson, pp. 74–95. Boulder, CO: Westview Press.
Goodall, J. (1973). Cultural elements in a chimpanzee community. In *Precultural Primate Behavior*, ed. E.W. Menzel, vol. 1, pp. 195–249. Fourth International Primatological Congress Symposia Proceedings. Basel: Karger.
Goodall, J. (1986). *The Chimpanzees of Gombe: Patterns of Behavior*. Cambridge, MA: The Belknap Press.
Guillaumet, J.-L. & Adjanohou, E. (1971). Le milieu naturel de la Côte d'Ivoire. *Mémoires ORSTOM*, 50, 157–264.
Haltenorth, T. & Diller, H. (1977). *Säugetiere Afrikas und Madagaskars*. Munich: BVL Verlagsgesellschafts Gmbh.
Hamilton, A. (1982). *Environmental History of East Africa: A Study of the Quaternary*. London. Academic Press.
Hauser, M.D. (1993). Cultural learning: Are there functional consequences? *Behavioral and Brain Sciences*, 16, 524.
Heyes, C.M. (1993). Imitation, culture and cognition. *Animal Behaviour*, 46, 999–1010.
Ingold, T. (1993). A social anthropological view. *Behavioral and Brain Sciences*, 16, 526–7.
Kawai, M. (1965). Newly acquired precultural behavior of the natural troop of Japanese monkeys on Koshima islet. *Primates*, 6, 1–30.
Kortlandt, A. & Holzhaus, E. (1987). New data on the use of stone tools by chimpanzees in Guinea and Liberia. *Primates*, 28, 473–96.

Kummer, H. (1971). *Primate Societies: Group Techniques of Ecological Adaptation.* Chicago: Aldine.
Kummer, H. & Goodall, J. (1985). Conditions of innovative behavior in primates. *Philosophical Transactions of the Royal Society of London, series B*, **308**, 203–14.
Marchesi, P., Marchesi, N., Fruth, B. & Boesch, C. (1995). Census and distribution of chimpanzees in Côte d'Ivoire. *Primates*. (in press).
Mason, W.A. (1978). Social experience and primate development. In *The Development of Behavior: Comparative and Evolutionary Aspects*, ed. G.M. Burghardt & M. Bekoff, pp. 233–51. New York: Garland STPM Press.
McGrew, W.C. (1974). Tool use by wild chimpanzees in feeding upon driver ants. *Journal of Human Evolution*, **3**, 501–8.
McGrew, W.C. (1992). *Chimpanzee Material Culture: Implications for Human Evolution*. Cambridge: Cambridge University Press.
McGrew, W.C. & Tutin C.E.G. (1978). Evidence for a social custom in wild chimpanzees? *Man*, **13**, 234–51.
McGrew, W.C., Baldwin, P.J. & Tutin, C.E.G. (1979). Chimpanzees, tools and termites: Cross-cultural comparisons of Senegal, Tanzania and Rio Muni. *Man*, **14**, 185–214.
Nishida, T. (1973). The ant-gathering behavior by use of tools among wild chimpanzee of the Mahale Mountains. *Journal of Human Evolution*, **2**, 357–70.
Nishida T. (1987). Local traditions and cultural transmission. In *Primate Societies*, ed. B.B. Smuts, D.L. Cheney, R.M. Seyfarth, R.W. Wrangham & T.T. Strusaker, pp. 462–74. Chicago: University of Chicago Press.
Nishida, T. & Hiraiwa, M. (1982). Natural history of a tool-using behavior by wild chimpanzees in feeding upon wood-boring ants. *Journal of Human Evolution*, **11**, 73–99.
Nishida, T. & Uehara, S. (1980). Chimpanzees, tools and termites: Another example from Tanzania. *Current Anthropology*, **21**, 671–2.
Nishida, T., Wrangham, R. W., Goodall, J. & Uehara, S. (1983). Local differences in plant-feeding habits of chimpanzees between the Mahale Mountains and Gombe National Park, Tanzania. *Journal of Human Evolution*, **12**, 467–80.
Olson, D. & Astington, J. (1993). Cultural learning and educational process. *Behavioral and Brain Sciences*, **16**, 531–2.
Rahm, U. (1971). L'emploi d'outils par les chimpanzés de l'ouest de la Côte d'Ivoire. *La Terre et la Vie*, **25**, 506–9.
Rogoff, B., Chavajay, P. & Matusov, E. (1993). Questioning assumptions about culture and individuals. *Behavioral and Brain Sciences*, **16**, 533–4.
Russon, A.E. & Galdikas, B.M.F. (1995). Constraints on great ape imitation: Model and action selectivity in rehabilitant orangutan (*Pongo pygmaeus*) imitation. *Journal of Comparative Psychology*, **109**, 5–17.
Savage, T.S. & Wyman, J. (1843/44). Observations on the external characters and habits of *Troglodytes niger*, Geoff. and on its organization. *Boston Journal of Natural History*, **4**, 362–86.
Savage-Rumbaugh, E.S., Rumbaugh, D.M. & Boysen S. (1985). The capacity of animals to acquire language: Do species differences have anything to say to us? *Philosophical Transactions of the Royal Society of London*, **308**, 177–85.
Segall, M., Dasen, P., Berry, J. & Poortinga, Y. (1990). *Human Behavior in Global Perspective: An Introduction to Cross-Cultural Psychology*. New York: Pergamon Press.
Struhsaker, T.T. & Hunkeler, P. (1971). Evidence of tool-using by chimpanzees in the Ivory Coast. *Folia Primatologica*, **15**, 212–9.
Sugiyama, Y. (1981). Observations on the population dynamics and behavior of wild chimpanzees

at Bossou, Guinea, 1979–1980. *Primates*, 22, 435–44.
Sugiyama, Y. (1985). The brush-stick of chimpanzees found in south-west Cameroon and their cultural characteristics. *Primates*, 26, 361–74.
Sugiyama, Y. (1990). Local variation of tools and tool use among wild chimpanzee populations. In *The Use of Tools by Human and Non-Human Primates*, ed. A. Berthelet & J. Chavaillon, pp. 175–87. Oxford: Oxford University Press.
Sugiyama, Y. & Koman, J. (1979a). Social structure and dynamics of wild chimpanzees at Bossou, Guinea. *Primates*, 20, 323–39.
Sugiyama, Y. & Koman J. (1979b). Tool-using and -making behavior in wild chimpanzees at Bossou, Guinea. *Primates*, 20, 513–24.
Sumita, K., Kitahara-Frisch, J. & Norikoshi, K. (1985). The acquisition of stone-tool use in captive chimpanzees. *Primates*, 26, 168–81.
Teleki, G. (1974). Chimpanzee subsistence technology: Materials and skills. *Journal of Human Evolution*, 3, 575–94.
Thorpe, W. (1956). *Learning and Instinct in Animals*. London: Methuen.
Tomasello, M. (1990). Cultural transmission in tool use and communicatory signalling of chimpanzees? In *"Language" and Intelligence in Monkeys and Apes: Comparative Developmental Perspectives*, ed. S.T. Parker, & K.R. Gibson, pp. 274–311. New York: Cambridge University Press.
Tomasello, M., Davis-Dasilva, M., Camak, L. & Bard, K. (1987). Observational learning of tool-use by young chimpanzees. *Journal of Human Evolution*, 2, 175–83.
Tomasello, M., Kruger, A. & Ratner, H. (1993a). Cultural learning. *Behavioral and Brain Sciences*, 16, 495–511.
Tomasello, M., Savage-Rumbaugh, S. & Kruger, A. (1993b). Imitation of object related actions by chimpanzees and human infant. *Child Development*, 64, 1688–705.
Tutin, C.E.G. & Fernandez, M. (1983). *Recensement des Gorilles et des Chimpanzés du Gabon*. Centre International de Recherches Médicales de Franceville, Gabon.
Visalberghi, E. (1988). Responsiveness to objects in two social groups of tufted capucin monkeys (*Cebus apella*). *American Journal of Primatology*, 15, 349–60.
Visalberghi, E. & Fragaszy, D. (1990). Do monkeys ape? In *"Language" and Intelligence in Monkeys and Apes: Comparative Developmental Perspectives*, ed. S.T. Parker & K.R. Gibson, pp. 274–73. New York: Cambridge University Press.
Whiten, A. (1989). Transmission mechanisms in primate cultural evolution. *Trends in Evolution and Ecology*, 4, 61–2.
Whiten, A. (1993). Human enculturation, chimpanzee enculturation (?) and the nature of imitation. *Behavioral and Brain Sciences*, 16, 538–9.
Whiten, A. & Ham, R.(1992). On the nature and evolution of imitation in the animal kingdom: Reappraisal of a century of research. In *Advances in the Study of Behavior*, ed. P.J.B. Slater, J.S. Rosenblatt, C. Beer & M. Milinski, vol. 21, pp. 239–83. New York: Academic Press.
Whiten, A. & Milner, P. (1984). The educational experiences of Nigerian infants. In *Nigerian Children: Development Perspectives*, ed. H. Valerie Curran, pp. 34–73. London: Routledge & Kegan Paul Ltd.
Whitesides, G. H. (1985). Nut cracking by wild chimpanzees in Sierra Leone, West Africa. *Primates*, 26, 91–4.
Wynn, T. (1993). Layers of thinking in tool behavior. In *Tools, Language and Intelligence: Evolutionary Implications*, ed. K.R. Gibson & T. Ingold, pp. 389–406. Cambridge: Cambridge University Press.

19

On the wild side of culture and cognition in the great apes

SUE TAYLOR PARKER AND ANNE E. RUSSON

The various contributors to this volume concur with other primatologists that, barring inappropriate rearing conditions, great apes display symbolic cognitive skills similar to those human children achieve between 2 and 3 years of age (for a summary, see Chapter 1). Here and elsewhere they, along with other investigators, have demonstrated that great apes attain these levels of thought in a variety of arenas including physical and logical knowledge as well as social knowledge (e.g. see Chapters 4, 5, 8, 9, 10, 12, 13, 14, 15, 17, and 18). In their capacity to attain symbolic levels of thought across domains they apparently differ from other non-human primates including cebus monkeys (see Chapters 2, 3, and 12).

As evidence of the mental accomplishments of great apes has accumulated, more students of wild populations have begun to apply the term culture to the local traditions of chimpanzees (e.g. see Chapters 10 and 18) and more students of captive great apes have begun to speak of "enculturated apes" (e.g. see Chapters 11, 13, and 17). Although controversial, these usages have historical precedents.

In the 1950s, Japanese primatologists began to describe the social transmission of new behaviors in Japanese macaques as "cultural" (Itani, 1974; Itani & Nishimura, 1973; Kawai, 1965; Kawamura, 1959; Menzel, 1973). Since then many primatologists have presented evidence of social traditions in wild or semi-wild populations of chimpanzees, macaques, and baboons (e.g. Burton, 1992; Cambefort, 1981; Huffman, 1984; McGrew, 1992; Nishida, 1987; Wrangham *et al*, 1994; and Chapters 10 and 18).

Whether or not we grant cultural capacities to nonhuman primates hinges on our definition of culture. If we look to anthropological definitions as a guide, however, we find that anthropologists themselves do not always agree on definitions of culture (for reviews, see e.g. Freilich, 1989; Keesing, 1974; Kroeber & Kluckholn, 1952; McGrew, 1992; Quaitt & Reynolds, 1992). Keesing (1974) divided anthropologists into those who see cultures as adaptive systems and those who see them as ideational systems. The view that cultures are adaptive systems based on technology has been popular among archeologists and anthropological ecologists (e.g. Harris, 1979; Steward, 1955; White,

1959), while the view that cultures are conceptual systems, dependent on language, has been popular among linguistic, cognitive, and psychological anthropologists (e.g. Goodenough, 1981; Wallace, 1961). Those who emphasize adaptation agree that "The ideational components of cultural systems may have adaptive consequences – in controlling population, contributing to subsistence, maintaining the ecosystem, etc." (Keesing, 1974, p. 76). Those who emphasize ideational components are divided by a paradox in these models – "on the one hand, of cognitive reductionism that misses . . . the code of cultural meanings and conventions; and on the other, of a spuriously autonomous . . . and uniform world of cultural symbols freed from the constraints of the mind and brain by which cultures are created and learned" (Keesing, 1974, p. 88). Keesing suggests a definition of culture that reconciles the two opposing views of adaptation versus ideology by emphasizing the interplay between species-specific cognition and representations of the world:

> Such a conception of culture as an idealized body of competence differentially distributed in a population, yet partially realized in the minds of individuals, allows us to bring to bear a growing body of knowledge about the structure of the mind and brain and the formal organization of intelligences. Even though no one native actor knows all of the culture, and each has a variant version of the code, culture in this view is ordered not simply as a collection of symbols . . . but as a system of knowledge, shaped and constrained by the way the human brain acquires, organizes, and processes information and creates "internal models of reality"
> *(Keesing, 1974, p. 89).*

Although many cultural anthropologists think of culture as a uniquely human (or hominid) phenomenon that creates a unique "cultural niche" (e.g. Fox, 1972), others have taken the lead of primatologists distinguishing "pre-" and "proto-" levels of culture in nonhuman primates (e.g. Hallowell, 1959; Kroeber, 1928; Wallace, 1961). We also see some common ground in the definitions of anthropologists and biologists: Kummer, the Swiss ethologist, has defined culture as "behavioral variants induced by social modification, resulting in individuals who will in turn modify the behavior of others" (Kummer, 1971, p. 15). Bonner, the American biologist, has defined culture as "the transfer of information by behavioral means, most particularly by the process of teaching and learning" (Bonner, 1980, p. 9). In his comparative study of chimpanzee populations, McGrew (1992), the American psychologist/anthropologist, has abstracted from the anthropologist Kroeber six criteria for recognizing culture in other species: innovation, dissemination, standardization, durability, diffusion, and tradition, and added two of his own, nonsubsistence and naturalness (occurring in the wild). While we agree with McGrew that we should grant cultural capacities to species that meet all of these criteria, we believe these rather cryptic criteria need expansion and further explication.

COMPARATIVE DEFINITION OF CULTURE

One of the primary reasons for defining culture in broad comparative terms is to facilitate understanding of the evolution of cultural capacities. Because we want to understand how cultural capacities arose during evolution, we seek a definition of culture that focuses on parallels between human and nonhuman cognition and traditions (e.g. Gibson & Ingold, 1993; Hallowell, 1959; Quiatt & Itani, 1994; Wallace, 1961). Several considerations are pertinent to devising a comparative definition of culture appropriate to nonhuman primates:

1. Because culture as it is defined for humans hinges on language and other cognitive abilities that are unique to our species, we want to broaden the concept to embrace nonhuman abilities that could have set the stage for the emergence of "full-blown" cultural systems.
2. At the same time, a wider definition should remain compatible with definitions of culture as it occurs in full-blown human form, if it is to serve the aims of comparative analysis and evolutionary reconstruction.
3. Just as human cultures are defined as intraspecific phenomena whose variation occurs across structured social groups, so nonhuman cultures should be defined as intraspecific phenomena that vary across groups.
4. Just as anthropologists have sought concepts of culture capable of accommodating the dual phenomena of culture change and stasis and the role of contact in culture change (e.g. Steward, 1955; Wallace, 1961), so primatologists need a comparative concept that accommodates the dual phenomena of stasis and change and the role of intergroup contact in change.
5. Cultural capacities vary along a continuum within related species; just as various early hominid species displayed "tool cultures" of varying complexity (e.g. Schick & Toth, 1993; Wynn, 1989), so nonhuman primates display differing degrees of cultural capacity.
6. Cultural adaptations are generated by "cognitive mechanisms" that support social transmission and storage of information. These mechanisms include observational learning (particularly imitation), teaching, and symbolic communication (e.g. Bonner, 1980; Donald, 1991; Heyes, 1993; Kummer, 1971; Tomasello et al., 1993).
7. The cognitive processes underlying "cultural adaptations" are a central focus in comparative studies of human cultures (e.g. Fox, 1972; Goodenough, 1981; Wallace, 1961) and therefore should be a central focus in comparative studies of cultural capacity (Donald, 1991; Quiatt & Itani, 1994).
8. Social transmission of representations of local knowledge from individuals of one generation to those of the next generation within a group – enculturation – has been distinguished from more general processes of socialization which are common to all individuals in a species (Mead, 1963), and from processes of acculturation of immigrants into the local knowledge of a new group. So, for consistency of usage, enculturation in primates should be defined as an intraspecific phenomenon.

In light of these considerations, we recommend the following comparative functional definition of cultures, which incorporates core features from cultural anthropology, psychology, and biology: **Cultures are representations of knowledge socially transmitted within and between generations in groups and populations within a species that may aid them in adapting to local conditions (ecological, demographic, or social).** These representations of knowledge depend upon various cognitive processes of individual learning and invention as well as social processes that transmit information in the population over time.

Cognitive bases for cultural adaptations

In our comparative discussion of nonhuman cultural capacities we follow the lead of earlier investigators in focusing on the mechanisms of social transmission of information, particularly on imitation and teaching. In order to identify taxonomic differences in degrees and kinds of cultural adaptation, we compare anthropoid species in terms of the cognitive abilities that underlie and support their various forms of information acquisition and transmission.

Current evidence suggests that primate species differ in the processes they have available for transmitting information socially (e.g. see Chapters 2, 3, 7, 14, and 16). Social transmission of information can be engendered by a variety of processes including low-level sensorimotor processes (e.g. such experientially based processes as stimulus or local enhancement, social facilitation, observational conditioning, or matched dependent learning), middle-level sensorimotor processes of kinesthetic-visual matching of familiar schemes, more advanced sensorimotor processes of kinesthetic-visual matching of novel schemes (for recent discussions, see Galef, 1988; Mitchell, 1993; Moore, 1994; and Chapter 7), or by primitive symbolic processes.

The cognitive capacities of monkeys seem to support social transmission primarily through experientially based processes (e.g. Huffman & Quiatt, 1986; Mineka & Cook, 1988; Visalberghi & Fragaszy, 1990; Chapters 2 and 3) and possibly through simple kinesthetic-visual matching. In contrast, the cognitive capacities of great apes appear to support social transmission of information through true imitation (e.g. Custance & Bard, 1994; Russon & Galdikas, 1993; Tomasello *et al.*, 1993; Whiten & Custance, 1994) and even demonstration teaching (e.g. see Chapters 13 and 18).

We use a Piagetian developmental framework as a means for conceptualizing these species differences in a cognitive and therefore cultural capacity (see e.g. Jolly, 1972; Parker, 1990). The utility of Piagetian and other frameworks from developmental psychology for comparative work derives from the epigenetic nature of the levels or stages of development, i.e. the fact that each succeeding stage within a given domain is constructed from the behaviors and abilities of the preceding stage and hence cannot occur without achievement of the prior stage (e.g. Jolly, 1972; Parker, 1990). It is important in this context, to emphasize that stage designations refer to the highest level abilities of individuals at a given age rather than typical performances (e.g. Piaget, 1970).

On the basis of Piagetian sensorimotor stages, we distinguish simple imitation, characteristic of human imitation during sensorimotor stages 3 and 4 (beginning at about 8 months in humans), from true imitation, characteristic of human imitation during sensorimotor stages 5 and 6 (beginning at 12–24 months; Piaget, 1952, 1962). Simple imitation involves reenactment of observed actions already in an individual's behavioral repertoire; true imitation involves matching of novel observed actions (see Mitchell, 1987, for a different scheme). According to Piaget this matching occurs by imitative recombination and differentiation of existing schemes[1].

In addition to showing true imitation, great apes also show a rudimentary capacity for pretend play (performing sensorimotor schemes out of context) characteristic of human children of 2–3 years of age during the symbolic subperiod of the preoperations period. We call this simple pretend play to distinguish it from more complex levels of pretend play that involve more orderly and exact replications of reality using collective symbols, characteristic of children from 3 to 6 or 7 years during the intuitive subperiod of the preoperations period, (Piaget, 1962; for a related typology, see Bretherton, 1984).

In addition to Piaget's model, we also use Caro & Hauser's (1992) comparative typology of levels of teaching: opportunity teaching, coaching, and demonstration teaching. It is notable that demonstration teaching, their highest level, seems to depend on the capacity for simple pretend play (see Chapter 16).

Piaget's sensorimotor model implies that the generation of new behaviors depends upon several factors including the number and kinds of schemes, their patterns of maturation, the breadth of application of existing schemes in new contexts or toward new goals, and the potential for differentiating and coordinating new and old schemes into new higher order schemes, either through imitation or invention. In comparative terms, this suggests that a species potential for generating new schemes is set in part by species-typical repertoires of sensorimotor schemes and in part by the degree of mobility or flexibility those schemes have to recombine (e.g. Parker, 1977). Cognitive differences between monkeys and great apes seem to arise both from differences in the size of the repertoire of schemes and from differences in the mobility or recombinability of schemes (e.g. Gibson, 1990). Species differences in cognition may also be influenced by such factors as species-typical temperaments, relationships, and attention structures (e.g. Cambefort, 1981; Chapters 6 and 7).

In the following section we review some reports of "cultural adaptations" in monkeys and apes focusing on the roles that imitation, teaching, and pretend play may have in social transmission of information. Although our discussion focuses on the highest levels of imitation and teaching demonstrated by a species, we recognize that anthropoid primates acquire skills through a variety of species-typical cognitive processes of different levels of complexity, ranging from relatively simple experiential processes to higher level symbolic processes (e.g. Goodall, 1986; Tomasello, 1990; Chapter 7). We also recognize that higher level abilities in great apes and humans coexist with and depend upon lower level ones.

Intimations of culture in monkeys

Japanese primatologists were the first to characterize social transmission of information in monkey troops as "cultural." Kawamura introduced this terminology in a controversial paper he gave at a meeting of the Anthropological Society of Nippon and the Japanese Society of Ethnology in 1955 to describe invention and social transmission of local food habits within troops of Japanese macaques through "imitation" (Itani & Nishimura, 1973). When Kawamura published his paper on "subcultural behavior" ten years later, processes of cultural transmission were being studied in approximately 20 troops of Japanese macaques, of which he himself was studying 11 (reported in Itani & Nishimura,

1973). The primary technique in these studies was to introduce new foods to different troops and track the rejection or acceptance and subsequent propagation of the new eating and/or feeding habits. Introduced foods included candy, wheat, and sweet potatoes. Significant differences among troops were reported in the nature and speed of propagation of new feeding habits within the troops. These differences were confounded by differences in introduced foods and associated inventions as well as troop structures. Studies of cultural or subcultural transmission have continued to be a major focus of research among Japanese primatologists since 1955. In this period a few Western primatologists have reported "cultural" transmission of other nonfeeding behaviors in Japanese macaques (e.g. Huffman, 1984, 1994) and other macaque species (e.g. Burton, 1992).

Also in the 1950s, K.R.L. Hall, the South African primatologist, did experimental studies of the acquisition of feeding habits among wild and captive Cape baboons (Hall, 1962) that paralleled those that Japanese primatologists did on Japanese macaques. More recently another South African primatologist has carried out a comparative experimental study of "cultural transmission" of feeding habits on wild baboons and vervet monkeys (Cambefort, 1981).

Hall (1962) described "observational learning (imitation)" of which foods to eat by a 3 year old male human-reared baboon he released into a Cape troop. As this juvenile followed members of the troop and watched them digging and eating, he repeated their actions and expanded his diet to include the items he saw them eating. Hall also described intent watching by another juvenile of an adult female who was digging up bulbs and roots. Hall (1962) and others (e.g. Altmann & Altmann, 1970) reported that baboons frequently sniff the mouths of other troop members to discover what they are eating. Cambefort (1981) described species differences in rates of discovery and propagation to new palatable and unpalatable foods that were hidden near "cues." In both Cape baboons and vervet monkeys, juveniles showed the highest frequency of discovery, but the difference in rate of discovery by juvenile baboons as compared to other age–sex classes was much greater than among vervets. Likewise, the propagation of the new feeding habit was much more rapid – the juveniles acting as models for all the other age–sex classes – among baboons than among vervets among whom propagation was more or less homogeneous and random (Cambefort, 1981). Strum (1975) has described increased hunting in a baboon troop as a cultural tradition, though such ecological factors as increased prey densities may have played a major role in the increase.

Although observational learning is important in the identification and acceptance of new foods, most observers agree that most new behaviors in monkeys seems to be acquired in large part through individual experience – ecological and social – and/or simple imitation rather than through direct social transfer in the form of true imitation and teaching (e.g. Galef, 1990; King, 1994; Visalberghi & Fragaszy, 1990). Teleki's (1975) description of the inability of termite-loving baboons to imitate chimpanzees termite-fishing with tools provides a striking example of the limited imitative abilities of monkeys. As we might expect, given their limited sensorimotor intelligence, their smaller brains, and their relatively brief period of immaturity as compared to great apes, monkeys show limited cultural potential. They seem to generate novel behaviors primarily through the application of existing schemes in new contexts rather than production of new schemes through coordination and differentiation of existing ones.

Incipient cultures in great apes

Whereas most students of monkey populations have avoided the concept of culture, students of great apes, particularly students of wild chimpanzees, have enthusiastically adopted the concept (see Chapters 10 and 18). In contrast to the focus in studies of macaques and baboons on social transmission of acceptance and/or preparation of experimentally introduced foods, the primary focus in studies of chimpanzees has been on regional differences versus local similarities in selection and preparation of indigenous foods and their social transmission (as well as social transmission of nonfeeding behaviors: e.g. Goodall, 1973; McGrew, 1992; Nishida et al., 1983; Wrangham et al., 1994).

In a systematic effort to control for ecological determinants, McGrew (1992) compared food species and potential raw materials for tools available to chimpanzees in six key sites in Africa. He thereby demonstrated the occurrence of differences in diet, tool types, and tool uses among chimpanzees in different regions even when the same plant and animal species are present in both regions. Conversely, he demonstrated similarities in diet, tool type, and tool use among chimpanzees in adjacent groups in the same region.

Boesch (1992, and Chapter 18) has concentrated on mechanisms of social transmission of tool use in nut-cracking within a group of Western chimpanzees. He has shown compelling evidence for demonstration teaching of nut-cracking techniques to juveniles by mothers and reciprocal imitative modeling by the juveniles of their mothers' technique. Matsuzawa & Yamakoshi (Chapter 10) have described an apparent case of social transmission of nut-cracking technique from one group to another through observational learning and imitation of an immigrant female by indigenous juveniles.

As indicated previously, demonstration teaching by great ape mothers seems to depend upon the capacity for simple pretend play which entails representing social roles of others (Byrne & Whiten, 1988; Povinelli et al., 1992; Premack & Woodruff, 1978; Chapter 14) as well as related motivational or attentional factors. Reciprocally, the capacity of juvenile great apes to benefit from demonstration teaching seems to depend upon true imitation. These interlocking abilities, combined with their more extended periods of immaturity, give great apes a greater capacity for social transmission than monkeys.

It is important to emphasize that great apes apparently display true imitation relatively rarely compared to human infants (Russon & Galdikas, 1993, 1995; Whiten & Byrne, 1991; Whiten & Ham, 1992; Chapter 14). This could be because imitation is notoriously difficult to recognize in the wild. Moreover, failures to elicit imitation in great apes in laboratory settings may be artifacts of experimental designs that ignore developmental and motivational constraints on imitation: true imitation develops after the age of 3 years and is concentrated on replicating patterns of behavior that are modeled by close associates and which are at the leading edge of the imitator's abilities. (Russon & Galdikas, 1995; Chapter 7). The fact that the capacity for true imitation first emerges during the juvenile period means that great apes respond to demonstration teaching late in the long process of acquiring new skills.

The apparent rarity of imitation in great apes, then, is consistent with its later development in these species as compared to humans. It is also consistent with the more limited behavioral repertoire of great apes compared to human infants, particularly in terms of vocal schemes (e.g. Chevalier-Skolnikoff, 1976) and combinatory skills (e.g. Chapter 12). Predictably, given the apparent rarity of imitation, simple pretend play has been reported only a few times in great apes, primarily in cross-fostered individuals (Patterson & Cohn, 1994; see Chapters 7, 13, 14, and 16). Nevertheless, these few cases are compelling because they have been reported in all species of great apes.

The few examples of demonstration teaching in great apes suggest that it supports true imitation in several ways at several points in the acquisition process (see Boesch, 1991): demonstration assists novices in mastering difficult constituents of behavioral strategies and at more complex levels, and it assists them in segmenting complex behavioral strategies into manageable constituents. Demonstration teaching seems to boost the effectiveness of true imitation by helping novices to analyze behavioral strategies as they progressively construct more complex skills.

Analytical skills are central to the sort of proto-constructive cognition that characterizes great apes, at least when complex strategies are built up by hierarchically coordinating simple abilities and subroutines (e.g. Gibson, 1990; Greenfield, 1991). This picture is consistent with studies of imitation in rehabilitant orangutans that suggest that what is acquired in true imitation is novel applications and/or novel arrangements of known elements (Russon & Galdikas, 1993). This suggests that true imitation offers great apes a means for transferring skills between problems: imitation may constitute a first step in freeing skills from contextual ties, making them "mobile" (Piaget, 1962) and "transportable" for application to new problems (Meltzoff, 1988). This step leads into the transition from sensorimotor to symbolic thought.

The contribution of true imitation and demonstration teaching to the local knowledge of great apes may arise from their coupling within the context of apprenticeship for technical skills (Boesch, 1991; Chapters 16 and 18). The incipient apprenticeship for nut-cracking that Boesch describes in Western chimpanzees provides the best example of enculturation in nonhuman primates, i.e. of social transmission of "local knowledge" from one generation to the next. It is interesting to note that primatologists have so far used the term "enculturation" to refer to social transmission of information from humans to great apes during cross-fostering (see Chapters 11, 13, and 17) rather than to refer to social transmission of information within a species in the wild as we propose.

Cultures of early hominids and modern humans

Although the cognitive capacities of early hominids are notoriously difficult to assess, the relatively simple nature of their tool technologies as compared to those of later hominids suggest that they displayed less advanced forms of cognition than do modern humans (e.g. Parker & Gibson, 1979; Schick & Toth, 1993; Wynn, 1989). Anthropologists typically mark modern human origins by the emergence of such symbolic representations as art and statuary as well as occupation of a greater range of habitat types and the emergence of more rapidly changing, more complex regionally variable tool technologies (e.g. Mellars & Stringer, 1984). This landmark highlights the gap between the highest abilities of great apes (rudimentary symbolic abilities) and the highest abilities of modern humans (hypothetical-deductive reasoning and propositional language). From a Piagetian perspective this gap corresponds to the developmental distance between the early preoperational, symbolic abilities of 3 year old children, and the formal operations of adolescents; that is, the developmental period that occurs between the onset of complex pretend play and the achievement of formal operational reasoning.

We set the point of cognitive divergence of the early hominids from their common ancestor with great apes as being coincident with the move from simple to complex pretend play because simple pretend play seems to be the highest cognitive achievement of the great apes (see Chapters 13 and 14). The onset of pretend play is significant because it is the first step in the transformation from sensorimotor to conceptual representation in human children. Pretend play is closely allied to language in that it is a form of "event representation" (Bretherton, 1984; Nelson & Gruendel, 1986) in which children represent their own behavior and that of others. Developmentally, they enact increasingly complex social routines practicing a variety of social roles. Pretend play thereby develops representation of the roles of self and other as part of the representation of events. In other words, pretend play opens the door to various forms of symbolic representation, including language and more complex forms of teaching that are crucial to the cultural capacity of modern humans.

Our review suggests some guidelines for characterizing levels of cultural adaptation in living primates on the basis of increasing levels of cognitive sophistication as diagnosed by frameworks from human development. Specifically, we follow the tradition of distinguishing the following cultural levels in anthropoid primates:

1. Preculture, typical of monkeys and other social mammals, which involves limited social transmission of representations of local knowledge based on a variety of mechanisms including simple imitation and other early sensorimotor abilities.
2. Proto-culture, typical of chimpanzees, which involves social transmission of representations of local knowledge based on true imitation, simple pretend play, and limited demonstration teaching characteristic of late sensorimotor-early symbolic level cognition.
3. Ur-culture, typical of Middle Pleistocene hominids, which involved social transmission of more complex representations of local knowledge based on substantial symbolic abilities, characteristics of late-preoperations and early concrete operations.
4. Eu-culture, typical of modern humans, which involves propositional language and declarative planning based on formal operational reasoning (Parker & Milbrath, 1993).

Although we agree with Donald (1991) that in comparing species "cognition is the most important dimension along which cultures are distributed," our diagnosis of cultural levels differs from his in the characterization of the abilities of the great apes as having rudimentary symbolic abilities, and in the use of developmental models to distinguish levels of cognitive and communicative abilities.

In addition to characterizing the kinds and levels of cognition underlying social transmission of information in various anthropoid species and diagnosing the associated levels of cultural potential, we want to understand the ecological variables involved in the expression of cultural potential. In order to approach this issue we need to examine life history, social organization, and other ecological parameters.

AN EVOLUTIONARY-ECOLOGICAL MODEL FOR SPECIES DIFFERENCES IN CULTURAL POTENTIAL AND THEIR EXPRESSION

Because we are focusing on the adaptive significance of culture and cognition in the wild, we would like to understand the contexts in which cultural capacities evolved and the contexts in which they are expressed. In other words, we would like to know why some species display these capacities in the wild while others do not.

We begin our inquiry by noting that our comparative functional definition of cultures (representations of knowledge that are socially transmitted within and between generations in groups and populations of a species) has some important socio-ecological entailments. First, it assumes the existence of a stable, inter-generational community of individuals of the same species whose members interact persistently and regularly in a variety of natural or semi-natural settings. In other words, this definition of culture assumes the same conditions that are prerequisites to forms of reciprocal altruism (Trivers, 1971). Secondly, it implies the diffusion of information through dispersal or movement of adult or subadult members of at least one sex between groups (Kummer, 1971). Thirdly, it implies that the local adaptations have arisen through social transmission of novel behaviors invented by members of the group or through importation of novel behaviors into a group by immigrants from other local groups.

Given these entailments, we hypothesize that the cognitive and hence the cultural potential of a species is shaped by several factors: first, by its life history; secondly, by species-typical foraging and feeding strategies; and thirdly, by species-typical social organization. In the following sections we briefly consider each of these factors as it impinges on the cultural capacities of anthropoid primates.

Life history and cognitive parameters

Such life history characteristics as gestation period, brain size, age at weaning, age at puberty, age at first reproduction, and life span (all of which correlate with body size) set both the neurological and experiential parameters for cognitive development. Since cognitive potential is set both by brain size and by length of immaturity, it is greater in the slowest developing, largest brained species (i.e. the great apes) than in the faster developing, smaller-brained species (i.e. lesser apes and monkeys).

All the great apes share common life history features of prolonged gestation, infancy and juvenility, long life span, and large brain size. These features, which reflect their recent common ancestry, ally them with humans and distinguish them from their next closest relatives, the lesser apes and the Old World monkeys, who have shorter gestation periods, shorter infancy and juvenile periods, shorter life spans, and smaller brains (see life history table for primates in Harvey et al., 1987).

We argue that the similar life histories of great apes are the basis for their common cognitive abilities and common cultural potentials. The foraging and feeding strategies and social organization of great apes, in contrast to their life histories, have diverged significantly during their evolution. Consequently great apes differ considerably in these features. We suggest that these differences in feeding strategies and social organization help to explain differences among the great apes in the expression of their cognitively shaped cultural potential. We also suggest that differences in feeding strategies and social organization in Old World monkeys help to explain species differences in the expression of the lesser cultural potential of these taxa.

Species-typical foraging and feeding strategies among great apes and other anthropoid primates

Dietary niches carry entailments that arise from differences in the spatial and temporal distribution of food species. Chimpanzees and orangutans (as specialists on widely dispersed, seasonal fruit sources) are forced to disperse and travel widely in large home ranges. Bonobos have smaller home ranges, and gorillas (as specialists on abundant herbaceous materials) are able to travel together relatively short distances in relatively small home ranges. These differences in ranging patterns in turn constrain their mating systems, social interactions, and social organizations (Wrangham, 1987), which in turn influence the expression of their cultural potential.

A common evolutionary phenomenon may help to explain differences in related primate species: when a group of sister species (known as an adaptive array) arises from a common ancestor, they tend to "radiate" into a variety of specialized feeding niches and associated body sizes which reduce their potential competition with one another. Typically, adaptive arrays among the primates, e.g. the great apes, the lesser apes, and, within the Old World monkeys, the macaques, the baboons, and the guenons, display a range of body sizes associated with a range of dietary specializations: larger-bodied species within an array relying preferentially on less nutritious, lower grade foods such as mature leaves, pith, etc. (herbivory); medium-bodied species relying on a mixture of lower grade and higher grade foods such as fruits and leaves (frugivory) or fruit, leaves, insects, and small animal prey (omnivory); the smallest species relying more on higher quality more nutritious foods such as insects (insectivory), and/or gums, honey, and nectar (e.g. Fleagle, 1988).

This pattern is repeated to some degree in the great apes: the smallest species, bonobos and chimpanzees, being omnivorous, having the greatest range of dietary items, and relying most extensively on fruits, nuts, insects, and other animal foods; the largest species, gorillas, being herbivorous, having the most restricted range of dietary items, and relying most extensively on leaves and pith; mid-to-large size species, orangutans, being primarily frugivorous and having a dietary range that is apparently closer to that of chimpanzees than gorillas in the variety of plants consumed (e.g. Fleagle, 1988; Richard, 1985). Obviously, competition has been a greater factor in niche separation among African apes than between them and the more geographically distant Asian apes.

Because they feed at at least two trophic levels, omnivorous species such as chimpanzees are better able to exploit a variety of habitats than species with narrow dietary niches. Consequently their home ranges often embrace mosaic habitats that include both rainforest and woodland savanna. Exploitation of woodland savannas brings increased seasonality of resources and hence increased dependence on embedded food sources during dry seasons. Embedded foods such as social insects (termites, ants, and bees), roots, fossorial animals, and hard shelled fruits and nuts are available to species who can extract them with powerful teeth and hands or with the aid of tools. Omnivorous extractive foraging with anatomical manipulators is common among orangutans and baboons while extractive foraging with tools is common among chimpanzees (Gibson, 1986; Parker & Gibson, 1977, 1979).

Populations of an omnivorous species who live in different areas are likely to eat somewhat different diets: first, because they are more liable to encounter local microclimates and associated variations in the relative abundance and/or occurrence of food species; and secondly, because they eat so many food species that they can often afford to be selective. Whereas conditions that vary between regions are likely to generate differences in resource exploitation strategies between populations living in these different regions (regional populations), similar conditions within regions are likely to favor similar patterns among local groups within a regional population.

Given the breadth of their feeding niche and the dispersion of their food sources during the dry season, chimpanzees more than any other species of great apes, show both regional variation and local convergence in feeding strategies. Regional populations of Eastern chimpanzees living in noncontiguous areas characterized by different climates display considerable differences in their diets in relation to the number of potential food items in common (e.g. Nishida *et al.*, 1983). Local groups within regional populations show much less variation. Likewise, regional populations show greater variation in their use of medicinal plants whereas local groups within a regional population show greater similarity both in plant items and in consumption modes (Huffman & Wrangham, 1994). In contrast to chimpanzees, bonobos live exclusively in tropical rainforests. Although their diets are similar to those of chimpanzees, their habitats are richer and less variable and they are therefore able to find food within smaller home ranges. They also seem to rely more on fibrous foods and invertebrate animal foods, which are more abundant than the foods chimpanzees rely on when fruits are scarce (Nishida & Hiraiwa-Hasegawa, 1987). Mountain gorillas, being primarily herbivorous, experience high abundance and relatively little seasonal variation in their diets and hence are better able to find adequate foods within much smaller home ranges than are chimpanzees. Lowland gorillas, especially Western lowland gorillas, show somewhat greater dietary breadth (e.g. Tutin & Fernandez, 1983). Finally, orangutans, being primarily frugivorous and therefore experiencing seasonal variations in food availability, have larger home ranges than any other great ape species (e.g. MacKinnon, 1974).

Some parameters in species-typical social environments

We suggest that social transmission of information within and among local groups within a species is both facilitated and constrained by the nature of demography and the organization of social groups as well as by the life history and associated cognitive capabilities of the species. The birth interval as well as the length of infancy and juvenility determine the window of opportunity for intensive social learning. Demography determines the age spectrum of the group and hence the numbers of individuals in various age cohorts as well as their sex ratio. The composition of peer groups of young monkeys or apes varies with the birth interval, the seasonality of mating, and the size and composition of social groups. Immature macaques tend to have large cohorts of playmates as a consequence of living in large multimale/multifemale groups in which females have infants every year or so. Seasonal breeding and births in these species generate large cohorts of same-age infants (e.g. Lindberg, 1980). Immature chimpanzees, in contrast, tend to have small cohorts as a consequence of their mothers long birth intervals and of living in subgroups with their mothers and next oldest siblings (e.g. Goodall, 1986).

Other related parameters of subgrouping within social organization include specific age and sex related affinities and disaffinities (Kummer, 1971) and the phenomena of attention structures related to dominance hierarchies (Chance & Jolly, 1970). These factors, particularly "centripetal attention structures" in which all members of a troop orient toward dominant males, seem to increase the facility with which information is socially transmitted (Cambefort, 1981).

Evidence suggests that among anthropoid primates, variations in the following parameters of social organization determine the probability and speed of social transmission of information:

1. The nature of subgroupings of individuals within the larger group (mother–offspring, juveniles, males, or males versus females, or harem units) according to various affinities and disaffinities (e.g. Kummer, 1971).
2. The stability and/or instability of these subgroups over time (i.e. the changeability of associations).
3. The spatial contiguity or spread (i.e. contiguous groupings versus spatially spread out groupings).
4. The structure of attention (Chance & Jolly, 1970).
5. The nature of group continuity according to the patterns of philopatry and dispersal (i.e. philopatric matrilines and dispersing males, philopatric males and dispersing females, bisexual dispersal into new breeding units: Pusey & Packer, 1987).

In the following sections we briefly address some of the implications these parameters may have for social transmission of information.

The social organization of monogamous species such as the lesser apes restricts the opportunity for inter-generational information transfer to parents and juvenile offspring. Similarly, the semi-solitary social organization of orangutans, who apparently have no stable units beyond mother–offspring units, restricts even further inter-generational transfer of information. The somewhat larger social units that include one male and several adult females with their adult female kin and their immature male offspring, which are typical of guenons and many langur monkeys, have increased potential for inter-generational transfer of information among females with limited opportunities for observational learning within the group for young males. (Interestingly, immature males in species with this organization experience opportunities for social learning in the all male social groups they join before they begin cooperating and competing to take over harem units.)

Multimale/multifemale social organizations that are typical of macaques and baboons among Old World monkeys, cebus monkeys among New World monkeys, as well as chimpanzees and bonobos among the great apes, apparently provide the organization most conducive to social transmission of information. The nature of this potential however, apparently varies, with the attention structure of the troop (Cambefort, 1981). The sexes of the resident/philopatric members within a social group may also constrain and/or facilitate information flow. In most monkeys, female matrilines form the stable residential core of groups and most adolescent males leave their natal group (Pusey & Packer, 1987), while in chimpanzees and probably bonobos, male patrilines form such a core group and most adolescent females leave their natal groups (Nishida & Hiraiwa-Hasegawa, 1987; Pusey & Packer, 1986).

To the extent that female primates play a privileged role as conduits of information to their offspring, their dispersal out of one group into another group has greater potential for disseminating new behaviors to other groups through their children (see Goodall, 1973; Chapter 10). According to this notion, female dispersal in the context of large social groups opens up increased possibilities for social and cultural transmission of new technologies and behaviors among local groups given adequate cognitive abilities and appropriate ecological challenges. The fullest expression of this tendency can be found in chimpanzees. The potential for information flow is less in orangutans because both sexes disperse (female offspring forming home ranges close to those of their mothers, and males ranging more widely) and because orangutan social life is semi-solitary from the adolescent period onwards. Likewise, the tendency for both male and female dispersal amongst gorilla groups probably reduces the opportunity for continuity of traditions that is offered by a continuous philopatric core of a group with inter-generational continuity. (Interestingly, leaf-eating red colobus monkeys, one of the few Old World monkeys characterized by female dispersal and large social groups, show no evidence of social traditions probably because they lack the life history and feeding strategies that are associated with cultural adaptations.)

OF ALL THE GREAT APES, WHY DO ONLY CHIMPANZEES DISPLAY PROTO-CULTURE?

Given the foregoing patterns, we are in a better position to address the question of why – given the similar cognitive abilities of all the great apes – only the chimpanzees (and possibly bonobos (Thompson, 1994)) seem to exhibit regional cultures? As hinted in the previous section, we hypothesize that the more cultural nature of chimpanzees as opposed to gorillas and orangutans arises from the nature of their feeding strategies, their demography, and their social organization.

In addition to having the cognitive prerequisites for proto-culture (true imitation and demonstration teaching), chimpanzees use a feeding strategy that favors tool use in extractive foraging on a variety of embedded foods which vary locally (which favors apprenticeship in tool use). In addition, they live in large social groups connected through female dispersal, a pattern that favors local social transmission of traditions through demonstration teaching and true imitation. Although studies of bonobos have yet to reveal patterns of cultural transmission in the wild, preliminary evidence suggests that they engage in less tool-mediated extractive foraging than chimpanzees (see Chapter 9). This makes sense given that their diet and their rainforest habitat provide more abundant and consistent food sources (Nishida & Hiraiwa-Hasegawa, 1987; Chapter 9) thereby reducing their need to rely on embedded foods which require tool-mediated extraction.

Gorillas and orangutans, in contrast to chimpanzees, use feeding strategies that minimize tool use, and, in turn, minimize demonstration teaching of tool-mediated extractive foraging. Although both gorillas and orangutans engage in complex manipulation of food items (e.g. Byrne & Byrne, 1991; Rijksen, 1978) they are strong enough to extract embedded foods without recourse to tools (Parker & Gibson, 1979). Although gorillas have female dispersal, which favors transmission of local traditions through females to infants in adjacent groups, so far there are no reports of social traditions in wild gorilla groups. Orangutan females also disperse but they are semi-solitary after adolescence, displaying few social groupings beyond the mother–offspring unit. Given the common cognitive abilities of all the great apes, however, it seems likely that their common ancestor displayed feeding niches, social organizations and proto-cultural adaptations similar to those seen in modern chimpanzees (see Chapter 16) which favored apprenticeship through true imitation and demonstration teaching of tool-mediated extractive foraging.

Overall, existing reports of the adaptations of wild populations of great apes suggest that true imitation and demonstration teaching occurs primarily among chimpanzees. Expression of these capacities in the wild – which are common to all the great apes in captivity – seems to correlate with dependence on apprenticeship in tool-mediated extractive foraging on embedded foods associated with an omnivorous diet. It also correlates with life in large multimale/multifemale social groups characterized by female dispersal and male philopatry.

It is important to note, however, that the apparent uniqueness of chimpanzee adaptation may be an artifact of the larger database that exists for wild chimpanzees as compared to wild gorillas and orangutans. Assessment of the expression of cultural potential depends upon comparative studies of groups within a local region and of populations across regions which control for ecological and genetic differences. So far such studies have been done only on chimpanzees (McGrew, 1992; Chapter 18). Recent reports that describe consumption of ants and termites by gorillas (Tutin & Fernandez, 1983; Watts, 1989) and tool use in wild orangutans (van Schaik & Fox, 1994) suggest that dietary differences among the great apes may be less dramatic than they have seemed. These data suggest that future studies of intra-specific variations within gorillas and orangutans may reveal some proto-cultural adaptations in these species.

Given these caveats, we note that this brief analysis of the life histories, feeding strategies, and social organizations of the anthropoid primates suggests that the expression of proto-cultural adaptation in the form of local feeding traditions depends upon the concatenation of several factors: (a) cognitive abilities for imitative tool use; (b) large social groups characterized by centripetal attention structures and female dispersal; and (c) an omnivorous feeding niche characterized by highly seasonal foods some of which require tool-mediated extractive foraging. As far as we know now, of all the living anthropoid primates, only chimpanzees display all of these factors in combination.

The validity of this ecological model is supported by reports of precultural adaptations in macaques and baboons, who are omnivorous feeders engaging in extractive foraging (without tools) and who live in large multimale/multifemale groups with centripetal attention structures. Although these species are characterized by male rather than female dispersal, matrilines do split off to form new groups, which offers a means for social transmission across groups through females. The emphasis in the literature on precultural adaptations among Japanese macaques is probably an artifact of the provisioning done with this species. Our ecological model predicts that studies focused on social transmission should yield other examples of precultural adaptations among other similarly adapted species.

SUMMARY AND CONCLUSIONS

In this chapter we have defined culture in comparative terms that we hope are compatible with anthropological, biological, and psychological usages. Using a Piagetian framework to diagnose different levels of cognition in anthropoid species, we have explored the different mechanisms of social transmission of information supported by these levels. Based on these differences, we have proposed a classification of four levels of cultural adaptation in monkeys, great apes, early hominids, and humans. Finally, we have proposed an ecological/evolutionary model to explain the conditions under which anthropoid species express their cultural potential in the wild. This model, which is an extension and elaboration of the apprenticeship model, emphasizes the interplay of social organization and dietary niche in facilitating the expression of cognitive potentials associated with various levels of sensorimotor and symbolic intelligence and periods of immaturity.

ACKNOWLEDGEMENTS

This chapter represents an attempt to bridge conceptual gaps in uses of the terms culture, enculturation, knowledge, tradition, and imitation by anthropologists, biologists, and psychologists. The authors, who themselves come from different backgrounds – biological anthropology and psychology – have grappled with these difficulties in writing the chapter. We thank R. Thomas Rosin for his comments from the perspective of a cultural anthropologist.

NOTES

1 Following the lead of earlier investigators, comparative psychologists and comparative developmental psychologists have defined "true imitation" in at least two different ways: Galef (1988), Whiten & Ham (1992), and Russon & Galdikas (1993) define it as "one individual learning new behaviors demonstrated by another by observations of the demonstration" (Chapter 7). Parker (1977, 1993; Chapter 16) defines it in terms of Piaget's (1962) fifth and sixth stages of imitation in human infants and children, i.e. imitation of novel schemes through either trial-and-error matching of a model's behavior (fifth stage) or through interiorized (insightful) matching of a model's behavior. Although these two usages overlap to some degree they differ significantly in the specificity of their mechanisms. Likewise "simple imitation" is used in two slightly different ways in this chapter: first, in strictly Piagetian terms to refer to imitation of existing schemes, and secondly, in comparative psychological terms to refer to matching through a variety of mechanisms such as stimulus or local enhancement.

REFERENCES

Altmann, S. & Altmann, J. (1970). *Baboon Ecology*. Chicago: University of Chicago Press.

Boesch, C. (1991). Teaching among wild chimpanzees. *Animal Behaviour*, **43**, 530–532.

Boesch, C. (1992). Imitation as a measure of attribution. *Behavioral and Brain Sciences*, **15**, 149.

Bonner, J.T. (1980). *The Evolution of Culture in Animals*. Princeton, NJ: Princeton University Press.

Bretherton, I. (1984). Introduction: Piaget and event representation in play. In *Symbolic Play*, ed. I. Bretherton, pp. 3–41. New York: Academic Press.

Burton, F.D. (1992). The social group as information unit: Cognitive behaviour, cultural processes. In *Social Processes and Mental Abilities in Non-human Primates: Evidences from Long-term Field Studies*, ed. F.D. Burton, pp. 31–60. Lewiston, NY: The Edward Mellon Press.

Byrne, R.W. & Byrne, J. M. E. (1991). Hand preferences in the skilled gathering tasks of mountain gorillas (*Gorilla g. beringei*). *Cortex*, **27**, 521–546.

Byrne, R.W. & Whiten, A. (eds.) (1988). *Machiavellian Intelligence: Social Expertise and the Evolution of Intellect in Monkeys, Apes, and Humans*. Oxford: Clarendon Press.

Cambefort, J. (1981). A comparative study of culturally transmitted patterns and feeding habits in the chacma baboon. *Quarterly Review of Biology*, **67**, 151–74.

Caro, T.M. & Hauser, M.D. (1992). Is there teaching in nonhuman animals? *Quarterly Review of Biology*, **67**, 151–74.

Chance, M. & Jolly, C. (1970). *Social Groups of Monkeys and Apes*. New York: Dutton.

Chevalier-Skolnikoff, S. (1976). The ontogeny of primate intelligence and its implication for communicative potential: A preliminary report. *Annals of the New York Academy of Sciences*, **280**, 173–216.

Custance, D. & Bard, K. (1994). The comparative and developmental study of self-recognition and imitation: The importance of social factors. In *Self-awareness in Animals and Humans: Developmental Perspectives*, ed. S.T. Parker, R.W. Mitchell & M.L. Boccia, pp. 207–25. New York: Cambridge University Press.

Donald, M. (1991). *The Origins of the Modern Mind*. Cambridge, MA: Harvard University Press.

Fleagle, J. (1988). *Primate Adaptation and Evolution*. New York: Academic Press.

Fox, R. (1972). The cultural animal. In *Man and Beast*, ed. J. Eisenberg & W.S. Dillon, 1st edn, pp. 15–28. Washington: Smithsonian Institution Press.

Freilich, M. (1988). Introduction: Is culture still relevant? In *The Relevance of Culture*, ed. M. Freilich, pp. 1–26. New York: Bergin and Garvey.

Galef, B.G., Jr (1988). Imitation in animals: History, definition and interpretation of data from the psychological laboratory. In *Advances in the Study of Behaviour*, ed. J. Rosenblatt, R. Hinde, E. Shaw & C. Beer, vol. 6, pp. 77–100. New York: Academic Press.

Galef, B.G., Jr (1990). Tradition in animals: Field observations and laboratory analysis. In *Interpretations and Explanations in the Study of Behaviour: Comparative Perspectives*, ed. M. Bekoff & D. Jamieson, pp. 74–95. Boulder, CO: Westview Press.

Gibson, K. (1986). Cognition, brain size and the extraction of embedded food resources. In *Primate Ontogeny, Cognition and Social Behaviour*, ed. J.G. Else & P.C. Lee, pp. 93–105. Cambridge: Cambridge University Press.

Gibson, K. R. (1990). New perspectives on instinct and intelligence: Brain size and the emergence of hierarchical mental construction skills. In *"Language" and Intelligence in Monkeys and Apes: Comparative Developmental Perspectives*, ed. S.T. Parker & K.R. Gibson, pp. 97–128. New York: Cambridge University Press.

Gibson, K.R. & Ingold, T. (eds.) (1993). *Tools, Language and Cognition in Human Evolution*. Cambridge: Cambridge University Press.

Goodall, J. (1973). Cultural elements in a chimpanzee community. In *Precultural Primate Behavior*, ed. E. W. Menzel, vol. 1, pp. 144–84. Fourth International Primatology Congress Symposium Proceedings. Basel: Karger.

Goodall, J. (1986). *The Chimpanzees of Gombe: Patterns of Behavior*. Cambridge, MA: The Belknap Press.

Goodenough, W. (1981). *Language, Culture and Society*, 2nd edn. Menlo Park, CA: Benjamin/Cummings.

Greenfield, P. (1991). Language, tools and the brain: The ontogeny and phylogeny of hierarchally organized sequential behavior. *Behavioral and Brain Sciences*, 14, 531–95.

Hall, K.R.L. (1962). Numerical data, maintenance activities and locomotion of the wild chacma baboon, *Papio ursinus*. *Proceedings of the Zoological Society, London*, 139, 181–220.

Hallowell, A. I. (1959). Behavioral evolution and the emergence of the self. In *Evolution and Anthropology: A Centennial Appraisal*, ed. B. Meggars, pp. 36–60. Washington, DC: American Anthropological Association.

Harris, M. (1979). *Cultural Materialism*. New York: Random House.

Harvey, P., Martin, R. D. & Clutton-Brock, T. (1987). Life histories in comparative perspective. In *Primate Societies*, ed. B. Smuts, D. Cheney, R. Seyfarth, R. Wrangham & T. Struhsaker, pp. 181–96. Chicago: University of Chicago Press.

Heyes, C. (1993). Imitation, culture and cognition. *Animal Behaviour*, 46, 999–1010.

Huffman, M. (1984). Stone play of *Macaca fuscata* in Arashiyama B troop: Transmission of a nonadaptive behavior. *Journal of Human Evolution*, 13, 725–735.

Huffman, M. (1994). On the acquisition of newly innovated cultural behaviors in nonhuman primates: A case study of stone handling, a socially transmitted behaviour in Japanese macaques. Paper presented at the Conference on Social Learning and Tradition in Animals, Madingley, Cambridgeshire, UK.

Huffman, M.A. & Quaitt, D. (1986). Stone handling by Japanese macaques (*Macaca fuscata*): Implications for tool use of stone. *Primates*, 27, 413–23.

Huffman, M. & Wrangham, R. (1994). Diversity of medicinal plant use by chimpanzees in the wild. In *Chimpanzee Cultures*, ed. R. Wrangham, W. McGrew, F. de Waal, & P.G. Heltne, pp. 129–48. Cambridge, MA: Harvard University Press.

Itani, J. (1974). Communication systems among primates. *Annual Review of Social Psychology, Japan*, 15, 31–54.

Itani, J. & Nishimura, A. (1973). The study of infrahuman culture in Japan. *Precultural Primate Behaviour*, ed. E. Menzel, vol. 1, pp. 26–50. Fourth International Primatology Congress Symposium Proceedings. Basel: Karger.

Jolly, A. (1972). *The Evolution of Primate Behavior*. New York: MacMillan.

Kawai, M. (1965). Newly acquired pre-cultural behavior of the natural troop of Japanese macaques. In *Japanese Monkeys*, ed. S. Altmann, pp. 1–30. Alberta: University of Alberta Press.

Kawamura, S. (1959). The process of sub-cultural propagation among Japanese macaques, *Primates*, 2, 43–60.

Keesing, R. (1974). Theories of culture. *Annual Reviews of Anthropology*, 76, 73–97.

King, B. (1994). *The Information Continuum*. Santa Fe, NM: School of American Research Press.

Kroeber, A. (1928). Sub-human culture beginnings. *Quarterly Review of Biology*, 8, 325–42.

Kroeber, A. & Kluckhohn, C. (1952). *Culture: A Critical Review of Concepts and Definitions*. Papers of the Peabody Museum of American Archeology and Ethnology. Cambridge, MA: Harvard University Press.

Kummer, H. (1971). *Primate Societies*. New York: Aldine.

Lindberg, D. (ed.). (1980). *The Macaques*. New York: van Nostrand Reinhold.

MacKinnon, J. (1974). The behavior and ecology of wild orangutans (*Pongo pygmaeus*). *Animal Behaviour*, 22, 3–74.

McGrew, W. (1992). *Chimpanzee Material Culture. Implications for Human Evolution*. New York. Cambridge University Press.

Mead, M. (1963). Socialization and enculturation. *Current Anthropology*, 4, 184–8.

Mellars, P. & Stringer, C. (eds.) (1984). *The Human Revolution*. Princeton, NJ: Princeton University Press.

Meltzoff, A. N. (1988). Imitation, objects, tools and the rudiments of language in human ontogeny. *Human Evolution*, 3, 45–64.

Menzel, E. (ed.). (1973). *Precultural Primate Behavior*, vol. 1. Fourth International Primatology Congress Symposium Proceedings. Basel: Karger.

Mineka, S. & Cook, M. (1988). Social learning and the acquisition of snake fear in monkeys. In *Social Learning: Psychological and Biological Perspectives*, ed. T.R. Zentall & B.G. Galef Jr, pp. 51–73. Hillsdale, NJ: Lawrence Erlbaum.

Mitchell, R.W. (1987). A comparative-developmental approach to understanding imitation. In *Perspectives in Ethology*, ed. P.P.G. Bateson & P.H. Klopfer, vol. 7, pp. 183–215. New York: Plenum Press.

Mitchell, R.W. (1993). Mental models of mirror-self-recognition: Two theories. *New Ideas in Psychology*, 11, 295–325.

Moore, B. (1994). The evolution of imitative learning. Paper presented at the Conference on Social Learning and Tradition in Animals, Madingley, Cambridgeshire, UK.

Nelson, K. & Gruendel, J. (1986). Children's scripts. In *Event Knowledge*, ed. K. Nelson, pp. 21–45. Hillsdale, NJ: Lawrence Erlbaum.

Nishida, T. (1987). Local traditions and cultural transmission. In *Primate Societies*, ed. B. Smuts, D. Cheney, R. Seyfarth, Wrangham, & Struhsaker, pp 462–74. Chicago: University of Chicago Press.

Nishida, T. & Hiraiwa-Hasegawa. (1987). Chimpanzees and Bonobos: Cooperative relationships among males. In *Primate Societies*, ed. B. Smuts, D. Cheney, R. Seyfarth, R. Wrangham & T. Struhsaker, pp. 165–78. Chicago: University of Chicago Press.

Nishida, T., Wrangham, R., Goodall, J. & Uehara, S. (1983). Local differences in plant feeding habits of chimpanzees between the Mahali Mountains and the Gombe National Park, Tanzania. *Journal of Human Evolution*, 12, 467–80.

Parker, S.T. (1977). Piaget's sensorimotor series in an infant macaque: The organization of non-stereotyped behavior and intelligence in human and nonhuman primates. In *Primate Biosocial Development*, ed. S. Chevalier-Skolnikoff & F. Poirier, pp. 43–113. New York: Garland Publishing, Inc.

Parker, S.T. (1990). The origins of comparative developmental evolutionary studies of primate mental abilities. In *"Language" and Intelligence in Monkeys and Apes: Comparative Developmental Perspectives*, ed. S.T. Parker & K.R. Gibson, pp. 3–63. New York: Cambridge University Press.

Parker, S.T. (1993). Imitation and circular reactions as evolved mechanisms for cognitive construction. *Human Development*, 36, 309–23.

Parker, S.T. & Gibson, K.R. (1977). Object manipulation, tool use, and sensorimotor intelligence as feeding adaptation in cebus monkeys and great apes. *Journal of Human Evolution*, 6, 435–49.

Parker, S.T. & Gibson, K.R. (1979). A developmental model for the evolution of language and intelligence in early hominids. *Behavioral and Brain Sciences*, 2, 367–408.

Parker, S.T. & Milbrath, C. (1993). Higher intelligence, propositional language, and culture as adaptations for planning. In *Tools, Language, and Cognition in Human Evolution*, ed. K. Gibson & T. Ingold, pp. 314–44. Cambridge: Cambridge University Press.

Patterson, F. & Cohn, R. (1994). Self-recognition and self-awareness in lowland gorillas. In *Self-awareness in Animals and Humans: Developmental Perspectives*, ed. S.T. Parker, R.W. Mitchell & M.L. Boccia, pp. 273–90. New York: Cambridge University Press.

Piaget, J. (1952). *The Origins of Intelligence in Children*. New York: Norton.

Piaget, J. (1962). *Play, Dreams and Imitation in Children*. New York: Norton.

Piaget, J. (1970). *Genetic Epistemology*. New York: Norton.

Povinelli, D., Nelson, K. & Boysen, S.T. (1992). Comprehension of role reversal in chimpanzees: Evidence of empathy? *Animal Behaviour*, 43, 633–640.

Premack, D. & Woodruff, G. (1978). Does the chimpanzee have a theory of mind? *Behavioral and Brain Sciences*, 4, 515–526.

Pusey, A. & Packer, C. (1987). Dispersal and philopatry. In *Primate Societies*, ed. B. Smuts, D. Cheney, R. Seyfarth, R. Wrangham & T. Struhsaker, pp. 250–66. Chicago: University of Chicago Press.

Quiatt, D. & Itani, J. (eds.) (1994). *Hominid Culture in Primate Perspective*. Niwot, CO: University of Colorado Press.

Quiatt, D. & Reynolds, V. (1992). *Primate Behaviour: Information, Social Knowledge, and the Evolution of Culture*. Cambridge: Cambridge University Press.

Richard, A. (1985). *Primates in Nature*. New York: Freeman.

Rijksen, H.D. (1978). *A Field Study on Sumatran Orangutans (Pongo pygmaeus abelii Lesson 1927): Ecology, Behaviour and Conservation*. Mededelingen Landbouwhogeschool Wageningen. Wageningen, BV: H. Venman & Zonen.

Russon, A. & Galdikas, B.M.F. (1993). Imitation in free-ranging rehabilitant orangutans (*Pongo pygmaeus*). *Journal of Comparative Psychology*, 107, 147–61.

Russon, A. & Galdikas, B. M. F. (1995). Constraints on great ape imitation: Model and action selectivity in rehabilitant orangutan (*Pongo pygmaeus*) imitation. *Journal of Comparative Psychology*, **109**, 5–17.

Schick, K. & Toth, N. (1993). *Making Silent Stones Speak: Human Evolution and the Dawn of Technology*. New York: Simon Schuster.

Steward, J. (1955). *Theory of Culture Change*. Urbana, IL: University of Illinois Press.

Strum, S.C. (1975). Primate predation: Interim report on the development of a tradition in a troop of olive baboons. *Science*, **187**, 755–7.

Teleki, G. (1975). Primate subsistence patterns: Collector-predators and gatherer-hunters. *Journal of Human Evolution*, **4**, 125–84.

Thompson, J.A.M. (1994). Cultural diversity in the behavior of *Pan*. In *Hominid Culture in Primate Perspective*, ed. D. Quiatt & J. Itani, pp. 95–115. Niwot, CO: University of Colorado Press.

Tomasello, M. (1990). Cultural transmission in the tool use and communicatory signaling of chimpanzees. In *"Language" and Intelligence in Monkeys and Apes: Comparative Developmental Perspectives*, ed. S. T. Parker & K. R. Gibson, pp. 274–311. New York: Cambridge University Press.

Tomasello, M., Kruger, A. & Ratner, H. (1993). Cultural learning. *Behavioral and Brain Sciences*, **16**, 495–552.

Trivers, R. (1971). The evolution of reciprocal altruism. *Quarterly Review of Biology*, **46**, 35–57.

Tutin, C. & Fernandez, M. (1983). Gorillas feeding on termites in Gabon, West Africa. *Journal of Mammology*, **6**, 530–1.

van Schaik, C.P. & Fox, E.A. (1992). Tool use in wild Sumatran orangutans (*Pongo pygmaeus*). Paper presented at the XVth Congress of the International Primatological Society, Kuta-Bali, Indonesia, 3–8 August.

Visalberghi, E. & Fragaszy, D.M. (1990). Do monkeys ape? In *"Language" and Intelligence in Monkeys and Apes: Comparative Developmental Perspectives*, ed. S.T. Parker & K.R. Gibson, pp. 247–73. New York: Cambridge University Press.

Wallace, A.F.C. (1961). *Culture and Personality*. New York: Random House.

Watts, D. (1989). Ant eating behavior of Mountain gorillas. *Primates*, **30**, 121–5.

White, L. (1959). *The Evolution of Culture*. New York: McGraw Hill.

Whiten, A. & Byrne, R.W. (1991). The emergence of metarepresentation in human ontogeny and primate phylogeny. In *Natural Theories of Mind: Evolution, Development, and Simulation*, ed. A. Whiten, pp. 267–81. Oxford: Basil Blackwell Ltd.

Whiten, A. & Custance, D. (1994). Functions and mechanisms of imitation: Studies of monkeys, apes and human children. Paper presented at the Conference on Social Learning and Tradition in Animals, Madingley, Cambridgeshire, UK.

Whiten, A. & Ham, R. (1992). On the nature and evolution of imitation in the animal Kingdom: Reappraisal of a century of research. In *Advances in the Study of Behavior*, ed. P.J.B. Slater, J.S. Rosenblatt, C. Beer & M. Milinski, vol. 21, pp. 239–83. New York: Academic Press.

Wrangham, R. (1987). Evolution of social structure. In *Primate Societies*, ed. B. Smuts, D. Cheney, R. Seyfarth, R. Wrangham & T. Struhsaker, pp. 282–96. Chicago: University of Chicago Press.

Wrangham, R., McGrew, W., de Waal, F. & Heltne, P. (eds.) (1994). *Chimpanzee Cultures*. Cambridge, MA: Harvard University Press.

Wynn, T. (1989). *The Evolution of Spatial Competence*. Urbana, IL: University of Illinois Press.

Index

altruism, 81, 88
Anderson, J.R., 8, 28–30, 34–9, 44–6, 59, 74, 85, 214, 334–5, 356, 360, 409
Antinucci, F., 5, 10–12, 259–60, 262, 351, 362
anthropocentrism, 1, 6, 326–7
apprenticeship, 348–64, 415, 437, 444–5
  and tool use, 348, 357–64, 437, 444–5
  cognitive mechanisms
    demonstration teaching, 348–64, 437, 444–5
    true imitation, 172, 348–64, 415, 437, 444–5
    self-awareness, 348–64, 438
  definition, 348
  evolution, 348, 357–63
  function, 359–64
  in great apes, 357–64
  in humans, 357–8, 362
  in monkeys, 357–62
  see also social cognition, physical cognition
attachment, 243, 246–8
attention, 133–4, 138–9, 146, 243–5, 250–3, 379–82, 391, 442–3
  see also social cognition
Aureli, F., 97–100
autism, 184–6, 306–9, 315

baboons
  culture, 430, 435–6, 445
  development, 27–8
  feeding, 441, 445
  gaze & eye contact, 135, 141
  imitation, 406, 435
  object manipulation, 28–9
  phylogenetic relations, 440
  social organization, 443–5
  tool use, 28, 364
Baldwin, J.M., 275, 281, 335
Bard, K., 27, 140, 190, 193–4, 205, 235–40, 250–2, 293, 361, 379, 382
Bates, E., 57, 133, 146, 154, 382

Bayley Scales of Infant Development, 238–51
Beck, B., 24–5, 29–30, 58, 190, 193, 358
behavioral development
  human influences, 235–53
  in great apes, 27–19, 235–53
  in humans, 28, 236–7, 241–2, 250–3
  in monkeys, 27–8, 252
  methods, 236, 238–40
  neonatal, 26–7, 236–7, 250–1, 379
  neurobehavioral, 26–8, 236–7
  of emotions, 236, 246
  of gaze, 250, 253
  of manipulation, 28–9
  of temperament, 236–53
  rates, 27–8, 240–53
  stress influences on, 236–53
  relation to cognitive development, 27–9, 238, 250–3
behavioral tradition *see* traditions
Boesch, C., 6, 24, 29–31, 58, 74, 186, 203, 212–17, 228, 303, 342, 353–4, 360, 371–3, 377, 384–6, 404–10, 413–16, 419–21, 425, 436–7
Bonner, J.T., 405, 431–2
bonobos
  apprenticeship, 358–9
  attention, 382
  communication, 140, 200–6, 382–3
  culture, 443–4
  development, 28, 140, 198–200
  ecological cognition, 195–9
  enculturation, 310–11
  feeding, 191–2, 205–7, 440
  food niche, 440–1
  gaze & eye contact, 140
  habitat, 191, 196, 204, 440–1
  human influences, 373–88
  Kanzi, 204, 307, 376
  life history, 191

bonobos (*cont.*)
  logicomathematical cognition, 377
  mindreading, 388
  mirror-self-recognition, 377
  object manipulation, 28, 190–204, 222–3, 373
  phylogenetic relations, 13, 440
  pretense, 307, 376
  social cognition, 199–207
  social organization, 191, 443
  tool use, 31, 126, 190–207, 376, 385
  true imitation, 310–1, 351–2, 385–6
Boysen, S., 8, 11, 177–80, 302–4, 378
brain
  allometric scaling, 124
  as Turing machine, 124
  cerebellum, 126
  development, 26–8, 439–40
  evolution, 112, 124–5
  gut–brain tradeoff, 125–6
  in great apes, 26–8, 112, 124–6, 439–40
  in haplorhines, 26–8, 112, 125
  in hominids/humans, 112, 126, 439–40
  in monkeys, 26–8, 439–40
  neocortex, 24, 124–5
  size, 112, 124–6, 439–40
Brazelton Neonatal Behavoural Assessment Scale (NBAS), 27, 236–7, 379
Bretherton, I., 350, 352, 354–5, 438
Byrne, R.W., 6–9, 24, 38–9, 114–15, 119–20, 122–6, 157, 171, 302–17, 348, 363, 383, 436, 444

Call, J., 383, 385, 388
Cambefort, J., 430, 434–5, 442–3
Caro, T.M., 349, 352–4, 418, 434
Carpenter, C.R., 4, 11, 153–4, 157
causal cognition, 6–8, 24–6, 31, 46–7, 57–75, 122, 215–17, 257, 266–8, 373–7, 389–94
  causal contingency, 257
  causal relations, 31, 46–7, 57–8, 66, 69–75
  development, 266–8
  function, 266–8
  in great apes, 7, 122, 215, 267–18
  in humans, 266
  in tool use, 31, 47, 57–75, 190–207, 215
  levels, 7–8, 24–6, 31, 46–7, 69–75, 266–9, 272–5
  methods, 259–61
  see also object manipulation, physical cognition, tool use
Cebus
  apprenticeship, 360–2
  brain, 26–8
  causal cognition, 7, 26, 31, 34, 47, 66, 70–5, 267
  cognitive levels, 25–8, 31–4, 47, 58–9, 66, 70–3, 258, 265–72
  cognitive organization, 268–72
  convergent evolution, 23–5
  cooperation, 66, 74
  deception, 39–40, 43, 73
  development, 26–9, 47, 257–75, 362
  ecological cognition, 257–75
  feeding, 360–2
  habitat, 196
  imitation, 46–7, 64–7, 73, 362, 406
  insight, 30–4
  learning abilities, 25–6
  life history, 26–7, 363
  logicomathematical cognition, 24–6, 34, 47, 257, 265–72
  mindreading, 39–40, 43, 73–4
  mirror-self-recognition, 35–8, 47, 74
  object manipulation, 26–9
  perspective-taking, 43
  phylogenetic relations, 23, 358–9
  physical cognition, 26, 31, 46–7, 257, 267–72
  self awareness, 35–8, 47
  social organization, 443
  teaching, 74, 362
  tool use, 25, 28–34, 46–7, 57–75, 267, 359–62, 406
Cheney, D.L., 9–12, 38, 73–4, 83, 102, 301–5, 317, 342, 393
Chevalier-Skolnikoff, S., 26–7, 193–5, 205, 283, 351–2, 437
chimpanzees
  apprenticeship, 354, 359–61, 437, 444–5
  attention, 379–82, 436, 445
  behavioral development, 26–9, 145, 198–9, 217, 226–7, 235–53, 361, 416–19
  behavioral laterality, 215
  brain, 26–8, 111–12, 124–5
  causal cognition, 24, 31, 66, 71–5, 215–17, 267–8
  cognitive development, 23, 27–8, 64–7, 73, 145, 237–45, 250–3, 257–75, 361, 416–9
  cognitive levels, 3, 23–7, 31, 58, 71–3, 101–3, 257–75
  cognitive organization, 268–72
  communication, 139, 201–5, 382–4, 406, 419–24
  consolation, 81, 93–104

cooperation, 66, 74, 87–9, 100, 121, 387
culture, 3, 208–12, 224–8, 235, 250, 404–26, 430–45
deception, 39, 73, 102–3, 180–7, 302, 305, 384
ecological cognition, 257–75
empathy, 101–3, 304, 329–43
enculturation, 46, 143, 235–53, 310–11, 425–6
feeding, 111, 126, 186–7, 191–2, 205–7, 214–28, 360, 441, 444–5
food niche, 111–12, 125–6, 360, 440–1
gaze & eye contact, 136–43
habitat, 196, 212–14, 228, 408–14
human influences, 46, 143–5, 235–53, 351, 373–89
insight, 3, 24, 29–30
language, 4, 23, 178–87, 268–9, 330–2, 342–3
learning abilities, 25, 57–75
life history, 26, 362
logicomathematical cognition, 9, 265–72, 377–8
metarepresentation, 185–7
mindreading, 24, 39, 73–4, 102–3, 124, 185–7, 301–5, 388–9
mirror-self-recognition, 24, 35–8, 47, 74, 122, 293, 333–4, 342–3, 377
neonates, 27, 235–7, 379
numerical competence, 177–87, 378
object manipulation, 26–9, 373
ostension, 142–5
perspective-taking, 30, 102–3, 121, 301–5
phylogenetic relations, 3, 13, 111–12, 124–5, 358
physical cognition, 257, 268–72
population-specific behavior, 208, 212–28, 404–14, 424–6, 436
pretense, 306–7, 375–6, 436–7
rearing, 224–8, 235–53, 351
reconciliation, 81, 89–92, 103
role reversal, 102, 304, 354–5
secondary representation, 24
self, 24, 35, 38, 47, 325, 332–43
self-medication, 222, 441
social organization, 84–7, 100, 104, 119–20, 126, 443
teaching, 74, 354, 360–2, 386, 416–19, 436–7, 444–5
tool use, 3, 24, 28–31, 57–75, 111–13, 119, 196–206, 214–28, 351, 360, 376, 385, 404–25, 436, 441, 444–5
true imitation, 24, 43, 46, 64–7, 73, 119, 171, 226–7, 280, 293–5, 307–11, 335–7, 343, 351, 360–2, 384–6, 406–7, 415–16, 436–7, 444–5

cladistics, 13, 111–12, 357–9
classification, 377–8, 390–1
Clutton-Brock, T.H., 112–13, 124
cognition in primates, 2–9, 430, 433–4, 437
  development, 4, 8
  evolution, 4–5
  function, 4–5
  historical views, 2–6
  in great apes, 2–9
  in monkeys, 7–8
  levels, 3, 5, 7–9, 433–4
  mechanisms, 7, 430, 437
  socio-cultural influences, 10–1
  see also cognitive development, cognitive evolution, cognitive levels, cognitive organization, ecological cognition, social cognition
cognitive development, 5, 7, 10, 235, 240–3, 250–3, 257–95, 371–95, 433–4
  evolutionary perspectives, 10, 275–75, 318, 348–64, 433–4, 437–40
  differentiation, 433–4
  epigenetic models, 5, 10, 350, 433
  equilibration, 274
  hierarchization, 6, 268, 272–5
  human influences, 10, 240–2, 250–1, 278–95
  in great apes, 7–10, 66, 73–5, 198, 226–7, 240–2, 250–3, 257–75, 278–95, 310, 351–4, 361–2, 371–95, 425, 436–9
  in humans, 9–10, 57–8, 65–6, 73–5, 184–6, 240–2, 250–3, 257–75, 280–1, 284–5, 294–5, 305–18, 348–56, 433–4, 438–9
  in monkeys, 9–10, 47, 66, 257–75, 362, 439
  mental construction, 6
  methods, 10
  rates, 11, 26–8, 145, 258, 262, 265–75, 361–2, 436–7, 240–2
  rearing conditions, 4, 7–8, 278–95, 371–95
  recombination, 433–4, 437
  recursion, 272–3
  sequencing, 257, 265, 267–72
  symbolization, 430
  synchrony versus asynchrony, 10, 47
  see also individual abilities
cognitive evolution
  apprenticeship in, 348–64, 444–5
  continuity/discontinuity, 3–6, 14
  extractive foraging in, 5, 8–9, 12, 26, 112, 206–7, 348–65
  heterochrony in, 10–3, 258, 267–75
  hierarchization in, 6, 268, 272–5

cognitive evolution (*cont*)
  in great apes, 5, 8, 12–13, 112–13, 125–6, 348–65
  in hominids, 3, 12, 348, 359–64
  in primates, 112–13, 125–6, 206–7, 348, 360–4
  Machiavellianism in, 5, 83, 112
  recapitulation, 258
  see also individual abilities
cognitive gaps, 2–12, 24–47, 84–5, 103, 131, 136–7, 146, 186–7, 257–75, 300, 303–7, 311–12, 317–19, 392–5, 430, 433–4, 438
cognitive levels, 47, 215–17, 305–18, 373–6, 438
  developmental patterns, 11, 312–14
  see also individual abilities
cognitive linkages
  attribution and
    causal cognition, 74–5
    self-awareness, 312–13
  causal/physical cognition and
    cooperation, 74
    deception, 74
    language, 57
    social cognition, 73–5
    teaching, 74
    true imitation, 66, 226–7
  cooperation and
    true imitation, 66–7
  deception and
    metarepresentation, 184–6
    mindreading, 39, 102–3, 184–6, 301
    numerical competence, 180–6
    ostension, 145
    role reversal, 355
    self-awareness, 312–13
    true imitation, 283, 292, 295
  demonstration teaching and
    mirror-self-recognition, 355–7
    pretense, 355, 436, 438
    role reversal, 355, 436
    self-awareness, 357
    true imitation, 10, 353–7, 361, 407, 418, 436–7
  ecological cognition and
    language, 257–8, 268–70
    social cognition, 9–11, 112, 206–7, 295, 348–9, 357–64
  kinesthetic-visual matching and
    mirror-self-recognition, 280, 355
    planning, 280
    pretense, 280
    true imitation, 280, 293, 350–2

  language and
    logicomathematical cognition, 9, 378
    numerical competence, 178–87
    ostension, 145
    self, 329–33, 356
    teaching, 353
    true imitation, 153–5, 280, 283–5, 294–5, 351–2
  logicomathematical cognition and
    physical cognition, 257–8, 268–75
  metarepresentation and
    mindreading, 306, 311–19
    secondary representation, 308, 315–19
    true imitation, 308, 315–19
  mindreading and
    ostension, 145–7
    perspective-taking, 102–3, 304–5, 316
    pretense, 305–19
    secondary representation, 306, 311, 313–19
    self, 329–32, 334, 356
    self-awareness, 38, 312–13, 318
    true imitation, 123–5, 304, 307–11, 315–19, 352, 389
  mirror-self-recognition and
    self-awareness, 85, 312–13
    true imitation, 280–1, 291–3, 311–13
  perspective-taking and
    true imitation, 316–17
  pretense and
    secondary representation, 313–14
    self, 313, 438
    true imitation, 10, 307–9, 315–18, 350–2
  reconciliation and
    self-awareness, 313
  role reversal and
    true imitation, 304
  self-awareness and
    reciprocal altruism, 313
    true imitation, 318, 335–7, 343
  secondary representation and
    true imitation, 308, 315–19
cognitive organization, 7–10, 257–75, 300–19, 348–64, 371–95
  see also cognitive development, cognitive evolution, cognitive linkages, individual abilities
communication
  addressing, 133, 141–3
  attention, 133–4, 138–9, 146–7, 382–3
  coded versus inferential, 139, 144–5
  communicative intent, 131–48

Index   455

gaze & eye contact, 131–7
human influences, 382–3
intentional, 382–4, 391–3
referential, 142, 199–207
symbolic, 433
tool use in, 199–207
see also apprenticeship, attention, consolation, cooperation, culture, deception, language, ostension, reconciliation, social cognition, teaching
communicative simulation *see* true imitation
complementary role taking *see* role reversal
conflict resolution *see* postconflict behavior
consolation, 80, 93–104, 137
   and eye contact, 137
   definition, 81–3, 93
   evolution, 83
   function, 81–3, 93
   cognitive mechanisms, 84–5, 98–104
      mindreading, 102
      contagion, 83, 101–2
      perspective-taking, 102
   side-directed behavior, 93–4
   social mechanisms, 81–4, 90–3, 98–104
      altruism, 81
      dominance style, 100–1
      emotional contagion, 83, 101–2
      empathy, 83, 93, 102–3
      succorance, 81–3, 101–3
      sympathy, 83, 93
   see also reconciliation, social cognition
cooperation, 66–7, 73–4, 386–7, 391
   coalitions, 386–7
   evolution, 87
   human influences, 386–7
   in great apes, 386–7, 391
   role reversal, 387
   see also social cognition
cross-fostering *see* enculturation
culture, 4, 6, 11, 208–9, 235–6, 250–2, 278–9, 294–5, 371, 392–4, 404–26, 430–46
   and ecological cognition, 11
   and feeding, 434–6
   and tool use, 432, 436
   canalizing, 414–15, 425
   cognitive mechanisms, 11, 226–7, 278–9, 294–5, 406–8, 414–19, 423–5, 430–46
   criteria, 405–8, 431–5
   definition, 405, 430–2

dissemination, 208–9, 405–7, 419, 439
evolution in primates, 439–446
function, 430–1, 439–43
in great apes, 4, 6, 11, 208–9, 226–8, 392–4, 430–45
in humans, 431–2, 437–8
in monkeys, 430–1, 434–6, 445
innovation, 405, 408, 419–24, 439
levels of, 8, 431, 438, 445
methods, 208, 226–8, 407–26, 443
social mechanisms, 208–9, 405–7, 415, 419, 439–43
versus ecological learning, 407–14, 435–6, 439–43
see also apprenticeship, communication, enculturation, imitation, language, population-specific behaviors, pretense, social learning, teaching, true imitation
Custance, D., 152, 156, 171, 280, 293, 309–10, 337, 351, 356, 384, 406, 433

Damon, W., 326–8, 332–3, 356
Darwin, C., 3, 279
Dawkins, R., 119, 301, 303
deception, 6–8, 39–43, 102–3, 120–6, 184–6, 283–5, 292, 295, 301–2, 305, 383–4
   cognitive mechanisms
      associative, 40, 43, 123, 126
      executive control, 184
      metarepresentation, 184
      mindreading, 39, 102–3, 184–6
      perspective-taking, 39, 43
      planning, 123
   human influences, 383–4
   in autism, 184–6
   in great apes, 8, 39–40, 43, 102–3, 120–3, 184–6, 283–5, 292, 295, 302, 305, 383–4
   in humans, 184–6
   in monkeys, 122–3, 126, 301–2, 305
   levels, 7–8, 38–43, 102–3, 123–6, 184
   methods, 39
   strategic, 184–6
   see also communication, mindreading, social cognition
demonstration teaching, 292, 354–5
   cognitive mechanisms, 354–7, 436
   evolution, 357–62
   function, 354, 359–61
   in great apes, 292, 353–5, 359–62
   in humans, 353–4
   in monkeys, 353, 362

demonstration teaching (*cont.*)
  of tool use, 354, 357–9, 436
  see also apprenticeship, teaching, imitation
de Waal, F., 4, 6, 80–3, 88–93, 100–1, 137, 301, 379, 382–4, 387, 426

ecological cognition, 5, 257–75, 372–8
  development, 257–75
  elements, 259–68, 273
  evolution, 257–8, 270–75
  first & second order, 262–8, 273–5
  grammatical and proto-grammatical, 257–8, 268–9, 273
  hierarchization, 268, 272–5
  interconnections, 199–207, 271–75, 348–64
  levels, 257–8, 262–75
  in great apes, 113–19, 257–75
  in humans, 257–75
  in monkeys, 257–75
  methods, 259–61
  originalist & derivationist hypotheses, 257–8, 268–70
  recursiveness, 272–5
  see also cognition in primates, logicomathematical cognition, physical cognition, numerical competence, object manipulation, tool use
ecological learning *see* individual learning
  emotions, 308, 337–42
empathy, 83, 93–5, 101–3, 283, 304, 308, 313–15, 339–41
  cognitive mechanisms
    emotional contagion, 83, 85, 101–3
    mindreading, 102, 315
    self-awareness, 85
  definition, 83, 315, 339
  development, 83–5, 102–3, 314
  in great apes, 102–3, 283, 304, 308, 339–42
  in humans, 83–5, 314, 339–41
  in monkeys, 98–9, 102–3
emulation, 294, 317, 384–5, 390–1, 394, 406–7
enculturation, 11, 46, 143–6, 235, 250–2, 278–82, 294–5, 310, 384–6, 394–5, 425, 432, 437
  definition, 281
  versus cross-fostering, 281
  versus socialization, 281, 432
  see also culture
evolution of primates, 24, 301, 307–11, 318, 357–63
  adaptive radiation in, 363, 440

  convergent processes, 43, 358–60
  extractive foraging in, 26
  phylogenetic inertia in, 362
Exline, R.V., 131–6
experiential learning *see* individual learning
eye contact, 35, 131–48, 285
  and attention, 133–4, 138–45
  and pointing, 139–41
  functions, 35, 133–138, 146
  in great apes, 136–46, 285
  in humans, 131–4, 138, 144–6
  in monkeys, 35, 133–8, 141, 146
  in requests, 138–46
  ostensive, 133, 138
  see also communication, gaze, ostension

food processing
  cognitive complexity, 113–14
  cognitive mechanisms
    hierarchization, 115–19
    iteration, 114
    representation, 119
    teaching, 354, 359–62, 416–19, 444–5
    true imitation, 115–19, 126, 217, 226–7, 354, 359–61, 414–16, 444–5
  development, 115, 126, 217, 226–7, 348–64
  in great apes, 113–19, 126, 205–7, 214–22, 348–64, 405, 408–19, 424–5, 444–5
  laterality in, 114–18
  manual versus tool assisted, 113–19, 205–7, 348–65, 444–5
  see also object manipulation, tool use
Fossey, D., 126, 138
Fragaszy, D., 23, 26–8, 46, 307–9, 335, 350–1, 360–2, 406, 419, 424, 433–5

Galdikas, B., 119, 126, 160, 280, 293–4, 361
Galef, B.G., 152–9, 279, 293, 307–9, 350–1, 361, 406, 415, 419, 424, 433–5
Gallup, G., 24, 34–5, 81, 85, 92, 122, 311–12, 325, 333–4, 356, 376–7
Gardner, R.A. & Gardner, B.T., 8, 23, 43, 178, 332–3, 351
gaze, 134–5, 139–41, 146, 250, 253, 382, 391
gibbons, 29
Gibson, K., 5, 9–12, 23, 31, 58–9, 112–13, 125, 153, 190, 206, 257–8, 348, 358–61, 431, 434, 437, 441, 444
goal emulation *see* emulation
Gómez, J.C., 122, 132, 138–42, 146, 194, 205,

303–5, 315, 373, 382, 385, 393
Goodall, J., 5, 24, 29–30, 58, 72, 89, 136–8, 186, 193, 196, 200–1, 205–6, 217–19, 224–8, 360, 373, 384–5, 405–6, 410, 419–25, 436, 442–3
Gopnik, A., 281, 312, 334, 352, 357
gorillas
  apprenticeship, 358–9, 363
  brain, 34, 112, 124–6
  causal cognition, 26, 34, 122
  cognitive levels, 118–26
  communication, 382–3
  cooperation, 120–1
  culture, 444–5
  deception, 120–3
  development, 115, 122, 138, 145
  empathy, 343
  food niche, 112–13, 125–6, 440–1
  feeding, 112–17, 126, 444
  gaze & eye contact, 135–44, 388
  human influences, 145, 373–88
  Koko, 122, 283, 307, 329, 333, 354, 376
  language, 329–33, 343–3
  laterality, 114, 118–19, 123–5
  logicomathematical cognition, 377–8
  mental maps, 113
  mindreading, 121, 126, 305
  mirror-self-recognition, 35, 122, 333–4, 343, 356, 377
  object manipulation, 112–17, 122, 126, 363, 373
  ostension, 142–5
  perspective-taking, 121
  phylogenetic relations, 13, 111–12, 125, 358
  planning, 123
  pretense, 307, 376
  role reversal, 121
  self, 122, 325, 329, 333–43
  social cognition, 121–4
  social organization, 119–21, 126
  teaching, 354, 386
  tool use, 26, 29, 34, 60, 113, 122, 126, 196, 204–5, 362–3, 376, 444
  true imitation, 118–19, 123–4, 334–7, 343, 351, 363, 385
Gould, S.J., 13, 258, 271
Greenfield, P., 9, 12, 154, 190, 204, 268, 437
Grice, H.P., 132–3
Guillaume, P., 281, 293, 335

hand preference see laterality

Hart, D., 281, 293, 326–8, 332–4, 343, 356–7
Harvey, P.H., 112–13, 124, 439–40
Hauser, M., 349, 352–4, 418, 434
Hayes, K.J. & Hayes, C., 4, 24, 43, 280, 293, 295, 307–10, 338, 351, 373, 377–9, 382–5
heterochrony, 258, 270–2
Heyes, C., 153, 302, 305, 308, 312, 315–18, 406–7, 425, 432
hominids
  brain, 112, 125
  cognitive evolution, 12–13, 206–7, 348–9, 357–9, 362–4
  cognitive levels, 437–9
  cognitive organization, 12
  culture, 405, 437–9
  evolution, 5, 12, 111–12, 125, 206–7, 348–9, 362–4
  feeding, 186–7, 348–9, 359–64
  food sharing, 186–7
  niche, 12, 111, 349, 359–64
  technology, 207, 362
  tool use, 206–7, 348–9, 360–4
Huffman, M., 222, 422, 430, 433–5
humans
  apprenticeship, 348–64
  attention, 133, 146, 392–3
  autism, 184–6
  behavioral development, 27, 133, 138, 236
  brain, 112, 125–6
  causal cognition, 6, 71–5, 112, 257–8, 266–7
  cognitive development, 5–10, 27, 35, 58–9, 64–5, 71–3, 83–5, 112, 154–6, 184–6, 240–2, 257–75, 280–5, 294, 312–18, 333–5, 350–7, 433–8
  cognitive evolution, 3–13, 58, 112, 125–6, 152, 186–7, 206–7, 257–8, 270–5, 294, 357–64
  cognitive levels, 5–6, 12, 28–9, 38, 43, 64, 71–3, 83, 131, 144, 153–6, 184–6, 257–75, 280–5, 312–8, 350–6, 433–4, 438
  cognitive organization, 153–4, 257–75
  communication, 131–4, 138, 144–6, 250, 281
  culture, 6, 11, 187, 250, 278
  deception, 184–6, 283–5
  ecological cognition, 6, 257–75
  empathy, 83–5, 339–41
  enculturation, 250–2, 281, 295, 392–4
  gaze & eye contact, 131–46, 250–3
  kinesthetic-visual matching, 280–1, 350–1
  language, 43–4, 85, 134, 153–6, 187, 257–8, 268–9, 280, 283–5, 295, 329–33, 356
  learning abilities, 154

humans (*cont.*)
  logicomathematical cognition, 6, 257–8, 263–75
  metarepresentation, 146, 184–6, 312–15
  mindreading, 184–6, 315, 318
  mirror-self-recognition, 35, 38, 74, 85, 280–1, 284–5, 333–6
  neonates, 27, 43, 236–7, 250, 280–1, 315
  numerical competence, 184–6
  object manipulation, 28–9, 58–9, 139
  ostension, 131–4, 140–6
  perspective-taking, 284–5
  phylogenetic relations, 3–4, 13, 23, 112, 125, 152
  physical cognition, 257–8, 266–75
  pretense, 281, 284–5, 307, 315, 318, 354–5
  primary representation, 312–15
  rearing, 236–7, 250–2, 392–4
  reconciliation, 81, 137
  role reversal, 354–5
  secondary representation, 312–15
  self, 6, 35, 38, 81, 85, 325–6, 329–43, 356–64
  social cognition, 74–5
  teaching, 6, 155, 353–64, 418
  tool use, 6, 57–65, 71–5, 206–7, 222, 348–9, 359–64
  true imitation, 6, 43, 64–5, 73, 139, 152–6, 171–2, 250–1, 278–85, 295, 307, 315, 318, 335, 350–64, 407, 433–4
Humphrey, N., 5, 38, 83, 112, 207, 301–3, 379

imitation, 4, 6–7, 43–7, 64–7, 118, 152–72, 278–95, 304, 307–311, 315–19, 335–7, 349–56, 361–4, 384–6, 406–7, 414–16, 419, 423–5, 433–7
  and modality, 279–81, 285–95, 317, 350–6, 433
  cognitive mechanisms
    first order, 156–9, 350–1, 385, 433
    circular reactions, 281
    contagion, 351
    kinesthetic-visual matching, 280, 288–93, 433
    memory, 155
    stimulus enhancement, 46, 279, 309–10, 350, 361, 385, 406–7, 415–16, 424–5
    second order, 155–6, 280, 292, 308, 317, 350–1, 357, 433
  definition, 307, 349
  development, 278–95, 350–2
  human influences, 278–80, 294–5
  impersonation, 171–2, 294
  in great apes, 43, 46, 64, 118–19, 128, 152–72, 278–95, 226–7, 307–11, 315–19, 351–2, 384–6, 406–7, 414–16, 423–5
  in humans, 64, 350–1
  in monkeys, 46, 64, 351, 434–6
  levels, 7–8, 152, 155–6, 162, 166–7, 171, 279, 294, 308, 316–7, 350–2, 384, 433
  match, 154–5, 161–7, 286–94, 350–1
  mimicking, 171, 317, 384
  neonatal, 43, 250, 280–1, 315, 318
  purposes, 152, 155–9, 162, 170–1, 279, 283, 295, 415–16, 424–5
  rehearsal, 154–5, 162, 294
  self-imitation, 283
  social factors, 283, 294
  see also emulation, kinesthetic-visual matching, social cognition, social learning, stimulus enhancement, true imitation
imitative learning *see* true imitation
individual learning, 118, 157–9, 414–16, 423–5
  see also social learning
Ingmanson, E.J., 190–2, 199–200, 206
innovation, 419–24
insight, 3, 24, 29–34
instrumental cognition *see* physical cognition
intention, 131–3

Jolly, A., 6, 38, 83, 112, 194, 207, 301, 379, 433, 442

Kano, T., 126, 191–5, 199, 205–6, 222
Kawai, M., 5–6, 419, 430
Kellogg, W. & Kellogg, L., 4, 379
King, D., 353, 435
Klüver, H., 25, 58, 60
Köhler, W., 3, 24, 29–30, 58–60, 73, 122, 194, 205, 279, 387
Kortlandt, A., 212, 409
Krebs, J.R., 301–3
Kummer, H., 6, 11, 83–5, 88, 405–6, 419, 422, 431–2, 442
kinesthetic-visual matching, 280–5, 293–4, 433

Langer, J., 8–12, 47, 257–75
language, 4–8, 23, 153–5, 257–8, 268, 279–85, 292–5, 329–33, 342–3, 431–2
  development, 153–5, 279–85
  in great apes, 4–8, 23, 156, 178–87, 333
  in humans, 153–5, 333
laterality, 114, 118–19, 123–5, 215, 291–2
learning abilities, 4–5, 25–6, 47, 57–75
  associative, 25–6, 47

mediational, 5, 8, 25–6, 47
transfer index, 25
Leslie, A.M., 71, 184, 305–6, 312, 314
Lewis, M., 281, 334, 337–8, 356
logicomathematical cognition
  classification, 263–6, 377–8
  development, 257–66
  evolution, 257–58
  human influences, 378
  in great apes, 8, 265–6
  in humans, 257, 263–5
  levels, 8, 263–6, 268–70
  methods, 259–61
  object concept, 6, 26, 257, 373–4, 390–1
  operations, 263–8
  quantification, 178, 182–3, 263–6
  see also numerical competence, physical cognition

macaques
  behavioral development, 93, 119, 252
  causal cognition, 7, 26, 36, 267
  cognitive development, 257–75
  cognitive levels, 26, 258, 265–72
  cognitive organization, 268–72
  communication, 141
  consolation, 93, 96–104
  cooperation, 86, 88
  culture, 419, 430, 434–6, 445
  dominance style, 86, 100–1
  ecological cognition, 257–75
  emotional contagion, 98–9, 102–3
  empathy, 98–9, 102–3
  feeding, 364, 434–6, 445
  gaze & eye contact, 136–7, 141
  hierarchization/recursiveness, 272
  imitation, 46
  learning abilities, 25
  life history, 26–7
  logicomathematical cognition, 257–75
  matrilines, 86
  mindreading, 74, 102–3, 302–5
  mirror-self-recognition, 35, 85
  object manipulation, 26–9
  perspective-taking, 102–3
  phylogenetic relations, 440
  physical cognition, 257–75
  rearing, 252, 425
  reconciliation, 81, 89–93, 103
  role reversal, 302–5
  social organization, 84, 86, 100, 104

species variability, 89–91, 98–104
tool use, 7, 26, 34, 267
Matsuzawa, T., 7, 58, 178, 211–19, 226–8, 377–8, 436
McGrew, W., 7, 24, 29–31, 113, 126, 186, 192, 206–7, 211, 215, 224, 358–61, 376, 385, 405–10, 419–22, 430–1, 436, 444
McKinney, M., 13, 258, 271
Meltzoff, A., 43, 46, 65, 153–8, 170, 280–1, 292, 309, 315, 318, 335, 350–1, 355–6, 364, 437
mental attribution *see* mindreading
Menzel, E.W., 5, 102, 139, 250, 311, 376, 379, 382, 387, 430
metarepresentation, 305–6, 311–15
  definition, 312–14
  development, 312–14
  false belief, 305, 313–15
  see also perspective-taking, secondary representation
Miles, H.L., 178, 281–5, 294, 332–3, 351, 354–6, 373, 378–9, 383–4
mindreading, 6, 39–40, 43, 102–3, 301–9, 313–15, 388–93
  and autism, 306–8
  cognitive mechanisms
    associative, 389
    intentionality, 24
    metarepresentation, 314
    perspective-taking, 39, 43, 102–3, 302–4, 388
    secondary representation, 24, 314–15
  definition, 301, 314
  development, 304–5, 313–5
  evolution, 301
  false beliefs, 389, 391
  in great apes, 8, 24, 39, 121, 301–5, 314–15
  in humans, 304–8
  in monkeys, 39–40, 43, 302–5
  levels, 7, 302–4
  methods, 301–5
  see also social cognition
mirror behavior, 34–8, 311–12, 333–4, 356, 375–7
  and eye contact, 35–6
  and self-referencing, 334
  as tool use, 35, 376–77
  cognitive mechanisms, 36–8
  development, 35, 333–4
  human influences, 377
  in great apes, 34–8, 311–12, 333–4, 356, 377
  in humans, 35–8, 311, 333–4, 356
  in monkeys, 34–8, 311, 334, 356

mirror behavior (cont.)
  mirror-guided reaching, 36–8
  self-directed behavior, 35, 356, 377
  social responses, 35–6, 311
  see also mirror-self-recognition
mirror-self-recognition, 34–8, 74, 81, 85–6, 122, 280–5, 293, 311–13, 333–4, 355–6, 375–77
  cognitive mechanisms
    metarepresentation, 313
    secondary representation, 311
  development, 35, 85, 281, 283–5, 293
  in great apes, 85–6, 122, 283–5, 311–12
  in humans, 38, 281, 311
  in monkeys, 35–6, 85
  levels of, 35–8, 311–13
  mark test, 35, 122, 333–4, 356, 377
  methods, 35–8, 122, 333–4, 356, 377
  see also mirror behavior, self-awareness
Mitchell, R., 6, 10, 38–9, 85, 153, 156, 169, 279–81, 293, 302, 307, 312, 316–18, 326, 334–5, 349–51, 355–7, 433
Moerk, E.L., 154–6, 161–2
Moore, B., 152–3, 156–8, 315–17, 433

Nishida, T., 4, 11, 120, 195–6, 205, 211, 222, 387, 405–8, 419–24, 430, 436, 441–4
numerical competence, 178–87, 378
  and food sharing, 180–3, 186–7
  evolution of, 186–7
  human influences, 378
  in great apes, 178–83, 378
  in monkeys, 378
  in strategic deception, 180–7
  levels, 180–7
  methods, 179–80
  see also logicomathematical cognition

object manipulation, 8, 28–9, 193–5, 222–5, 373–6, 384–5, 390–1, 419–23
  cognition mechanisms, 8, 28–9
  development, 28–9
  ecological factors, 224
  human influences, 373–6
  in great apes, 28–9, 193–5, 222–5, 419–25
  in humans, 28–9
  in monkeys, 28–9
  in pretense, 373–6
  in symbolic communication, 419–22

population specific, 224
social factors, 424–5
tool use, relation to, 28, 190–3
see also causal cognition, physical cognition, tool use
object permanence, 6, 26, 373–4, 390–1
observational learning see true imitation
orangutans
  apprenticeship, 358–9, 363
  attention, 285
  brain, 125
  Chantek, 171, 278–95, 332, 354
  cognitive levels, 292–3
  communication, 140, 382–3
  culture, 278, 295, 444–5
  deception, 283–4, 292, 295
  development, 27, 278–95
  enculturation, 278–81, 294–5
  food niche, 440–1
  feeding, 119, 126, 444
  gaze & eye contact, 140
  habitat, 205
  human influences, 373–88
  true imitation, 119, 152–73, 278–95, 311, 335–7, 351–2, 363, 384–5, 407, 437
  kinesthetic-visual matching, 283–8, 293
  language, 278–95, 332, 343–3
  laterality, 291–2
  logicomathematical cognition, 293, 377–8
  mindreading, 140, 307, 315, 342, 388
  mirror-self-recognition, 35, 122, 283–5, 293, 333–4, 343, 377
  object manipulation, 119, 126, 190, 193–4, 205, 373
  perspective-taking, 283–5
  phylogenetic relations, 13, 111, 125, 358
  pretense, 285, 292, 376
  role reversal, 292, 295, 354
  self, 325, 332–43
  social organization, 442–5
  teaching, 292–5
  tool use, 29, 62, 190, 196–9, 204–5, 294, 362–3, 376, 385, 444
ostension, 131–48, 285
  and attention, 132–3, 141–2
  and communicative intent, 132
  and eye contact, 133, 138–46
  definition, 132
  development, 145
  human influences, 143–6
  in great apes, 138–46, 285

in humans, 131–4
in monkeys, 138, 141, 145
see also social cognition

Parker, S.T., 5–12, 23, 26, 31, 34–5, 58–9, 72, 85, 112–13, 125, 154–5, 172, 190, 206, 257–8, 280–1, 293–4, 312, 326, 334, 348, 351–2, 356–62, 433–8, 441, 444
Patterson, F., 122, 178, 311, 329, 332–4, 351–2, 356, 376–8, 383–5
Perner, J., 184, 305, 312–15
perspective-taking, 39, 43, 102–3, 285, 302–5, 316, 388
see also mindreading
physical cognition, 26, 31, 46–7, 57–75, 257–75, 373–7, 389–90
  development, 266–8
  functions, 266–8
  in great apes, 24–6, 31, 257, 267–8
  in humans, 257, 266
  levels, 47, 72–3, 373–6
  methods, 259–61
  object permanence, 373–4
  see also apprenticeship, causal cognition, ecological cognition, object manipulation, tool use
Piaget, J., 5–6, 11, 23, 26–31, 57–8, 72, 122, 152–3, 170–2, 200, 257, 266, 280–1, 292–4, 307, 350–5, 364, 373, 392, 433–4, 437
planning, 123, 281–5
pointing, 43–5, 139–41
population-specific behavior, 208, 214–17, 226–8, 404–26
  see also culture, behavioral traditions
postconflict behavior *see* consolation, reconciliation
Povinelli, D., 74, 85, 102, 121, 302–4, 334, 354–5, 363, 383, 387–9, 392, 395, 436
Premack, D., 5–9, 24, 39, 102, 121, 139, 178–80, 301–5, 314–15, 354, 371, 378, 383, 388–9, 436
pretend play, 280–1, 292, 306–8, 313–15, 350–2, 373–6, 390, 434, 438
  cognitive mechanisms
    imitation, 376
    kinesthetic-visual matching, 280
    pretense, 313–4
    secondary representation, 313–14, 434, 438
    true imitation, 281, 434
  definition, 306, 434
  development, 280–1, 306

human influences on, 373–6
in autism, 306
in great apes, 306–7, 313–15, 373–6, 434, 438
in humans, 306, 314–15, 434, 438
in monkeys, 307
levels, 434
methods, 306–7
object manipulation in, 306, 373–6
see also pretense
pretense, 8, 169, 279, 283–5, 305–8, 315, 350–2, 373–6
in great apes, 6, 169, 283–5, 315
levels, 6, 313–5, 434, 438
see also pretend play, social cognition

reconciliation, 80, 89–93, 137
  and deception, 92
  and eye contact, 137
  cognitive mechanisms
    mirror-self-recognition, 81, 92
    self-awareness, 81, 92
  definition, 80
  function, 80–1, 91, 103
  in great apes, 81, 89–92, 103
  in monkeys, 81, 89–93, 102
  in humans, 81
  species factors, 90–3
  see also social cognition
Reynolds, P., 362
Reynolds, V., 57–9, 222, 430
Ridley, M., 112, 357
Rogoff, B., 348, 418, 425
role reversal, 302–4, 354–5
  see also social cognition
Rumbaugh, D., 5, 38, 178, 378
Russon, A.E., 152, 159–66, 280, 283, 292–4, 311, 332, 335, 338, 351–2, 363, 373, 376, 384–5, 406, 433, 436–7

Savage-Rumbaugh, S., 140, 178, 190, 251, 268, 303, 307, 352, 376–7, 379, 383–8, 393, 425
Schaik, C. van, 361–2, 444
Schiller, P.H., 28, 30, 72, 376
second-order intentionality *see* mindreading
self, 6, 85, 325–343, 438
  criteria, 328–42
  cognitive mechanisms, 85
  components, 326–8
  definition, 85, 326

self (cont.)
  in humans, 343
  in great apes, 8, 343
  in monkeys, 343
  levels, 7–8
  methods, 328–42
  social factors, 85
  see also consolation, mirror-self-recognition, reconciliation, self-awareness, self-understanding, social cognition
self-awareness, 8, 34–8, 81, 85–6, 122, 308–9, 312–13, 327, 348–9, 354–6
  and empathy, 313, 339–41
  and mirror-self-recognition, 312–13, 334, 355–6
  and self-conscious emotions, 337–41, 356
  and self-referring, 332–3
  cognitive mechanisms, 355–6
  definition, 355
  development, 85–6, 356
  evolution, 357–9
  function, 359–61
  in autism, 308–9
  in great apes, 35, 38, 122, 311–12, 332–5, 338–41
  in humans, 38, 308–11, 332–5, 338–41
  in monkeys, 35, 311
  kinesthetic-visual matching, 355
  levels, 312, 355–6
  objective, 327, 332–41
  subjective, 327, 338–41
  see also mirror self-recognition, self
self-medication, 222–3
self-understanding
  and empathy, 339
  and language, 329–33
  in humans, 329–42
  in great apes, 329–42
  in monkeys, 329–32
  methods, 329–33, 339
  personal memories, 328–33, 339
  self-representations, 328, 331–4, 338–9
  self-theories, 328, 332, 338–9
  see also self
sensorimotor (Piagetian) abilities
  circular reactions, 27
  human influences, 373–4
  in great apes, 26–7, 59, 373–4
  in humans, 27
  in monkeys, 26–7, 31, 59
  mental representation, 26

Seyfarth, R.M., 9–12, 38, 73–4, 83, 88, 102, 301–5, 317, 342, 393
social cognition, 6, 38–46, 80–104, 378–89
  in great apes, 121–4, 199–20
  in primates, 301
  evolution, 83, 301
  methods, 83–4
  see also apprenticeship, attention, communication, consolation, cooperation, deception, imitation, mindreading, ostension, perspective-taking, pretense, reconciliation, role reversal, self, social learning, teaching, true imitation
social learning, 309, 384–6, 406–7, 414–16, 423–5, 442
  cognitive mechanisms, 384–6
  emulation versus mimicking, 384–5
  human influences, 384–6
  in monkeys, 46
  ontogenetic ritualization, 385–6
  see also apprenticeship, culture, imitation, individual learning, social cognition, stimulus enhancement, teaching, true imitation
social organization, 87–91, 442–3
social tradition see traditions
social transmission, 11, 432–9
  cognitive mechanisms, 433
  in great apes, 436–7
  in monkeys, 433–7
  levels, 433, 435–6
  social mechanisms, 422–3
  see also communication, culture, imitation, true imitation, kinesthetic-visual matching, social learning, stimulus enhancement, teaching
Sperber, D., 131–2, 139, 144–5
spider monkeys, 141
Spinozzi, G., 259–62, 265–7, 272, 377
stimulus enhancement, 46, 156, 279, 309–10, 350, 361, 406–7, 415–16, 424–5
  see also imitation, social learning, true imitation
Sugiyama, T., 211–19, 222, 379, 406–10, 413, 420–2
symbolic play see pretend play

tactical deception see deception
teaching, 6, 8, 348–55, 386, 391–3, 416–19, 434–7, 444
  coaching, 352–4, 386, 418, 434
  cognitive mechanisms, 353–5
  definition, 352, 416

demonstration, 353–4, 386, 418, 434–6
development, 354–7
human "teaching", 389–94
in great apes, 8, 353–5, 416–18, 434–7
in humans, 353–4, 418–19
in monkeys, 353, 362
of tool use, 436
opportunity, 352–4, 386, 391–3, 417–18, 434
scaffolding function, 392, 417–18
see also demonstration teaching, social cognition
theory of mind *see* mindreading
Thorpe, W.H., 153, 158, 279, 293–4, 406
Tomasello, M., 8, 24, 46, 123, 139, 143–5, 153, 156, 171, 224, 235, 279, 294, 305, 309–11, 317–18, 335, 350–2, 382–95, 405–7, 415, 421, 424–5, 432–3
tool-related behavior *see* object manipulation
tool use, 3, 5, 25, 29–35, 47, 57–75, 190–207, 211–29, 348–9, 358–65, 375–7, 390–1, 404–9, 424–5, 436–7, 444–5
  cognitive mechanisms
    associative, 30–1, 69, 71–2
    attribution, 74–5
    hierarchization, 217
    insight, 29–30, 34
    symbolic representation, 7–8, 31, 34, 47, 69, 71–2
  cooperative, 66–7, 74
  definition, 29, 57–8, 191, 348
  development, 28, 58–9, 65–6, 73, 197–9, 217, 349, 360–1
  ecological factors, 360–4, 444–5
  ecological versus social, 122, 195–207
  errors, 60, 67–8
  evolutionary significance, 58, 206–7, 348–9, 357–64, 444–5
  function, 359–61
  human influences, 376, 390–1
  in great apes, 3–8, 28–31, 34, 58, 64–75, 122, 191–2, 195–207, 211–29, 348, 351, 359–64, 404–19, 436–7, 444–5
  in humans, 57, 64–5, 68–72, 75, 359–64
  in monkeys, 7, 25, 30–34, 46–7, 59–60, 64–75, 359–62, 406
  in symbolic communication, 204, 206
  intelligent, 31, 58–9, 348, 358–64
  life history factors, 361
  methods, 61–71, 192, 214–15, 226–7
  metatool, 7, 31, 215
  population-specific, 211, 215, 219, 222–4, 228, 404–14
  social factors, 424–5, 444–5
  social transmission of, 404–29, 436
  tool set, 7, 30–1, 34, 214–17, 222
  see also apprenticeship, causal cognition, culture, demonstration teaching, food processing, mirror behavior, object manipulation, true imitation
trial-and-error learning *see* individual learning
Trivers, R., 88, 349, 439
true imitation, 43–7, 64–7, 118, 123, 152, 156, 161–72, 226–7, 279, 292–4, 304, 307–11, 315–19, 335–7, 348–64, 384–6, 391–2, 406–7, 414–16, 419, 423–5, 433–7, 442–4
  action level, 118, 123, 308
  cognitive constraints, 352–3
  cognitive mechanisms
    hierarchization, 119
    secondary representation, 315–18
    symbolic representation, 280, 292–3, 308
  deferred, 46, 155, 167–8, 280–3, 292, 351
  definition, 43, 118, 123, 153, 350–1, 406, 433
  development, 43, 278–95, 315, 318, 350–2, 384–6, 436–7, 442
  evolution, 318, 348, 357–62
  functions, 359–61, 406–7, 414–16, 419, 423–5, 433, 437, 444
  human influences, 43, 46, 278, 294–5, 310, 351, 384–6, 391, 437
  in autism, 315, 318
  in great apes, 43, 46, 64, 118–19, 123, 152–72, 226–7, 278–95, 307–11, 315–18, 335–7, 350–2, 358–63, 384–6, 406–7, 414–18, 433, 436–7, 444–5
  in humans, 64, 153–6, 171, 280–1, 307–8, 310, 313–18, 335–6, 350–2, 362
  in monkeys, 46, 64–7, 279, 305–9, 311, 335, 350–1, 358, 362, 406, 419, 433–6
  in nonprimates, 315–18
  levels, 118–19, 123, 126, 155, 166–7, 171, 308, 316–7, 350–2, 363
  matching, 154–5, 163–6, 296–92, 294, 351
  methods, 43, 46, 119, 156–9, 171, 226–7, 279–82, 285–8, 309–11, 350–2, 385, 406–7, 415–16, 425–6, 436–7
  of arbitrary acts, 43, 46, 283, 292–4, 310, 351, 384
  of communication, 43–6, 281–3, 292–3, 351, 385–6, 407, 423–5
  of object manipulation, 283, 288, 310–1, 351, 384–5

true imitation (*cont.*)
  of social roles, 302–4
  of tool use, 46, 64–7, 73, 226–7, 279, 283, 294, 302, 310–1, 349–51, 357–63, 385, 406–7, 414–18, 436–7, 445
  program-level, 119, 126, 171, 317, 363
  purposes, 154–5, 170–2
  rehearsal, 155, 167–70, 294, 351
  social constraints, 279, 283, 294, 406–7, 425–6
  speed of dissemination, 419
  versus individual learning, 119, 156–9, 414–15
  see also emulation, imitation, pretense, social cognition, social learning, stimulus enhancement
traditions, 5, 11, 208, 211–12, 226–8, 404–26, 430, 434–8, 444
  and culture, 11, 404–26, 430, in great apes, 11, 436–8, 444
  in humans, 437–8
  in monkeys, 5, 434–6
  see also population specific behavior

vervet monkeys, 301, 436
Visalberghi, E., 24, 31, 46, 59–60, 72–3, 157–8, 171, 279, 293–4, 307–9, 335, 350–1, 360–2, 406, 419, 424, 433–5
Vygotsky, L., 11, 257, 352, 371, 392

Werner, H., 170, 257, 271, 275
Whiten, A., 6–10, 24, 38–9, 120–3, 145–6, 152–4, 279–80, 293–4, 301–9, 312–17, 335, 348–50, 383–5, 395, 407, 418, 425, 434–6
Wood, D., 171, 353–4
Wrangham, R., 186, 222, 430, 436, 440
Wynn, T., 407, 432, 437

Yamakoshi, G., 211, 214, 222, 226
Yando, R., 172, 352, 354
Yerkes, R.M. & Yerkes, A.W., 3, 24, 58, 60, 222, 279, 333–9, 383
Yerkes Research Center, 235, 238–9

zone of proximal development, 352–3